S0-AJA-440

The Struggle for Workers' Health:

A STUDY OF SIX INDUSTRIALIZED COUNTRIES

Ray H. Elling

Baywood Publishing Company, Inc.
Farmingdale, New York

Library of Congress Catalog Card Number: 85-28739
ISBN: 0-89503-047-0

Library of Congress Cataloging-in-Publication Data

Elling, Ray H., 1929-
The struggle for workers' health.

Includes bibliographies and indexes.
1. Industrial safety – Europe. 2. Industrial safety – Germany (East) 3. Industrial safety – United States. I. Title.
HD7262.5.E85E44 1986 363.1'1'094 85-28739
ISBN 0-89503-047-0

Dedication

To the workers of the world
that they may unite to take control of their lives
and insist on safe and healthy work.

Acknowledgments

The field work for this study was conducted during my sabbatic leave in 1981. I would like to thank the University of Connecticut for this opportunity. The European Regional Office of the World Health Organization appointed me as Temporary Advisor and offered some assistance with travel. I am especially grateful for this sponsorship and support. No endorsement of this work is implied by this recognition. To the rank-and-file workers, union spokespersons, OSH experts, and others (see Methods Appendix) who agreed to interviews and supplied written information in each of the six study countries I owe a great deal and offer here my thanks. I am especially grateful to the key informants in each country who commented on draft chapters and sent additional information. It is a rare privilege to benefit from such warm and supportive, yet critical colleagueship. Finally, but by no means least, my warm thanks to Margit E. Elling for the long hours and excellent job of typing and manuscript preparation.

Preface

To some extent workers are faced with the non-choice "your job or your life" in almost every work setting in every land. This is a cruel choice. The worker and her or his family need sustenance. The worker must work. Yet the job may be dangerous–insidiously so, with long acting chemicals and stress; or acutely so, with dangers of explosions, fires, falls, cave-ins, electrocutions. At the same time, if a decent paying job is not guaranteed by the society, there may be thousands of other potential workers standing in line ready to take a hazardous, self demeaning job as a matter of survival–just to keep one's head above water.

The problems of work-related diseases and accidents are widespread and serious. In U.S.A. each year more people die from such causes than were lost in any year of war, except for one bloody year of the Civil War, and hundreds of thousands more are maimed or seriously disabled. With hundreds, if not thousands of new chemicals entering the workplace every year, and faster work paces often coupled with maintenance cut-backs "to meet the competition," these problems are getting worse in many places. Yet job deaths and disabilities are not equally widespread and serious in every country. And there are remarkable differences between countries in protective measures–both those directed specially toward the workplace and those which should be available (but usually are not) through primary health care.

What could explain such differences as the following? 1) In U.S.A. almost all industrial physicians (present in only a minority of workplaces) are hired and fired by management and very few have adequate preparation in occupational health; thus workers generally lack trust in these physicians and refer to them as "company quacks." In Sweden, a worker-union-controlled health and safety committee has the hiring and firing power as regards the occupational safety personnel and the national worker health and safety agency only recommends candidates who have met minimal qualifications; further, at least half of Swedish workers are covered by such services. 2) In U.S.A. workers are still fighting to achieve the "right to know" what substances they are working with, what the harmful effects might be and how to avoid problems. In Sweden this right is guaranteed by the Co-determination Law (joint union-management planning of production) and by the training of more than 110,000 union health and safety representatives with at least forty hours of hazard identification and control information. 3) The U.S.A. and Sweden are both poor in assuring that primary health care providers will recognize work-related disease and properly follow up the particular case as well as stimulate preventive efforts for other workers who may be similarly exposed.

As I began this work I did not know the specifics of such differences, but I expected that the strength of workers' movements, as well as the form of organization of the general health services would be related to the adequacy of worker protection (both on the job and in primary care). The nations chosen for study were selected to represent a range of such differences. Thus, in an effort to better understand how strong worker protection systems differ from weak ones, this volume reports and interprets a study carried out in six nations–Sweden, Finland, The German Democratic Republic, The Federal Republic of Germany, The United Kingdom of Great Britain and Northern Ireland, and the United States of America. The work involved interviews with reputational leaders of different interest groups as well as observations, extensive document study and correspondence with key informants. The work is limited by what one person could do during a sabbatic leave and follow-up months. Yet a clear picture seems to emerge. This picture should be confirmed or revised by cross-national in-depth historical case studies. But for now some guidance in this important matter can be given.

While it can not be said (contrary to my expectations) that the strength of the workers' movement is strongly determinant of how the general health services deal with work-related disease (Sweden's education of medical students is as poor in occupational health as is that in U.S.A.) it seems clear that the overall picture as regards measures to protect workers is much better where the workers' movement is stronger.

The main lesson to be drawn is that workers themselves must gain and make full use of "the right to know" what chemicals they are working with and take charge of their own health and working conditions. Since workers and their unions usually face the powerful opposition forces of the owners and managers and their experts, this is not an easy prescription. But the variations in the strengths of workers' movements revealed in this and other works speak strongly to the capacity of men and women of all ethnic backgrounds to organize for solidarity in the pursuit of justice, well being, and the realization of human potential.

Table of Contents

List of Figures

List of Tables

CHAPTER 1
Introduction

WORK AND HUMAN ORGANIZATION

> You become your job. I became what I did. I became a hustler. I became cold, I became hard, I became turned off, I became numb. Even when I wasn't hustling, I was a hustler. I don't think its terribly different from somebody who works on an assembly line forty hours a week and comes home cut off, numb, dehumanized. People aren't built to switch on and off like water faucets.
>
> (Roberta Victor, Hooker in *Working* by Studs Terkel, p. 102)[1]

> You can't take pride any more. . . . You're mass producing things and you never see the end result of it. . . . I got chewed out by my foreman once. He said, "Mike, you're a good worker but you have a bad attitude." My attitude is that I don't get excited about my job. I do my work but I don't say whoopee-doo. The day I get excited about my job is the day I go to a head shrinker. How are you gonna get excited about pullin' steel? How are you gonna get excited when you're tired and want to sit down? . . . Somebody has to do this work. If my kid ever goes to college, I just want him to have a little respect, to realize that his dad is one of those somebodies. . . .
>
> (Mike Lefevre, Steelworker in *Working* by Studs Terkel, pp. 1, 2, 6)

Work presents men and women with a central part of their identities. Yet, if the worker is not in control and is being used for another's gain, it is a dehumanizing, alienating experience. For millions of workers, there is an essential ambivalence. On the one hand, it is boring drudgery: "I do my work but I don't say whoopee-doo." On the other hand, one is somebody through his work: "I just want him (my, hopefully college-bound son) to have a little respect, to realize that his dad is one of those somebodies."

It is something of a rationalization for Roberta to say her work is not terribly different from Mike's. After all, his is an openly accepted occupation producing publicly valued steel products for the society. Hers is a clandestine "service." But at a deeper level, both people have become their work and both are prostitutes to a system which exploits them. Both sense that they are exploited, that they lack control. They are alienated from themselves and the structures which oppress them; they often feel helpless. As Mike put it,

> Hell, if you whip a damn mule he might kick you. Stay out of my way, that's all. Working is bad enough, don't bug me. . . . Who you gonna sock? You can't sock General Motors, you can't sock anybody in Washington, you can't sock a system . . . *But I gotta get it out.* I want to be able to turn around to somebody and say, 'Hey, fuck you?' You know? (Laughs) . . . So I find a guy in a tavern. To tell him that. And he tells me too. I've been in brawls. He's punching me and I'm punching him, because we actually want to punch somebody else (pp. 2, 3).

Work structures the society as well as its individuals. There are occupations which place the person in control of immense resources and thousands of other workers. There are other occupations which provide ample room for individual creativity. But the mass of workers have little power and little recognition for what they produce—even though the society's essential goods would not be made without their work. Mike, again:

> Somebody built the Pyramids. Somebody's going to build something. Pyramids, Empire State Building—these things just don't happen. There's hard work behind it. I would like to see a building, say, the Empire State, I would like to see on one side of it a foot-wide strip from top to bottom with the name of every bricklayer, the name of every electrician, with all the names. So when a guy walked by, he could take his son and say, "See that's me over there on the forty-fifth floor. I put the steel beam in." Picasso can point to a painting. What can I point to? A writer can point to a book. Everybody should have something to point to (p. 2).

Of course Mike's view is a very individualistic one. He has been brain washed (*socialized* is the sociological euphemism if you prefer it) by the dominant culture which upholds and is created by our form of political economy—a capitalist system based on private property, oriented toward competition in the marketplace. It is said to depend on individual initiative—in spite of the fact that the increasingly monopolistic form of capital makes massive power plays and fixed arrangements as regards markets and suppliers the real bases of a high rate of profit. In a collective society, Mike might point to all the buildings and the cars, etc. and speak with pride of his own work and that of his comrades. This may be an idealistic view, but it is a theoretical and (some would argue) partially achieved reality in some countries. As the pinion point for both personal and societal organization, work presents the primary arena in which the most vital struggles of society are engaged.

Sometimes, the struggle is highly personal, even secret, at least as far as the dominant forces in the situation are concerned. For Roberta, it is an *act*: "The call girl ethic is very strong. You were the lowest of the low if you allowed yourself to feel anything with a trick. The bed puts you on their level. The way you maintain your integrity is by acting all the way through" (p. 94). Mike has a number of ways of personal struggle. When he was younger and single, he punched back. He knocked a foreman off a shipping dock once after the foreman had put his hand on his shoulder and shoved him toward the next box to be loaded. Now he is older and has a family, so he limits his resistance to verbal exchange and secret, minor sabotage. A young college graduate foreman once told him to reply "Yes sir" instead of "Yeah." Mike exploded, "Who the hell are you, Hitler? What is this 'Yes sir' bullshit? I came here to work, I didn't come here to crawl. There's a fuckin' difference" (p. 3). Mike resisted and he lost. He got broke down a grade–25 cents an hour less. But he has this other more secret form of resistance. It is a mixture of seeking to enhance his identity and of resisting: "Sometimes, out of pure meanness, when I make something, I put a little dent in it. I like to do something to make it really unique. Hit it with a hammer. I deliberately fuck up to see if it'll get by, just so I can say I did it."

The struggle around work is most acute when it moves from the personal to the collective level. The supreme form of collective struggle on the workers' side is the strike, the shutting down of production. At this point, owners usually call in the power of the state in one form or another–the courts for an injunction; or the police or national guard to "keep order," allowing the hiring of scabs and the breaking of a strike. Examples are legion. One of the more heroic and exemplary moments of solidarity in U.S.A. labor history joined textile workers–men and women, Poles and Italians, Irish and Jews–in the "Bread and Roses Strike" of 1912 in Lawrence Massachusetts. The bosses used the police, beatings by thugs, and tried to pit the workers against each other along sex and ethnic lines by asserting that some were after the jobs of others. They also attempted red-baiting by playing the chords of a worker-antagonistic culture, calling the organizers "Marxists," "socialists," and worse. But the workers won their demands. Mrs. Mary K. O'Sullivan, first woman organizer of the AFL wrote, "Catholics, Jews, Protestants and unbelievers, men and women of many races and languages, were working together as human beings with a common cause. . . . It was the most unselfish strike I have ever known. . . ."[2]

The struggle can be won. But the experience in each country is different. Working men and women have been much more successful in some countries than in others. In Sweden, for example, the workers' movement is stronger than it is in the United States. I will come back again and again to discussing the varying strengths of workers' movements for it is the central thesis of this book that a stronger workers' movement will lead to more healthy and safe working conditions. Before exploring this hypothesis by examining the occupational safety and health (OSH) situation in six industrialized countries (Sweden, Finland, German Democratic Republic–G.D.R., Federal Republic of Germany–F.R.G., United Kingdom of Great Britain and Northern Ireland–U.K., and

United States of America–U.S.A.) it is important to recognize what part OSH plays in the struggle over work and to recognize the depth and extent of the OSH problem.

THE OCCUPATIONAL SAFETY AND HEALTH PROBLEM

Roberto Acuna is thirty-four-years-old, an organizer for the United Farm Workers of America.

> I saw the need to change the California feudal system, to change the lives of farm workers, to make these huge corporations feel they're not above anybody–I began to see how everything was so wrong. When growers can have an intricate watering system to irrigate their crops, but they can't have running water inside the houses of workers. Veterinarians tend to the needs of domestic animals, but they can't have medical care for the workers. They can have land subsidies for the growers, but they can't have adequate unemployment compensation for the workers. They treat him like a farm implement. In fact, they treat their implements better and their domestic animals better. They have heat and insulated barns for the animals, but the workers live in beat-up shacks with no heat at all.
>
> Illness in the fields is 120 percent higher than the average rate for industry. It's mostly back trouble, rheumatism and arthritis, because of the damp weather and the cold. Stoop labor is very hard on a person. Tuberculosis is high. And now because of the pesticides, we have many respiratory diseases.[3]

There may or may not be an epidemic of work-related diseases and injuries raging in the U.S.A. and other countries. It is hard to know for several reasons. This uncertainty is greatest for work-caused diseases. The situation is clearer for injuries obviously connected with the workplace, such as falls, fires, explosions, and so forth. These are acute, usually violent and undeniable: thus, they are better reported and recorded.[4]

Diseases, are more problematic for several reasons. First, there are the clinical difficulties in distinguishing work-related diseases from other or "ordinary" diseases. Occupational diseases cover a whole gamut from relatively acute skin diseases; to stress-induced coronary disease which may develop very gradually; to asthma-like conditions set off by metal dust, fumes, or other air pollutants which may be an on-again/off-again chronic condition; to asbestos-caused lung cancer and mesothelioma which may take fifteen to thirty-five years to show itself clinically from time of the worker's first exposure to asbestos.[5] The trouble for the clinician, especially one not trained to pay attention to OSH diseases and not able to take a work history as well as a family history, is that a metal-dust asthma looks like a pollen-caused asthma; a solvent-caused skin rash looks like a rash from some home cleaning agent; a lung cancer caused by asbestos looks like one caused by smoking.[6] If the work-caused diseases are not diagnosed as such by primary care personnel they will not be reported and recorded.

Second, primary health care (PHC) providers (the ones workers may and usually do go to when they first seek professional help) are not usually prepared at all or are grossly underprepared to recognize, properly treat and preventively pursue these problems for other workers who may be similarly exposed in the way the patient was.[7] This is one of the central concerns of this work which we will examine in later chapters in relation to each study country.

Third, the organizational and financial linkages between the general health services, especially at PHC level, the OSH standard setting and enforcement agency (OSHA, in the U.S.A. and its counterpart agencies in other countries) and any special OSH services at the workplace are seldom figured out in such a way as to identify, count, and properly pursue these problems. This is true even on a case base, to say nothing of proper epidemiologic surveillance and follow-up to protect other workers exposed to the same hazards which were associated with a case which may have been picked up at the PHC level of the general health services.[8]

Finally, since OSH problems are inextricably intertwined in the political economy of production, the gathering and use of information is affected by the conflicting interests of capital and labor. These interests and leverages range from capital's ability to move jobs (and sometimes hazards)[9] in a political economic world-system, say from Connecticut to Maine or to India,[10] to a local union's hesitancy to demand a new ventilation system at a time of a depressed economy for fear of losing jobs to Maine or India.

For all the above and other reasons, there is an abysmal inadequacy about the OSH information system, especially in U.S.A.[11] This inadequacy has reached the depths in the "asbestos cover-up."[12] We will not pause here to go into the fascinating, even if often disappointing story of OSH data development and use, including the political economy of scientific research in this sphere of concern. Chapter 4 takes up this problem in some detail. Here it is enough to know that we are often dealing with consensual estimates or even highly controversial figures in the OSH field.

Although the extent and depth and exact character of the OSH problem are not well known, there is every reason to believe that they are widespread, numerous, varied, and serious. Still-used, ten-year-old estimates for the U.S.A. place the number of deaths from work-related diseases at 100,000, while another 15,000 die annually from traumatic injuries on the job.[13] For a variety of reasons suggested above, these figures probably underestimate the number of such deaths.

An estimated 390,000 new cases of disabling industrial disease occur each year.[14] This estimate also understates the problem, in part because it is a projection based upon the California workers' compensation system. These systems are notorious for a variety of reasons, including nonrecognition of occupational disease by PHC providers, for their underreporting of work-related disease. In addition, it probably understates the problem because a number of prevalent diseases may be work-related, but are not yet recognized as such. Thousands of stress-related conditions, including numerous complaints of headaches and eye strain from workers facing video display terminals for eight hours a day,[15] should be counted and may, in fact, be related to the development of more serious problems. One especially provocative, carefully conducted national prospective survey of working men in Sweden showed that those with less self-determination at work (for example, they did not feel free to leave their jobs at an appropriate time for ½ hour once a week to take care of some personal business) were at much greater risk of suffering from coronary disease.[16] Just as coal miners in the U.S.A. succeeded in having a law passed to compensate them for black lung disease,[17] clerical, technical, and other workers (for example, air traffic controllers) exposed to stressful work conditions may succeed to have legislation passed to protect and compensate them.

> Stress has become the black lung disease of the technical classes, says labor organizer Stanley Aronowitz. Aronowitz thinks organized labor must begin addressing issues, like stress, that are of concern to white-collar employees in high tech industries.[18]

Estimates of the proportion of cancer stemming from the workplace range from 4 percent to ten times that. (This is an example of the kind of problem I try to untangle in Chapter 4.)[19] If one chooses 20 percent as a reasonable compromise estimate, then, with 450,000 some deaths expected from all cancers in the U.S.A. population in 1984,[20] there will be 90,000 work-related deaths from this set of diseases alone. It is not an incidental point that cancer death rates are much higher for minority workers who often get the dirtiest and most hazardous jobs, such as the predominantly black workforce placed over the coke ovens in U.S.A. steel mills.[21] The workplace substances clearly associated with cancer include asbestos associated with lung, mesothelioma, and gastro-intestinal cancers, benzine with leukemia; aniline dyes with bladder cancer; vinyl chloride with all cancers; but especially angiosarcoma of the liver; hydrocarbons with lung cancer. There are some 450 chemical substances listed by the International Cancer Agency in Lyon France as confirmed or highly suspect human carcinogens. Another 1800 chemicals have something reported in the literature which makes them suspect. These reports may range from short-term, live-cell, mutagenic tests such as the Ames Test, to animal studies, to a single unconfirmed epidemiologic study. In addition, there are problems of radiation for uranium miners and hospital X-ray technicians, as well as the possibility that microwaves may be suspect. For reasons of the number of substances and other hazards which have been shown to cause cancer, some only recently discovered or confirmed, the large number of other suspect substances, and the long latency period between initial exposure and the clinical manifestation of the disease (fifteen to thirty years for many cancers, though leukemia seems to have a shorter "incubation" period, perhaps only a few years) some analysts have asserted that we have only seen the tip of the iceberg. They predict a veritable epidemic by year 2000.[22] This does not bode well for the World Health Organization's push toward "Health for All by Year 2000" through improved primary health care (PHC).[23]

Currently there are some one quarter of a million chemicals used in all workplaces. Some 15,000 are in industrial use. Many have been shown to be hazardous not only for cancer, but for a host of other diseases–silicosis from silica dust, most notably among miners; berylliosis from beryllium; severe skin disorders from solvents; and so on. Only a small proportion of these chemicals have been adequately tested for human effects.

> The effects of the vast majority of substances in common use are not really known, and little is being done to expand our knowledge. Each year about 3000 new chemicals are introduced into industry, yet their effects on the workers who will use them are never tested.[24] Standards for allowable exposure exist for less than 500 of the toxic chemicals currently in use. These standards are supposed to be the maximum concentrations to which workers can be exposed for eight hours a day, every day, without developing disease. They are based on average exposures, which means it is not illegal to be exposed to higher levels at certain times so long as the average exposure is below the legal limit.[25] This limit is known as the *threshold limiting value*, or TLV.[26]

We may only have seen or become aware of the tip of the iceberg, not only of work-caused cancer, but of all kinds of life-threatening and debilitating occupational diseases. The reasons for thinking so include: 1) the difficulty of distinguishing work diseases from other diseases, 2) the lack of training for PHC providers as well as lack of motivational structures so they could and would report and more adequately treat and participate in the prevention of these problems, 3) the long latency period between exposure to hazards and the noticeable appearance of some work diseases, 4) the entry of phenomenal numbers of new chemicals and other hazards (for example, VDTs and low level radiation) into workplaces, 5) almost weekly discovery of previously unknown cellular, animal, or human toxic effects of both new and previously used chemicals, and 6) the frequent distortion and even suppression of information on work hazards.[27]

Figures on mortality, disability, days lost from work due to OSH problems, worker compensation, medical care costs, and other OSH costs, lead to value lost estimates (by who knows what calculus as regards the value of a worker's life)[28] of some $35 billion a year in U.S.A.[29]

Obviously, this is not a problem only for U.S.A. In Sweden, a much smaller industrialized country (some 8.3 million versus the U.S.A.'s 230 million population), which has proceeded further than the U.S.A. and most other countries in protecting workers from OSH problems,[30] it is still estimated that 300 workers die each year from work-related causes, some 120,000 are hurt at or enroute to or from work, and some 2,000 are disabled. As the source recognizes, "These figures do not include injuries resulting from stress, monotony or a steadily increasing work pace."[31]

In the U.K., a medium-sized country among those compared in this work (56 million population) and the starting place of industrial capitalism,[32] progress has been made from the ravages of sweat shops and child labor of early capitalism to the social welfare state of modern Britain. Still, in 1980 it was estimated that over 1400 people die every year as a result of occupational accidents and diseases and at least 300,000 suffer serious injury while at work. "In economic terms it has been estimated that the cost of this to the nation is in the order £2 million per annum. The cost to those involved, however, can never be calculated."[33]

Where are rates of work-caused injuries, illnesses and deaths lowest and where are they highest among our six study countries? OSH information systems are so poor and problems attached to making estimates so many (see Chapter 4) that this work more or less avoids the problem. Instead, this book focuses on organizational arrangements and other measures to protect the safety and health of workers at work. If I were forced to give crude estimates of effectiveness, I believe that the strongest protection measures go along with the lowest rates. For example, the G.D.R.'s OSH protection system is much stronger than the F.R.G.'s and information comparing them shows the G.D.R. to have lower occupational accident, illness, and death rates. The figures cited above, if calculated out to yield crude work-caused deaths per 100,000 population (not age, sex, or work-type controlled) show fifty for U.S.A., 3.6 for Sweden, and just 2.5 for U.K. By these calculations, the U.S.A. crude work-related death rate is fourteen times the Swedish rate and twenty times the U.K. rate! Taking other information into account, I estimate that workers suffer the fewest work-related injuries, illnesses, and deaths per 100,000 in Sweden and the G.D.R.; Finland seems to fall in the middle along with the U.K.; the F.R.G. is next; and the U.S.A. has the highest rates. But, again, the reader is cautioned that these are only best guesses.

The OSH problem may be most severe and varied in underdeveloped countries undergoing industrialization in agriculture and other sectors. In these countries, workers often carry a burden of malnutrition and infectious disease which presumably makes them all the more susceptible to toxic hazards at work. The problems are especially severe in capitalist-oriented countries with repressive regimes which have not adopted protective standards, do not enforce any they may have, and do not permit unions.[34]

THE HISTORY OF OSH

In the first century AD, Pliny the Elder wrote about a lung disease suffered by slaves who worked in asbestos mines and described a crude respirator devised for their protection.[35] In the sixteenth century, Paracelsus studied a number of diseases, for example, what he called "Phthisis," which he connected with such occupations as mining and smelting. Agricola's *De re metallica*, including observations on the hazards of mining and metallurgy, was also written in the sixteenth century. But the starting point of the field of occupational health or industrial medicine is usually identified with the publication in 1700 of Bernardino Ramazzini's (1633-1714) monumental treatise, *The Diseases of Workmen*. His life-long study of the hazards and diseases associated with the medieval crafts contributed undertakings and principles still referred to by industrial physicians of this day. In the 1770s Sir Percival Pott documented the first links between work and cancer by revealing the high rate of scrotal cancer among London's chimneysweeps. It is a sad fact that people are still dying from cancer due to work exposures to the same substances, hydrocarbons.[36] Even though knowledge which would protect workers may be in hand, it is often not brought to bear. One of the goals of the present work is to understand and suggest ways of correcting this tragic situation.

The conditions which early industrial capitalism foisted upon its workers and their families were anything but idyllic. Engels described the horrors of sweat shops, child labor, and the undernourished, overcrowded, impoverished, disease-ridden working class.[37] Long before his studies, public outcry and the fear of explosive unrest had led to the Elizabethan Poor Laws and the important Poor Law Reform Bill of 1834. Eventually, an increasingly organized labor movement brought about changes in the mines and mills. But these were piecemeal. The U.K.'s Factory Act of 1844 required "Certifying Factory Surgeons" whose sole function was to verify that children were old enough to work. In 1855 other functions were added including investigation of industrial accidents. The first Workman's[38] Compensation Act of 1897 led some larger employers to appoint physicians to protect themselves against claims (not primarily to protect the workers). Thus, as in other countries, the first industrial medical services were in large plants. In 1906 the first modern medical case of a death from asbestos was presented by a British physician, Montague Murray. In 1924, another British physician, Dr. Cooke, reported several deaths due to asbestos among relatively young women who had sorted the fibers in one factory. In a few years, regulations for improved ventilation were instituted in Britain. Thus, protective standards were adopted piecemeal too, usually a new standard for any disaster followed by public outcry. Today, the United Kingdom's OSH standards manual is a massive catalogue which has grown like Topsy. Only the person deeply steeped in its history and meaning can find his way through it. Certainly, it is beyond most ordinary mortals, including most workers. The development of a national health insurance system in 1911 and then, following WW II, the first National Health Service in a capitalist country[39] also form part of the background of workers' health, especially as regards the relation of PHC to workers' health. But these developments and other aspects of the OSH system are reserved for discussion in Chapter 11 on the U.K. Here, before looking briefly at OSH developments in U.S.A., I will simply note that OSH laws in the U.K., the first industrial capitalist nation, were not brought together in a reasonably comprehensive manner until the Health and Safety at Work Act of 1974.[40]

Industrial capitalism developed somewhat later in U.S.A. Public concern, particularly for safety problems where women and children were employed in the textile mills, led in 1867 to the first department of factory inspection. Ten years later Massachusetts passed the first workplace safety law; it re-required guards for textile spinning machinery. The largest labor organization, The Knights of Labor, demanded in the 1870s and 1880s "the adoption of measures of providing for the health and safety of those engaged in mining and manufacturing, building industries, and for indemnification . . . for injuries."[41] After 1900, when labor unrest over hazardous conditions was joined by public outcries over a number of especially notable tragedies, most industrialized states in the U.S.A. went beyond the minimal laws they had to deal with work hazards. For example, the New York State Factory Investigating Commission was formed in the aftermath of the Triangle Shirtwaist Fire. In this disaster, 145 of 500 workers, mostly young Jewish and Italian immigrant women, burned to death, were suffocated, or lost their lives jumping from the building (the company was located on the upper three floors of a ten-story building in New York City). Part of the problem was that most of the doors were blocked or locked or only opened inward.[42]

The state workers' compensation systems came in after 1910 when the National Association of Manufacturers and the National Civic Federation (dominated by large corporate members) began to lobby for such measures. While these measures gave some aid to injured workers (the focus was almost entirely on injuries, with little or no recognition of work-caused disease—a characteristic which predominates to this day)[43] their main purpose was to protect owners from unpredictable and increasingly large and successful suits brought by injured workers and their families. Thus following adoption of the workers' compensation laws, the worker could no longer sue his or her employer.[44] The employer had a predictable "cost of doing business" with possibly lower outlays than continuing suits may have entailed. In no state were the benefits close to 100 percent of lost wages.

In 1913 the National Safety Council was organized and launched its "safety first" educational campaign. This may have done some good in the way of raising awareness and concern for accidents, but it led to a "blame the victim" approach to worker education—the focus being on the "accident prone" or careless worker, rather than on hazards such as slippery floors or unguarded machinery which could be engineered out of the situation.

One of the true heroes of the field, Dr. Alice Hamilton, began her life's work in Chicago. She conducted the famed "Illinois Survey" of conditions in the lead industry. Her work was for the most part unsupported and informal–that is, conducted by herself without official or industry sponsorship. It took her to many parts of the industrialized U.S.A. One of her studies focused on mercury poisoning among workers in the felt hat industry in Danbury, Connecticut. In 1919, she became the first woman Harvard faculty member when the industrial hygiene program was formed. Her autobiography is an inspiring document.[45]

Pressures to conserve and enhance the labor force–especially the labor shortages of the two World Wars–led to renewed emphasis on workplace safety. In addition, such outrageous disasters as the Gauley Bridge "incident" in West Virginia, periodically raised public consciousness and brought about specific protective measures, or, as in this episode, simply the formation of an industry-sponsored protective association which evolved into the Industrial Health Foundation (IHF). In this disaster, more than 470 men died and 1,500 were severely disabled. Workers, mostly black and poor, desperate for a job during the Great Depression, signed on for fifty cents (later thirty cents) per hour. They were herded with their families into chicken-coop-like ten-by-twelve-foot shacks with twenty-five to thirty people in each.

> The water tunnel was being cut through almost pure silica, and the dust was so thick that workers sometimes could see barely ten feet in the train headlights. Instead of waiting thirty minutes after blasting, as required by state law, workers were herded back into the tunnel immediately, often beaten by foreman with pick handles.[46]

Aside from the formation of the Air Hygiene Foundation under auspices of the Mellon Institute of Pittsburgh (later the AHF evolved into the IHF) not much happened as a result of the public hearings and outcry over this disaster. The West Virginia workers' compensation law was amended to include silicosis as compensable, but workers could collect benefits only if they had been working for two years or more on the job which exposed them to silica dust. This made sure that none of the Gauley Bridge survivors would qualify, nor would any future such "incident" lead to compensation.

The close link between the state of a political economy and OSH is revealed by this experience when contrasted with measures taken during wartime when labor was in short supply:

> The failure of the horror at Gauley Bridge to precipitate any effective, broad-based action against the scourge of work accidents and diseases was a clear indicator of the low status of the job safety and health issue during the Depression years. The advent of World

> War II brought a revival of interest in the problem as the federal government sought to reduce occupational accidents that might adversely affect the war effort.[47]

It was not until 1970, following the Farmington, West Virginia deaths of seventy-eight miners in 1969 and passage of the landmark Coal Mine Safety and Health Act by the U.S.A. Congress (see note 17) that broad national OSH legislation was passed in the form of the Williams-Steiger Occupational Safety and Health Act, or OSHA. "The passage of the OSH Act had a tremendous significance both as a watershed in the struggle over reforms in occupational health and as a demonstration of the vitality of the new forces that had entered this struggle in the 1960s."[48] Joining the long but not until then very successful struggle of organized labor were the forces of the civil rights movement; a minority workers' movement; the environmental health movement; the women's movement (both of these last representing more middle-class forces) and the Anti-Vietnam War movement which to some extent tended to unify anti-establishment forces opposed to authority gone wrong. OSHA came into being in a flurry of activity in the 1970s which leveled off in a declining economy and came to a standstill or even a roll-back under the Reagan Administration.[49] More of the OSH history and the struggle surrounding it in U.S.A. will be presented in Chapter 12.

It is more than a matter of causal curiosity that broad, integrative, OSHA-like legislation was passed in the late 1960s or early 1970s in all six study countries. This common experience (even though the laws themselves reflect very significant differences) will be taken up in Chapter 3 along with especially notable differences between the six countries. I conclude this introductory history segment by noting that these and other countries have much to learn from one another. Unfortunately, the transfer and use of life-protecting knowledge across borders has not often occurred in an expeditious, effective way in the past—witness the discovery of Asbestosis in 1926 in Britain and the failure of the U.S.A. to adopt an asbestos standard until 1970. The relevant knowledge for saving workers' lives is not just technical. It is also sociological knowledge concerning the ways in which workers' movements have organized in a general way to struggle for improved OSH as well as specific arrangements they have achieved for OSH services on the job and other conditions. It is in part to encourage the exchange of such knowledge between workers' movements and professional and technical people concerned with improving OSH in different countries that the present work is written.

ELEMENTS FOR ENHANCEMENT OF OSH INCLUDING PHC

When conditions affecting OSH are to be examined and compared, it is helpful to share an awareness of the main elements of an OSH system. I have already suggested that the broad forces of a nation's political economy functioning within and influenced by the capitalist world-system play a determining role—such matters as the level of monopoly and competition within an industry; how easily parts of production can be transferred to low-wage, nonunion, underdeveloped countries; the level of unemployment; and so on. These elements will be further discussed in Chapter 5 on framework. Here, it is important to identify the elements more directly connected to protecting workers' health.

In each country, workers' struggles have led to the establishment of some kind of labor inspectorate, some branch of which focuses on safety and health issues. Sometimes these are located in the Ministry of Labor or its equivalent, sometimes in the Ministry of Health or its equivalent. Traditionally these agencies have concentrated on safety matters—the hazards of fire, explosion, electrical shock, falls, being drawn into a machine, having a finger or hand or limb cut off by a machine, and so forth. Increasingly, in recent years, in part due to such landmark epidemiologic work as that of Irving Selikoff and his colleagues on asbestos and cancer,[50] and much subsequent work on the large number of other toxic substances,[51] standards have been passed with the prevention of work diseases in view. Standards are *more* or *less* adequate. For example, under the OSHA standard for noise, *ninety* decibels average over an eight-hour exposure, it has been estimated that 18 percent of workers would go deaf at this level over a working lifetime. The Swedish standard is eighty-five decibels, less than half the U.S.A. level or more than twice as protective (because noise is measured on a ratio scale, with a doubling at about every three dB). Once in place, standards must be enforced and the number, qualifications, and mandate of inspectors as well as their reception in plants, all relate to their ability to assure compliance.

Just as a standard may not be protective enough, the lack of a standard is no assurance that a substance is safe. In the movie, "Song of the Canary,"[52] workers at an Occidental Petroleum chemical plant in California discover they are becoming sterile from working with the pesticide, DBCP. There was no standard. An emergency standard was adopted after many workers were affected at many production and use cites across the country.

Perhaps the primary element for improved OSH is worker education and control. A good deal more will be said about this. Here it is only necessary to note that without knowledge of the hazards they are exposed to, knowledge of the likely effects, and knowledge of how to protect themselves, including knowledge of how to organize, for example, to achieve improved OSH provisions in a contract, workers are helpless to protect themselves. It may be that someone on management's side will tell them to wear a respirator or other personal protective equipment. But they may not know that engineering controls or administration controls (for example, rotating jobs to achieve lower exposures) are preferable and personal protective equipment a last resort or only a temporary or supplemental measure. Also, they may not realize that the respirator has to fit correctly and be technically effective for the particular hazard.

This key element connects to a set of others in the shop. Is there a well organized health and safety committee with a strong worker voice? Are there trained full-time OSH union representatives free to walk around and look for and correct hazards? Is there an OSH medical, nursing and safety engineering service with other appropriate personnel (industrial hygienists, stress-aware psychologists, ergonomicists) who are responsive to workers' concerns and needs? These elements vary markedly from one study country to another.

Is there a comprehensive and effective information system? Such a system should include epidemiologic and hazard surveillance, an inventory of hazardous materials, their distribution, and those exposed, regular reporting and follow-up of work-related clinically discovered work disease cases, and special studies of old and new hazardous materials and conditions.

Finally, since they are so important in first identifying work disease (or failing to identify it—all too often the case), in properly treating a case and following up to assure proper work placement and other matters, including proper compensation if due, and in providing information for proper surveillance and follow-up of other workers who may be similarly exposed, what OSH preparation do PHC providers have and what mandate and organizational arrangements do they function under? In a regionalized national health service which has PHC units at the periphery of the system assigned geographically, one might expect PHC personnel to become familiar over time with the workplaces and hazards in their areas even though they have not had a thorough introduction to OSH problems in their training. The NHS in the U.K. should offer this advantage, at least theoretically,[53] over the come-and-get-it, market-oriented system in the U.S.A.

The present work can only reflect on the general situation of each of these components in the six study countries. I did give special attention to worker education and control over OSH conditions. And, since my interest was especially strong in how the general health services, especially at PHC level, back-up and support workers' health (or fail to do so) I concentrated considerable attention on the medical, health services aspects of OSH. Thus, I gave less attention to safety aspects and the work inspectorate. These aspects are very important, so they were not ignored. I am simply alerting the reader to the main emphasis of this volume which is on worker involvement and OSH services in the plant and the OSH-PHC connection. However, I do try to describe the general situation as regards the work inspectorate and the information system and other aspects of the OSH system in each country.

THE APPROACH TO STUDY COUNTRIES

A more detailed statement of methods is included in the Appendix. Here I simply want to present the logic of choice of the study countries, indicate my auspices and general approach to each country, and suggest both limitations and strengths of this work.

Since my focus was on the OSH-PHC connection, I wanted to study countries with quite differently organized health systems. Also, since I had a general working hypothesis that a strong workers' movement, both historically and currently, would be associated with a better OSH system as well as better OSH-PHC connections, I wanted to study countries with workers' movements with evidently different strengths and forms. Although the OSH problems of the peripheral and semi-peripheral countries (the underdeveloped countries)[54] are very serious, perhaps more serious than in core capitalist nations (see note 34), it seemed best to limit the study to industrialized countries for comparability of problems and personal reasons of logistics and feasibility. Thus the study countries provided a range of comparisons suggested by Table 1. More detail will be given in Chapters 7-12 on each country. Obviously, the categories are meant to be suggestive. Thus, it is not certain that the workers' movement in the U.K. today under Margaret Thatcher and the Torries is much stronger than the workers' movement in F.R.G., but its history suggests it should be stronger.

The kinds of indicators one has in mind when making these tentative classifications include whether labor has had its own political party, whether that party is or has been in power, the proportion of the workforce which is organized, whether the labor law provides for closed shop organizing (single union, complete membership if a majority vote it in) etc. The G.D.R. is a special case without truly independent unions but the state is said to function as a "workers state" under guidance of the Communist Party.

I did the main field work for this volume during my sabbatic leave in the Fall of 1981. I have stayed in touch with key informants in each country since and have received both their critiques of early drafts describing the OSH situation in their countries and updates on major developments through Spring, 1984.

The work was done without a special grant. However, after presenting my plan to a number of senior staff at the European Regional Office of the World Health Organization in Copenhagen, I did receive some travel support and the sponsorship of the organization in the form of my serving as a Temporary Advisor. This sponsorship probably did not matter much in most countries. I knew colleagues with interests close to mine in each country and these colleagues helped arrange interviews and gain access to written materials.[54] But this sponsorship was very important for my visit to G.D.R. Without it, I do not believe I would have got as thorough an overview as I did under the very willing and supportive auspices of the Ministry of Health for which I am most grateful. Although I prepared a 223-page preliminary report for WHO/EURO, the present work is of a different character and WHO/EURO has no responsibility for anything contained herein.

I believe the work has some of the strengths of an "in-depth" contrasting case study in the field of cross-national study of health systems, a notion discussed further in Chapter 2. I believe the reputational sample (see Methods Appendix)

Table 1. Study Countries by Strength and Form of Workers' Movement and Structure of the Health Service System

Structure of Health Services	*Strength and Form of Workers' Movement*			
	Independent Union Movements			*Workers' State*
	Weak	*Medium*	*Strong*	
Segmented national health insurance (Medicare)—Primarily market oriented	United States of America			
National health insurance—primarily market oriented	Federal Republic of Germany			
Mixed public-private, planned market system		Finland		
National health service, regionalized and geographically targeted at periphery		United Kingdom of Great Britain and Northern Ireland		
National health service, regionalized, but not geographically targeted at periphery			Sweden	
National health service, regionalized and geographically targeted at periphery				German Democratic Republic

includes key spokespersons from major relevant interest groups in each country. Obviously, there are a number of weaknesses. However careful I was to take complete notes during (or immediately following) interviews, however diligent I was about collecting a complete set of relevant documents in each country, this is still the work of one person who had the time and means to visit each country for only about two weeks. Even if I have poured diligently through the mass of materials and tried to keep in touch with key informants, it must be seen as an exploratory, hypothesis generating work. Where I guess that something may be one way rather than another, but I am uncertain, I will say so. Nevertheless, some important differences seem so clear and documentable and some findings seem to be so internally consistent with the rest of the "case" that I have been emboldened to present conclusions and recommendations as well as general impressions relevant to a sociological understanding of workers' safety and health and the role of PHC in support thereof.

NOTES AND REFERENCES

1. Several quotes will be used from this marvelous book to help put some ordinary people behind the otherwise abstract ideas presented here. Studs Terkel, *Working*. New York: Avon Books, 1975.
2. William Cahn, *Lawrence 1912, The Bread and Roses Strike*. New York: Pilgrim Press, 1980, quote from p. 216.
3. Terkel, op. cit., pp. 30 and 36. On the generally oppressive and diminishing character of work for others' interests see André Gorz, *Farewell to the Working Class, An Essay on Post-Industrial Socialism*. Translated from the French by Michael Sonensher. Boston: South End Press, 1982 (original 1980).
4. Even with these events there is disagreement, especially when death is not involved. For example, in classes conducted by a New Directions Program for Worker Education in OSH which I directed for some five years (1978–1983) we often heard examples such as a worker being brought back in the afternoon with a cast on his leg after having suffered a broken leg in the morning. The worker may be put at a desk to stamp cards or answer the phone or do nothing. The point is that for insurance purposes and safety competition between branches, the local manager does not report an accident resulting in "days lost from work." In his critical analysis of workers' health and safety, Navarro puts the number of deaths from accidents, fires, etc. at between 14,000 and 28,000: Vicente Navarro, The Health of Working America in *Medicine under Capitalism*. New York: Prodist, 1976.
5. A good, but not exhaustive listing of hazardous occupations, work-related diseases and known and suspected dangerous substances is: Marcus M. Key, *et al.*, (ed.). *Occupational Diseases, A Guide to Their Recognition*. Washington, D.C.: U.S. Public Health Service, National Institute for Occupational Safety and Health, revised edition, June 1977.
6. Barry S. Levy and David H. Wegman, Recognizing Occupational Disease in Levy and Wegman (eds.), *Occupational Health, Recognizing and Preventing Work-Related Disease*. Boston: Little Brown and Co., 1983.
7. Barry S. Levy, The Teaching of Occupational Health in American Medical Schools, *Journal of Medical Education 55*:18–22, January, 1980.

8. In this regard, some nations have better health services organizational structures than others. Most national health service systems are regionalized in such a way that the most local PHC service units have a geographically defined population to serve. In such situations, whether or not PHC providers have received adequate preparation in OSH, one expects these local service units to become familiar over time with the workplaces and sorts of problems they produce. While it may help, this is not usually a complete solution to the OSH-PHC problem. See Julian Tudor Hart, Primary Care in the Industrial Areas of Britain, Evolution and Current Problems, *International Journal of Health Services 2*:349–365, 1972; also James McEwen, *et al.*, The Interface between Occupational Health Services and the National Health Service, *Public Health* (London) *96*:155–163, 1982.
9. Barry I. Castleman, How We Export Dangerous Industries, *Business and Society Rev.*:7–14, Fall, 1978; also Castleman, Double Standards: Asbestos in India, *New Scientist* pp. 522–523, 26 February, 1981.
10. Elsewhere I have attempted to summarize the essential elements of this world-system as found in the extensive work of Immanuel Wallerstein, André Gunder Frank, and others, and to apply it to a range of international health problems: Ray H. Elling, The Capitalist World-System and International Health, *International Journal of Health Services 11*:21–51, 1981.
11. Daniel M. Berman, The Official Body Count, Chapter 2 in *Death on the Job, Occupational Health and Safety Struggles in the United States.* New York: Monthly Review Press, 1978. See also the excellent chapter on differential diagnosis and counting of work-related disease The Incidence of Occupational Disease, in Peter S. Barth with H. Allan Hunt (eds.), *Workers' Compensation and Work-Related Illnesses and Diseases.* Cambridge, MA: MIT Press, 1980.
12. One of the more glaring examples of the misuse of OSH information is the asbestos cover-up by Johns Manville, other asbestos manufacturers, perhaps the U.S.A. government, and in some instances, where unions feared plant closings, even unions: Samuel S. Epstein, The Asbestos 'Pentagon Papers', pp. 154–165 in Mark Green and Robert Massie (eds.), *The Big Business Reader, Essays on Corporate America.* New York: Pilgrim Press, 1980. This cover-up now forms the base of thousands of actually and potentially successful third-party suits against Manville and other suppliers of asbestos (the worker compensation laws prevent workers from suing their own employers). Judge Urges Manville Have 20-Year Fund for Payment of Future Asbestos Claims, *The Wall Street Journal*, October 26, 1983, p. 16. Also, Harry F. Rosenthal, U.S. Charges Asbestosis Cover-up, *The Hartford Courant*, September 17, 1983, p. A.4.
13. Nicholas A. Ashford, *Crisis in the Workplace, Report to the Ford Foundation.* Cambridge, MA: MIT Press, 1976.
14. *President's Report on Occupational Safety and Health*, 1972.
15. In the case of VDTs, physical ergonomic and radiation effects may be creating these problems as well as stress. Howard Hu, Other Physical Hazards and their Effects, pp. 219–233 in Levy and Wegman (eds.), op. cit. (note 6) esp. pp. 234–235.

16. Robert Karasek, *et al.* Job Decision Latitude, Job Demands, and Cardiovascular Disease: A Prospective Study of Swedish Men, *American Journal of Public Health 71*:694–705, July, 1981. The influence of this factor of self-determination at work held up statistically in spite of information of other serious risk factors such as diet and smoking being controlled in the study. The general sphere of stress and democracy at work has received a lot of attention in Sweden. We will come back to this concern in Chapter 7 on Sweden. There is also some work on this sphere of concern in the U.S.A.: James S. House, *Occupational Stress and the Mental and Physical Health of Factory Workers*. Ann Arbor: Survey Research Center, Univ. of Michigan, 1980.
17. The Black Lung Association in Fayette and Kanawha counties in West Virginia succeeded in getting a state law after 42,000 of 44,000 miners went out in a wildcat strike to demand a broad health and safety reform bill in 1969. "When 78 miners were killed in an explosion at Consolidated Coal's huge No. 9 mine in Farmington, later that year, union and public pressure finally forced the passage of the historic Coal Mine Safety and Health Act by Congress." This bill was the immediate forebringer of the OSHA law of 1970 about which more later. Molly Joel Coye, Mark Douglas Smith, and Anthony Mazzocchi, Occupational Safety and Health, Two Steps Forward, One Step Back, pp. 79–106 in Victor W. Sidel and Ruth Sidel (eds.), *Reforming Medicine, Lessons of the Last Quarter Century*. New York: Pantheon Books, 1984, quote from p. 83.
18. Stress Concerns Labor Organizer, *The Hartford Courant* January 30, 1984, p. C7. Connecticut and some other states have passed laws to compensate police and firemen for coronary disease which is defined by law as related to the high stress of these jobs. Industry, fearing a broad precedent, has attempted to overturn these laws.
19. Health Education and Welfare Secretary, Joseph Califano, citing a federally sponsored study estimated in 1978 that between 20 and 40 percent of cancer deaths were caused by on-the-job exposures. Daniel Berman, *Death on the Job Occupational Health and Safety Struggles in the United States*. New York: Monthly Review Press, 1978, p. 46. Later an industry-sponsored study came up with the 4 percent estimate: R. Doll and R. Peto, The Causes of Cancer: Quantitative Estimates of Avoidable Risk of Cancer in the United States Today, *Journal of the National Cancer Institute 66*:1191 ff, 1981. Samuel Epstein, Author of *The Politics of Cancer*. New York: Anchor, 1979, said on the McNeil-Lehrer News Hour on March 2, 1984 that worker exposure to just six chemicals will account for one-third of cancers in future years.
20. American Cancer Society, Facts and Figures, 1984, (pamphlet). The cancer death rate per 100,000 population has been rising from 143 in 1930 to 152 in 1940 to 158 in 1950 to 169 in 1980 (p. 3).
21. American Cancer Society, Cancer Facts and Figures for Black Americans, 1979.
22. Irving J. Selikoff and E. Cuyler Hammond, Environmental Cancer in the Year 2000, pp. 687–696 in *Proceedings from the 7th National Cancer Conference*, Los Angeles, 1973. New York: American Cancer Society.

23. *Alma-Ata 1978, Primary Health Care*. Geneva: WHO, 1978. For a critique of the PHC ideology see Vicente Navarro, A Critique of the Ideological and Political Positions of the Willy Brandt Report and the WHO Alma Ata Declaration, *Social Science and Medicine 18*:467–474, 1984.
24. In the meantime, in U.S.A., the Toxic Substances Control Act passed in 1976 has mandated that new chemicals for industrial use be put through a battery of different kinds of short-term (live-cell mutagenic tests and several variants thereof) as well as certain animal tests. A chemical may be refused if found too hazardous by these means. However, these tests do not assure that workers will show no effects, nor do they cover the quarter million chemicals already in use.
25. These 500 standards were simply taken over in the OSHA laws as it came into force in 1970. They came from industry-sponsored industrial hygiene research conducted through such organizations as the American National Standards Institute. The unfortunate thing is that they only reflect acute effects—rashes, headaches, other frank, quickly noticed symptoms or diseases. They do not reflect long-acting chronic effects such as cancer or arthritis or coronary disease. Only one new standard was adopted as part of the OSHA law, that for asbestos.
26. Jeanne M. Stellman and Susan M. Daum, *Work Is Dangerous to Your Health*. New York: Vintage Books, 1973, quote from pp. 155–156.
27. In Chapter 4 and elsewhere I will go into such cases as the "Asbestos coverup" and the struggle for the workers' "Right to Know" what chemicals they are working with, what the possible effects are, and how to protect themselves. This right is still not assured in U.S.A. Associated Press, OSHA Prepares Controversial Rule on Disclosing Workplace Chemicals, *The Hartford Courant*, November 22, 1983, p. A4. Also, OSHA Issues 'Right to Hide' Rule, State Right to Know Laws Threatened, *CACOSH* (Chicago Area Council on Occupational Safety and Health) *Health and Safety News* November/December, 1983, p. 1.
28. The Ford Motor Company estimated a person's life on average was worth $500,000 when they did a "cost-benefit analysis" (your cost, my benefit) of the predicted loss of life from placing the Pinto's gas tank in the rear, unshielded, versus the costs of redesigning the car. Mark Dowie, Pinto Madness, *Mother Jones* pp. 18–24ff, September/October, 1977. The figure for human life yielding this estimate of $35 billion costs due to OSH problems in the U.S.A. was probably not that high.
29. Ashford op. cit. (note 13) esp. Sect. 314 beginning p. 83.
30. Steven Kelman, *Regulating America, Regulating Sweden: A Comparative Study of Occupational Safety and Health Policy*. Cambridge, MA: MIT Press, 1981. Kelman documents many of these achievements, though I disagree sharply with his cultural explanations (the "overhead state," a notion of respect for authority stemming from the days of absolute kingship; and the idea of accommodationist institutions). These ideas ignore the strength of the workers' movement in Sweden and the gains won in sometimes severe, sharp and prolonged struggle. A paper of Vicente Navarro's documents this history and destroys the illusion of Sweden as a country of

perpetual labor peace: The Determinants of Social Policy, A Case Study: Regulating Safety and Health at the Workplace in Sweden, *International Journal of Health Services 13*:517–561, 1983.

31. Occupational Safety and Health in Sweden, Stockholm: The Swedish Institute, no date (probably 1979), quote from p. 1.
32. Wallerstein argues that the capitalist mode of production began in the world-system controlled by Portugal. But this was a largely slave-worked agricultural and extractive economy. The Laws of Enclosure and other conditions generated the first surplus urban population to be put to work as "wage slaves" under industrial capitalism in England. Imanuel Wallerstein, *The Modern World-System: Capitalist Agriculture and the Origins of the European World-Economy in the Sixteenth Century*. New York: Academic Press, 1974.
33. Trades Union Congress, Workplace Health and Safety Services, TUC Proposals for An Integrated Approach, London: TUC Publications, 1980, quote from p. 5 (booklet).
34. Ray H. Elling, Industrialization and Occupational Health in Underdeveloped Countries, *International Journal of Health Services 7*:209–235, 1977.
35. Donald Hunter, *Diseases of Occupations*, 4th edition. Boston: Little, Brown, 1969, p. 972.
36. The tragic fact that workers, more usually black workers, are dying from exposure to hydro-carbons emanating from the coke ovens around Pittsburgh long after these substances were first found to cause cancer is examined in Joseph K. Wagoner, Occupational Carcinogenesis: The Two Hundred Years since Percival Pott, pp. 1–3 in *Occupational Carcinogenesis*, special issue of the *Annals of the New York Academy of Science 271*:1–3, 1976.
37. Today people aware of the contribution of women to the labor movement insist on the term "Workers'" (instead of "workman's") Compensation."
38. Bismarck adopted the first national health insurance system in the world in 1883. It was the socialist-oriented Soviet Union which adopted the first National Health Service following the revolution of 1917.
40. More of the OSH history for each study country will be presented in the country chapters.
41. Charles Levenstein, A Brief History of Occupational Health in the United States, pp. 11–12 in Levy and Wegman, op. cit., quote from p. 11. I am indebted to this piece and to Daniel Berman's work, *Death on the Job*, for much of the U.S.A. history of OSH presented here and in more detail in Chapter 12 on U.S.A.
42. Berman, op. cit. (note 11): 9.
43. Peter S. Barth with H. Allan Hunt, *Workers' Compensation and Work-Related Illnesses and Diseases*. Cambridge, MA: MIT Press, 1980.
44. The many recent successful suits brought against Johns Manville Company and other asbestos manufacturers by workers who have suffered from asbestosis and cancer are third-party suits against suppliers of a hazardous material not the workers' employers. Judge Urges Manville Have 20-Year Fund for Payment of Future Asbestos Claims, *The Wall Street Journal* p. 16, October 26, 1983.

45. Alice Hamilton, *Exploring the Dangerous Trades*. Boston: Little, Brown, 1943. See also Barbara Sickerman, *Alice Hamilton: A Life in Letters*. Cambridge, MA: Harvard University Press, 1984.
46. Berman, op. cit. (note 11): 27. One observer reported, "Three hours was the standard elapsed time between death in the tunnel and burial. In this way the company (Rhinehart-Dennis, affiliated with Union Carbide) avoided the formalities of an autopsy and death certificate. It was estimated that 169 blacks ended up in the field, two or three to a hole." Joseph A. Page and Mary-Win O'Brien, *Bitter Wages, Ralph Nader's Study Group Report on Diseases and Injury on the Job*. New York: Grossman, 1973, p. 161. Today as Union Carbide attempts to promote an image of being a "safe company" in the face of Bhopal and harmful gas leaks in Institute, West Virginia, it is important to remember its nefarious role in the Gauley Bridge disaster of 50 years ago.
47. Page and O'Brien, op. cit. (note 46): 63.
48. Coye, *et al.*, op. cit. (note 17): 81.
49. Philip J. Simon, *Reagan in the Workplace: Unraveling the Health and Safety Net*. Washington, D.C.: Center for Study of Responsive Law, 1983.
50. Irving J. Selikoff, Cancer Risk of Asbestos Exposure, pp. 765–784 in Howard H. Hiatt, *et al.* (eds.), *Origins of Human Cancer*. New York: Cold Spring Harbor Laboratory, 1977.
51. Richard R. Monson, *Occupational Epidemiology*. Boca Raton, FL: CRC Press, 1980.
52. The title is taken from the fact that coal miners used to take small birds into the mines. If the bird stopped singing, the miners knew they had to escape from the area as methane gas was threatening their lives. The dilemma of modern workers is that there usually is no early warning system—except other workers who die or show symptoms from some previously unknown hazard.
53. André Gunder Frank, *Capitalism and Underdevelopment in Latin America*. New York: Monthly Review Press, 1969. Also, Walter Rodney, *How Europe Underdeveloped Africa*. Dar es Salaam: Tanzania Publishing House, 1972.
54. I leave unnamed all interviewees and key informants to protect them and to leave me completely free to describe and analyze the situation in each country as best I can. Nevertheless, I owe a great debt to all who cooperated. And all who were asked did cooperate except for one or two persons in each country whose schedules simply could not be reconciled with mine. Even in some of these cases, I conducted a partial interview by phone.

CHAPTER 2
Workers' Health in the Context of Cross-National Study of Health Systems (CNSHS)

WHY STUDY DIFFERENT HEALTH SYSTEMS?

At the most abstract level, philosophers of science argue that we cannot "know" anything except by comparison. Similarly, Eastern religions teach that beauty and ugliness, good and evil, the moral and immoral, the Yin and Yang, are closely interwoven, really two sides of a single nature and cannot be understood or appreciated one without the other.[1]

Since every known people everywhere, at all times, has had some way to confront the threats of illness and death, there is much to be learned about our own "health" (really medical care) system or non-system, as some would argue, by comparison with more highly planned systems.

The approach might be stoic or minimal, as among certain tribes of American Plains Indians who attempted to repress all recognition of illness and saw the only honorable death as one occurring in combat with an enemy. Among these people, even deep depression, which could lead, in another culture, to suicide, might be denied as an illness and converted into aggression against the enemy:

> If a warlike Crow Indian became depressed, he would chant the songs and wear the garb that marked him as a 'Crazy Dog,' pledged to die fighting the enemy within the season. His will to self-destruction was thus achieved through homicide, a means acceptable to his society and akin to that sometimes suspected by psychologists in the bravado of racing drivers or fighter pilots.[2]

Or the approach to illness and death might be as elaborate and complex as among those in a variety of mixed ethnic groups in the worker and peasant classes of Latin America who see any illness as an imbalance between hot and cold caused by ingestion of a hot or cold substance.[3] Sometimes this approach is complicated by elements of voodoo. And usually there is at hand some facet of the most elaborate and complex approach of all–modern or cosmopolitan medicine. This facet may be a clinic, a family planning program, a public health immunization or malaria control program, or, in urban centers, a hospital.

How is a people's approach to health and illness determined? Much of the work on this question in the past has reflected the cultural concerns of anthropologists–the cross-cultural study of health and illness conceptions and behaviors. Recently, those who worked from this perspective have begun to realize that social structural forces–the very establishment of modern facilities, or the support by the government of one traditional medical group or another–bring about changes in health orientations and behaviors. But these changing perspectives often remain at a micro-cosmic level–the structural changes going on in the immediate area of a village one is studying and the concomitant changes in health and illness beliefs, values, attitudes, and behaviors.[4] Or, even if a broader perspective is adopted, including national political forces, these may be treated as relatively free-standing or capricious aspects of some leader or party, without clear recognition of the intertwining of political and economic forces often extending into and stemming from an extranational capitalist political-economic world-system.[5]

As adequate health and health care have come to be more and more regarded as fundamental and universal human rights,[6] the popular demand for improved care has taken the form of political pressure for improved health systems in nearly all of the world's 150 or so nation-states. It is within the nation-state that political-economic forces fashion national budgets, including funds for health manpower, facilities, and services. Thus, while it remains important to study cross-cultural variations in orientations toward illness and ways of dealing with it (also, many such variations can be identified within most countries) an overriding importance must be given to the political economy of the nation-state as influenced by the world-system in determining the nature of health services.

This perspective will be elaborated in Chapter 5 on framework and will pervade the other chapters of this volume.

Although the approaches to illness and death across the world are many and the mixes between forms of traditional and cosmopolitan medicine are often marvelous for a U.S.A. social scientist or health administrator to behold, one phenomenon stands out: the increasing belief in the power and value of modern medicine and consequent rising demand for it, however partially and inadequately it is conceived. This belief may be wrong, or partially wrong, for cosmopolitan medicine has its critics. These range from: (1) those like McKeown who say the efficacy of mechanistic modern medicine in relation to changing population survival has not been demonstrated, whereas the case for changing nutritional, sanitation, and living conditions generally is strong;[7] (2) to those like Powles who say it may be effective, but a large part of this is due to the placebo effect–people who believe in science or modernism believe in cosmopolitan medicine and are helped by it, while those who believe in traditional explanations and approaches and seek that kind of help are aided equally well;[8] (3) to those like Illich who find the iatrogenic effects of modern medicine and its assertion of professional control over individual, social and (from Illich's view as a former priest) religious problems downright dangerous and harmful.[9] In short, there is plenty of ambiguity in relation to which the full range or forces defining any social reality can operate.[10] And the dynamic cultural hegemony, established by and in turn reinforcing the capitalist political-economic world-system appears to be heightening the demand for curative, high-tech cosmopolitan medical care almost worldwide.[11]

At the same time that the demand for modern medicine (or some conception of it) is rising nearly everywhere in the world, so are the costs. Dr. Halfdan Mahler, the director general of the WHO has portrayed the situation this way:

> The challenge is immense if the World Health Organization's definition of health be taken seriously, this definition which was adopted following the most destructive war the world has ever seen. As you know, it considers health to be something different from the absence of disease and infirmity, and speaks instead of a state of physical, mental, and social well being. What is more, the definition considers health to be a right for everybody without discrimination. All this, though solemnly accepted by the member states of WHO, may seem like a bad joke when set against the realities of the health scene today. On the one hand we have the persistence in tens of millions of people of such diseases as cholera, malaria, and onchoceriasis. On the other we have gigantic modern machinery geared to the treatment of the whole range of disease up to the point of obfuscating the distinction between life and death. In one sophisticated city, more than 70 percent of all so-called health expenditure is used on people who are going to die within the next twelve months. In another highly developed country, a network of sophisticated renal dialysis centers catering to the needs of a few hundred patients with chronic kidney disease was given priority over a network of rural health centers catering to the needs of hundreds of thousands of women and children. These examples reflect, in my opinion, an obsessional concern of the medical profession—my own profession—with marginal disease and tend to pervert the very concept of health.
>
> Countries further down the developmental scale are busy imitating this kind of perversion. In a developing country, which constitutionally declares health a universal human right, you find in one province 80 percent of the health budget being used to support one teaching hospital, whereas in outlying parts complete coverage is supposed to be achieved by one general purpose dispensary for half a million people.
>
> These examples are not exceptional, they are ordinary, and the costs are spiraling. While over the last twenty years prices in many developed countries rose by 50–100 percent the cost of medical care rose by 300–500 percent. While in the same period the population in

> such countries increased by 10–20 percent, hospital utilization rose by three and a half times that figure. Paradoxically, with such stupendous cost increases and disease addiction has gone increasing dissatisfaction with the health care system on the part of its consumers, because serving champagne to the few while the many do not get their daily bread is hardly promoting confidence.
>
> The general picture in the world is of an incredibly expensive health industry catering not for the promotion of health, but for the unlimited application of disease technology to a certain ungenerous proportion of potential beneficiaries and, perhaps, not doing that too well either.[12]

The dissatisfaction Dr. Mahler writes of, stems not only from demands for more curative-oriented, high-tech care being frustrated by higher costs. Nor is it due solely to the two class systems of care found in many countries—one for the elite and another for the workers and farmers and the unemployed. There is an ambivalence, for the same care which is increasingly demanded has become hierarchized, bureaucratized, and depersonalized, leaving individual patients without any sense of control. In response to all these concerns, the recent, surprisingly successful formation of the People's Medical Society in U.S.A. takes on significance as a move to gain greater control and alleviate these conditions.[13] Even national health service systems which cover all costs through taxes or some sort of social security may reflect this problem. Such complaints have been made about the NHS in the U.K. And, while recognizing significant accomplishments in health in the U.S.S.R., Navarro's central critique of the Soviet system is directed toward the need for greater democratization and politicization of this too highly centralized, expert-determined system.[14]

One of the more important consequences of this situation for health services administrators and planners is that political and governmental leaders of the 150 or so nation-states of the world, faced with rising demand and rising costs, are beginning seriously to seek ways of rationalizing health systems.[15] The pressure is on, so to speak, not only in the U.S.A., but in varying ways in nearly every country in the world. This is not to say that the health care systems of all countries, or even all industrialized countries, are becoming alike (the convergence hypothesis). Different systems reflect rather remarkable variations in these problems and the ways they are dealt with. Thus cross-national comparisons offer much potential insight and chances for creative health system development.

The pressure of rising demand and costs *has* led to a search for alternatives. One important work searches out so-called "promising" alternatives for developing countries—that is, those alternatives which are said to be: (1) effective in meeting "basic health needs" (i.e., adequate food supply and nutrition, pure water for drinking, proper sanitation, adequate shelter and clothing, immunizations, family planning, and front-line acute medical care); (2) supportable within the present and expected resource levels of the country; and (3) socio-politically capable of being widely adopted throughout the country within a period of a few years. China, Cuba, and Tanzania are described along with projects in India, Venezuela, Niger, and some other countries.[16]

Another important work surveys the systems in twenty "developed" countries in Western Europe, the U.S.A., and the Soviet Union and suggests there is much to learn by cross-national studies with regard to four managerial-type problems: "(1) getting better information for managing available resources; (2) eliminating incentives to waste of resources; (3) coordinating interdependent services more closely; and (4) improving the assignment of responsibility for the way resources are used, over both the shorter and the longer term."[17]

In the meantime WHO convened the Alma-Ata Conference in 1978. Representatives from 155 nations plus the family of United Nations agencies solemnly undertook to pursue "Health for All by Year 2000" as *the* social goal of the UN system and its member States. Primary health care (PHC) was agreed to be the means by which this goal is to be pursued.[18]

PHC was seen both first as a broad thrust of the system toward anticipating and dealing with health problems as close to their place of occurrence as possible and secondly as a set of eight minimal actions to be accomplished at the periphery of the system:

> . . . education concerning prevailing health problems and the methods of preventing and controlling them; promotion of food supply and proper nutrition; an adequate supply of safe water and basic sanitation; maternal and child care, including family planning; immunization against the major infectious diseases; prevention and control of locally endemic diseases; appropriate treatment of common diseases and injuries; and provision of essential drugs.[19]

Less well known was the identification at Alma-Ata of certain national contextual conditions as prerequisites for effective support of PHC, primarily: authentic citizen participation in health affairs as well as other decision-making in the society; and social and economic justice. Subsequent developments have pointed to the need to move beyond broad philosophic statements and conduct sets of down-to-earth case studies so as to better understand in practical terms supports for and pursuit of PHC (or their lack) in different systems.[20] The expectation is to thereby enhance the prospects of health for all. This goal needs all the enhancing it can get in a world which spent over 660 billion dollars in 1982 on armaments and military forces while thirty children died every minute for want of food and inexpensive vaccines and medicine.[21]

It is within this "Zeitgeist" that the increasing scholarship and research in the field of cross-national study of health systems (CNSHS) is to be seen.[22]

But there are sharp limits to this search for alternatives. One is in the area of methods and data which I will develop further in Chapter 4, with special reference to OSH data problems and services. Another is the relative lack of work of a continuous, systematic sort—that is, examining over time, with a common, clearly stated framework and approach to gathering of information, the ways nations organize scarce resources for health. There is still no center or network of centers for this purpose anywhere in the world, although some beginnings have been made.[23]

But the greatest limitation of a search for alternatives stems from the nature of health systems in their societies. From the perspective adopted here, the very search for alternatives has a "managerial" cast to it which tends to disembody health systems or parts of them from the socioeconomic, political, cultural, and epidemiological contexts in which they are embedded. Navarro illustrates the problem in the introduction to his historical analysis and critique of the Soviet health system:

> This focus on the health sector without analysis of the socioeconomic system that determines it assumes an autonomy and near independence of the health sector that is both unempirical and unhistorical. Moreover, it leads to conclusions that are also empirically invalid and ineffective policy-wise. This is most clear in international health studies, where pieces of and individual experiences in the health sector are usually perceived as meriting "export" to other countries. A recent example of this is the present attention given to the Chinese barefoot doctors' experience. Much has been written concerning that interesting use of personnel in the People's Republic of China. Consequent to that interest, many national and international agencies, including the World Health Organization, are encouraging the export of that specific Chinese experience to other countries. Oblivious of China's socioeconomic determinants that may explain the success of the barefoot doctors, those exporters have tried to transform what is basically a political phenomenon into a mere managerial one. Not surprisingly, the export of the Chinese barefoot doctors' experience has failed to grow in those countries where the socioeconomic determinants were different ones.[24] Indeed, that experience assumes and presumes a series of economic and political parameters that determine its success. The purpose in citing this example and many others is not to question the value of international studies, but rather to stress the great importance of analyzing the nature of the health sector within a country's broader economic and political parameters which determine it.[25]

To illustrate further, some years ago it was pointed out that general hospitals in Sweden receive patients about as sick as is the case in the U.S.A. and discharge them about as well, and the Swedish hospitals do so, on the average, with only slightly more than half the personnel per patient than in U.S.A. hospitals.[26] If one understands that rising hospital costs contribute a major portion of increasing health care costs and one realizes that approximately three-fourths of hospital costs are attributable to personnel, a saving of close to one-half in personnel amounts to a very significant practical as well as theoretical matter. Does this difference really exist? And what is involved? A later study further documented the difference in terms of personnel hours per patient with diagnosis controlled (thus ruling out the longer average stay in Swedish hospitals which is largely attributable to the long-term care function being performed in the general hospital rather than in separate institutions).[27] But to the second question we have only speculative answers–perhaps it is the greater cultural homogeneity in Sweden; or a clearer focus on medical care goals as oposed to hotel functions; or a vigorous or more devoted work ethic; or less "defensive medicine" (fewer tests, etc.) since there are many fewer malpractice suits; or the fact that two or three physicians staff a ward (rather than a dozen or so community-based practitioners plus house staff) thus cutting complexity of treatment styles and communication; or some combination of these or other hypotheses. Whatever the explanation, the difference is real and remains for further study to be clearly understood.

But suppose it was found that the relative efficiency of Swedish hospitals is primarily attributable to the staffing of wards with two or three full-time salaried physicians and associated simplification of treatment styles and effectiveness of communication. What part of this pattern could we envision transferring so as to affect U.S.A. health policy and decision making? Probably very little under our dominant pattern of office-based physicians who are oriented toward a market and regard the hospital as their "workshop" (as opposed, say, to conceiving it as the hub or subcenter of an extended and coordinated health system as one can find in well-regionalized systems in some other countries).[28]

As the Royal Commission in New Zealand said in the opening chapter of its 1972 report on social security, ". . . social security administrators are everywhere prisoners of history, tradition, and political will."[29]

Still, it is not correct to be totally discouraging and relativistic. Comparison–implicit or direct–is the source of all improved understanding. Thus, there are several points to make about the gains to be realized from CNSHS even in the present underdeveloped state of this field.

First, we can see our own situation in high relief when regarding it through the reflection of experiences in other countries. For example, although there are one or two other countries which spend as much as the some $1300 per person spent in the U.S.A. on all health care in 1983 (Sweden is one, in spite of the greater efficiency of its hospital personnel–underlining once again the need to understand the whole system rather than examining some facet in isolation), more critical and possibly creative approaches to health services organization may be stimulated by the realization that many other countries attempt to cover all parts of their population with adequate health care for less than one-hundredth of this amount.[30]

A comparison with other industrialized countries is perhaps more relevant to U.S.A. experience. In late 1977 when he visited the U.K. to study the National Health Service, the then U.S.A. Secretary of the Department of Health, Education and Welfare, Califano, remarked to the media that costs for all hospital care in the U.S.A. were rising at $1 million an hour. About 55 percent of this was inflation; the rest, added services and additional population coverage. The U.S.A.'s total health-medical care bill had been rising well above the general inflation rate for more than fifteen years when, in 1980, it rose the most it had risen in that period of time. At the same time (1980) the Gross National Product (GNP) declined in a generally depressed world economy. In that year, health and medical care costs consumed 9.4 percent of GNP, or $1,067 per person. The U.K by contrast spent just 6.6 percent of its smaller GNP on all health and medical care for a total of $608 per person–just 57 percent of what was spent in the U.S.A. that year. By 1983, these costs had reached 11 percent of GNP in U.S.A. or more than $1300 per person. This was double the 5.5 percent of the smaller GNP spent on health in the U.K. that year. At the same time, using a varied set of indicators, Robert Maxwell reported the health level to be more favorable in U.K than in U.S.A. This picture of more benefit for less money began to attract attention even among conservative business circles in U.S.A.[31] This is not to imply that all is well with the National Health Service in the U.K. (there are waiting times for elective surgery as well as other problems)[32] but it seems logical that there would be clear cost saving as well as effectiveness advantages from such a generally well-regionalized NHS system.[33] Are there new types of health man and woman power we should develop? New systems of finance, of organization, such as a regionalized NHS with OSH given full attention–something like the Dellums Bill which has been in the U.S.A. Congress for years? Or should our efforts concentrate on broad societal transformation with the health services but a part of our concerns? Probably we cannot have a fully regionalized health services system giving full attention to workers' health without thorough going social transformation which brings about equity in distribution and control of productive as well as consumer resources.

A second gain to be realized from CNSHS may be in the form of principles. For example, we gain insight if we examine how it is that China, a nation with an estimated gross national product per person per annum of $160 in 1972 (one thirty-fifth of our GNP per capita for that year–$5,590) can, by all reports,[34] provide coverage with something approaching adequate health care to its 942 million people, nearly one-quarter of the world's population.[35] There are many principles and possible lessons to speculate about in this case. But one quite clear principle evinced in the system of barefoot doctors (so called because in many rural places they remove their footcover to join their peasant comrades in planting rice, etc., and thereby stay close to the people and their problems)[36] and public involvement in health work is that health problems should be anticipated and handled in preventive or other fashion as close to their place of occurrence as possible. Instead, more expensive and less effective systems such as in U.S.A. seem to detach every problem from the people and shunt it up the hierarchy of increasingly specialized and expensive services and facilities. We could not apply this principle in the U.S.A. by developing a vast neighborhood-based network of barefoot doctors. But we might find other ways more appropriate to our circumstances—that is, assuming significant change in our political-economic structure which would place a high priority on adequate primary and preventive care for all our people, transforming the class-based, racist, sexist society we now have[37] into one of more equal participation and sharing.

A third potential gain from CNSHS is in the identification of functional equivalents in other systems which could suggest improved approaches in the country of concern. (CNSHS should be a resource to all countries and is not only a matter of applying inferences to U.S.A. health policy and decision making.) The major difference between most other countries and the U.S.A., that is, of a health-system focus versus an organizational or institutional focus, throws many concerns, such as coordinating services within local and regional areas, into a new light. Or, another example, interfacing with various publics, especially regulatory agencies such as PSROs (Professional Standards Review Organizations), is hardly a concern in some other systems where our fetish for controlling costs by mechanisms for evaluating the appropriateness of care is replaced by a much more closed-ended, single-stream financing system and assignment of responsibility within such defined limits to a relatively highly regarded and trusted team of professionals. (The U.K. may offer such an example, though some research to document and further explore this seeming contrast is called for.)

Finally, and most important, through the kind of contrasting sociopolitical-historical study of different systems suggested here, we can identify and hopefully avoid transformational mistakes. That is, those of us who believe that health and health services will be enhanced through democratization and thorough politicization of the system, including the health services (rather than leaving everything to elite "professionals," "experts," and "managers") can learn what errors to avoid by comparison of the courses of different countries. As Navarro puts it:

> . . . one aim of this volume is to single out those forces that appeared in the Soviet Union's experience which may replicate themselves—both in liberating and oppressive dimensions—in other socialist experiences as well. Indeed, in any liberating force there is always, by definition, the possibility of the opposite. This is why I dedicate this work to the Cuban Revolution, a most important liberating force in today's world, which is in the very new and initial stages of creating a new society and a new human being.
>
> And it is in that enormous liberating potential that there is also the possibility of its opposite; and the strengthening and weakening of those opposites depend on the forces here defined. It is in this

> dialectical situation, as the Cuban people of course are aware, that the analysis of other socialist experiences is of paramount importance.[38]

Although it is an immensely complex matter with solutions varying necessarily with historical, socioeconomic, political, cultural, and epidemiological conditions present in each country, the central health services problem can be phrased as how to cover all parts of each nation's population with adequate health care at costs which the country itself can sustain. Since nations are the largest more or less autonomously acting social units (within limits placed upon them by the capitalist political economic world-system) able in varying degrees and in different ways, according to the phase of the struggle for human liberation they are in, to marshal the resources and political will to attack this problem, it is within these units that the worldwide health services drama is being played out. Consequently, it is on these units and on subnational, regional (more or less complete) health systems that I will focus.

The special concern here is for workers' health as this is addressed (or not) by the general health services within the political-economic context of each nation functioning within the world-system. Unfortunately, this problem has received little attention from sociologists.[39] The same is true for health planners, students of international health, occupational health specialists, union people specially concerned for OSH, government officials, WHO leadership, and various kinds of scholars who have devoted their efforts to comparative health systems. In short, the problem of OSH and its relations to PHC has been almost completely ignored.

THE STUDY OF OSH WITHIN THE CNSHS FIELD

Within the field of Cross-National Study of Health Systems (CNSHS), the study presented here represents a mix between focusing on a facet of health systems and attempting to understand overall health systems in their political-economic, cultural, and historical contexts. It is this latter approach which I have elsewhere recommended and for which I attempted to suggest the main points of the in-depth, contrasting case studies method, the method generally followed in this work.[40]

Available work in CNSHS is not adequate. Our understanding of different systems of organizing scarce resources for health within varying political-economic, sociocultural, and epidemiologic contexts is all too anecdotal.

In part, this situation is due to the difficulty in using available data and to the cost, organizational complexities, and other problems of achieving strict comparability even when fresh data are gathered across cultures and systems. In part, there has been a lack of adequate framework which I take up in Chapter 5. And there has been a lack of clarity as to the basic logic of design—the country comparisons or contrasts being undertaken and the reasons for them. This last problem relates to framework, but in the first chapter, somewhat in anticipation of the framework relevant to this study, I have given the rationale for selecting the six study countries included in this work.

Aside from mere anecdotal accounts, there have been impressive attempts to carry out strictly comparable studies of several medical care systems, but these studies have only examined some facet of the medical care system, such as financing,[41] utilization of services,[42] definition and treatment of schizophrenia,[43] dental caries and their treatment,[44] surgeons and operations,[45] hospital staff-to-patient ratios,[46] home care and care of the elderly,[47] maternal and child care,[48] and the place of the pharmaceutical industry in health insurance systems.[49] These works do not provide an adequate overview of key elements in the organization of the different systems so as to confirm or contribute to our sociological imagination as to what is going on.[50]

The "WHO International Collaborative Study of Medical Care Utilization" deserves special discussion in this category. After more than ten years of careful planning; frequent meetings of the seven national research teams to achieve comparability of data gathering approaches in twelve sites within the seven countries–Argentina, Canada, Finland, Poland, the U.K., U.S.A., and Yugoslavia–analysis of data; write up; and waiting for publication; a mammoth volume appeared with thirty authors and five consultants listed in addition to the two editors.[51] The work sheds considerable descriptive light on the range and concentration of levels of medical care use of several sorts–hospitals, physicians, and other use–with its seventy-three tables, 200 figures, and thirteen appendices. But there have been sharp criticisms of the work. One is its cost in money, personnel, and time, considering the product and its usefulness. Another is that the twelve sites chosen for study are not representative of the seven countries. In Canada, for example:

> What remains unclear from this extensive household survey, at least at the time of this writing, is how the communities or regions were selected in the first instance and to what extent they represent the national experience of the seven countries involved. It is unlikely, for instance, that the two rural districts in the Lower Fraser Valley of British Columbia, Grande Prairie, Alberta or the region of North Battleford in Saskatchewan portray the full scope of health services or how they were used in the three western provinces, their major cities, or the rest of Canada. Despite its sophisticated methodology, at no point in the analysis is a rationale given of how these four Canadian sites were selected. Throughout the interim report and the published papers, the assumption prevails that these communities represent Canadian experience as a whole, a premise which is at best tenuous.[52]

Another criticism has been that such a study is ahistorical and static, not reflecting change in use in relation to changing financing patterns, availability of facilities and services and a changing medical-cultural hegemony as to what is the "best" or proper thing to do about an illness. The exclusion of traditional or "nonscientific" forms of service use from the study underlines this point and constitutes another questionable aspect of the study. Finally, and most important, while remarkable variations in use of services between the twelve study areas in the seven countries are described, very little attention was paid to deliberately contrasting the overall health systems in these countries and their underlying political-economic differences which would have lent greater understanding. In short, one is inclined to read of the differences in use and associated patient and medical setting characteristics and say, "So what?" This shortcoming is highlighted by examining Appendix A. "Description of Selected Characteristics of the Study Areas and Their Health Services System." The health services are described from official documents as if one could take such descriptions at face value. Thus, for example, on paper, the Argentine system seems a good deal more generous and complete in its coverage of all the people than it in fact is in that highly stratified country at the time under a rightist, military, capitalist government; yet there is no discussion of this discrepancy in Appendix A or in the disembodied discussion of survey data in the text.

Another type of work provides useful descriptions of different health care systems, but even these are limited because the authors employ largely unstated frameworks in abstracting from the total reality of countries and their health care systems. Generally, these publications are of two types. One is an edited collection or compilation by different authors employing a variety of unstated frameworks.[53] Another type has the merit of describing several national health systems from one (or at least fewer) perspectives, since only one author is involved, but the framework is still only implicit. That is, it is not made clear why the aspects which are described are considered important and why some other considerations, such as political-economic history and background, are left out.[54] A recent contribution improves slightly on this category. Although it is an edited volume with different authors covering the health services in fourteen countries, it has the merit of the editor having required the authors to cover certain topics–social and political context, medical practice, hospitals and other health care institutions, medical education, nurses and other health personnel, ambulatory care, health costs and funding, and public health services.[55] However, this work has other serious shortcomings. Statements are often offered with no citation of relevant research. For example, in the editor's own chapter on U.S.A., there are at least three glaring examples (paraphrasing): 1) hospital boards are elected by nonrepresentative constituencies and business and banking and lawyers and professionals predominate but it is widely felt that these boards represent community interests (no research cited though there is work and most of it would dispute this "wide feeling"); 2) racism used to be a problem in admission to medical schools but racial barriers are completely down (this is outrageous and no work is cited); 3) it is simply not known how much unnecessary surgery is done in U.S.A. (Bunker's work comparing U.K. and U.S.A. could have been cited–see note 45). In general, the work is purely descriptive and uncritically so. It lacks any analytic framework and does not take advantage of material in the volume itself to make cross-system comparisons; thus it is not comparative as the title would suggest.

In still another category is work describing and perhaps attempting to analyze a single system. Because some systems, such as that in the People's Republic of China, seem promising enough in terms of meeting the basic health needs of all the people at costs that can be sustained and still allow for the pursuit of other priorities, a rich body of work has begun to develop.[56] But the "tightness" of articulation between socioeconomic-political structures and elements of these health systems have still not been explored by systematic contrasts with other systems. Although it has the kind of manipulative, managerial ring to it of which I am critical, the following statement by the executive board of WHO, in referring to the kind of health system in the People's Republic of China, has the merit of recognizing the need for improved understanding of this embeddedness:

> It could be said that the way in which such a service (as that in China) could be run is already known and that what is lacking is a national will and a manner of overcoming the entrenched opposition of organized medicine. While these reasons for lack of change may be valid, there are few or no examples where a change in emphasis of the type and degree required has been introduced within an existing health service without a preceding change in social policies. This link between health service and political structure is not so intimate that the health services cannot change separately and independently within most sociopolitical systems. However, the manner in which health services change in different systems and under different circumstances, the process of change, and the dominating constraints making change difficult, are largely unknown. The Board would consider useful the collection of a body of knowledge relating to this question, but WHO should give even higher priority to participating in and documenting such changes as they may occur, in order that these experiences can be made widely available and the lessons learned can be used by others.[57]

Finally, a few valuable descriptions and analyses of different systems exist which are systematic in employing a common, explicit framework,[58] and there is a need for more such work. However, the reasons for comparing different countries in these works are usually unclear or have to do simply with curiosity, convenience, or sponsorship. In short, the logic of comparison is not spelled out or is lacking. With some important exceptions, these works lack adequate breadth in their frameworks, focusing as they usually do mainly on financing, structure, and numbers of personnel internal to the health services rather than developing a historical view of the political economy within which the health services are to be understood.[59]

I have proposed in-depth studies of contrasting national health systems to (1) make the logic of design and country selection explicit, (2) allow qualitative understandings where non-comparable discrete data are a problem, and (3) facilitate theoretically guided broader understanding of health systems and their contexts in place of the limited views one gains through purely descriptive studies, or studies of facets of systems which may isolate information on survey scales but do not develop the meaning context within which the information is gathered.

The basic idea of this approach is to identify contrasting cases (countries) on conceptually interesting grounds and study them in depth. Such studies may be carried out through the use of historical records, available current data and reports, and, if resources and logistic and access conditions allow, field studies. Such field studies may range from the extensive, almost ethnographic type, to a shorter period during which strategically selected persons are interviewed and situations observed. The field work for this study was of this briefer sort, but included the gathering of historical information and documents.

Elsewhere I have considered the differing kinds of contrasting cases.[60] In the present work, I do not focus on contrasting health outcomes (though there seem to be remarkable differences between the study countries as regards incidence of occupational accidents—the work-related disease data are also suggestive, but there are too many uncertainties in such information to concentrate on these differences). Instead, the study countries were picked to reveal notable differences in strengths and forms of workers' movements, as well as varying types of health service systems, since these dimensions were thought to have an important relationship to OSH conditions and services and to arrangements for PHC.

CROSS-NATIONAL STUDIES OF OSH

The sad truth is that any kind of comparative or contrasting study of provisions for workers' health is very rare. This is rather startling when we consider that the health of the basic productive forces of societies is being thus ignored.

The Ford Foundation Report, *Crisis in the Workplace*,[61] included a chapter of descriptive material on several European OSH systems. But these were very partial or segmental descriptions often out of context. It was useful simply to give some "flavor" of other systems and suggest that there are significant variations between countries.

Many countries issue descriptive material concerning their own OSH systems. We will find specific references to examples in the later chapters on the study countries. There are also some more analytic studies of OSH struggles and conditions in particular countries. There is notable work which I will have occasion to cite again on F.R.G. (West Germany),[62] Italy,[63] Sweden,[64] and Cuba,[65] and certain others.[66] Valuable as these works are they do not provide comparisons or contrasts except casually or by implication, though the work by Navarro comparing Sweden and U.S.A. is an important exception (note 64).

The International Labor Organization (ILO) provides a very valuable bibliographic service on OSH. Many issues of these CIS Abstracts and Bibliography include descriptions of single OSH systems and sometimes descriptions of particular facets such as OSH education for technical personnel in several countries. For example, CIS Abstract November '81-1499 in Vol. 8 No. 5 (1981) notes that the 19th International Congress on Occupational Health was held in September 1978 in Dubrovnick Yugoslavia and Vol. 5 of the Proceedings has fifty-nine papers under various headings including "occupational health in various countries." Again, such material can be a valuable descriptive resource. WHO also issues descriptive material on different systems from time to time,[67] and general guideline statements.[68]

Elsewhere I have attempted to provide an overview and analysis of the especially severe OSH problems in underdeveloped countries.[69] There are many facets to this problem, including the export or development of hazardous industry in more easily exploited countries (because there are no labor unions, there are repressive regimes, etc.)[70] and the special exploitation of women workers.[71]

There is one somewhat outdated comparative survey of OSH regulatory standards in industrialized countries.[72] A primarily descriptive but very informative work on a closely related subject provides information on the health and safety regulation of biotechnology in the European Communities, U.K., France, F.R.G., Switzerland, Japan, and U.S.A.[73] Among the rare comparative works on OSH are three unpublished papers. One by Goldsmith compares three socialist-oriented countries: Cuba, G.D.R. (East Germany), and the U.S.S.R.[74] Although the author visited Cuba, the rest of the work is based on available materials.

A study by Atkins compares OSH cancer hazard policy, standards and enforcement in U.S.A. and U.K.[75] The author was a medical student at the University of Connecticut at the time and did this work as his required research project. His observations and impressions will be important for later chapters on the U.S.A. and U.K. One of his main impressions is that management and labor are much more adversarial as regards OSH matters in U.S.A. as compared with the "work it through" approach in U.K.

A very informative comparison of OSH regulation in Austria, F.R.G., and Sweden concludes:

> These countries are characterized by organized labor and related political power which have contributed much to the job safety and health practices there; the balance of power among government,

> industry, and labor has forced an openness and a necessity for communication which is more visible than in this country [U.S.A.]. The appearance is one of a system whose components are more willing to be guided by mutual needs and reciprocal demands.[76]

The emphasis here on the beneficial impact of organized labor's political power on OSH comes closer to my own understanding than the interpretation Kelman places on his comparison of OSH in Sweden and the U.S.A. (see below).

A recent Ph.D. thesis compares OSH and environmental policies of major multi-national chemical firms based in U.S.A. and U.K. One conclusion is: "The industry resistance to regulation, in both countries, also impedes the quality and delivery of corporate medical services . . . corporate neglect, secrecy, and passivity in response to environmental and health and safety considerations continues in the United States and Great Britain."[77]

The major systematic work in this field is Kelman's study comparing OSH regulation and enforcement in Sweden and the U.S.A.[78] While this work is valuable in providing information of a comparative sort, it is its own strong recommendation for further, theoretically guided work. The author seems to have had no guiding framework; or was guided implicitly by an idealistic notion that seemingly free-floating cultural differences explain everything. Having correctly perceived a more accommodating approach to OSH enforcement in Sweden (whereas nearly every OSH matter is disputed heftily in the courts and otherwise in U.S.A.) Kelman asserts that Swedes have a near reverence for things authoritative (stemming he suggests from the days of absolute kingship) thus they do not question and argue but simply carry standards out. There are other aspects to this argument and I will have occasion to come back to this work. But it is all too idealistic and simplistic. His most important failing is that he appears to dismiss, even ignore, the major differences in the character and strength of the workers' movements in Sweden and U.S.A. and the bitter struggles Swedish workers have waged in years gone by against management exploitation. As Navarro puts it, "The most important weakness of Kelman's work (and that of many other analysts of social policy in the U.S.) is the absence of any conception of social class as a determinant of power, including state power."[79] I will try to correct this weakness in Chapter 5 on Framework and elsewhere throughout this work.

There are other cross-national studies which relate more to the social relations of production and the organization of workers' movements, than they do OSH problems per se. Some of these works are very germane to this study,[80] but are better discussed in Chapter 5 on framework.

NOTES AND REFERENCES

1. Modern particle physics has come to share some of these perspectives in its consideration of positive and negative, image-like "particles" (which

themselves may be interactions of electro-magnetic fields, rather than "hard" particles) which erase each other when they meet, emitting photons of light in the process. Gary Zukan, *The Dancing Wu Li Masters, An Overview of the New Physics*. New York: Bantam Books, 1980.

2. Merwyn Susser and Wilbur Watson, *Sociology in Medicine*, 2nd edition. London: Oxford University Press, 1971, p. 89.
3. Alan Harwood, The Hot-Cold Theory of Disease: Implications for Treatment of Puerto Rican Patients, *Journal of the American Medical Association 216*:1153–1158, May 17, 1971.
4. See George Foster's report on his own shift in perspective: Medical Anthropology and International Health Planning, *Medical Anthropology Newsletter* 7:12–18, May 19, 1976.
5. John M. Janzen clearly identifies changing national political forces as determining the rise and fall of different traditional medical systems in Zaire. But he does not go the further step of laying out the political economy of these moves other than to record the change from colonialist rule to national rule without considering whether the national government is a client, neocolonialist government functioning in the interests of expropriating multinational corporations. John M. Janzen, The Comparative Study of Medical Systems as Changing Social Systems in the special issue of *Social Science and Medicine 12* No. 2B:121–129, April 1978, Charles Leslie (ed.). For some work employing Wallerstein's conception of "core" and "peripheral" nations in the world's capitalist political economy in an analysis of exploitative relations in favor of the "core" nations, see Christopher Chase-Dunn, The Effects of International Economic Dependence on Development and Inequality: A Cross-National Study, *American Sociological Review 40*:720, 738, 1975; and Richard Rubinson, The World Economy and the Distribution of Income within States: A Cross-National Study, *American Sociological Review 41*:638–659, 1976.
6. In 1948, the U.N. Universal Declaration of Human Rights stated: "Everyone has the right to a standard of living adequate for health and well-being of himself and his family, including . . . medical care . . . and the right to security in the event of sickness." As quoted in Milton I. Roemer, *Comparative National Policies on Health Care*. New York: Marcel Dekker, 1977, p. 233. Aside from the satisficing or Rawlsian ethics implied in this phrasing, the emphasis is clearly on greater equity in health and health care. A more complete egalitarian conception is found in the preamble to WHO's constitution published in 1974: "The enjoyment of the highest attainable standard of health is one of the fundamental rights of every human being without distinction of race, religion, political belief, economic or social condition."
7. Thomas McKeown, A Historical Appraisal of the Medical Task, pp. 29–57 in Gordon McLachlan and Thomas McKeown (eds.), *Medical History and Medical Care*. London: Oxford University Press, 1971.
8. J. Powles, On the Limitations of Modern Medicine, *Social Science, Medicine, and Man 1*:1–30, 1973.
9. Ivan Illich, *Medical Nemesis*. London: Calder and Boyars, 1975.

10. Sherif's early work using the autokinetic effect and group and authority influences in the definition of norms and standards concerning movement of a pinpoint of light in a dark room provides the experimental base for this conception. A more general formulation is found in Peter L. Berger and Thomas Luckmann, *The Social Construction of Reality*. New York: Doubleday, Anchor, 1976. But this phenomenological approach needs to be qualified, for social life is not as ambiguous as a dark room. Realities do present themselves even though varying interests with control over power resources may create ambiguities for purposes of obfuscation and exploitation. See Richard Lichtman, Symbolic Interactionism and Social Reality: Some Marxist Queries, *Berkeley Journal of Sociology 15*:75, 94, 1970. I will come back to the range of social forces affecting perceptions of OSH problems and the gathering and use of OSH data to create a dynamic cultural hegemony which most accept and some struggle to change.
11. I will come back to the notions of the cultural hegemony and cultural politics in Chapter 4, and elsewhere. Here I might note that more and more work points to the dependence, or partial dependence, even of socialist-oriented states on the core capitalist states. See Christopher K. Chase-Dunn (ed.), *Socialist States in the World-System*. Beverly Hills: Sage Publications, 1982. The 28 billion dollar debt Poland owes to Western banks would be only the most obvious example.
12. Halfdan Mahler, The Health of the Family, Keynote Address, in *Health of the Family*. From the International Health Conference sponsored by the National Council for International Health, Washington, D.C., October 10, 1974, p. 2.
13. 10 Surgeons to Stay Away From, *People's Medical Society Newsletter 4*(2): 1, October 1983.
14. Vicente Navarro, *Social Security and Medicine in the USSR: A Marxist Critique*. Lexington, MA: Lexington Books, 1977.
15. Anne R. Somers, The Rationalization of Health Services: A Universal Priority, *Inquiry 8*:48-60, March 1971; David Mechanic, Ideology, Medical Technology, and Health Care Organization in Modern Nations, *American Journal of Public Health 65*:241-247, March 1975.
16. V. Djukanovic and E. P. Mach, *Alternative Approaches to Meeting Basic Health Needs in Developing Countries: Report of a Joint UNICEF/WHO Study*. Geneva: WHO, 1975.
17. Robert Maxwell, *Health Care, the Growing Dilemma: Needs Versus Resources in Western Europe, the U.S. and the U.S.S.R.* New York: McKinsey and Co., 1974, quote from p. 41. This work has been updated and deepened with the number of countries limited to ten. I comment on it further on in this chapter (see note 58).
18. *Alma-Ata 1978 Primary Health Care*. Geneva: WHO/UNICEF, 1978.
19. From the Declaration of Alma-Ata, Roman VII, Point 3 (op. cit., see note 18, p. 4). Unfortunately, there was no explicit recognition of occupational disease and injury though they are in principle included in the sixth and seventh and perhaps other of the minimal eight PHC actions.

20. Bogdan Kleczkowski, Ray Elling, and Duane Smith, *Health System Support for Primary Care*. Public Health Monograph Series, No. 80. Geneva: WHO, 1984.
21. Ruth Leger Sivard, *World Military and Social Expenditures, 1983*. Washington, D.C.: World Priorities, 1983. An estimate for 1984 is that more than 1,000 billion dollars will be spent worldwide for arms and armies.
22. I have provided elsewhere a two-volume annotated bibliographic introduction to the field: Ray Elling, (1) *Cross-National Study of Health Systems: Concepts, Methods and Data Sources* and (2) *Cross-National Study of Health Systems by Country and World Regions, Including Special Problems*. Detroit: Gale Research, 1980.
23. The Health Services Research Center, Graduate School of Business, University of Chicago, under Odin Anderson's direction, has done several cross-national studies, mainly comparing U.S.A. and Swedish experience. An exchange research-study program has been established between UCLA School of Public Health and the Department of Social Medicine, University of Copenhagen, under Erick Holst's direction in Copenhagen. Jan Blanpain directs an Institute for European Health Services Research at the Catholic University of Leuven, Belgium. This center focuses on the Western European EEC countries. I have served for the past ten years as coordinator of an informal, *defacto* consortium, the Northeast Program for Cross-National Study of Health Systems (NEP-CHSHS), joining the interests and talents of scholars with a variety of perspectives and disciplines in several nearby universities. Timothy Empkie, Assistant Professor, Department of Family Medicine, Brown University in Providence, RI now serves as coordinator of NEP-CNSHS.
24. H. A. Ronaghy and S. Solter, Is the Chinese 'Barefoot Doctor' Exportable to Rural Iran? *Lancet 1*:1331–1333, June 29, 1974.
25. Vicente Navarro, op. cit. (note 14), p. xviii.
26. Paul Lembcke, Hospital Efficiency: A Lesson from Sweden, *Hospitals 33*: 34–38, 92, April 1, 1959.
27. Egon Jonsson and Duncan Neuhauser, Hospital Staffing Ratios in the United States and Sweden, in Ray Elling (ed.), *Comparative Health Systems*, supplemental volume of *Inquiry 12*:128–137, June 1975.
28. Ross Danielson, *Cuban Medicine*. New Brunswick, NJ: Transaction Books, 1978.
29. As quoted in Derrick Fulcher, *Medical Care Systems: Public and Private Coverage in Selected Industrialized Countries*. Geneva: International Labour Office, 1974.
30. John Bryant, *Health and the Developing World*. Ithaca, NY: Cornell University Press, 1969.
31. B. Newman, Britain Healthier for Less Money, *The Wall Street Journal* Vol. CCI, No. 28, February 9, 1983, pp. 1 and 27. The conservative Prime Minister, Margaret Thatcher, who seemed to threaten to privatize and dismantle the NHS as she came into office, seemed to have realized both the cost saving advantages and political popularity of the NHS. She is quoted in this article as saying, "The NHS is safe with us." It should not be imagined

that this change came about just out of the goodness of her heart. Organized labor and coalitions of patient groups, health workers, and left-liberal political activists have been carrying on a vigorous struggle to save the NHS. See, for example: the politics of health group, Cuts and the NHS, pamphlet number two, London: The Group (Blackrose Press) no date (approx. 1980).

32. Ellie Scrivens and Walter W. Holland, Inequalities in Health in Britain. A critique of the report of a research working party. *Effective Health Care 1*: 97–108, 1983.
33. I attempt to trace the historical roots of the idea of regionalization in health and offer a ten-point model of what I believe to be a fully developed concept of regionalization in Ray Elling, *Cross-National Study of Health Systems, Political Economies and Health Care*. New Brunswick, NJ: Transaction Books, 1980, esp. pp. 100–101.
34. Victor Sidel and Ruth Sidel, *Serve The People: Observations on Medicine in the People's Republic of China*. Boston: Beacon Press, 1973. Also Shahid Akhtar, *Health Care in the People's Republic of China: A Bibliography with Abstracts*, Ottawa: International Development Research Center, 1975. An update of the Sidel's book is: Ruth Sidel and Victor W. Sidel, *The Health of China, Current Conflicts in Medical and Human Services for One Billion People*. Boston: Beacon Press, 1982.
35. The Environmental Fund Washington, D.C., estimate for mid-1975. By 1980, China's population was recorded as 1,006,000,000. Ruth Sivard, op. cit. (note 21) Table II, p. 34.
36. Ray Elling, Notes on a Discussion with a Chinese Barefoot Doctor, *WHO Dialogue* (Geneva) *14*:9–12, October 1973.
37. Leslie A. Falk, The Negro American's Health and the Medical Committee for Human Rights, pp. 70–88 in Ray Elling (ed.), *National Health Care: Issues and Problems in Socialized Medicine*. New York: Lieber-Atherton, 1973. Charles R. Greene, Medical Care for Underprivileged Populations, *The New England Journal of Medicine 282*:1187–1188, May 21, 1970. S. Leonard Syme and Lisa F. Berkman, Social Class, Susceptibility, and Sickness, pp. 398–405 in H. D. Schwartz and C. S. Kart (eds.), *Dominant Issues in Medical Sociology*. Reading, MA: Addison-Wesley, 1978. Barbara Ehrenreich, Gender and Objectivity in Medicine, *International Journal of Health Services 4*:617–623, 1974.
38. Vicente Navarro, *Social Security and Medicine in the U.S.S.R.*, op. cit. (note 14) p. xxi.
39. To my surprise and disappointment I learned that the papers session I organized for the Medical Sociology Research Committee of the International Sociological Association's 10th World Congress of Sociology meeting in Mexico City in August 1982 was the first sociological papers session on OSH ever held anywhere.
40. Op. cit. (see note 33). Obviously "in-depth" is severely qualified by time and resources. Both were quite limited in this effort (see Methods Appendix). However, the goal of achieving an overall qualitative understanding, or "Verstehen" as Max Weber introduced it into sociological methods, is also an essential part of "in-depth" and that goal has guided this work.

41. Brian Abel-Smith, *An International Study of Health Expenditure and Its Relevance for Health Planning*, WHO Public Health Papers no. 32. Geneva: WHO, 1967.
42. Many utilization studies could be cited. The best known and the most extensive involving field research in twelve places distributed in seven countries, is Robert Kohn and Kerr L. White (eds.), *Health Care, An International Study: Report of the WHO International Collaborative Study of Medical Care Utilization*. London, New York, Toronto: Oxford University Press, 1976. A cross-national comparison limited to hospital use and employing secondary data is Ragnar Berfenstam, Cross-national Comparative Studies on Hospital Use, *World Hospitals 9*:143–149, October 1973.
43. WHO, *The International Pilot Study of Schizophrenia*. Geneva, 1973. On efforts to develop reliable cross-national measures of mental illness diagnoses, see Norman Sartorius, The Cross-National Standardization of Psychiatric Diagnoses and Classifications, pp. 73–81 in M. Pflanz and E. Schach (eds.), *Cross-National Socio-medical Research*. Stuttgart: Thieme, 1974. (Notes and references, pp. 94–97.)
44. Lois K. Cohen and David F. Barmes, International Collaborative Study of Dental Manpower Systems in Relation to Oral Health Status, *Social Science and Medicine 8*:325–327, May, 1974.
45. The United States, with about twice as many surgeons per population unit as in England and Wales, has about double the number of operations per population unit, J. P. Bunker, Surgical Manpower: A Comparison of Operations and Surgeons in the United States and in England and Wales, *New England Journal of Medicine 282*:135–144, 1970. See also Marlene Lugg, A Comparison of Hospital Morbidity Statistics: Western Australia; The United States; England and Wales; Scotland. Paper presented at the American Public Health Association Meetings, Chicago, November 1975.
46. Jonsson and Neuhauser, op. cit. (note 27).
47. J. C. Brocklehurst, *Geriatric Care in Advanced Countries*. Baltimore: University Park Press, 1975. Virginia Little, Factors Influencing the Provision of In-Home Services in Developed and Developing Countries, Paper presented at the 10th International Congress of Gerontology, Jerusalem, June 1975.
48. Suzanne Arms, *Immaculate Deception: A New Look at Women and Childbirth in America*. Boston: Houghton Mifflin, 1975. This book is trenchantly critical of the over-eager, hospital-based operative intervention during childbirth in the U.S.A. medical marketplace, and contrasts this with the much more home-based approach in the Netherlands. On child care in developing countries, see David Morley, *Paediatric Priorities in the Developing World*. London: Butterworth, 1973.
49. Pharmaceutical Manufacturer's Association, National Health Program Survey of Eight European Countries (Washington, D.C.: the Association, 1970).
50. C. Wright Mills, *The Sociological Imagination*. New York: Grove Press, 1959.
51. Kohn and White, *Health Care*, op. cit. (note 42).
52. Robin F. Badgley, International Health Research Methods: Footnotes on Canadian Health Services, pp. 128–140 in Pflanz and Schach (eds.), *Cross-National Socio-medical Research*, op. cit. (note 43) quote from p. 132.

53. A work of mine including descriptions and analyses of "developed" and underdeveloped countries as well as studies of special problems falls in this category: Ray H. Elling (ed.), *Comparative Health Systems*, supplemental vol. of *Inquiry 12* (June 1975). Some of the major works in this group are: I. Douglas-Wilson and Gordon MacLachland (eds.), *Health Services Prospects: An International Survey*. Boston: Little, Brown, 1973; John Z. Bowers and Elizabeth Purcell (eds.), *National Health Services: Their Impact on Medical Education and their Role in Prevention*. New York: Josiah Macy, Jr. Foundation, 1973; John Fry and W. A. V. Farndale (eds.), *International Medical Care*. Wallingford, PA: Washington Square East Publishers, 1972; Stanley R. Ingman and Anthony E. Thomas (eds.), *Topias and Utopias in Health: Policy Studies*. The Hague: Mouton: Chicago: Aldine, 1975. This last is more innovative than most collections in including works mainly from a dependency theory or conflict perspective.
54. Derick Fulcher, *Medical Care Systems: Public and Private Health Coverage in Selected Industrialized Countries*. Geneva: International Labour Organization, 1974; Allan Maynard, *Health Care in the European Community*. Pittsburgh: University of Pittsburgh Press, 1975; Michael Kaser, *Health Care in the Soviet Union and Eastern Europe*. Boulder, CO: Westview Press; London: Croom Helm, 1976.
55. Marshall W. Raffel (ed.), *Comparative Health Systems, Descriptive Analyses of Fourteen National Health Systems*. University Park, PA: Pennsylvania State University Press, 1984. The countries included are Australia, Belgium, Canada, China, Denmark, England, France, Germany (West), Japan, Netherlands, New Zealand, Sweden, U.S.S.R., and U.S.A.
56. The most noted work is Victor W. Sidel and Ruth Sidel, *Serve the People*: op. cit. (note 34). This has been updated after the recent modernization, anti-cultural-revolution trend, Sidels, op. cit. (note 34). An extensive and valuable annotated bibliography is: Shahid Akhtar, *Health Care in the People's Republic of China, A Bibliography with Abstracts*, IDRC-038e. Ottawa; International Development Research Centre, 1975. Works on other frequently described systems are cited and annotated in Chapter 3 of Ray H. Elling, *Cross-National Study of Health Systems: Countries, World Regions, and Special Problems*. Detroit: Gale Research, 1980. The sections are: 3.1 Canada; 3.2 Cuba; 3.3 People's Republic of China; 3.4 Sweden; 3.5 United Kingdom; 3.6 U.S.S.R.
57. Organizational Study on Methods of Promoting the Development of Basic Health Services, Annex 11 to Official Records of the World Health Organization, no. 206. Geneva: WHO, 1973, p. 110. My position is that health services *are* highly dependent upon and determined by their political economic contexts and cannot be changed separately in any significant way as this WHO statement asserts.
58. These works are cited and annotated in Chapter 4 of the first volume of my annotated bibliographic introduction to CNSHS (see note 22 above). Notable among them are: Milton I. Roemer, *The Organization of Medical Care under Social Security*. Geneva: International Labour Organization, 1969; E. Richard Weinerman with Shirley B. Weinerman, *Social Medicine in*

Eastern Europe: The Organization of Health Services and the Education of Medical Personnel in Czechoslovakia, Hungary, and Poland. Cambridge, MA: Harvard University Press, 1969; Odin W. Anderson, *Health Care: Can There Be Equity? The United States, Sweden, and England.* New York: John Wiley and Sons, 1972. A recent work, sharing a good deal of the perspective of this volume, which compares Sweden, United Kingdom, Soviet Union and the People's Republic of China in relation to the United States health system issues is Victor W. Sidel and Ruth Sidel, *A Healthy State: An Internationl Perspective on the Crises in United States Medical Care.* New York: Pantheon Books, 1977. Another fairly recent work with similar concerns, but a much more limited, managerial perspective, but one which nevertheless spells out a comparative framework: Joseph G. Simanis, *National Health Systems in Eight Countries*, DHEW Publication no. (SSA) 75-11924. Washington, D.C.: GPO 1975. Also Robert J. Maxwell, *Health and Wealth, An International Study of Health Care Spending.* Lexington, MA: Lexington Books, 1981 includes ten industrialized countries, but might also be considered as a study of a single facet of these systems as the focus is so heavily on costs. Most of the data are from 1975. The ten countries are: Australia, Canada, France, Italy, Netherlands, Sweden, Switzerland, U.K., U.S.A., and West Germany. The most recent work in this group focuses on health planning in four countries but gives enough about the overall health systems and national contexts to be quite valuable. It also includes a stimulating Appendix reviewing other work in CNSHS. Unfortunately, it lacks explicit framework and methods statements and seems not to recognize class-struggle as the motor of change. Victor G. Rodwin, *The Health Planning Predicament, France, Quebec, England, and the United States.* Berkeley, CA: University of California Press, 1984.

59. Sidel and Sidel, *A Healthy State*, op. cit. (note 58). The dedication of this book suggests the breadth I have in mind: "For Dr. Salvador Allende Gossens and other health workers in Chile who were killed or are in exile because they knew that the only way to protect and promote the health of a nation is to redistribute wealth and power among its people—'El Pueblo unido jammas sera vencido.'"
60. For an approach to country selection based on contrasting "health outcomes" while countries are similar in overall resource levels, see Ray Elling and Henry Kerr, Selection of Contrasting National Health Systems for In-depth Study, pp. 25–40 in Ray Elling (ed.), *Comparative Health Systems*, supplemental volume of *Inquiry 12*, June 1975. For a consideration of other kinds of contrasts see *Cross-National Study of Health Systems, Political Economies and Health*, op. cit., (note 33) esp. pp. 61–64.
61. Nicholas A. Ashford, *Crisis in the Workplace.* Cambridge, MA: MIT Press, 1976.
62. One of the most notable is Hans-Ulrich Deppe, Work, Disease, and Occupational Medicine in the FRG, *International Journal of Health Services 11*: 191–205, 1981.
63. Giorgio Assennato and Vicente Navarro, Workers' Participation and Control in Italy: The Case of Occupational Medicine; and Giovanni Berlinguer, Work

and Health in Capitalist Societies: Some Italian Experiences, in *Work and Health*, Health Marxist Organization, Packet No. 5 (New Haven) no date, probably 1980, (reprinted in V. Navarro and D. Berman (eds.), *Health and Work under Capitalism*, Farmingdale, NY: Baywood, 1983); also Michael R. Reich and Rose H. Goldman, Italian Occupational Health: Concepts, Conflicts, Implications, *American Journal of Public Health 74*:1031–1041, 1984.

64. Vicente Navarro, The Determinants of Social Policy. A Case Study: Regulating Health and Safety at the Workplace in Sweden, *International Journal of Health Services 13*:517–561, 1983.
65. Robin Alexander and Pamela K. Anderson, Pesticide Use, Alternatives and Workers' Health in Cuba, *International Journal of Health Services 14*: 31–41, 1984.
66. R. Charles Clutterbuck, The State of Industrial Ill-Health in the United Kingdom, pp. 141–151 in Navarro and Berman, op. cit. (note 63); *Health Services in the USSR, Report of a Study Tour*, Public Health Papers, No. 3. Geneva: WHO, 1960 includes a chapter on OSH and an annex giving an organization chart for such a service; R. Repond, Industrial Health Services in the USSR, *World Health 20*:18–24, March 1969; William Y. Chen, Relation of Occupational Health to General Community Health, pp. 247–260 in J. R. Quinn (ed.), *China Medicine As We Saw It* NIH Pub. No. 75-684. Bethesda, MD: John E. Fogarty International Center, 1974. The Navarro and Berman work also has chapters on several countries other than those already noted, op. cit. (note 63). Few of the planning initiatives studied by Rodwin took OSH explicitly into account. Quebec did give it special attention, but created new structures rather than relating OSH clearly to local health and social service center, op. cit. (note 58) esp. p. 138.
67. Report of a Working Group on Evaluation of Occupational Health and Industrial Hygiene Services, Stockholm, 15–17, April 1980. Copenhagen: WHO/EURO. One of the more valuable pieces is the informal appendix to the above report giving fairly full descriptions of the OSH systems in eight European countries. This document was not officially released as figures could not all be verified and some information was out of date. It simply served as background for the Working Group.
68. Health Aspects of Wellbeing in Working Places. Report on a WHO Working Group. EURO Reports and Studies No. 31. Copenhagen: WHO/EURO.
69. Ray Elling, Industrialization and Occupational Health in Underdeveloped Countries, *International Journal of Health Services 7*:209–235, 1977.
70. Barry I. Castleman, The Export of Hazardous Factories to Developing Nations, pp. 271–308 in Navarro and Berman, op. cit. (note 63); Barry I. Castleman, Double Standards; Asbestos in India, *New Scientist 26*:522–523, February 1981.
71. Barbara Ehrenreich and Annette Fuentes, Life on the Global Assembly Line, *Ms* pp. 53–59 ff, January, 1981.
72. B. Holmberg and M. Winell, Occupational Hygienic Standards—An International Comparison, *Arbete och Halsa* (Stockholm) *13*, 1976.

73. Alan M. Fox, *et al.* International Health and Safety Regulation of Biotechnology. Prepared for the USA Congress, The Office of Technology Assessment. Washington, D.C.: Kaye, Scholer, Fierman, Hays and Handler, July 15, 1983.
74. Frank Goldsmith, Occupational Safety and Health in Cuba, the German Democratic Republic, and the Soviet Union, Ithaca, NY: New York State School of Industrial and Labor Relations, Cornell University, 1977. Reproduced.
75. Elisha H. Atkins, Regulation of Occupational Carcinogens in the United States and Great Britain. The Politics of Science, University of Connecticut Medical School, 1977. Reproduced.
76. Patrick J. Coleman, Regulating Job Safety and Health in Austria, West Germany and Sweden. A Report to the German Marshall Fund of the United States, 1979. Reproduced.
77. Jane Ives, A Cross-Cultural Comparative Study of the Regulation and Administration of Occupation Safety and Health Care Policies in the Workplace: the Chemical Industry in the United States and Great Britain. Unpublished Ph.D. dissertation submitted to London University, 1984, quote from the abstract.
78. Steven Kelman, *Regulating America, Regulating Sweden: A Comparative Study of Occupational Safety and Health Policy*. Cambridge, MA: MIT Press, 1981.
79. Navarro, Health and Safety at the Workplace in Sweden, op. cit. (note 64): 523.
80. An especially important article is Michael Burawoy, Factory Regimes Under Advanced Capitalism, *American Sociological Review 48*:587–605, October, 1983. This builds on Harry Braverman's, *Labor and Monopoly Capital*. New York: Monthly Review Press, 1974 and other work. For an early, less critical view, see Harold L. Wilensky and Charles N. Lebeaux, *Industrial Society and Social Welfare*. New York: Russell Sage Foundation, 1957.

CHAPTER 3
An Overview of OSH and OSH-PHC Relations in Study Countries and Elsewhere

A GROWING AWARENESS AND CONCERN

In each of the study countries–Sweden, Finland, G.D.R., F.R.G., U.K., and U.S.A.–there was a veritable rash of OSH legislation and activity during the late 1960s and 1970s. This spurt of activity and legislation concerning OSH in the late 1960s and early 1970s may have been limited to industrialized nations,[1] but extended far beyond the study countries in focus here. An overview of eight Western European countries in 1975 suggested many recent accomplishments building upon earlier developments.[2] The OSHA law of 1970 in the U.S.A. was especially important. The background of its development was outlined in Chapter 1. But such developments were not limited to the capitalist industrialized nations. Also in the socialist-oriented Eastern European countries, there were great strides taken in OSH matters in this period. We will see this illustrated in the chapter on the G.D.R. Here I cannot go into all these developments. Details will be given for each study country in those chapters (7-12). Here it is enough to take note of this widespread phenomenon as part of a raised consciousness and awareness of OSH. This "movement" must be understood within the dynamics of each country's political economy. And the history of struggle over these issues reaches far back in each case, usually to the origins of workers' struggles against the brutalizing conditions of early industrial capitalism.[3]

One especially fascinating commonality across most of the countries I studied and other countries was the existence of what in the U.S.A. are known as COSH groups, Councils on Occupational Safety and Health. The earliest was the Chicago Area COSH. Now there are some forty such groups. In New England there is Mass. COSH, RI COSH, and in 1981 ConnectiCOSH was formed. Typically, these are collectives of activist unionists, sometimes with official union backing, and medical technical, legal, social welfare, social science students, and experts ready to attack specific OSH problems at local level and act politically on several levels.[4] In Denmark there is the "Aktionsgruppen Arbejdere Akademikere" which has done significant work on epoxy, the still existent problems of chimney sweeps (200 years after Percival Pott's work) and other problems. In the U.K. there is the "Work Hazards Group," part of the British Society for Social Responsibility in Science." In Finland, there is the "Front for Health" which focuses on workers' health and safety but takes up other health issues too. One fascinating project involved publication and wide dissemination of a set of hazard standards acceptable to workers and much more stringent than those established through the noted, and well funded Finnish National Institute for Occupational Health and Safety.[5] In Sweden, a group called SARB was successful for a time in the 1970s,[6] but seems to have gone out of existence as the official system met most workers' demands and expectations, and even the Swedes began to worry about jobs and employment in the downturn of the world economy. The whole Solidarity movement in Poland was very much alive when I was conducting my field work. I visited Poland for several days around the time of the one-hour general strike on October 28, 1981. Occupational health issues were very central concerns.[7] I have also heard of such groups in Norway and Italy. I did not find one in either of the Germanies. But there has been one in an incipient form in Frankfurt, F.R.G. where activist professors and students advise I.G. Metall and other unions without having formed an organization. The existence of these COSH-type groups in many countries suggests several lessons.

First, the existence of such groups even in, so called, social welfare states (which are still capitalist states) suggests that the state apparatus, even if it does not serve directly and exclusively as the "executive committee" of the ruling capitalist class, as Lenin asserted,[8] does *not* meet the full demands of workers and their representatives. Thus in mediating and controlling the class struggle, as Mandel and others have suggested,[9] the state tends to support the interests of the ruling capitalist class. This point is important for understanding the weakness of OSHA in U.S.A. as compared with OSH provisions in countries where the workers' movements are stronger.

Second, and obvious perhaps on other grounds as well, the existence of COSH-type groups, reinforces our conviction that OSH problems are not mere technical problems. They do involve often exquisitely complex and detailed technical matters. But these matters quickly become issues for local worker-management negotiation and for class struggle on the wider scene.[10] Thus OSH problems are preeminently political-economic issues and must be examined in such a framework.

Third, these COSH-type groups universally have a left-oriented political composition, though they seem to include a wide range of such views and party interests; thus they tend to be a unifying force, rather than sectarian. The point is that OSH problems, because of their centrality in the control of production, as well as their straight forward, literally vital impact on the well being of workers, present an excellent arena in which to educate and organize, raise workers' consciousness, and generally advance the class struggle. It is also an excellent forum for those wishing to see advances in social medicine and public health, though, curiously, this linkage between health activitists and activist workers and unions has been very weak, except for those involved in COSH-type groups.[11]

Fourth, the existence of these spontaneously, independently developed groups in several countries (I could find no sign of systematic mutual support and communication between such groups in different countries which I think is unfortunate for the working-class struggle for improved OSH) confirms the widespread growing concern for OSH, especially in the late 1960s and 1970s.

In the U.S.A., this ferment has, in addition to the OSHA law, had the following components and effects:

> Unions, often under pressure from their memberships, are forming health and safety departments and teaching their members to identify and correct hazards. Workers and progressive scientists, lawyers, and health specialists are organizing and working together through grassroots health and safety coalitions, such as the Chicago Area Committee for Occupational Safety and Health Urban Planning Aid (Now Mass COSH) in Boston, and the Philadelphia Area Project on Occupational Safety and Health. Medical researchers, led by Doctors Irving Selikoff and Sam Epstein, have formed the Society for Occupational and Environmental Health, which admits nonprofessionals as members; and the general public's awareness of working conditions has increased through a few books and mass media programs. Within this context it is possible to point to specific advances for working people: the establishment of their rights under OSHA; the general acceptance of higher and more realistic estimates of the size of the problem; the new questioning of the traditional role of the company doctor; the new interest in occupational disease (in addition to the earlier almost exclusive focus on trauma from falls, fires, explosions, and other accidents); the increased number of collective bargaining and research initiatives in health and safety; and coalfield gains regarding black lung compensation.[12]

This study will not attempt an explanation of this widespread "movement or ferment."[13] But I find it fascinating and some speculation may be in order. While OSH problems have been with us for a long time,[14] a more vigorous organized labor entered upon struggle to gain the worker's "right to know" what substances they were working with and the hazards involved in the 1960s and early 1970s when each country's economy and the world-economy were in a boom phase and jobs were not scarce. One long-term Swedish student of and contributor to OSH relates these developments to the ferment of the Vietnam War years and the growing dissatisfaction of workers and others with the dangers and diminished human dignity placed upon them by arbitrary, often irrelevant and exploitative authorities:

> The sixties ended with unrest and violence in many parts of the industrial world. The political uprising of 1968 in Prague was forcefully terminated in August. In France de Gaulle was shaken in his presidential chair by students riots. In the United States police fired at students in university campuses. Sweden too had its share of political unrest with student occupation of official buildings, and, for us, an unusual and serious strike in the mines. Oddly enough it was not so much for money that the miners went on strike, but for human dignity and better working conditions.[15]

In the U.S.A., as already noted in Chapter 1, the background to the passage of OSHA included not only the anti-Vietnam War movement, but a civil rights and minorities movement, a women's movement, and a reinvigorated labor movement. Capital was on the defensive, even though its champion, Richard Nixon, was in power. But just how this "climate of the times" could have carried to all five other study countries as well as other industrialized nations I do not know. Perhaps the cultural hegemony which dynamically supports and is formed by the capitalist political-economic world-system also carried with it much of the counter culture of resistance forces from one core country to another as well as to semi-peripheral and peripheral countries.[16]

But there were certainly other influences than just those (if any) stemming from U.S.A. As suggested earlier, OSH developments have to be understood primarily in terms of the struggles in each country. In Sweden, for example, workers' impatience with the "labor peace" in the late 1960s. This "peace" had been achieved through the "social contract" of Social Democratic government. This contract left control of the means of production in the hands of capital for forty years and thereby led to deteriorating work conditions as Capital sought to improve its rate of profit through administrative and technical rationalizations of the production process, while labor achieved only marginal gains in terms of social welfare benefits. Workers' growing impatience was expressed in wildcat strikes and other ways. These workers' rebellions of 1969–1970 against rationalized working conditions took the organizations of both labor and capital by surprise. The response to this ferment in the realm of OSH was a set of very impressive additional gains for labor.[17] Ferment—whether of U.S.A., anti-establishment inspiration or locally initiated—was certainly involved in the heightened concern and action for OSH in the late 1960s to early 1970s. I also think there was some impact from the epidemiologic studies of Selikoff and others suggesting that the size of the OSH disaster from asbestos and other exposures was more, much more, than working people, their representatives, and conscionable researchers, scholars, and the public would stand for. But on top of these factors, the high employment rate and generally high gear of the economies in those years, made it more possible for workers and unions to stand up for the right to know and effective standards and enforcement.

CONTRASTS IN OSH PROVISIONS

This cross-nationally observable "movement" by no means suggests equal success. In fact, in the U.S.A. a rather weak OSHA act passed under Nixon in 1970,[18] has been further weakened under Reagan.[19] By contrast, in Sweden where some 95-98 percent of blue-collar workers are unionized; some 70-75 percent of all other workers, including professional, scientific, and technical workers are organized (while only 18.8 percent of U.S.A. workers are currently unionized);[20] where labor takes a broad socialist approach to social policy (whereas U.S.A. labor has been caught in a narrow Gomperism since the late 1800s) and there has been a labor government in power for over forty-five years (while the U.S.A. has never had a credible labor party) a much stronger set of laws and overall OSH system were put in place. To illustrate the enormous differences, see Table 2, as adopted from Navarro.[21]

I cannot discuss all of these items here. More will be said in the chapters on Sweden and U.S.A. But to illustrate and underline the differences, let us just elaborate on two of the items. The first item, "Right to stop work" applies to the individual worker who perceives a hazard on his or her job. In Sweden, the worker can stop work without fear of firing or other reprisal until a union health and safety representative gives assurance that there is no problem or a plan is agreed whereby the worker will be adequately protected and the problem corrected. In the U.S.A., this right was granted in an extremely limited way through an OSHA citation appealed by a construction company to the Supreme Court. A worker had refused to work on a night time construction job when there was no safety net underneath the scaffolding. He was fired. A co-worker continued to work and was killed in a fall. The court ruled that since it was "after hours" and OSHA could not be called to settle the safety of the situation, the worker had the right to refuse to work the job without punishment. Otherwise (except when OSHA is unavailable to settle differences) the worker in U.S.A. has no right to refuse to work in the face of hazards, no matter how severe or obvious!

There is a more important meaning to this right which Table 2 in itself does not reveal.

In Sweden some 110,000 union health and safety representatives have been trained with forty hours or more of instruction in recognizing and controlling hazards. These OSH Representatives have the right *by law* to stop the production process in their plants when they see a work hazard of any sort. It can be an acute hazard of an electrical short and possible fire or even a long acting hazard such as a carcinogen, the effects of which might not show up for fifteen

Table 2. Rights of Safety Delegates and Safety Committees in Sweden and in the United States for the Majority of the Working Population

	Sweden	*United States*[a]
Right to stop work in the face of perceived hazard	+	partial
Right to refuse work in the face of perceived hazard	+	partial
Right to bargain on technical and other aspects of work process	+	–
Right to be informed about changes in work process and materials used	+	–
Right to bring in consultant (paid by management)	+	–
Right to have workers' rights publication	+	–
Right to have mandatory training for all new workers	+	–
Right to strike (local unions) over OSH conditions	+[b]	+[b]
Right to OSH consultant's services from federal agencies	+	only for management
Right of interpretation of labor law	+	–
Right of veto power over changes in work process	–	–
Right to do actual monitoring of OSH conditions at work	+	–
Right of access to monitor information and to observe monitor	+	+
Right to be trained as monitor	+	–
Right to be informed about hazards	+	partial
Right to be informed about potential hazards	+	partial
Right to be informed automatically about OSH hazards	+	variable
Right to have access to OSH studies and reports	+	+
Right to call inspector	+	+
Right of access to medical records of individual worker	+	+
Right of access to medical records (aggregate information) by unions	+	–
Control over occupational health services	+	–
Hiring and firing of physicians and engineers	+	–
Inspector's right to stop work	+	–

[a]For further information on workers' rights in the United States, see C. Seltzer, Surveying and Analyzing the Field of Employees' Rights Related to Occupational Disease, U.S. Department of Labor, 1982.

[b]If approved by agreement.

to thirty years. The production is stopped until a representative of the work inspectorate (the equivalent of OSHA which is called ASV, *Arbetarskyddsverket*, or Worker Protection Board) arrives and explains to the OSH Representative's satisfaction (1) there is no problem; it only looks like a problem; or (2) there is a problem and here is how you and your co-workers can work safely until the problem is removed in the following way, with a plan spelling out this way. There is nothing like this in the U.S.A.[22]

With regard to the second right to be discussed here, "Control over occupational health services," (third from the bottom of Table 2) workers in U.S.A. see the industrial physician as untrustworthy and usually refer to him as "the company quack." They or their representatives (the union, when there is one) have no control over the hiring or firing of the industrial medical and safety engineering personnel (if there are any). In Sweden, even though the management has veto power by virtue of controlling the purse strings,[23] there is a majority of workers representing the union on the health and safety committee and it is this committee which decides the hiring of the chief industrial physician and chief safety engineer after screening candidates submitted by ASV. Firing is also in the committee's hands, though this is more difficult because of the Security of Employment Act, designed to protect all workers.

The variations one can observe in OSH conditions and provisions between countries are no doubt primarily explained by the particular economy and history and current struggles between labor and capital within each country (and between labor and state production managers in socialist-oriented countries). However, these variations are also determined or importantly influenced by a country's economy and particular firms interacting with the world-system (see note 16). Thus if the world economy is depressed, as it has been under modern stagflation,[24] and workers and their representatives (unions and a political party, if they have one) have not achieved a system of taking up the slack of unemployment through public works, or in other ways,[25] it will be much easier for capital to get rid of sick or complaining workers and replace them from the reserve army of the unemployed than, let us say, to invest in a new ventilation system.

Another general effect of the world-system, whether depressed or not, stems from the general division of labor between core nations and semi-peripheral and peripheral nations. Depending upon the skill level of a country's labor force and of particular firms (it is harder to break down complex, skilled tasks and "taylorize" them for displacement to cheaper, less well-organized labor in semi-peripheral and peripheral countries, though computerization and robotization are rapidly changing the boundaries of these limits)[26] workers' bargaining position over OSH conditions may be weakened by threats of plant closings or displacement of jobs to other countries. Some analysts assert that such displacements of jobs and whole production units are occurring expressly to avoid OSHA and environmental controls.[27] Indeed, attempts by comprador bourgeoisies in semi-peripheral and peripheral nations to attract outside investment on this basis can become quite crude. For example, the State of Mexico placed the following newspaper advertisement:

> Relax. We've already prepared the ground for you. If you are thinking of fleeing from the capital because the new laws for the prevention and control of environmental pollution affect your plant, you can count on us.[28]

Other students of the hazard export situation (or hazard development, since a hazardous factory may be set up in a semi-peripheral or peripheral country without having first been in operation in a core nation) assert that references to OSHA and environmental pollution laws are only being used by capitalist owners as threats to weaken enforcement. These analysts argue that, in fact, production is being moved for other reasons—primarily cheap, easily exploited labor, lower taxes, ease of access to raw materials, and so on.[29] In any case, whether OSHA-type laws are a factor or not, the worldwide taylorization of labor and displacement of jobs to the semi-periphery and periphery of the world-system is a widespread and growing phenomenon.[30] This is bound to differentially affect the bargaining positions of workers' movements with regard to OSH concerns according to how exposed a country is to this kind of displacement. Such exposure can lead to the rise of a new despotism and very harsh working conditions indeed.[31]

Of course, workers will resist hazardous conditions as they become more aware. Such resistance may even break out in underdeveloped countries where unemployment is high. In Ghana, a French owned factory, L'Air Liguide, was taken over by the workers from August to October 1983. While most workers in core industrialized nations who work with carbide (a byproduct of acetylin production) are protected by special clothing and have access to a sink and shower, the workers in Ghana were without any protective clothing, goggles, gloves, or aprons and carried carbide in open pans in front of them or on their heads. In contact with moisture in the atmosphere it is a severe irritant and can cause blindness. Union leaders had been dismissed for demanding protective clothing and medical check ups.[32] The take over brought changes.

Beyond such general effects of the world-system on whole nations, there are direct effects of competition from firms producing the same or a very similar product in another country. Thus, while I was in Sweden, a dispute broke out between activist workers and the union in a small plant producing fire-safe boxes and files for business offices. The activists were backing the plant physician who wanted immediate changes to set up site-specific ventilation for the welding booths, while the union sided with the owners, arguing that a company in F.R.G. which produced approximately the same products could win out on the world market if costs in the Swedish plant went up too much; thereby the Swedish workers might loose their jobs. Your job or your life. However, strong Sweden is in OSH protection, it has not completely done away with this inhuman choice in every circumstance.

There is a range of fascinating variations in OSH measures to be observed between countries. These suggest different histories and traditions but also current differences in political economies and strengths of workers' movements. Only a few are mentioned below to stimulate the reader's curiosity. I also mention some differences to point up dimensions of a model to be elaborated in Chapter 6.

Before presenting this range of OSH differences between study countries, let us examine some general socioeconomic and health characteristics of these six countries.

GENERAL SOCIOECONOMIC AND HEALTH CHARACTERISTICS OF STUDY COUNTRIES

More detail will be presented on each study country in the country chapters (7-12). This is especially so for historical background. Thus the reader is cautioned that the cross-sectional comparisons presented in Table 3 and in this section are merely offered as introductory characterizations and not fully developed contextually placed information. Thus, for example, Finland has the lowest density of persons per square kilometer of area in Table 3, first column, but a considerable part of the country is covered with water, swamps, and forests. Thus the density of living arrangements is not much different from that of Sweden, though it remains the least densely settled and most rural of the study countries.

Or, to take another example closer to our concern with determinants of OSH conditions and provisions, the population figure for the G.D.R. becomes especially meaningful only when one knows that (1) some 3 million people fled before the fortified wall separating G.D.R. from F.R.G. was built in 1961 and some 50,000 more fled after the wall was built; (2) the population growth rate has been very low; (3) the push toward industrialization has been vigorous. Thus there is a serious labor shortage and OSH protection is given high priority and is consequently strong.

From our general knowledge of the world situation we can see from the second column in Table 3 that our study countries are all very wealthy. Their 1980 GNP per Capita ranged from $7,226 in G.D.R. to $13,962 in Sweden, while the world as a whole had $2,619 GNP per person and the two-thirds of the world in peripheral and semi-peripheral states had just $794 GNP per person. One of the poorest countries, Chad, had just $113 and Bangladesh had just $129 GNP per person.

Table 3. General Socio-economic and Health Characteristics of Study Countries, 1980[1]

Study Country	*Population Area (sq. kms), Density (persons per sq. km)*	*GNP per Capita U.S.A. $*	*Percent of Population Urban*	*Percent of Exports Made up of Manufactured Goods*	*Public Expenditures per Capita U.S.A. $*			*Population per Physician*	*Population per Hospital Bed*	*Infant Mortality Rate*[2]	*Life Expectancy*[3]
					Military	*Education*	*Health*				
Sweden	8.3 million 450,000 18	13,962	83	NA	459	1,350	1,035	450	68	7	75
Finland	4.8 million 337,000 14	10,333	59	NA	170	585	410	530	64	8	73
German Democratic Republic (G.D.R.)	16.7 million 108,000 155	7,226	76	70.7[4]	360	304	341	490	94	12	72
Federal Republic of Germany (F.R.G.)	61.6 million 249,000 247	13,399	85	88.9	434	616	884	440	87	13	73
United Kingdom of Great Britain (U.K.)	56.0 million 244,000 230	9,213	91	77.5	478	494	443	620	121	12	73
United States of America (U.S A.)	227.7 million 9,363,000 24	11,347	73	60.8	632	571	439	550	171	13	74

[1] Data for first 10 columns (except columns 3 and 4) taken from Ruth Sivard, *World Military and Social Expenditures, 1983*. Washington, D.C.: World Priorities. Columns 3 and 4 (percent urban and percent of exports manufactured) are from *World View 1982*. Boston: South End Press, 1982. Data for 1980 unless otherwise indicated.

[2] Deaths under one year per 1,000 live births.

[3] Expectation of life at birth—males and females combined.

[4] "Machinery and equipment" 55.8 plus; "Consumer goods" 14.9; "raw materials" 29.3. The G.D.R. does not report "Food and agriculture," "Raw materials," and "Manufactured goods" as do the other study countries for which *World View, 1982* gave data.

Table 3 continued below

Study Country	Type of Health Services System	Specialty OSH Services	Strength and/or Type of Workers' Movement	Political Economy and Percent Unemployed in 1981
Sweden	National Health Service with universal coverage. Quite fully regionalized PHC not geographically targeted. Some self pay.	Well developed, especially for large and medium sized plants by agreement between employers and workers' unions. Strong worker control.	Strong. 95% or more of blue collar workers organized; some 75% of white collar workers organized. Labor govt. in power for over forty-five years. Strong labor laws.	Capitalist social welfare. Relatively decentralized, concerted authority structure. Comparatively equitable distribution of income. Two % unemployed.
Finland	National Health Insurance system covering all who do not seek private care on their own. A dualistic system. Not well regionalized. PHC not geographically targeted.	New law requires services for all workers, tooling up to prepare manpower now underway. Strong back-up from National OSH Institute. Some worker control; not as strong as in Sweden.	Moderately strong. Some 80% of workers organized. No labor government. Paternalistic tradition. Labor laws not as strong as in Sweden.	Capitalist, quasi-social welfare. Somewhat more centralized authority structure than Sweden; less concerted. Sharp disparities in income. Five % unemployed.
German Democratic Republic (G.D.R.)	National Health service with universal coverage and almost no private elements. Quite fully regionalized. PHC geographically targeted.	Very fully developed in law and actuality for large factories; important variations for medium and small plants and agriculture. Formal worker control.	Strong. A workers' state. Communist party gives essential direction; weak in the sense that unions do not exercise an independent voice.	Socialist oriented (or state capitalist). Highly centralized, relatively concerted authority structure. Comparatively equitable distribution of income. Zero % unemployed.
Federal Republic of Germany (F.R.G.)	National Health Insurance system. Universal coverage. Some private elements. Little regionalization. PHC not geographically targeted.	Limited to prevention. Only developed for large plants. Recent law requires increasing coverage for medium plants. Little worker control.	Weak (some 30% of workers are organized). No labor government in power. Weak labor laws.	Capitalist. Decentralized, federated authority structure, somewhat divided. Sharp income disparities. Eight–nine % unemployed.
United Kingdom of Great Britain (U.K.)	National Health Service with universal coverage. Fairly well regionalized. PHC geographically targeted. Cutbacks leading to increasing privatization.	Mainly developed for large plants. Partially trained GPs may serve other plants. No requirement for such services in law. Little worker control.	Medium. Some 55% of workers are organized. Labor government has been in power, but not in recent years. Labor laws moderately strong. Labor under attack by conservative government.	Capitalist, social welfare. Relatively decentralized increasingly fractionated authority structure. Sharp income disparities. Twelve % unemployed.
United States of America (U.S.A.)	Mainly market-oriented, private pay system with some federal public programs, including compulsory national health insurance for those 65 and older (Medicare). No regionalization. PHC not geographically targeted.	Poorly developed. Some large plants have well trained physicians but almost all OSH services under exclusive management control. No requirement for such services in law.	Very weak. Some 19% of non-agricultural workers are organized. No labor government has been in power. Very weak labor laws. Union-busting and give-back contracts rampant.	Capitalist. Decentralized, federated authority structure, relatively fractionated. Very sharp disparities in distribution of income. Ten % unemployed.

The study countries are all quite highly industrialized and urbanized. While figures for Sweden and Finland were not given in *World View, 1982* for the percent of exports made up by manufactured goods (Column 4 in Table 3) both countries are highly industrialized, Sweden more so than Finland. However, both also export food products and raw materials, especially wood products, though these usually have value added through complex industrial processing, as, for example, in the largely automated newsprint paper mill I saw in Finland. This continuous mill turns wood pulp into 833 cm-wide giant rolls of paper at the rate of 1 km a minute! One of their large customers is *The London Times*.

The U.S.A. is most burdened by military expenditures among study countries with 5.6 percent of GNP going for this sad end. Britain and G.D.R. are not much better with 5.2 percent and 5.0 percent respectively. F.R.G. spends just 3.2 percent of its GNP for military purposes with Sweden at the same level, 3.3 percent. Finland which balances nicely between "East" and "West" in its economy and other ways, as we shall see in Chapter 8, spends only 1.6 percent of its GNP on the military.

In education, Sweden is far and away the one devoting the greatest share of resources to this sector of societal concern: 9.7 percent of GNP. The others are in a similar range from a low of 4.2 percent in G.D.R. to a high of 5.7 percent in Finland. The 5.0 percent of GNP spent in U.S.A. on education through public channels may particularly underestimate the level of support for education as more of the educational system is private in U.S.A. than in other study countries.

This last point regarding public versus private expenditures is especially relevant for the health sector in U.S.A. and Finland and to some extent F.R.G. These countries, especially the U.S.A., have important proportions of the health sector supported by non-public funds, while the health systems of Sweden, G.D.R. and U.K. are almost entirely publicly supported. Thus the figure in column seven of Table 3 must be understood as public support for health care and not all support. More discussion is given in the country chapters (7-12). I will also reserve discussion of the health systems for those chapters.

As regards general health indicators (infant mortality and life expectancy) the study countries are within the same general range. However, the positions of G.D.R. and U.K. are notable in comparison to those for U.S.A. and F.R.G. when it is realized that they do at least as well with fewer overall resources and considerably less spent on health (when both public and private expenditures for health are included). More on this later.

The content of the last three columns of Table 3 will be elaborated, in part in the next section of this chapter on variations in OSH conditions between study countries, and significantly in later chapters (7-12). As regards the political economies, all except G.D.R. are not only capitalist, but partake of the character of "late" or "monopoly" capitalism functioning in a world-system. Such systems of production and distribution are characterized by ever larger and more powerful agglomerations or centers of control of capital resulting from both horizontal integration (similar firms being swallowed up by the larger ones) and vertical integration (sources of supply as well as marketing outlets increasingly brought under control of the parent firm or holding group). This development is most obvious in U.S.A. where there are consequences also for the social relations of production,[33] organization of community life,[34] and other effects such as the fiscal crisis of the state and cutbacks in support for human services.[35] But capitalism being a world-system, its monopoly character and effects are observable in important ways in every study country. This statement is understandable when one realizes the size and power of the modern capitalist enterprise. In a listing of countries of the world according to GNP and companies according to sales (in billions of U.S.A. $s) the U.S.A. is first, the U.S.S.R. second, but Exxon with $80 billion sales ranks in the top fifteen, ahead of Sweden, Finland and G.D.R., and below U.K. and F.R.G. General Motors follows closely after and British Petroleum is in the top fifty along with the other "seven sisters" (the major oil firms, including Exxon).[36] This does not mean that there are no small competitive firms. In each study country, there are. It simply means that monopoly capital lends the major impetus to and essentially structures the system. Thus, for example, with regard to OSH, one may find, curiously enough, that very large firms may object less to certain OSHA-type regulations than do smaller firms. This can be because the large ones have strong control over their prices and simply add OSH costs to "the costs of doing business" and to the price of their products. Small firms, facing stiffer competition as to prices may find it harder to add OSH protections and services. If they should go under doing so, this would not displease the larger ones. They would be happy to swallow the littler ones.

The one exception among our study countries is the G.D.R. This is a socialist-oriented "workers' state" (some would argue state capitalist).[37] Firms are public and are designated VEB, Volkseigenebetriebe (people's own firms or works) rather than Limited, Incorporated, or, as in F.R.G., Society with Limited Liability (G.m.b.H., Gesellschaft mit beschraenkter Haftung). The implications for OSH of there being no major private productive property in G.D.R. will be discussed in Chapter 9 as will evidences of there still being effects from competition in the capitalist world-system. Although different as regards class relations and control of the means of production, the G.D.R. is as industrialized as some of the other study countries (Finland for example). In fact, even though it has but 16.7 million people, it is estimated that it has risen since WW II to eighth position in the world in industrial production.

OSH VARIATIONS ALONG DIMENSIONS OF A MODEL

In Chapter 6 I will develop the reasons for certain key concerns as regards provisions for OSH and I will define these concerns as dimensions of an ideal model. These concerns have to do with *policy, sponsorship and control, education, organization, information systems, and financing*. Here I will briefly cite some examples of major differences along these dimensions to give the reader some taste of what is involved and how great the differences can be. At this point, OSH provisions are cited somewhat out of context, only to highlight differences between study countries. A fuller understanding will be gained in the country case studies themselves (Chapters 7-12).

At the *policy level*, the clear priority given to workers' health in Sweden contrasts sharply with the situation in U.S.A. Many illustrations could be given. In U.S.A., the OSHA-funded New Directions Program for worker education in OSH has been a great boon. We have had such a program in Connecticut for five years (Reagan budget cuts resulted in a much diminished program in the last two years and an end to it in 1983). Each year we have trained some thirty-five "Trainers," that is workers with enough training and interest to go on to train other workers with their union's assistance. But in Sweden, more than 110,000 such people have been prepared with forty hours or more of training and many other educational efforts are undertaken. Sweden has 8 million people. Connecticut has 3 million people. If we were to have done as well, we would have had to have the funds and staff to prepare 41,250 Trainers. We have prepared some 175.[38]

The cynicism with which OSH has been treated in U.S.A. would have been intolerable in most other study countries in modern times.[39] The Gauley Bridge disaster which occurred in the depths of the Great Depression in the early 1930s would be hard to match for its perfidy and reckless disregard of workers' lives in any time and place (see Chapter 1). The asbestos cover up is another story which could perhaps only have occurred in U.S.A. This ranged over a period of thirty-five to forty years until the present court cases and other events revealed the full story of workers tricked into unknowingly sacrificing their health and lives over the long run in favor of short term gains for capital.[40] The famous "Guenther memo" brings us to the era of OSHA. In this action, Mr. Nixon's Director of OSHA, George C. Guenther:

> . . . reassured his superiors in the Department of Labor that "no highly controversial standard (i.e. cotton dust, etc.) will be proposed by OSHA or by NIOSH," that the political registration (party affiliation) of all field Regional Administrators was being checked, and that he would submit OSHA's hiring plan to the Republican National Committee and to the Committee to Re-Elect the President.[41]

The cut backs in OSHA and weakened enforcement under the Reagan administration are only the latest betrayals of the working class and evidence of deep cynicism concerning OSH at the policy level in U.S.A.[42] This low priority for OSH at policy level can not be found in the other study countries.

With regard to *sponsorship and control* of OSH services, there are also striking differences. Again, in Sweden, the OSH physicians and engineers in the workplace are hired and fired by the Health and Safety Committee controlled by a majority of union members, though management is represented and controls the purse strings. In G.D.R., OSH physicians are neither employed by management, nor unions, but by the Ministry of Health and report to the chief district doctor within this quite completely regionalized National Health Service system. In U.S.A., as we know, when there are any OSH physicians or other experts, they are employed by and report to management. There is, by the way, no OSH services standard under OSHA, that is, there are no requirements as to what health services must be provided to U.S.A. workers! Another especially interesting contrast is the centrality of the insurance agencies, the Berufsgenossenschaften in F.R.G. By contrast, U.S.A. insurers responsible for worker compensation payments have only lately entered the field.

As regards *education*, I have already pointed out one of the most important variations, namely in worker education in Sweden and U.S.A. This dimension connects with PHC and the general health services as well. While there usually are no OSH hours required for U.S.A. medical students the G.D.R requires thirty-six hours. F.R.G. requires none but students know there will be several OSH questions on their national board exams. There is also the matter of how OSH specialists will be prepared. Several countries are coming to or have a two-tiered system: several years on top of medicine for researchers-teachers and lesser preparation on top of medicine for plant physicians. Finland has adopted a law requiring OSH services for every workplace: thus they suddenly need hundreds of new OSH physicians at least minimally trained.

Organizational arrangements for OSH and OSH-PHC seem to reflect the most variation. I have already mentioned the important point that regionalized NHS systems tend to have geographic assignment of PHC services which can lead PHC providers to learn about workplace hazards in their service area. The U.K. and G.D.R. exemplify this among the study countries. Sweden has an NHS too; however, it lacks clear geographic assignments at the periphery. Finland, on the public side of its services–it is really a dual public-private system–shares most NHS characteristics. F.R.G. has an NHI, but its health services are organized largely by market forces as is true for U.S.A. health services. More of the details of organizational arrangements for PHC-OSH linkage will be mentioned in country chapters (7-12).

The problem of *information systems* for OSH surveillance has been mentioned. The sharpest contrast is between the U.S.A.'s chaotic, almost nonexistent system and the G.D.R. where a recent national law, 1981, requires a detailed survey, and systematic reporting and recording and follow-up of hazards in every workplace.

While the general OSH information system in U.S.A. is poor, it may be one of the best as regards special epidemiologic and other OSH studies, though in testing of chemicals it does not have a strong showing.[43]

Financing also shows fascinating differences. In Sweden, OSH services in the plant are paid 50 percent by the NHS and 50 percent by the employer. The employer's funds are said to be for prevention, the NHS funds for treatment. In the F.R.G., OSH services are paid 100 percent by the employer, and, by law (pushed through by organized medicine) required to be limited to preventive efforts so that in-plant OSH physicians will not take any patients away from community practitioners (though this is not the publicly expressed rationale).

NOTES AND REFERENCES

1. There has not been enough work on underdeveloped countries for me to be sure that OSH action and legislative developments in the late 1960s and early 1970s were limited to industrialized nations. Certainly my own literature survey of underdeveloped countries suggested disastrous OSH conditions and few positive developments. Ray Elling, Industrialization and Occupational Health in Underdeveloped Countries, *International Journal of Health Services* 7:209–235, 1977. However, it seems likely that there were important OSH actions in socialist-oriented countries in the underdeveloped parts of the world. There have been disappointing reports from China, for example, but when one considers the utterly debased situation of workers and peasants before the revolution, first, the very great advances in general living and health conditions have also advanced workers' health, and, second, some reports indicate major achievements in OSH: William Y. Chen, Relation of Occupational Health to General Community Health, pp. 247–257 in J. R. Quinn (ed.), *China Medicine As We Saw It* DHEW Pub. No. (NIH) 75-684. Bethesda: Fogarty Int'l. Center, 1974; also David C. Christiani, Occupational Health in the People's Republic of China, *American Journal of Public Health* 74:58–64, January, 1984. The same story could be told for Cuba, and more recently for Nicaragua which is placing great emphasis on protecting the health of workers and peasants just four years after the success of the armed struggle.
2. Foreign Experience and Its Relevance for the United States, Chapter 11, pp. 500–519 in Nicholas A. Ashford (ed.), *Crisis in the Workplace: Occupational Disease and Injury, A Report to the Ford Foundation.* The countries about which some information is given are Belgium, France, Netherlands, Sweden, Finland, Federal Republic of Germany, United Kingdom, and Italy. However, there are no specific references for this section and statements are sometimes incorrect. For example, it is said that in Sweden (as in France, Netherlands, and Belgium) "the occupational health service is almost exclusively preventive; treatment may be given only in the case of emergency," (p. 508). In fact, because half the payment for OSH services in Sweden stems from the National Health Services, about half the effort is devoted to treatment and there is considerable discussion around the issue. Among my study countries, it is only the F.R.G. which strictly limits OSH services to prevention. The main documented detail of this overview chapter

by Ashford covers the Robens Report in the U.K., Safety and Health at Work, 1972, an important example of the kind of activity and legislative ferment I am referring to.

3. Frederick Engels, *The Condition of the Working Class in England in 1844.* New York: J. W. Lovell Co. 1887 (English translation, 1st edition 1845 in German). Norman Thomas, *Human Exploitation in the United States.* New York: Frederick A. Stokes, 1934.
4. Daniel M. Berman, Organizing for Job Safety and Health, A Conversation about COSH Groups. Unpublished paper, February 22, 1978; Charles Levenstein, Leslie I. Boden, and David Wegman, COSH: A Grass-Roots Public Health Movement, *American Journal of Public Health 74*:964–965, 1984.
5. Terveysrintama (Health Front) *Työsoujelunormil* (Work Protection norms). Tampere, Finland: Kirjapaino Sanan Tie, 1977. Study Circles in local unions were encouraged. In one instance, the workers in a foundry read and discussed the heat standard–take a fifteen-minute break every hour if the temperature is over 100 degrees Fahrenheit–and decided they should follow it. The management complained to the National Institute for Occupational Health and Safety. The Institute sent two inspectors who could also not stand to stay in the heat continuously so the workers won and a new official standard was *de facto* adopted. This example highlights how OSH problems connect very directly with control of the production process; thus enter centrally into the struggle between capital and labor.
6. An impressive book was produced, mixing, in simple, straightforward terms, the technical and political-economic action aspects of both physical and psychosocial problems at work. SARB *Arbetsmiljo På Vems Villkor? En Bok Om Risker I Arbetet.* Stockholm, 1977.
7. Denis Macshane, *Solidarity, Poland's Independent Trade Union.* Nottingham, England: Spokesman, 1981, esp. Point 16 of the Gdansk Agreement: "To improve working conditions and the health services so as to ensure better medical protection for the workers."
8. V. I. Lenin, *The State and Revolution.* Peking: Foreign Languages Press, 1976.
9. Ernest Mandel, The State in the Age of Late Capitalism, Chapter 15 in Mandel (ed.), *Late Capitalism.* London: Verso, 1975. See also Fred Block, Marxist Theories of the State in World-System Analysis, pp. 27–37 in B. H. Kaplan (ed.), *Social Change in the Capitalist World Economy.* Beverly Hills: Sage, 1978.
10. A classic case of the interweaving of OSH science and politics in U.S.A. is found in Paul Brodeur, *Expendable Americans.* New York: Viking, 1973. This describes the process in which the Asbestos Standard was adopted as the only new standard in the OSHA Act of 1970. First, twelve fibers of 5 microns length per cubic centimeter of air over an eight-hour working day was considered; then 5 and then 2. The law adopted 5 with a shift to 2 in 1974. Subsequent evidence has led to an emergency standard of .01 fibers. (For example, the finding of Selikoff's Mt. Sinai Hospital team that families who had no asbestos exposure but moved into houses previously occupied

by asbestos workers began to show higher than expected rates of asbestos-caused lung and other diseases). But this has been challenged by industry and set aside. Hearings on a new standard are now underway. Joann S. Lublin, OSHA Proposal to Set Tougher Standard for Asbestos Exposure Offers 2 Choices, *The Wall Street Journal*, p. 8, April 10, 1984.

11. There are very notable and laudable exceptions. Dr. Lorin E. Kerr has devoted a lifetime to workers represented by the United Mine Workers and has been honored by the APHA for his work. The epidemiologic work of Dr. Irving J. Selikoff and Dr. Thomas F. Mancusso, faithfully carried through in spite of attacks by management forces is especially noteworthy. But many social medicine health planners concerned with achieving national health service systems, etc. seem to have ignored improved health on the job. One of the best illustrations of this point is that to this day the largest employer in all of Europe, namely the NHS in Britain, does not have an occupational health service of its own and hospitals and other health facilities are not exactly healthy places to work.
12. Daniel M. Berman, *Death On the Job, Occupational Health and Safety Struggles in the United States*. New York: Monthly Review Press, 1978, quote from pp. 35–36.
13. Technically speaking, I cannot document that it was a cross-national social movement with identifiable participants stimulating each other across national boundaries. Thus, it might better be termed a "ferment" having possibly independent, even if similar, origins in each country, that is, similar in the sense of class interests involved, and so on. For a discussion of the sociology of social movements, especially in relation to the physically disabled, including moves for improved workers' compensation, see Elliott A. Krause, The Political Sociology of Rehabilitation, pp. 201–221 and Constantina Safilios-Rothchild, Disabled Persons' Self-Definitions, pp. 39–56 in Gary O. Albrecht (ed.), *The Sociology of Physical Disability and Rehabilitation*. Pittsburgh: University of Pittsburgh Press, 1976.
14. The tragic fact that workers, more usually black workers, are dying from exposure to hydro-carbons emanating from the coke ovens around Pittsburgh long after these substances were first found to cause cancer is examined in Joseph K. Wagoner, Occupational Carcinogenesis: the two hundred years since Percival Pott, pp. 1–3 in *Occupational Carcinogenesis*, special issue of the *Annals of the New York Academy of Science 271*:1–3, 1976.
15. Ricardo Edstrom, unpublished conference paper, no date, probably 1980.
16. The idea of the world-system and its supporting cultural hegemony will be developed in Chapter 5. Here it may suffice to mention the work of Immanuel Wallerstein, Andre Gunder Frank, and others on the world-system. Immanuel Wallerstein, *The Capitalist World-Economy*. Cambridge: MA: Cambridge University Press, 1979; André Gunder Frank, *Crisis In the World-Economy*. New York: Holmes and Meier, 1980; Christopher K. Chase-Dunn (ed.), *Socialist States in the World-System*. Beverly Hills: Sage, 1982. And that of Antonio Gramsci on hegemony: The Modern Prince in Q. Hoare and G. Norwell (eds.), *Prison Notebook*. London: Laurence and Wishart, 1972.

17. Vicente Navarro documents the working class struggles in Sweden and gains in OSH: "The determinants of Social Policy, A Case Study: Regulating Safety and Health at the Workplace in Sweden, *International Journal of Health Services 13*, pp. 517–561, 1983. See also Birger Vicklund, The Politics of Developing a National Occupational Health Service in Sweden *American Journal of Public Health 66*, pp. 535–537, 1976.
18. Joseph A. Page and Mary Win-O'Brien, Introduction by Ralph Nader, *Bitter Wages.* New York: Grossman, 1973, especially The Battle for OSHA, Chapter 8. "The legislative history of the Occupational Safety and Health Act of 1970 (OSHA) provides more than just an arresting glimpse at the process by which laws are drafted and enacted. It also removes much of the gloss from the Nixon administration's blue-collar strategy. For while the President and Vice President were making rhetorical pronouncements that purported to support working class aspirations, the Nixon administration was struggling to perpetuate a bill that would furnish no more than cosmetic protection for employees in an area that concerned them in the most vital ways imaginable" (p. 167). Once passed, Nixon's campaign staff sought to use promises of selective enforcement to raise funds (the famous "Gunther Letter").
19. Philip J. Simon, Foreword by Ralph Nader, *Reagan in the Workplace: Unraveling the Health and Safety Net.* Washington, D.C.; Center for the Study of Responsive Law, 1983.
20. Under the pressures of monopoly capital and its union-busting in the currently depressed world-system, some observers see a near dissolution of organized labor in the U.S.A. or at the very least the need for a new strategy. Stanley Aronowitz, *Working Class Hero: A New Strategy for Labor.* New York: The Pilgrim Press, 1983.
21. Navarro, op. cit. (note 17) Table 9, p. 553.
22. Of course in unionized shops in U.S.A. (a minority) there is usually something in the contract about workers or union stewards being able to "red tag" a piece of broken equipment so they can stop using it and maintenance can find and repair it. Sometimes workers "red tag" dangerous machines which are otherwise working. But this is far from satisfactory. It is surreptitious, does not carry the sanction of law and does not cover all hazardous situations.
23. In many situations, the Swedish unions have found a way around this limitation by having a pre-budgeted approach to OSH; thus within the agreed budget, the union and its majority of representatives on the shop's health and safety committee have control.
24. This is only one of several conditions making for a crisis in the world-system. André Gunder Frank, op. cit. (note 16). Also see the appreciative review of this work of Frank's: Gregory Shank, The World Economic Crisis According to Andre Gunder Frank, *Contemporary Marxism* No. 5, pp. 147–153, Summer 1982. Also Albert Bergesen (ed.), *Crisis in the World-System.* Beverly Hills: Sage, 1983; and E. A. Brett, *International Money and Capitalist Crisis: The Anatomy of Global Disintegration.* Boulder, CO: Westview, 1983.

25. Recently, in F.R.G., the major union, I.G. Metall, carried out a strike of steelworkers and other metal workers for a thirty-five-hour week instead of forty hours with no loss of pay as a way of creating more jobs in a situation where there is over 9 percent unemployment. Metalworkers Threaten Strike in West Germany, *The Hartford Courant*, April 18, 1984, p. A17. The final result was a compromise between thirty-five and forty hours.
26. Susan Jonas and Marlene Dixon, Proletarianization and Class Alliances in the Americas, *Synthesis 3*:1–13, Fall, 1979.
27. Barry I. Castleman, The Export of Hazardous Factories to Developing Nations, *International Journal of Health Services 9*:569–606, 1979.
28. As quoted in Richard J. Barnet and R. E. Muller, *Global Reach*. New York: Simon and Schuster, 1974, p. 345.
29. Charles Levenstein and Stan Eller, 'For Example' Is No Proof, *Business and Society Review 33*, Spring, 1980.
30. F. Froebel, J. Heinrichs, and O. Kreye, *The New International Division of Labor: Structural Unemployment in Industrialized Countries and Industrialization in Developing Countries*. London: Cambridge University Press, 1980 and the "Review" of this work by Andre Gunder Frank in *Synthesis 3*(1):41–44, 1979.
31. Michael Burawoy, Factory Regimes Under Advanced Capitalism, *American Sociological Review 48*:587–605, October 1983.
32. Obi Bini (pen name) Anatomy of a Takeover, *Multinational Monitor 5*(4): 20, April, 1984.
33. Harry Braverman, *Labor and Monopoly Capital*. New York: Monthly Review Press, 1974.
34. Especially as regards local community power structures and decreasing face to face contact between power figures and people. For example, in the film, "Controlling Interest," the workers and people (including small business and local government people) in Greenfield Massachusetts receive a set of demands through the local manager from the headquarters of a corporation based in New Jersey. These demands include: work for no wage increases over the next four years even though there is raging inflation; lower our taxes; and build us a new plant, or we will move to South Carolina or Taiwan (paraphrased). There is a struggle and the workers and town keep the jobs but more or less loose. Unfortunately, liberal sociologists seem to have ignored this aspect of community power. And Marxist sociologists torn between the arguments of structuralists, who ignore power figures to focus on systems, and instrumentalists, who focus only on state power as an active agent or tool in class struggle, seem to have ignored local community power structures altogether. At the national level, see G. William Domhoff, *Who Rules America?* Englewood Cliffs, NJ: Prentice-Hall, 1967. Also G. William Domhoff (ed.), *Power Structure Research*. Beverly Hills: Sage, 1980.
35. Ray Elling, The Fiscal Crisis of the State and State Financing of Health Care, *Social Science and Medicine 150*:207–217, 1981.
36. Marlene Dixon, Dual Power: The Rise of the Transnational Corporation and the Nation-State, *Contemporary Marxism* No. 5, pp. 129–146, Summer, 1982.

37. For example, see Navarro's analysis of the Soviet Union which could also be applied to G.D.R.: Vicente Navarro, *Social Security and Medicine in the USSR: A Marxist Critique*. Lexington, MA: D. C. Heath, 1977.
38. For this and other aspects of the New Directions Program in Connecticut, including its links to the Yale Occupational Health Clinic, see Ray H. Elling, Worker-based Education, Organization and Clinical-epidemiologic Support for Occupational Safety and Health: The New Directions Program in Connecticut, in *Proceedings of the ILO-WHO meeting on Education for Occupational Health*, 16–19 August 1981, Sandefjord, Norway. Geneva: ILO, 1982.
39. Whether there were major differences in the "rawness" of early industrial capitalism as it developed in each country is debatable. The conditions in sweat shops, for women, for children and for men workers were everywhere "far from idyllic." (See note 3.)
40. Samuel Epstein, The Asbestos 'Pentagon Papers'," p. 154–165 in Mark Green and Robert Massie, Jr. (eds.), *The Big Business Reader, Essays on Corporate America*. New York: Pilgrim Press, 1980.
41. Molly Joel Coye, Mark Douglas Smith, and Anthony Mazzocchi, Occupational Health and Safety: Two Steps Forward, One Step Back, pp. 79–106 in Victor W. Sidel and Ruth Sidel (eds.), *Reforming Medicine, Lessons of the Last Quarter Century*. New York: Pantheon, 1984, quote from p. 86.
42. Philip J. Simon, *Reagan in the Workplace: Unraveling the Health and Safety Net*. Washington, D.C.: Center for Study of Responsive Law, 1983.
43. National Research Council: Only a few Chemicals Tested for Health Effects, *The Nation's Health* (Official Newspaper of the American Public Health Association), pp. 1 and 3, April, 1984.

CHAPTER 4
The Social Context and OSH Information

The numbers are far too large to accept that why we don't really know how many men and women are being killed or seriously disabled each year by their work, in which work settings, from which hazards and why adequate protective measures are not taken. We have to understand the structure of interests and flow of influence in the OSH information system as well as the technical problems of disease perception, hazard identification, sampling, measurement and other aspects of research and surveillance. Here, I first take up the structure of interests which provides the overall atmosphere, or in Gramsci's terms, the hegemony within which OSH problems are perceived or not and acted upon or not. Chapter 5 will elaborate the idea of hegemony and its supporting and supported political economy. In the second part of this chapter, I consider the individual's perception and use of information within the context of dominant influence. Finally, the chapter considers some especially problematic aspects of the OSH information system. While most examples are drawn from U.S.A. experience, the principles apply and the problems are found to one extent or another in other study countries.

THE CLASS-BASED NATURE OF OSH INTERESTS IN CAPITALIST SOCIETIES

> . . . I submit that if a million people in the so-called middle or professional class were dying each decade of preventable occupational disease, and if nearly four million were being disabled, there would long ago have been such a hue and cry for remedial action that if the Congress had not heeded it vast numbers of its members would have been turned out of office. Paul Brodeur, author of *Expendable Americans*[1] testifying before the House of Representatives Select Subcommittee on Labor, Washington, D.C., May 22, 1974.

Paul Brodeur made this statement after twenty months of tracking down the details of the asbestos story, especially the setting of asbestos standards under OSHA and the tragedy of several hundred workers' deaths in the Tyler Texas asbestos plant jointly owned by Pittsburgh Plate Glass (PPG) Company and Corning Glass Works. Brodeur's story of information and Dr. Grant's role *vis-à-vis* the Tyler Texas plant follows. Among many astounding and outrageous experiences Brodeur had which led him to understand the class-based bias of information standards, enforcement, and protection in the OSH field were his efforts to interview Dr. Lee B. Grant, who became Medical Director for PPG in 1965. Dr. Grant was a retired colonel who had been Chief of Aerospace Medicine for the United States Air Force Logistics Command.

In Chapter 1 it was noted that British clinicians and epidemiologists had identified asbestosis in the 1920s. It should also be recalled that since 1963 Dr. Irving Selikoff had been reporting very disturbing findings in the literature concerning his follow up studies of asbestos workers in New Jersey. A major study of his with Dr. E. Cuyler Hammond, vice president for epidemiology and statistics of the American Cancer Society, had been reported to the AMA annual convention in June 1963 and was published in *JAMA* in the Spring of 1964.[2] This study not only associated exposure to asbestos with asbestosis but with lung cancer and mesothelioma (an extremely rare cancer of the lining of the lung). Further, an October 1964 Conference of the New York Academy of Sciences on the Biological Effects of Asbestos alerted hundreds of scientists and industrial medical people to the dangers. And in 1966 the Academy sent out thousands of copies of its conference report.

Thus when Dr. Grant decided in 1966 to have a resurvey of the Tyler plant done to update a 1963 survey which had found, in a number of its samples, asbestos dust levels above the then accepted threshold limit value (TLV) of 5 million particles per cubic foot of air,[3] he should have arranged for medical exams including lung function tests and chest x-rays as well as an industrial hygiene survey. The survey, however, was only a study of air samples from sixteen different areas of the plant. The investigator, a safety and industrial hygiene engineer from PPG itself, reported in November 1966 that the TLV was exceeded in seven of the sixteen samples. In three air samples, the count was 20 million or more fibers per cubic foot of air. He made the same mistake as in the 1963 survey, that is, he used a general dust standard as if it were an asbestos standard (see note 3). Again, improved ventilation was recommended. Dr. Grant took no immediate action. In December of 1966, Dr. Grant commissioned Dr. George A. Hurst of the East Texas Chest Hospital to do a medical survey of the workers. The study was to cost $4200. But he later cancelled this in favor of another industrial hygiene survey done by Dr. Lewis J. Cralley, based in the U.S. Public Health Service's Division of Occupational Health in Cincinnati. It turns out that this seemingly independent agency was also industry dependent, for it had no legal authority to enter plants and gather information. It could only carry on its studies by invitation. This survey, dated March 27, 1968 reported to the works manager, with a copy to Dr. Grant, high proportions of air samples over the TLV. It used both a standard method of counting fibers and a new method in use in the U.K. (which later became standard in U.S.A). But again no explicit health warning was issued and the report ended with "Your cooperation in this study is sincerely appreciated and the data gained from your plant are of considerable value."[4] Dr. Grant took no special action, nor is it known if he offered the works manager a medical evaluation of the report. Whether he did or not, two years later the works manager himself died of mesothelioma.

Dr. Grant never got back to Dr. Hurst to have a medical study done and Dr. Cralley and his associates in the Division of Occupational Health of the USPHS never got around to doing the medical part of their survey which had been part of their protocol. Nor did Dr. Grant push them to do it.

Following the passage of OSHA by the U.S.A. Congress on December 29, 1970 and its effective date of April 28, 1971, union and other pressure brought about a NIOSH health hazard evaluation of the Tyler plant between October 26 and 29, 1971. This was headed by Dr. William M. Johnson, who had just completed a two-year program in occupational health at Harvard School of Public Health. He was backed up by Dr. Joseph K. Wagoner who replaced Dr. Cralley on August 1 as Director of the New Division of Field Studies and Clinical Investigations. In preparation for the study, Dr. Johnson had called Dr. Grant and was told that there wasn't much of a problem at Tyler "because the place was so dusty that people didn't stay around there long enough to get sick." Dr. Grant also said there were no plans to improve the ventilation.

> A thick layer of dust coated everything—from floors, ceilings and rafters to drinking fountains. As we walked through the interior, we saw men forking asbestos fiber into a feeding machine as if it were hay. They obviously had no idea of the hazard involved.[5]

Among other crimes, no one had told the workers of the hazards involved!

Later, when the depth of the tragedy was beginning to be known and the company began to worry about liability, Paul Brodeur sought to interview Dr. Grant. He had been denied this possibility on March 20, 1972 in a phone call from the president of Pittsburgh Corning, who said he was doing so on advice of legal counsel. But Brodeur hoped to succeed by speaking directly to Dr. Grant. Thus on November 12th he approached Dr. Grant at the end of a program of the American College of Preventive Medicine. Dr. Grant was outgoing president of the College and had presided over the program. He was on the platform of a ballroom in Atlanta where the American Public Health Association was holding its one hundredth Annual Meeting. Brodeur reminded him of his earlier attempts to speak with him and of the PPG president's refusal. He then asked Dr. Grant if there would be any time in the next day or two that he could talk with him. "I'm afraid I can't," was the reply. "In this instance, it's a question of the patient's rights." For a moment, Brodeur thought he had not heard right. Then it hit him. Dr. Grant was referring to the company as his patient!

Brodeur: "Do you mean Pittsburgh Corning?"

Dr. Grant: "Why, yes. If they don't want me to talk with you, there's nothing I can do."

Brodeur: "But isn't the patient all those men who worked in the Tyler plant?"

Dr. Grant: "Well, in the larger sense, of course, that's probably true. And, now, if you'll excuse me, I have some business to attend to."[6]

There are many lessons which could be drawn from this experience. One is that the most exquisite technical details (such as the general dust standard not being at all adequate as an asbestos standard) can matter a great deal. The ignoring or misuse of these details can result, as in this case, in damaged health and loss of life of hundreds of workers. A second point is that the workers and their representatives (the union) were not informed of these details and had little or no control over the gathering or flow of information. Nor were the workers even informed of the hazards–other than to be told to wear (often ill-fitting) respirators without being told exactly why and without being informed that wearing respirators should only be a temporary measure while preventive engineering solutions such as dust exhaustion and collection through an effective ventilation system are put in place. A corollary to the lack of information and control on the parts of workers and their representatives is the near complete control and even owning of the experts by the bosses–to the point where the corporate medical director thinks of the company as his patient! Another lesson is that as long as it is backed by power and position, expertise can be stupidly incompetent (not necessarily deliberately evil) and yet be accepted as authoritative and even be rewarded (Dr. Grant was made president of the American Academy of Preventive Medicine) before being exposed in its bias to protect the interests of the capitalist owner. A final point is the way in which government OSH officials (with some admirable and important exceptions, such as Drs. Wagoner and Johnson) pretend to act independently but in fact usually end up serving the interests of capital over labor. This of course is in the U.S.A. where the working class has not succeeded to mobilize the kind of strength which would put a labor government in power.

Brodeur also followed the role of different interests and the scientific studies they paid for and controlled in the setting of very weak asbestos standards by OSHA. First the recommendation of the American Conference of Governmental Industrial Hygienists (another industry-supported group, in spite of its name) was adopted making the "initial" standard twelve fibers of 5 microns length or more per cubic centimeter of air.[7] Then on August 3, 1971, Dr. Irving Selikoff wrote to Secretary of Labor Hodgson with copies to heads of several unions urgently requesting a revision. He noted that the British standard of two fibers was less than one-fifth the U.S.A. standard and that his work and that of Dr. Hammond indicated that thousands of preventable workers' deaths would occur at the twelve fiber level. Only on December 7, after the Industrial Union Department of the AFL-CIO sent a letter urging the Secretary to declare an emergency standard of two fibers, did the Secretary of Labor declare an emergency standard of five fibers.

The fur really began to fly, so to speak, when on March 14, 1972 the Department of Labor convened public hearings on a permanent standard for occupational exposure to asbestos. After an exhaustive review of the literature, NIOSH had issued a criteria document recommending a two fiber standard. Citing Selikoff's and Hammond's and others' work, the Industrial Union Department of the AFL-CIO testified that the standard should be made two fibers within six months, equal to what the British standard had been for some years, then within two years be lowered to one fiber, and eventually to zero exposure. The industry-paid consultants cited other studies and testified in favor of five fibers. One of the studies was from the U.K. and had actually served as the main basis for the British Occupational Hygiene Society's recommendation of two fibers which was adopted by the British Inspectorate of Factories in 1968. The study was done by two company paid experts, Dr. John F. Knox, Chief Medical Officer for Turner Brothers and Dr. Stephen Holmes, the company's industrial hygienist. These two evaluated x-rays of workers from the Rochdale Asbestos plant who were employed for ten years or more and still employed as of June 30, 1966.[8] With a finding that only eight of 290 workers showed asbestotic changes, they asserted that four or five fibers would be safe. The British Occupational Hygiene Society decided to play it safe and halved the low level making it two fibers. U.S.A. industry's experts, including Dr. Holmes, instead of playing it safe, urged a standard of five fibers.

After these hearings ended, an article by Dr. Hilton C. Lewinsohn appeared in the April 1972 issue of the *Royal Society of Health Journal*; Dr. Lewinsohn had replaced Dr. Knox as corporate physician of Turner Brothers. He had given chest x-rays to essentially the same men Knox and Holmes had studied (though some had left). He used new criteria for judging the x-rays. In the category of long term employed, twenty years or more, essentially the same workers included from this group in the study four and a half years earlier, he found that almost 40 percent had scarring consistent with asbestosis! This was ten times as much disease as was found before and called for a much lower standard than two fibers. In spite of this and other pertinent information being brought to his attention, Secretary Hodgson and Assistant Secretary George Guenther, Director of OSHA, issued the new permanent standard on June 6, 1972: it would remain five fibers for four more years, two fibers taking effect on July 1, 1976.

Since then, OSHA, forced by new and frightening evidence of the potency of asbestos in causing cancer, announced on April 10, 1984 that it would hold public hearings on a new standard. It had attempted earlier to adopt a new emergency standard of 500,000 fibers per cubic meter of air (the current two fiber level per cubic centimeter being equivalent to 2 million fibers of 5 microns length or more per cubic meter of air), but industry blocked that move by court action which decided that OSHA had not followed regular rule-making procedures including public hearings. The present public hearings are to consider, either 500,000 or 200,000 fibers per cubic meter of air. OSHA estimates that 375,000 U.S.A. workers are exposed to asbestos, including 49,000 above the 500,000 level and 4,000 above the 200,000 but below the 500,000 level.[9] Guess how capital and labor will line up in these hearings.

After assiduously following the Tyler Texas story and the story of standard setting, Brodeur concluded:

> By the late spring of 1973, I had long since come to the conclusion that there would be no quick end to the resistance of the medical-industrial complex to action that would ameliorate the plight of workers exposed to toxic substances, or, for that matter, to the capacity of many key government officials to react with timidity and deceit whenever they were required to make decisions regarding occupational-health problems which might run counter to the interests of the corporate giants that had been supplying money and manpower to political administrations for decades. One of the chief difficulties in overcoming the traditional business-as-usual approach to industrial disease was simply that public opinion could not easily be aroused against the delayed carnage occurring in the workplace. In short, unlike the casualty figures of the Vietnam war, which for years had been reported weekly from a specific geographical area, specific or dramatic reporting of casualties from industrial disease could never be provided for they were occurring years after the onset of exposure to toxic substances and among men and women who had been working in literally tens of thousands of shops and factories in every state of the union. For example, who would be likely to remember forty years from now that several hundred men in the hill country of East Texas who had died of asbestosis, lung cancer, gastrointestinal cancer, and mesothelioma had once been employed at a small insulation plant in Tyler owned by Pittsburgh Corning?[10]

Brodeur may have been even more outraged if he had learned of the full extent of the asbestos cover up. Actually, he did note that the story actually began when in 1918 American and Canadian insurance companies started refusing to insure asbestos workers. But he did not learn of Johns Manville and other manufacturers deliberately keeping hazard information from workers for thirty-five years.[11]

We must await the discussion of framework in Chapter 5 before we can gain a more complete understanding of these class-based forces in capitalist societies functioning within the world-system. Here, assuming that there are such forces (and I have only given from Brodeur's careful work a few illustrations of what could be mounds of such evidence) I want to lay out in the next section how such a hegemony could affect at social psychological level a person's selective perception, gathering, and use of OSH information. Then in the last section of this chapter I want to apply this understanding to highlight some of the more problematic aspects of OSH data and information systems.

THE SOCIAL PSYCHOLOGY OF OSH INFORMATION

To understand the functioning of dominant class interests in the individual OSH physician's mind, the perceptions of plant owners, managers and foremen, and OSH inspectors and sometimes workers and even union representatives (if they have not broken out of the hegemony and put on a new set of glasses for viewing the world, a form of struggle which is always present to some degree—a point, to which I will return) I have to introduce the element of ambiguity.

Ambiguity is present in matters of health and sickness and in the definitions of work hazards as in all aspects of social life.[12] The dynamics of social life could be said to revolve around the element of ambiguity. Is an act good or bad? Is this painting a work of art or a piece of trash? Is this chemical carcinogenic? Only in animals? Also in humans? Does this x-ray show asbestotic hardening or are those shadings imperfections in the x-ray? Is chyrsotile asbestos benign while the real dangers of lung cancer and mesothelioma as well as asbestosis stem from crocidolite or amosite forms (differently shaped fibers are involved in these forms) of asbestos?

Playing on this last ambiguity, Dr. George W. Wright, a long-time paid medical consultant for Johns-Manville, testified at the OSHA asbestos standard hearings of March, 1972, by citing a study of miners in Quebec exposed only to chrysotile.[13] He asserted that mesotheliomas were virtually absent in workers exposed only to chrysotile—the type which accounts for 95 percent of the world's production and the type which Johns-Manville mines, uses, and sells almost exclusively. "Mesothelioma appears to be predominantly linked with exposure to crocidolite or amosite," Dr. Wright declared. "Therefore, both of these types of asbestos should be controlled more stringently than is chrysotile."[14] Of course adoption of a split standard would add immense complexity to the counting of fibers in every air sample and therefore to enforcement, a situation which would not be displeasing to the asbestos manufacturers. But it would be very disturbing to workers, their representatives, and medical and scientific people supporting workers' interests. Thus Dr. Nicholson, co-worker in the Mt. Sinai, New York group of OSH epidemiologists assembled by Dr. Selikoff, later submitted for the record a statement which documented that the problems found in shipyard and insulation workers from exposures before WW II and even through 1950 could not be attributed to crocidolite or amosite since only a very tiny amount of this material had been used in the manufacture of insulation materials. Therefore, chrysotile *is* a serious hazard.[15]

Assuming that Drs. Grant (see beginning of chapter) and Wright are not deliberately lying or evil men, how could they see and do as they did? Their sources of pay and recognition, their positively valued significant others, those from whom they receive self-approval–probably the most general and powerful form of human motivation–would be most likely to support perceptions and behavior of this sort, rather than perceptions and behavior which would highlight damage to workers from conducting business as usual.

In a famous laboratory experiment, Sherif created a situation of near total ambiguity–a totally dark room except for a pinpoint of light at one end of the room.[16] Because of the autokinetic effect (due to very slight movement of the eyeballs) the point of light appeared to move, even though it was stationary. Subjects were asked how far the light moved from side to side. The subject might say, "Three inches." Then the experimenter could introduce various types of social influences into the situation. Most notably, for our purposes here, figures of authority could be introduced–let us say someone introduced as a Professor of Psychology with much experience making these judgments. If this person, whom the subject had no reason to mistrust, and from whom the subject wished to receive self approval, said, "Oh, no. This light moves at least six inches," the subject was faced with a dilemma. His or her own senses had suggested three inches. But here is a quite believable and authoritative person doubling that estimate and furthermore it is an authority figure whose approval one would like. A situation of cognitive dissonance has been created for the subject–two creditable assessments which must be resolved. After some consideration and hesitancy the subject might say, "Well, maybe it moves four and a half inches, but I don't think it moves six, does it?"

The experiment can be varied to reflect varying social influences–group influences, for example, and groups composed in different ways. A group with women or minority members or others whose opinions were not highly valued in the hegemony of the time might not influence the subject as much as groups composed of high status members.

Now let us translate this back into the OSH situation. Suppose we envision Dr. Grant reporting to a group of top management people in Pittsburgh Corning. He has had air samples taken at the Tyler Plant. Some of the samples are over the accepted (industry-sponsored general dust) standard of the time–five million particles per cubic foot of air. He has not bothered to have lung function tests or x-rays done on the workers. But Dr. Grant who has retired into OSH work, so to speak, from an earlier career in the Armed Forces, may not be as abreast of asbestos health hazard information as he should be. Perhaps he hasn't read the Selikoff-Hammond article in the *Journal of the American Medical Association*. Or, even if he has, the management group, his bosses (though he also takes guidance from his professional association, the Occupational Medical Association, a group of physicians having similar kinds of bosses) complain that their rate of profit is not what they had hoped in comparison with the company's competitors and investing in new ventilation will hurt this picture. Furthermore, as Dr. Grant himself has rationalized, the workers don't stay around long enough in those dusty conditions to actually get sick (though no one has actually asked the Director of Personnel for figures on the average or range of length of employment of the workforce at Tyler). Furthermore, the plant and equipment are getting old and it probably won't be used for more than a few years more, so why invest a lot of money in ventilation now? So, the decision is to do nothing; just continue to watch the situation.

We must keep in mind that decisions affecting each specific hazard situation are taken within the broader hegemony (which varies as we will see later from country to country) in which certain major arrangements are already defined beyond question. For example, it is hard for workers or their representatives to raise questions because the law says the plant is private property. It "belongs" to the owners of the company and their agents, the managers and foremen, are in charge. To an alien being, this arrangement might seem very strange. This being could also see that the values of products are produced by the workers and hundreds of workers and their families have their livelihood from what they do. But that is the way it is.[17]

SOME PROBLEMATIC ASPECTS OF OSH INFORMATION

Some might object that class-based bias and the influence of the dominant hegemony may apply to less than perfectly trained industrial physicians who are in effect captives of management (Dr. Grant rationalized his captivity by defining the company as his patient) but it could not apply to scientific research in this field.

One need only go back to the young Arrowsmith in Sinclair Lewis' novel of the same name to realize that the influences of competition—not in the marketplace but in the rush to publish—can lead to the acceptance of shortcuts even in bio-medical laboratory research. Arrowsmith, recently out of medical school and still idealistic, joins a research laboratory and finds to his dismay that his boss, the head of the institute, has engaged in shoddy practices, if not downright fudging of the data.

Elsewhere, I have drawn on experience while working in the Research Division of WHO Headquarters to identify sorts of sociopolitical influence which might and often do affect decisions at each stage of the research process.[18] In analyzing any particular research one can usefully envision a two-dimensional chart. Along one side are the different levels of social organization, reaching from the individual researchers to the sponsorship and control structures of the project and the kinds of influences which might be involved at these different levels of organizations. For example, How is the research team composed? Is there an advisory board and does it include workers or their representatives? And so on. Along the other dimension are the phases in any research process–the way the problem is defined; the logic of the design (that is, the comparisons to be made); the sample of subjects and controls; the research instruments (questionnaires with questions posed one way or another or air samplers placed near the workers' breathing zones or not); data processing; data manipulation; presentation of data (for example, disaggregated tables or presentation with certain columns and or rows combined); use of the study (for example, to support an asbestos standard of five fibers of 5 microns length per cubic centimeter of air or a standard of zero detectable exposure). The suggested exercise is to consider each level of social organization and likely influences in relation to each phase of the research process.

For example, why did Dr. McDonald, whose study Dr. Wright cited in support of a weak asbestos standard, include on an equal basis in his study sample many workers with little exposure to airborne fibers (those working in open-air pits with a wet mining process) along with only a few with heavier exposures (those in the processing mill)? Also, why did he not acknowledge in his publication that his Institute received funding indirectly from Johns-Manville?

The awareness of scientific knowledge production as just another arena in which social processes of production are at work has been growing.[19] The journal, *Science for the People*, is explicitly devoted to exposing class-based bias in science. Kotelchuck's analysis of industry-sponsored research on asbestos appeared in that publication (see note 11). Analyses of specific OSH hazard researches and investigations have appeared such as those on nuclear power risks,[20] and workers exposed to so called, low-dose radiation.[21] An extensive analysis of class-based interests in cancer research includes considerable material on the pitfalls of company-sponsored research on carcinogens and work exposures, including such "tricks" as (1) failing to include in epidemiologic studies early retirees (who may be sick) and deceased workers, and (2) ignoring "the healthy worker effect" in reports of lower rates of disease and death than would be expected based on rates in the general population (which is on average sicker than a population which can work).[22] A general consideration of the ways in which OSH-related medical research and knowledge serves ruling class interests in capitalist societies has been offered.[23]

Fortunately, the hegemony is never complete. With the year 1984 just behind us, the projections of Orwell's book regarding the growth and frightening prospect of the ruling class using the media and secret, as well as overt, surveillance for mind control give us pause. But, first, there are limits to the ambiguity in most situations. No matter how great the authority, one cannot get the subject in Sherif's dark room to say with conviction that the point of light moves three feet (unless of course the person feels under duress and in fear and does not view lying about the movement of a point of light as a matter of vital principle).[24] Second, not all the science and authority are on the side of the bosses. In Chapter 3 I made note of COSH-type groups, many functioning in U.S.A. and several in other countries. These groups include scientists and medical people employing their knowledges and skills on behalf of working men and women. Also many independent investigators—the works of Selikoff, Hammond, Nicholson, and others were cited in relation to the asbestos story—pursue their work with workers' interests clearly in mind and full cognizance of the power of dominant forces to challenge their work, so they are especially careful to do good work.[25] With such resources, and a growing awareness and concern, the ordinary worker may challenge "the powers that be":

> The docs keep telling me there's nothing wrong with the place where I work. I guess they're supposed to know it all because they've had a lot of education and everything. I'm no expert like they are, but I sure as hell know there's something wrong in that mill and the other guys are saying the same thing. One thing I know for sure—that place is killing us.[26]

In fact, the hegemony can be turned around. This of course takes a broad and deep running societal transformation to accomplish. In the process of such a struggle science and human liberation can go hand in hand.[27] I have only touched upon the ways scientific research is often used to uphold ruling-class interests. There would be many more examples and general points to make if space allowed.[28] But there are many other problematic aspects of OSH information systems. Here I will again only touch on three more problematic aspects—the day to day OSH surveillance and reporting systems; the problem of clinical diagnosis and reporting or lack of reporting of OSH disease (most usually this concerns PHC providers) and the workers' right to know.

A country's ongoing, day-to-day OSH surveillance and reporting system has or should have a number of facets to it. I will say more about this in Chapter 6 in which an ideal model for OSH and OSH-PHC is developed. For the present, let us simply note that the standards covering occupational hazards are one part. The inspection apparatus is another closely related part as is the means of enforcement to correct violations. Another is any ongoing and continuous survey of hazards and workers exposed, including medical follow-up of exposed workers and follow-up of hazard removal or protection. Finally, there is (or should be) reporting of accident and illness statistics by industry and specific type of work so that concerned workers and OSH personnel can identify high risk areas and undertake special studies or prevention efforts if the problems are generally understood but not fully under control. All of these facets may and usually do have class-based problems of inadequate or biased information associated with them. We cannot examine all these facets here.

Because of its importance for gaining an overall appreciation of a country's OSH situation, and especially prevalent problems, let us here focus on the statistical system by industry and type of job.[29] I have already noted in Chapter 1 how we really do not know the dimensions of the OSH problem and as late as 1984 are more or less stuck (at least in U.S.A.) with ten-year-old crude estimates of some 100,000 workers dying yearly from work-related disease and another 15,000 from trauma due to accidents on the job. Let us see why this is so.

In 1908 Frederick Hoffman noted the lack of any national survey or statistical system for reporting industrial accidents. He went on to use death certificates to estimate that there were between 15,000 and 17,500 deaths annually among male workers due to accidents in "dangerous industries or trades." In 1915, apparently including women workers, he raised his estimate to 25,000. But after 1919, the Labor Department issued estimates based on the worker compensation systems which had been established in most states as a means of compensating workers and their families in a minimal way while allowing employers to avoid being sued for damages through the courts. Suddenly the estimates of deaths due to work accidents dropped to 9,392. By this kind of estimate they rose to 12,531 by 1927. Late in the 1920s the National Safety Council (NSC) took over the issuance of these estimates. Its interest was to show that workplaces were becoming ever safer. Between 1963 and 1971 the estimates ranged between 14,100 and 14,500 deaths from work accidents. These were achieved by allocating in some unstated way to "work, home, and public" the deaths from all accidents after subtracting motor vehicle accidents which were reasonably accurately reported.

> A spokesman from the NSC statistical department admitted that there was no standard written procedure used to compute the industrial accident death total. 'It's basically guesswork,' he said, 'nothing like what I learned in school; but you get used to it after you've been here awhile.'[30]

To understand why state worker compensation systems offered poor bases for making estimates of work-related accidental and disease deaths it may be enough to cite the example of Missouri. Projecting from the experience of other states, a special survey, and national estimates, Berman observes: "It is likely that more than half the deaths caused by job accidents and almost all the deaths from occupational disease go uncompensated in Missouri, so, in other words, up to 95 percent of the deaths caused by working conditions in Missouri never reach the compensation system."[31] In 1968, 115 work-related deaths were reported; yet the true number was anywhere from 380 to 2,780.

Reports of work accidents not involving a death are even worse. Employers, hoping to keep insurance costs low, often encourage safety programs which focus on getting the employee back to work without it becoming a "reportable lost time accident," one which causes death, permanent disability or at least a full day's inability to work. In our OSH education program for workers we have been given numerous examples of workers injured in the morning and back in the afternoon perhaps wearing a cast doing some meaningless paper shuffling or answering a phone.[32] In addition, accident reporting systems keep being revised, often narrower definitions, so there is little chance of making meaningful comparisons from one era to another.

For all these and other reasons, in this study I do not focus much on statistics concerning occupational accidents and disease for comparing the six study countries. To illustrate the problem, while the U.S.A.'s definition of a work-related accident has tended to get narrower over the years, in the G.D.R., Sweden, and Finland a work-related accident includes motor vehicle or pedstrian or other accidents on the way to and from work!

While the reporting of work accidents is bad, especially in U.S.A., the reporting of occupational disease is abysmal in most countries.

A special medical and industrial hygiene survey of workers in a sample of workplaces with fewer than 150 employees in the Pacific Northwest section of U.S.A. found 451 cases of "probable occupational disease." This was defined as "manifestations of disease . . . consistent with those known to result from excessive exposure to a given injurious agent; this injurious agent is present in the patient's working environment and significant contact in course of usual duties is likely." Of this total, 399 (88.5 percent) were absent from the employer's log (required by OSHA regulations) and were missing in compensation records. The authors of this Seattle study concluded that there is a "vast reservoir of unreported occupational disease" in the working population.[33] This survey could take no account of diseases related to work stress nor of those due to any of the thousands of untested chemicals in the workplace. In the Missouri compensation system cited earlier, only 1,003 illness cases were reported in 1968—most were dermatitis severe enough to call for compensation. From other experiences, anywhere from 4,000 to over 120,000 cases probably occurred.

There are many reasons why occupational disease goes unreported. Employers discourage any industrial health and safety personnel they have (and in U.S.A. and some other study countries there may not be any such personnel, especially in medium to small-sized plants) from identifying disease with the workplace to avoid bad morale, a poor public image, and paying higher worker compensation insurance rates. But since workers often go to a community physician when they feel sick and first seek professional help, the main problem rests with poorly trained, little interested PHC personnel.[34] In part the lack of interest may stem from differences in class background between workers and physicians. Few physicians come out of working-class families. Also, very few have worked in a factory, even for a summer job. Finally, there is the problem of ambiguity in occupational disease. It usually looks like another "ordinary" disease.[35] An asthma due to chronic exposure to metal dust looks like any other asthma. Without proper training to be alert and ask the right questions and do the right tests, and without the interests and orientations within the hegemony of capitalist-based medical systems to pursue the worker's interest as a first priority, the problem usually goes unrecognized, improperly treated, and unreported.

An important place to begin correcting many problematic aspects of OSH information systems would be to put more and more adequate information in the hands of workers and their representatives. Training courses are only a part of this picture. Since this is such a key matter, I will not go into it here in detail, but will examine the workers' right to know in country chapters (7-12) and elsewhere. Here it is important to recognize that this is a problem in every study country but in some much less so than others. To illustrate the extremes, let us highlight, just two countries. In Sweden, the Law of Co-determination places unions into the production planning process along with management, so there is no question of the workers' right to know. Whether information is always passed along effectively may be another matter, though training efforts are also extensive in Sweden. In U.S.A., the national law is so poor that state efforts have been undertaken. But the, so called, "trade secrets" ideology is still being used by management forces to fight off individual state's worker-right-to-know bills. Recently, a revised national law has been proposed which threatens stronger state laws.[36] Worker training efforts have been minimal in U.S.A.

To better understand how such class-based biases can become central defining characteristics of OSH information systems, we need a framework with which to understand OSH as part of a country's political economy within a competitive capitalist world-system. The next chapter presents this broader framework.

> This hearing could have been held more than two years ago, when our department and affiliates first questioned the exclusion of known carcinogens from the interim standards promulgated under the Occupational Safety and Health Act. The consequence is that, as a result of two years of unjustifiable exposure to carcinogenic agents, regardless of anything done now, hundreds and perhaps thousands of men and women can be expected to experience agonizing death from cancer in the next two decades. Sir, in a truly civilized society we would hold personally responsible those who participated in this crime, both the callous political creatures and the cancer peddlers who bartered moral and statutory obligations. In a just society they would now be undergoing rehabilitation in a penal institution. Instead they walk freely—some of them are or have been in this room—as if evil is its own reward. (Sheldon Samuels, OSH Head for the Industrial Union Department, AFL-CIO, testifying before an OSHA hearing on standards for fourteen known human carcinogens (in addition to asbestos) September 14, 1973).

NOTES AND REFERENCES

1. Paul Brodeur, *Expendable Americans*. New York: Viking Press, 1974. The *Newsletter* of the Association of Trial Lawyers of America wrote of this book: "This splendid exposé uncloaks the total callousness, stupidity and deceit of the medical-industrial complex consisting of company doctors, industry consultants and key occupational-health officials at various levels of the state and federal governments."
2. Irving J. Selikoff and E. Cuyler Hammond, Asbestos Exposure and Neoplasia. *Journal of the American Medical Asociation 18*:22–26, 1964.
3. The 1963 survey itself offered a classic example of management-oriented bias. It was conducted by the Industrial Hygiene Foundation which as we saw in Chapter 1 grew out of the Gauley Bridge tragedy as an entirely industry-financed protective device. In its report to Pittsburgh Corning there was no mention of a health hazard. "Incredibly, the authors of the foundation's report appear to have based their judgment on the assumption that the threshold limit value of five million particles per cubic foot meant five million asbestos fibers, whereas the proponents of the threshold limit value had intended it (back in 1946 when it was adopted by another industry-backed voluntary association) to apply to *all* the particulate matter–fibrous and nonfibrous–in a given cubic foot of air. To understand the magnitude of this error, it should be noted that the study upon which the standard was based had been made in the winter of 1935-36 in four asbestos-textile plants where asbestos fibers were found to constitute about 10 percent of the total amount of airborne dust. The asbestos fibers in the airborne dust measured in the Tyler plant by engineers of the IHF ranged from a low of 29 percent to a high of 56 percent. The foundation, however, reported the percentages as if they were of little or no consequence, and contented itself with making recommendations for better housekeeping, better ventilation equipment, and improved maintenance of the ventilation system." Brodeur, op. cit. (note 1):10–11. Facts in the case of the Tyler plant and Dr. Grant are all drawn from Brodeur.
4. Brodeur, op. cit., p. 20.
5. From an interview of Dr. Johnson, Brodeur, op. cit., p. 48.
6. Brodeur, op. cit., p. 245.
7. A cubic centimeter is a small thimblefull of air. A micron is one-thousandth of a centimeter; 5 microns is the size which can be seen through an ordinary microscope; thus allowing enforcement. It should be noted that for every fiber of this size or larger there are hundreds of smaller fibers in the same thimblefull of air. Even ignoring this fact, a worker who inhales about eight million cubic centimeters of air in a working day takes in 40 million fibers of 5 microns or longer at the five fiber standard and 16 million at the two fiber standard.
8. These details are reported in Brodeur, op. cit., pp. 174–176. Note that by this method they missed workers who had died or retired early for health reasons or left for other reasons.
9. Joann S. Lublin, OSHA Proposal to Set Tougher Standard for Asbestos Exposure Offers 2 Choices, *The Wall Street Journal, 8*, April 10, 1984.

10. Brodeur, op. cit., pp. 229–230.
11. David Kotelchuk, Asbestos–Science for Sale, *Science for the People*, pp. 9–11, September, 1975. Samuel S. Epstein, The Asbestos 'Pentagon Papers' pp. 154–165 in Mark Green and Robert Massie, Jr. (eds.), *The Big Business Reader, Essays on Corporate America*. New York: Pilgrim Press, 1980.
12. Sandor Kelman, The Social Nature of the Definition Problem in Health, *International Journal of Health Services 5*:625–645, 1975.
13. John Corbett McDonald, Mortality in the Chrysotile Asbestos Mines and Mills of Quebec, *Archives of Environmental Health 22*:677–686, June 1971. One problem with this study was that many, perhaps most, workers included had worked in open-air pits in wet-rock mining; thus had little exposure to airborne fibers. Also, Dr. McDonald's Institute of Occupational and Environmental Health at McGill University was supported by the Quebec Asbestos Mining Association of which Johns-Manville was the dominant member. But the author did not acknowledge this in the article. See further Samuel S. Epstein, *The Politics of Cancer*. Revised and expanded edition. New York: Doubleday, Anchor Press, 1979, pp. 85–86.
14. As quoted in Brodeur, op. cit., p. 126.
15. Brodeur, op. cit., p. 129.
16. Muzafer Sherif, Group Influences upon the Formation of Norms and Attitudes, pp. 219–232 in E. Maccoby, *et al.*, (eds.), *Readings in Social Psychology*, 3rd edition. New York: Henry Holt, 1960. See also Peter Berger and T. Luckmann, *The Social Construction of Reality: A Treatise in the Sociology of Knowledge*. New York: Doubleday, 1966.
17. In U.S.A. the notion of private property was specifically upheld in OSH matters when the Supreme Court supported an employer who insisted that "his" factory was protected by the Fourth Amendment to the Constitution against unreasonable search and seizure when OSHA sought to enter on an unannounced inspection. Henceforth, OSHA must get a search warrant before they can enter "his" factory where some 7,000 people are employed or any other factory where the owner-employer objects to an inspection. (The Barlow decision.)
18. Ray Elling, Political Influences on the Methods of Cross-National Sociomedical Research, pp. 144–155, references pp. 170–172 in Manfred Pflanz and Elisabeth Schach (eds.), *Cross National Sociomedical Research: Concepts, Methods, Practice*. Stuttgart: Georg Thieme, 1976.
19. Alan Young, Rethinking Ideology, *International Journal of Health Services 13*:203–219, 1983.
20. Bruce L. Welch, Nuclear Power Risks: Challenge to the Credibility of Science, *International Journal of Health Services 10*:5–36, 1980.
21. Theodor D. Sterling, The Health Effects of Low-Dose Radiation on Atomic Workers, A Case Study of Employer-Directed Research, *International Journal of Health Services 10*:37–46, 1980.
22. Epstein, op cit. (note 13).
23. Vicente Navarro, Work, Ideology, and Science: The Case of Medicine, *International Journal of Health Services 10*:523–550, 1980.

24. This subject could be explored much further than we are able to do here. In an experiment almost as famous as Sherif's, Asch found that people could be convinced to agree to the equality of two lines of different length, but only slightly different, while lines of quite different length were never agreed to be equal. Solomon E. Asch, The Doctrine of Suggestion, Prestige, and Invitation in Social Psychology, *Psychological Review 55*:250–276, 1948. On the other hand, Milgram found that people were willing to give unbelievably (presumed real but actually absent) high electric shocks to an actor-subject simulating pain in an experiment under no more authority than the experiment itself as "a scientific study." Stanley Milgram, *Obedience to Authority: An Experimental View*. New York: Harper and Row, 1974. The widespread use of actual torture has been documented by Amnesty International. It is hard to say at just what points individuals and groups and masses will stand and fight against obvious wrongs and injustices. The comforting thing is that we know it happens and in labor history has happened many times, sometimes in circumstances calling for great sacrifice and courage.
25. This point is not often recognized or given enough emphasis. In work I have done with colleagues in the Connecticut New Directions Program for Worker Education in OSH, we have often been told by experienced union people that it is very bad and dangerous to "go off half cocked." The point is that if a union local is to get a new ventilation system or other advance through bargaining, they need very solid information to back up their claims. If they do not have it, they not only lose the particular battle, they open up the movement at large to lessened credibility. An excellent survey of the problems and possibilities in the field of OSH epidemiology is Richard R. Monson, *Occupational Epidemiology*. Boca Raton, FL: CRC Press, 1980.
26. Cancer patient and steelworker from the Bethlehem Steel Corporation mills, Sparrow Point, Maryland, 1978, as quoted in Navarro, op. cit. (note 23): 523.
27. Rita Arditti, Pat Brennan and Steve Cavrak (eds.), *Science and Liberation*. Boston: South End Press, 1983.
28. The movie, "Song of the Canary," brings out that a study, commissioned by Dow Chemical Company, was done on rats exposed to the pesticide DBCP some sixteen years before it was found that workers at Occidental Petroleum were becoming sterile from working with DBCP. The study had found that the rats developed testicular atrophy. But Dow never passed the study along to Occidental until four years before the union paid to have sperm tests done on the workers and the problem was discovered. For those four years, the Manager at Occidental, says "We didn't pay much attention to the Tolefson study because I asked our scientists and gee, we never thought testicular atrophy in rats necessarily meant there'd be human infertility." Chuck Levenstein is working on an important history of bysinosis. In Chapter 3 of his forthcoming book he lays out the sad tale of company-sponsored research on that problem.
29. Most of the illustrations in this part are drawn from Chapter 2, The Official Body Count, pp. 38–53 (notes pp. 204–210) in Daniel Berman, *Death on*

the Job, Occupational Health and Safety Struggles in the United States. New York: Monthly Review Press, 1978.

30. Ibid, p. 40.
31. Ibid, p. 47.
32. Berman cites the case of a worker at Rohm and Haas chemical plant in Passadena, Texas: "I got a skin irritation from benzene. My hands swelled up. Oh, they were huge. Well they brought me back out there and set me at the desk in front of the telephone . . . about two weeks to keep from listing a lost-time accident. . . . They'll bring the people back in on stretchers if they have to." op. cit., p. 42.
33. As cited in Berman, op. cit. (note 29):46–47, Original: "Seattle Pilot Study," Department of Environmental Health, School of Public Health, University of Washington, Seattle, no date (probably 1977). Also note that 16, or 1 percent, said "Yes" when asked if they'd had a serious accident or injury on the job in the last twelve months. "Serious" was defined with examples.
34. It was pleasing to see recognition of this problem in a recent, widely distributed medical newspaper: Mary Ann Khan, Community MDs must recognize Occupational Ills, *Medical News and International Report 8*: (1 and 8), April 16, 1984. The article notes that only half of U.S.A. medical schools include OSH lectures, some of these on an elective basis.
35. Barry S. Levy and David H. Wegman, Recognizing Occupational Disease, pp. 29–39 in Levy and Wegman (eds.), *Occupational Health*, Boston: Little Brown, 1983.
36. State Laws threatened; OSHA Issues 'Right to Hide' Rule, *CACOSH Health and Safety News 1*, November/December, 1983. For example, "lets the companies decide what to put on or leave off the label, and has a huge loophole for 'trade secrets'." For a detailed study of the repression of workers rights at American Cyanamide, Occidental Chemical, the Burnside Foundry, and J. P. Stevens cotton textiles, see Mary Gibson, *Workers' Rights.* Totowa, NJ: Rowman and Ollanheld, 1984.

CHAPTER 5
Framework

INTRODUCTION

When Marx and Engels wrote that "class struggle is the motor of history,"[1] they anticipated by some thirty years, albeit in a general way, Bismarck's adoption in 1883 of the first national health insurance system in the world in order to "cut the legs off the socialist workers' movement."[2] At the time, in that phase of competitive, industrial capitalism (as distinct from the dominant mode of monopoly capitalism today) Bismarck's liberal opposition, who were fighting for the sacred principles of individualism and laissez-faire and against state intervention, "called Bismarck a socialist–Bismarck who was out to destroy the socialist movement."[3] Bismarck of course was very modern, ahead of his time, so to speak, in anticipating the moves by which the welfare state could keep within bounds the demands of the organized sector of the working class so as to preserve the essential conditions for what Marx called General Capital (not necessarily each firm or business but capital in general). It was Bismarck's view, expressed as early as 1849, in reaction to the widespread but abortive revolutionary attempts in Europe in 1848, that "the social insecurity of the worker is the real cause of their being a peril to the state."[4] His answer to this threat when he became "the Iron Chancellor" was to adopt various palliative social welfare measures, while preserving conditions for profitability and leaving ownership and control of the means of production in the hands of individual capitalists or groups of capitalists and the capitalist ruling class in general.

This historical vignette includes nearly all the major elements needed for understanding the struggles around workers' health and safety: a more or less organized working class represented by the unions and the workers' party; a resisting capitalist ruling class employing various means to enhance or at least protect its position; a state apparatus which serves as one of these means (but also mediates or brokers the struggle by responding to workers' demands); and ideology or a cultural hegemony ("individualism" and so on) which can powerfully bias this class-based system as we saw in the last chapter in the definitions placed on everything from accidents ("it's the careless worker's fault") to epidemiologic data ("these asbestos miners—most of whom worked in the open air and with a wetted-down mining process, though this part is not reported—showed few disease effects, so five fibers of five microns length per cubic centimeter of air is a strong enough standard). To these elements we need only add the historical phases of capitalism and the form of its key enterprises (multinational corporations) competing through technological change, and other ways, within the world-system, to complete the broad outlines of a framework for understanding the generation and handling of OSH problems in different countries. Each of these elements requires discussion before we can examine health and welfare developments, including OSH developments, as intermediate outcomes in a continuing class struggle.

SOCIAL CLASSES AND CLASS CONFLICT

Part of the ideological attack of modern-day apologists for capital's exploitation of the working class is to pretend that class interests and therefore classes themselves are tending to disappear. This attack is best identified in the "end of ideology," a sort of technology-based homogenization supposed to characterize modern-day capitalism as "post-industrial."[5] But while technological changes may allow increasingly massive agglomerations of capital to compete with each other, they don't necessarily bring improvement in living conditions or class status for workers. A recent cartoon illustrates this: In the first frame, a woman worker sitting in front of a keyboard and video display terminal (VDT) says, "I made $12 an hour in the steel mill until I was laid off" (women and minority workers are usually among the last hired and first fired). In the second frame, she says, "So—I took computer classes and now I earn $6 an hour." Next she says, "Soon a more sophisticated computer will eliminate my present job." In the final frame she says, "I've achieved state-of-the-art unemployment!"[6]

Capitalism is a highly dynamic or dialectic system but until there is serious societal transformation, the dynamics do not involve the erasure of classes though there may be basic changes in the class structure. To identify and examine technological and other changes does not at all imply a disappearance of classes—except in the ideological statements of system apologists. But Braverman, considering the changing social relations of production in modern-day monopoly capitalism, finds an increasing proletarianization of certain categories of clerical and technical workers formerly associated with the petite bourgeoise.[7]

While I could review studies of concentration of wealth, concentration of ownership, and power structure to demonstrate the existence of a ruling bourgeoisie, a working class and associated classes (petit bourgeoisie, intellectuals, and popular masses) and their interests and class behaviors,[8] I do not have to reach beyond the OSH arena of concern to illustrate the existence of classes, class interests and class behaviors—often vicious class behaviors:

> If a factory worker drives his car recklessly and cripples a factory owner, the worker loses his license to drive, receives a heavy fine, and may spend some time in jail. But if a factory owner runs his business recklessly and cripples 500 workers with mercury poisoning, he rarely loses his license to do business and never goes to jail. He may not even have to pay a fine.[9]

In fact, it is in the factories and other workplaces that we can see most clearly the existence and functioning of the class structure. For here all pretense of "democracy," otherwise presented as a covering facade for the society at large through the constitution, Bill of Rights, and electoral process, is abandoned. Here, what Marx called the "dictatorship of the bourgeoisie" holds full sway:

> Try putting thirteen little pins in thirteen little holes sixty times an hour, eight hours a day. Spot-weld sixty-seven steel plates an hour, then find yourself one day facing a new assembly-line needing 110 an hour. Fit 100 coils to 100 cars every hour; tighten seven bolts three times a minute. Do your work in noise "at the safety limit," in a fine mist of oil, solvent and metal dust. Negotiate for the right to take a piss—or relieve yourself furtively behind a big press so that you don't break the rhythm and lose your bonus. Speed up to gain the time to blow your nose or get a bit of grit out of your eye. Bolt your sandwich sitting in a pool of grease because the canteen is ten minutes away and you've only got forty for your lunch-break. As you cross the factory threshold, lose the freedom of opinion, the freedom of speech, the right to meet and associate supposedly guaranteed under the constitution. Obey without arguing, suffer punishment without the right of appeal, get the worst jobs if the manager doesn't like your face. Try being an assembly-line worker.[10]

The ruling bourgeoisie and their managers have set up an authoritarian bureaucratic structure through which they control the means and processes of production. The other primary class—the working class—only controls its labor power. It does not own or control the means or processes of production.[11] Even the working class' control over the offering or withholding of its labor power may be weak or strong, depending upon level of unemployment (how large the threat of job loss is from the "reserve army of the unemployed"), similar other factors (such as the ease with which capitalism can export plants and jobs to cheaper labor markets elsewhere in the world-system) and strength of the workers' movement (level of organization and degree of solidarity).

"Well, there is no problem," the conservative observer might say, "the classes are in perfect harmony, forming an almost symbiotic system in which one produces, largely through muscle power, and is rewarded therefore and the other manages largely through head work and is rewarded more handsomely."[12]

But there is a problem. The classes are locked in constant conflict with each other. This conflict, as already noted, is the motor of history. Why is class conflict inherent in the system? How is this struggle carried on? And how does it relate to OSH concerns? These are the central questions for the remainder of this chapter, indeed, for the rest of this book.

Conflict is inherent in the system for two main reasons. First, the owners and controllers of the means of production, that is, the capitalist class, are a leisure class, and, as such, are nothing but a burden on the workers and the rest of society.[13] They have managers and technocrats to run affairs for them. There is of course a great deal in the cultural hegemony of capitalist society—phrases such as "intellectual and moral leadership"—which attempts to justify the existence and "rights" of this class. But they do nothing essential to the production process.

Even if they are actively engaged in major allocative decisions, as in the instance of David Rockefeller when he still headed Chase-Manhattan Bank and traveled around the world making major loans to some of the most despotic governments, there is nothing, other than the misallocation of capital, which worker control would not accomplish. Thus, objectively speaking, there is no justification for the existence of the capitalist class. Second, the surplus upon which the capitalist class depends in order to lead its often sumptuous existence is extracted from the working class which only has its labor power to sell. Depending upon the form and strength of its organization and local conditions of competition between clusters of capital in the world-system, workers may receive little more for their labor, including the unpaid labor of wives and mothers,[14] than is essential to reproduce themselves.

> . . . the relationship from which in one case a common interest in preserving and extending a particular economic system and in the other case an antagonism of interest on this issue can alone derive must be a relationship with a particular mode of extracting and distributing the fruits of surplus labour, over and above the labour which goes to supply the consumption of the actual producer. Since this surplus labour constitutes its life-blood, any ruling class will of necessity treat its particular relationship to the labour process as crucial to its own survival; . . .[15]

To survive, capital must compete in ever larger clusters or agglomerations in the world-system, not just to realize a profit, but to realize the highest rate of profit possible. Before discussing further the class struggle and the ways it is carried on and how this struggle and these ways affect OSH, we need to understand the basic economic elements in the rate of profit upon which the class structure and class conflict are based. In the formula $p = s/c + v$ the rate of profit (p) is equal to the surplus value (s) of products workers produce, divided by constant capital (c) plus variable capital (v). Appelbaum summarized the nature of this central relationship:

> Marx analyzed the value of commodity production in terms of three elements: constant capital (c), the value of the means of production used up during the production process (primarily the depreciated value of machines, buildings, and raw materials); variable capital (v), the value of the labor power applied to the production process (primarily the wage bill); and surplus value (s), the value of unpaid labor appropriated by the capitalist during production process (workers' labor time beyond that which is socially necessary to sustain the standard of living of the working class). Surplus value is the key to capitalist economic production. It is the source of all profits, including those which are reinvested in enhanced productive capacity (capital accumulation).[16]

In the early phase of industrial capitalism, the vanguard of the working class in the form of unions and left political parties had not been formed and working class consciousness and resistance was quite low.[17] Thus the condition of labor and the life of the working class in general were quite raw. Marx quoted Sir F. M. Eden:

> It may, perhaps, be worthy the attention of the public to consider, whether any manufacture, which, in order to be carried on successfully, requires that cottages and workhouses should be ransacked for poor children; that they should be employed by turns during the greater part of the night and robbed of that rest which, though indispensable to all, is most required by the young; and that numbers of both sexes, of different ages and disposition, should be collected together in such manner that the contagion of example cannot but lead to profligacy and debauchery; will add to the sum of individual or national felicity?[18]

Marx went on to quote Fielden who was more explicit:

> . . . Many, many thousands of these little, hopeless creatures were sent down into the north, being from the age of seven to the age of thirteen or fourteen years old—overseers were appointed to see to the works, whose interest it was to work the children to the utmost, because their pay was in proportion to the quantity of work that they could exact . . . cruelties the most heart-rendering was practiced upon the unoffending and friendless creatures who were thus consigned to the charge of master manufacturers; they were harassed to the brink of death by excess of labor . . . were flogged, fettered and tortured in the most exquisite refinement of cruelty; . . . they were in many cases starved to the bone while flogged to their work and . . . even in some instances were driven to commit suicide. . . . The

> profits of manufacturers were enormous; but this only whetted the appetite that it should have satisfied, and therefore the manufacturers had recourse to an expedient that seemed to secure to them those profits without possibility of limit; they began the practice of what is termed 'night working,' that is, having tired one set of hands, by working them throughout the day they had another set ready to go on working throughout the night; the day-set getting into the beds that the night-set had just quitted, and in their turn again, the night-set getting into the beds that the day-set quitted in the morning. It is a common tradition in Lancashire, that the beds *never get cold.*[19]

The history of progress in modern society is the history of the struggles of the workers' movement against such conditions and in general against the expropriation by the capitalist class of the surplus which workers produce.

Most directly, class struggle bears upon the conditions of work. Among the great achievements must be counted the eight-hour-day and child labor laws. But this struggle also carries into the realms of social welfare, health conditions, and health services, as well as other spheres.[20] The development of the first national health insurance in F.R.G. in 1883 in response to the demands of the socialist workers' movement has been noted. The National Health Service (NHS) in the U.K. must also be counted as a great achievement of the British working class.[21] Achievements in the OSH realm occupy a special place in this struggle, for they link both conditions of work and health provisions. But most important of all, they bear directly upon control of the work process—a central threat to capitalist control of the means of production. This is a key point I will come back to.

While class struggle shares many key aspects across nations, it is not everywhere uniform, nor are the achievements of the working class equal in all nations, not even in all industrialized nations:

> Indeed, labor movements have historically viewed social services (including health) as part of the *social wage*, to be defended and increased in the same way that *money wages* are.[22] In fact, Wilensky[23] has shown how the size of social wages depends, in large degree on the level of militancy of the labor movements. Thus, contrary to popular belief, the scope of social benefits in terms of social services is broader in France and Italy, than in Scandinavia or even in Britain. One indicator of this is the percentage of the GNP spent on social security which, in 1965, was 17.5 in Sweden, 14.4 in the U.K., 12.6 in Norway, 13.9 in Denmark and only 7.9 in the U.S.[24] And I attribute this to the higher militancy of the unions in those countries and to the existence of mass socialist and communist parties (whose platforms are at least in theory, anti-capitalist) that force upon the state an increase of social wages. The practical absence of a comprehensive coverage for social benefits in the U.S.A. is also undoubtedly due to the lack of an organized left party.[25]

The particular form which the class struggle takes within a country determines (along with the country's position in the world-system) the character of capitalism in that country. This statement applies to the intra-capitalist class struggles as well as to the interclass struggles between the working and bourgeois classes. To ignore these particular struggles would lead to gross generalizations. Even though two states are both capitalist, one governed by a fascist fraction of the capitalist class may be very different and more repressive of workers' interests than one governed by a social democratic fraction among the bourgeoisie.

NEO-IMPERIALIST MONOPOLY CAPITAL AND THE WORLD-SYSTEM

The works of Wallerstein and others are key in laying out the development of the capitalist world-system.[26] Works of others fill in more of the needed framework as regards health and health services in capitalist societies.[27] Here I can only attempt, first to highlight in somewhat static fashion the key elements of the neo-imperialist phase of monopoly capital in the current world-system, and, second, consider state action and other tools of the class struggle as these bear in a general way on OSH (more details of the struggles in each country will be found in country chapters, 7-12). The reader should realize that the current neo-imperialist phase of monopoly capital and socialist revolution is highly fluid and dynamic and that it has evolved through a number of phases to the current set of crises.

Frank nicely summarizes the phases of capitalist world-system development including strengthening of the core nations and active underdevelopment of many parts of the periphery.[28] Neo-imperialism has brought the expansion of wage slavery and world Taylorization of labor (the fitting of machines, human motion, and time for greater productivity) to replace slavery 'pure and simple.'[29] Monopoly capital in the form of large, powerful multi-national finance, industrial and agri-business corporations knows no particular national home.[30] Although we still talk of U.S.A.-based, Japanese, etc. corporations, perhaps this attitude only reflects the fact that certain corporations have learned how and have developed traditions of "working with" or manipulating the state apparatus in a given country.

The vast, unequal exchange and differential capital accumulation as between core and periphery are indexed by some figures for the U.S.A. From 1950 to 1965, according to the U.S.A. Department of Commerce, some $9 billion of investment flowed from the U.S.A. to poor underdeveloped countries while $25.6 billion in profit flowed out of these countries to the U.S.A. for a net return of $16.6 billion.[31] As regards Latin and South America, Navarro presents this picture:

> . . . the top 298 U.S. based corporations earn 40 percent of their entire net profit overseas, with their rate of profit from abroad being much higher than their domestic rate. Actually this rate of profit for these global corporations is even higher in the underdeveloped world, resulting in huge net outflow of capital from those countries back to the U.S. American corporations, for example, made direct investments in the Latin American continent of U.S. $3.8 billion during the period of 1950–1965, while extracting $11.3 billion, for a net flow of $7.5 billion back to the U.S.[32]

For France, one detailed calculation puts the annual flow of capital from Senegal alone to French financial and industrial corporations at some 20 billion CFA francs over a ninety-year period.[33]

Even a predominance of indigenous capital or local ownership does not give control over the flow of capital. Ownership and control are not equivalent. With indirect rule, the core-nation capitalist can reap huge profits even when investing in mines, etc. which have been nationalized by a bureaucratically based national bourgeoisie.[34] This observation offers a prelude to understanding the increasingly close intertwining of state finance and private monopoly capital as discussed below.

Elsewhere, I have examined the effects of the functioning of the world-system on a number of aspects of international health: poor general health levels in peripheral and semiperipheral nations, especially rising infant mortality rates and extremely high industrial accident rates in countries such as Brazil; commerciogenic malnutrition, including baby bottle disease; dumping and the exploitative sale in peripheral and semiperipheral countries of drugs, pesticides and other products which have been found to be hazardous and have been banned or restricted in core nations; genocidal and other threatening approaches to population control; the export to and/or development of hazardous and polluting industries in the periphery and semiperiphery; the export of human experimentation; the sale of irrelevant, high medical technology to countries lacking basic public health conditions and measures; the "brain drain" of medical and other experts from periphery and semiperiphery to core; and medical imperialism.[34] The general point is that uneven relations of exchange are created between core and other nations and exploited by increasingly mammoth multi- or metanational corporations of finance, industry or agribusiness.[35]

While jobs may be lost to the cheaper, less organized labor markets in the semiperiphery and periphery, it is important to realize that business also competes across core nations. The example of the increasingly challenged position of the U.S.A. auto industry has been cited. Such competition is a key characteristic of capitalism; in fact, the motor of the growth of its production-distribution entities into giant multinationals ("If you don't grow, you die," says the Board Chairman of a multinational in the film "Controlling Interest")—has very direct implications for OSH. For example, even in Sweden, where the workers' movement is extensively and well organized, the management of a medium-sized company cites possible loss of jobs by referring to the competition of a firm making essentially the same product in F.R.G. as the reason not to invest in an improved ventilation system at this time. Burawoy finds evidence of a "new despotism" in production processes influenced by state-supported "production politics" based upon the leverage of crisis and competition within the world-system.[36] The recent epidemic of cutbacks, takebacks, and union-busting in U. S.A. can be understood in light of this perspective.

KEY AGENTS OF MONOPOLY CAPITAL AND THE WORLD-SYSTEM

While this world-system describes relationships, including health relationships, on a world scale and between nations, it is important to realize what the several operative agents or vehicles of the world-system are. National states, their component and ruling class structures, multinational corporations, and a supporting cultural hegemony are among the most general political economic forms to keep in mind. As regards the health scene, there are more particular agents of the world-system, such as international health agencies; foundations; national bilateral "aid" programs; a variety of polluting and exploiting industrial firms, including agribusiness, commercial baby food suppliers, purveyors of chemical fertilizers and pesticides, and sellers of population control devices; and a medical cultural hegemony supportive of the activities of these agents on the world scene and in particular nations and locales (see note 34).

In this section, I offer only brief discussion of the general forms–nation states, classes, multinationals, and the cultural hegemony attendant upon and supportive of the world-system.

Strong central nation states arose as key mediating and organizing forces in the class struggle attendant on the rise and expansion of capitalism. In Lenin's view, "The state is a product and manifestation of the irreconcilability of class contradictions. The state arises where, when and to the extent that class contradictions objectively cannot be reconciled."[37]

The Armed Forces, police, legal mechanisms and conceptions, such as "private property," a key component of the cultural hegemony, hold the nation together in favor of the capitalist ruling class, the bourgeoisie, which controls the means of production and realizes the benefits of surplus extracted from the working class. The state in a socialist-oriented country functions on behalf of a different dominant class as our examination of the G.D.R. suggests.

In addition to raw force and the subtleties of a world view with which to legitimize the organized crime of capitalism, there are of course fiscal arrangements, service programs, including health programs, etc. which the state supports. In general, the main functions of the capitalist state are:

1. Provision of those general conditions of production which cannot be assured by the private activities of the members of the dominant class.[38]
2. Repression of any threat to the prevailing mode of production from the dominated classes or particular sections of the dominant classes, by means of army, police, judiciary and prison system.
3. Integration of the dominated classes, to ensure that the ruling ideology of the society remains that of the ruling class, and that consequently the exploited classes accept their own exploitation without the immediate exercise of repression against them (because they believe it to be inevitable, or the "lesser evil," or "superior might," or fail even to perceive it as exploitation).[39]

Gramsci made a distinction between class rule (dominio) and "hegemony." "Class 'rule' is expressed in economic relations, while 'hegemony' encompasses the complex interlocking of political, social and cultural relations of domination."[40] He laid great emphasis on the integrative function, for class domination based solely on repression "would be tantamount to an untenable state of permanent civil war."[41] But "integration" must be understood in terms of joining the exploited classes, not with each other, but with the interests of the ruling class. Indeed a large part of the cultural hegemony is directed toward dividing people against one another–racism, sexism, hierarchism, professionalism, etc.–precisely so they do not realize their common exploitation and decide to take state power into their own hands. Since much of the struggle for OSH is carried out through consciousness raising and information (the workers' "right to know" about the chemicals and other possibly hazardous substances they work with, for example) I will come back to the ruling cultural hegemony, and the struggle against it as relates to OSH issues.

Multinational finance, industrial, and agribusiness corporations are today's most powerful agents for extending the business of core nations into underdeveloped peripheral and semiperipheral nations (see note 30). In a ranking of nation states according to GNP and corporations according to sales, the U.S.A. is first and the U.S.S.R. second, but Exxon is in the top fifteen ahead of G.D.R. and on a par with India and the Netherlands![42] Several of the multinationals each control more resources than the twenty-five poorest nations of the world taken together.[43] Often these economic units are backed up by political forces in the form of armed forces, undercover agencies such as the CIA, etc., especially in times of 'crisis', i.e. when, as in the case of a duly elected socialist-oriented government coming to power in Chile, multinational control appeared to be slipping away.[44] These multinational entities singly or in concert (e.g. the Trilateral Commission which joins many of the ruling class interests of capitalist Europe, Japan and the U.S.A.[45]) manipulate national governments and officials. The Lockheed pay-offs are only the most widely known example. But, further, they encourage and carry the cultural hegemony of the core nations into the peripheral and semiperipheral nations where client elites (comprador bourgeoisies) more or less do their bidding under the legitimation of these conceptions while repressing organized labor and otherwise holding working and peasant classes in check.[46]

Let us now examine some of the concomitants of the functioning of this world-system as regards state actions and financing in general, before going on to consider the implications of the state and the cultural hegemony for OSH.

STATE ACTION AND FINANCE UNDER MONOPOLY CAPITALISM

Recall here the three basic functions of the capitalist state as quoted above from Mandel. I believe he is correct to distinguish the repressive from the integrative functions. O'Connor lumps these into one category; thus he sees two basic state functions: (1) assisting with the establishment and maintenance of profitable conditions so as to allow a competitively favorable rate of capital accumulation; and (2) legitimizing, through culture (law, education, the media, arts, etc.) as well as force, the governing apparatus and bourgeois ownership of the means of making profit. What kinds of expenditures does the capitalist state engage in to fulfill these basic functions? In Marxist conceptual economic terms (the actual expenditures as reported by nation-states will not be accounted for in these terms) state expenditures may be categorized, corresponding to the two broad state functions, as either *social capital* or *social expenses.*[47]

As we discuss these types of state expenditures it is necessary to recall the basic elements in the rate of profit so we might understand how such state financing can affect the profits of monopoly capital working hand in glove with the state apparatus.

State expenditures can be direct but most often are indirect in support of constant capital. This may be called *constant social capital*–roads, railways, education (for example, teachers and equipment needed to reproduce and/or expand workers' technical and skill levels) etc. Such social investments consist of relatively constant (even if increasing) expenditures as well as special projects and services which will increase a given amount of labor power and thus the rate of profit. Currently, for example, both the Democrats and Republicans seem to be preparing "the public" for some form of state support for "reindustrialization"–that is, taxpayers' buying of new equipment and technology to refurbish a steel industry and other industries from which high profits were drained while basic constant capital was left to deteriorate.[48] In the meantime, of course, monopoly capitalists have simply transferred a good deal of their investment funds to more profitable plants in Japan, Germany, etc. and will be glad to retransfer back if enough pressure can be brought to bear to make reindustrialization the public's tax problem and not theirs.[49]

State expenditures for *variable social capital* lower the costs of labor replacing or reproducing itself thus decrease what must be spent by the bourgeoisie to maintain an adequate supply of labor and thereby increase the rate of profit. Social insurance and many public health programs, insofar as they are truly effective, provide examples of such social consumption.

Both social (i.e., state rather than private) constant capital expenditures and social variable capital expenditures enhance the supply and strength of capital. The second major category of state expense, *social expense*, is a drain on capital and profits (if the taxes are paid out of potential profits, rather than workers' potential wages). But these expenditures are necessary to keep the system pasted together. They go mainly to the (in economic, not moral terms) surplus population—the unemployed, persons on welfare, the retired, etc. From time to time these expenditures will be cut in the interests of a higher profit rate for capital, but then pressure may mount, including national protest marches, such as The Poor People's March on Washington, D.C., and such funding will be restored. An historical analysis of this phenomenon suggests that activity on the streets, voting with one's feet, so to speak, has been a more effective way of restoring social expense funding than has the election of minority candidates as mayors, congressman, etc.[50] Probably there is this effect because street action gives the possibility of raised working class consciousness and sustained collective action; thus it is more frightening to the bourgeoisie than electoral politics in which each person goes into the ballot box as an individual.

Any state expenditure is likely to be in part both for social capital and social expense. Thus some transfer payments such as social insurance are intended to make reproduction of the work force less costly, while income subsidies for the poor may be to pacify and control part of the surplus population.[51] As O'Connor puts it:

> . . . precisely because of the social character of social capital and social expenses nearly every state expenditure serves these two (or more) purposes simultaneously, so that few state outlays can be classified unambiguously. For example, freeways move workers to and from work and are therefore items of social consumption, but they also transport commercial freight and are therefore a form of social investment. And, when used for either purpose, they may be considered forms of social capital. However, the Pentagon also needs free-ways; therefore they in part constitute social expenses. Despite this complex social character of state outlays we can determine the political-economic forces served by any budgetary decisions, and thus the main purpose (or purposes) of each budgetary item.[52]

What have been the major developments of the capitalist system as they bear upon state financing and as this has in turn affected capitalist development? Space does not permit a full answer. Clearly, the role of the state in funding early extractive expeditions (e.g., the Spanish search for gold in the New World) which led to the beginnings of capital accumulation and capitalist production arrangements, differed from the state in the Mercantilist period or from the later phases of *laissez-faire* competitive capitalism and the most recent phase of imperialist monopoly capital (see note 26).

With the rise of industrial capitalism, it was the destruction of the feudalistic absolutist state and its "interference" in the free production and exchange of goods which marked the success of the bourgeoisie. However, in the transition from competitive to monopoly capitalism, the narrowing of the state's role began to turn around with both the bourgeoisie's subjective attitude toward the state and the state's objective function altered. The state came to play a greater and greater role in safeguarding investments in the colonial period through military means. These were meant to guard against imperialist rivals as well as potentially rebellious colonial peoples and, not incidentally, provided an added sector for capital investment and accumulation.

Other associated factors led to an ever increasing role of the state. Under the influence of the working class movement, universal suffrage was won and powerful working class parties appeared in some countries. These parties functioned largely as integrative forces offering the illusion of equal political power to the worker, while the same unequal appropriation of the surplus he and she produced went on economically. They did lead to some concessions in social welfare and health which might best be seen in the above terms both as social expenses (to "cut the legs off" the socialist worker's movement as Bismarck understood) and as investments "in assuring the physical reconstitution of its labour-force where it was endangered by super-exploitation."[53] Sometimes feeling itself threatened by working class participation in the government (that is participation in its own exploitation) the bourgeoisie nervously resorted to military dictatorships, bonapartism and fascism.

In late capitalist development there is a further extension of the basic state functions:

> It is a consequence of three main features of late capitalism: the shortening of the turnover-time of fixed capital, the acceleration of technological innovation, and the enormous increase in the cost of major projects of capital accumulation due to the third technological revolution, with its corresponding increase in the risks of any delay or failure in the valorization of the enormous volumes of capital needed for them. . . . The result of these pressures is a tendency in late capitalism towards an increase not only in State economic planning, but also in State socialization of costs (risks) and losses in a steadily growing number of productive processes. There is thus an inherent trend under late capitalism for the State to incorporate an ever greater number of productive and reproductive sectors into the "general conditions of production" which it finances.[54] Without such a socialization of costs, these sectors would no longer be even remotely capable of answering the needs of the capitalist labour process.[55]

A tremendous concentration of capital occurs, leaving competition alive, even intensified, but between ever larger units. Thus in 1955 the bottom U.S.A.

Table 4. State Expenditures as a Proportion of Gross National Product and Gross Domestic Product, U.S.A. and Germany 1913-1970

Gross National Product U.S.A.		*Gross Domestic Product Germany (F.R.G. only, after 1948)*	
Year	*Percent*	*Year*	*Percent*
1913	7.1	1913	15.7
1929	8.1	1928	26.6
1950	24.6	1950	37.6
1960	28.1	1961	40.0
1970	33.2	1969	42.0

company on the Fortune 500 list had sales of $49.7 million. In 1979, with adjustments for inflation to have comparable dollars, the sales figure for the bottom company was $133 million, an increase of 268 percent.[56]

Thus with extended functions both in social investment and social control, the fraction of total capital which is redistributed, spent and invested by the state, steadily increases. This is seen for the U.S.A. and Germany (the Federal Republic only after 1948) in Table 4 (source: Mandel).

But all the state financing of social controls (social expenses) and other measures are not enough to avert the continuing internal contradictions and mounting crises.

Several conditions are at the root of these mounting crises. First, there is a continuing private appropriation of profit even though more and more capital costs are socialized. For example, ask yourself, "Who will get the profits, if any, from the Chrysler bail out–the stockholders or the tax payers who put up the loan?" Or, to whom do the profits on government cost-plus-profits contracts for arms manufacture go? Thus what O'Connor calls a 'structural gap' exists in which state expenditures have a tendency to outrun the means of financing them, though raging inflation has evolved as one way to discount current commitments with less valuable future funds. Second, with the increasing importance of state financing to the capitalist's success, there is chaos in government, reflecting the pressures and deals of special interests. Third, an increasing sophistication of the working class concerning the various ways in which business and the state rob them of the surplus they produce leads to a heightened class struggle and ever more complicated, subtle and sometimes ardent ways of revising the cultural hegemony to keep the working class from "breaking out" of the "voluntary" prison they are in. One of the latest and most potent efforts in this regard is represented by former Treasury Secretary Simon's book, "A Time For Truth" (as if it was not always a time for truth). A recent editorial advertisement by a major arms manufacturing corporation draws out several of the 'anti-government' themes of the book (but obviously it is anti-government only in certain social welfare directions, not against partnership for arms manufacture). The major theme with tax cutting implications is (quoted in the advertisement as taken from Simon):

> What we are seeing in America today is government dedicated to . . . the chopping down of those who produce wealth and the transfer of that wealth from those who have earned it to those who have not.[57]

Who is allied with whom and who are the enemies in this formulation? The capitalists and workers are put forth as partners and the enemy is that part of government which drains funds from these productive people to support the surplus population–the unemployed, persons on welfare, the elderly. Who next? Children? Maybe college students and professors who don't study useful, technical, direct production-related things? The objective facts are that the surplus produced by the workers is being largely appropriated by the bourgeoisie with increasing assistance from government.[47]

Perhaps other forces contributing to the mounting crises should be mentioned but space does not allow this. Mandel, recognizing also the world-system, summarizes the crisis atmosphere as stemming both from enhanced core-nation class struggle occasioned by cut backs stimulating working class consciousness, and from increased external struggle as exploited classes in peripheral countries seek to escape core nation domination–Cuba, China, Vietnam, Laos, Iran, Angola, Mozambique, Zimbabwe, Ethiopia, Nicaragua, El Salvador, etc.

As the state no longer can fully meet its major functions all at once, quick budget shifts and other characteristics of 'crisis management' develop. Thus the fiscal crisis of the state entails a management-leadership crisis.

Let us now turn from considering the implications of state action and financing for late capitalism, to considering the implications of class struggle within this system for OSH, especially in terms of the state and the cultural hegemony as they bear upon OSH. I will then conclude the chapter by offering a working hypothesis for comparing OSH in our study countries.

THE STATE, THE CULTURAL HEGEMONY AND OSH

The general dynamics I am concerned with are suggested by Stark:

> For Marx in *Capital*, the factory is less a place than a process and one not necessarily limited to the physical plant. With respect to epidemics, Marx argues that the capitalist's "greed" for profits leads him to so increase the worker's hours of labor (to increase "absolute surplus value") that laborers are driven below biological subsistence

> and cannot reproduce themselves as a class. Health deteriorates, class struggle intensifies, and epidemics break out. Then, the state responds to the crisis—with the Factory Acts, for example—to offset class struggle and protect the interests of landed and industrial capital as a whole.[58]

In the face of an actual or anticipated crisis caused by heightened or new attempts by the bourgeoisie at exploitation for increased profit and the response of the working class to these exploitative moves, the state may intervene to quiet the situation down and preserve the system, rather than risk a revolution through which the working class would assume state power and order affairs in its own interest.

There may not have been any very evident crisis of such vital proportions on the immediate horizon in the late 1960s and early 1970s[59] when much of the important OSH legislation (such as OSHA in the U.S.A.) was adopted in all our study countries. But the unsettled, high employment times of the Vietnam War era, when workers could be more demanding without fear of being fired from their jobs, makes it plausible that state intervention and concessions occurred in the OSH realm as in others to avert even more serious advances by the workers' movement. I will examine these advances and their shortcomings in subsequent chapters. For the moment, we might think of such state measures as rear guard actions protecting as much of ruling class interests as possible, depending upon the strength of the workers' movement.

But before such concessions are made, or even as an all pervasive influence affecting the concessions themselves, there is a powerful tool, the cultural hegemony, which tends to capture the language of the struggle and all that surrounds it in a net of meaning supportive of the ruling class.

> . . . One concept of reality is diffused throughout society in all its institutional and private manifestations, informing with its spirit all taste, morality, customs, religions, and political principles, and all social relations, particularly in their intellectual and moral connotations . . .
>
> It follows that hegemony depends on much more than consciousness of such interests on the part of the submerged classes. The success of a ruling class in establishing its hegemony depends entirely on its ability to convince the lower classes that its interests are those of society at large—that it defends the common sensibility and stands for a natural and proper social order.[60]

Thus very basic things such as the notion of factories as the "private property" of some individual capitalist or group of "owners" (who may do absolutely nothing in the actual production process) go unquestioned and are even confirmed in the new laws. In the U.S.A., for example, a factory owner in Idaho, Mr. Barlow, challenged OSHA's right under the 1970 law to come unannounced on inspection into "his" factory as a violation of the fourth amendment to the Constitution against unreasonable search and seizure of private property. The case went all the way to the U.S.A. Supreme Court where it was agreed that Mr. Barlow was right. Henceforth OSHA had to get a warrant from the court whenever an owner objected to their entry. While some people argue that this did not slow OSHA down much (others would say they never were that aggressive) since warrants were generally freely issued by judges presented with evidence of reasonable suspicion of workplace hazard violations (usually a signed formal complaint by some employee or the union was adequate) the case illustrates the power of such basic, unquestioned notions as "private property" in the realm of OSH struggles as well as the political-economically supported culture generally.

Other examples bearing directly on OSH concerns could be given by the dozens if space allowed and this were our main purpose. Only a few additional illustrations will be offered.

The sexism often separating men and women workers from joining in common struggle, entails many subtle but powerful notions. Sometimes supposed "female dexterity" is used to assign women to very detailed but boring and stressful and low paid work (forgetting that most watchmakers and micro-surgeons have been males). On the other side, men are encouraged to act "macho" and view each other as "sissies" if they hesitate to lift over 90 lbs. routinely on a shipping dock or in a warehouse or other workplace. This is very convenient for management who want higher productivity and would just as soon ignore OSH standards if higher profits could thereby be realized.

Some of the more powerful notions of the hegemony regulating or undercutting the workers' movement are indeed the most subtle. Some ideas come into the very consciousness of self and mode of expression. Even in a recession, people are heard to say, with a note of self depridation "I lost my job." Hardly ever do they say, "The company fired me because the economy is going sour."[61]

There are many ruling class apologists among the intellectual class who focus on technology, "the technological imperative," and industrialization as if these were automatic processes with lives of their own outside of any human determination and control. Workers themselves and their representatives (unions) may unthinkingly adopt such ideas and blame the pressure they face on such inanimate figments of the bourgeoisie's imagination. Thereby they are completely distracted from the essential issue–control over the means and processes of production. Technology does not have to be an inhuman thing–only when its sole purpose is to enhance the rate of profit for the owner does it take on this character.

Also, by struggling within the OSH realm only over chemical and other hazard standards, workers and their representatives are distracted once again from the key struggle. It is important to seek improved standards and more healthful and safe working conditions. But the key thing is to gain control over the means of production and the production processes. Within workers hands, given adequate information and expertise, as well as socialist control of the State and world-system to eliminate harmful competition, the implements and materials of production are not likely to be put to poisoning or otherwise harming workers.[62]

Even the workers' means of organizing through unions is undercut in many ways by the hegemony. Through the media, in schools, in almost every forum, speaking now of U.S.A. where the hegemony against unions is particularly strong, the image conjured up is of a fat, cigar-smoking, heavily dark stubbled union boss with his hand in the till of the pension fund creating rip off opportunities for the mafia. It is rarely mentioned that the Teamsters Union (the one most closely associated with this image) has a strong democratic reform movement alive within it. Nor is it mentioned that the bonding fees for bank executives have risen astronomically because of the high rate of "white collar" crime among them. Nor is it mentioned that no official of the United Auto Workers has ever been indicted for any crime.

At a conference on Video Display Terminals given by the New Directions Program for Worker Education in OSH of Connecticut in 1982, a scientist from Sweden who was to deliver an address on the subject added this to the chairperson's introduction of him: "I just want to add that I am an active member of the union in my country and I am proud of this." His American audience, even though many were union members, seemed mildly shocked and then broke out in spontaneous applause for this small but significant break in the hegemonic net of anti-union meaning surrounding them in U.S.A.

This story illustrates that the hegemony, although powerful, is never complete, that ordinary people can and do fight against it. The main way is through organization. Thus unions, labor parties and other forms of organization such as COSH-type groups are very key to our concerns.

A WORKING HYPOTHESIS

Considering the earlier presented history of OSH, including such incredible events as the Gauley Bridge disaster, the asbestos cover up, and the grudging yielding of improved OSH laws, within a worldwide competitive class-based system, dominated by the interests of the capitalist class, it seemed to me that the history and form and current strength of the workers' movement would be a major factor in determining the adequacy of OSH conditions in each country.

As I began this study I applied this idea both to special in-plant OSH services and conditions and to the linkages or backup between OSH and PHC. There will be reason to modify this conception as we come to interpreting the findings from the six case studies. But in general this conception has served as a useful guide to my work.

The strength of workers' movements is not a highly quantifiable phenomenon. Also, although there appear to be long-term trends within countries, there are ups and downs, so the readings of the moment must be qualified by historical study. Nevertheless, in spite of complexities and difficulties, it seems clear that high demand for labor (low unemployment), a strong labor or Left party, especially one which is in power or has held power; a high proportion of the labor force organized in effective unions; labor laws which provide for closed shops and other effective organizing; and a broad social philosophy or set of goals on the part of organized labor are all key indicators of the strength of workers' movements. These are important tools in carrying on the class struggle as it bears upon OSH. I have inquired into these and related characteristics of the workers' movements in each study country as a way of clarifying the gains as well as the yet-to-be-gained ground in the struggle for improved OSH.

NOTES AND REFERENCES

1. Karl Marx and Frederick Engels, *The Communist Manifesto*. New York: International Publishers, 1960 (original 1849).
2. As quoted in Henry E. Sigerist, From Bismarck to Beveridge, Developments and Trends in Social Security Legislation, *Bulletin of the History of Medicine 13*:365–388, April, 1943, quote and source, p. 380.
3. Ibid, p. 380.
4. Ibid, quote and source, p. 376.
5. Daniel Bell, Notes on the Post-Industrial Society (II), *The Public Interest*. No. 7:102–118, Spring 1967. There are many other such "harmonizing" attempts. The whole form of social science serves this function when it divides itself into various competing "disciplines," most of which ignore or revise social class into "cultural groupings," and treat politics separately from economics. Political science then becomes the study and analysis of relatively free floating (presumably potentially equal) interest groups, all examined as if there were no class base to the structure within which political struggle is carried out. Navarro examines a number of other homogenizing efforts, including consumerism, pluralist theories of power structure, social welfarism, and, curiously, some on the Left: the Frankfurt School (Habermas, et al.); Marcuse; and Aronowitz (the latter's historical analysis of the U.S.A. workers' movement remains, nevertheless, very important to this work–see Chapter 12 on U.S.A.). Vicente Navarro, Social Class, Political Power, and the State and their Implications in Medicine, *International Journal of Health Services* 7:255–292, 1977.
6. Fred L. Pincus, Students Being Groomed for Jobs that Won't Exist, *Guardian* (New York): 7, cartoon by Carol Simpson, May 9, 1984.
7. Harry Braverman, *Labor and Monopoly Capital*, New York: Monthly Review Press, 1974. Navarro offers a helpful definition: "Social relations of production are the relations which exist in a given process of production between the owners of the means of production and the producers, a relation which

depends on the type of ownership, possession, capacity for allocating and designing those means of production, and the use of the products of that process of production. Forces of production are the forces, instruments, labor, and knowledge which are organized to produce goods and services in a society. How the forces of production are organized, designed, and interrelated is determined by the social relations of production." Vicente Navarro, Work, Ideology, and Science: The Case of Medicine, *International Journal of Health Services 10*:523–550, 1980, quote from note 2, p. 536.

8. Much of this information is contained in Domhoff's work. See especially G. William Domhoff, *Who Rules America?* Englewood Cliffs, New Jersey: Prentice-Hall, 1967. Also, G. W. Domhoff (ed.), *Power Structure Research*. Beverly Hills: Sage, 1980. See also Frank Ackerman, *et al*, The Extent of Income Inequality in the United States, pp. 207–217 in Richard C. Edwards *et al.*, (eds.), *The Capitalist System, A Radical Analysis of American Society*. Englewood Cliffs, New Jersey: Prentice-Hall, 1972. Recently after public handwringing during the Detroit depression of 1979–1982 over the high price of U.S.A. labor and the supposedly resulting poor competitive position of the U.S.A. auto industry in its own market—*vis-a-vis* Japanese and other foreign-made cars (including Saabs and Volvos from Sweden and Mercedes and BMWs from F.R.G., to mention main competitors among our study countries)—U.S.A. auto workers, even though organized in the United Auto Workers Union, were forced to take cutbacks. Now it comes out that auto company executives, with concurrence of their boards have been paying themselves massive salary increases. Ford's chairman, Philip Caldwell got $1.4 million in salaries and bonuses in 1983! General Motors' top man, Roger Smith, got a 171 percent increase in annual pay! In addition, hundreds of millions of dollars were spread among 5,800 executives whose bonuses for a good year averaged $31,000!

"This contrasted sharply with the $640 bonus paid with fanfare earlier this year to GM workers, who had each given up about $3,000 in the 1982 United Auto Workers' Agreement." Ann DeCormis "Record profits, fat payoffs in 'poor' auto industry," *Guardian* (New York) (May 16, 1984: 8. Worrying that the legitimacy of the system might crack with the exposure of such disparities, even the most conservative columnists got in the act of criticizing: George F. Will, "Pay Those Auto Execs What They're Worth—and Nothing more," *The Hartford Courant* May 10, 1984: C9. When Caldwell said in reply to the furor that executive salaries and bonuses amounted to only $4 of the cost of each car and truck, Will quotes "one of the wisest Republicans in Washington" who said, dryly, that this is "a typical Republican statistic." But such disparities are nothing really new. See Paul Blumberg, "Another Day, Another $3,000: Executive Salaries in America," pp. 316–334 in Mark Green and Robert Massie, Jr. (eds.), *The Big Business Reader, Essays on Corporate America*. New York: The Pilgrim Press, 1980.

For an examination of classes and the concentration of power in the hands of the capitalist class in the U.K. see Ralph Miliband, *The State in Capitalist Society*. London: Weidenfeld and Nicholson, 1970. Even in

Sweden, which some portray as socialist, though it remains quite clearly a capitalist state, workers' sons and daughters made up less than nine percent of the top political-bureaucratic levels in 1961: G. Therborn, Power in the Kingdom of Sweden, *International Socialist Journal Y.2*:51–63, 1965.

9. A. Miller, The Wages of Neglect, *American Journal of Public Health 65*: 1217–1220, 1975. The author was then head of the United Mine Workers of America.
10. Michael Bosquet, The Prison Factory, *New Left Review 73*:23, 1972 as quoted in Vicente Navarro, "Work, Ideology, and Science: The Case of Medicine," *International Journal of Health Services 10*:523–550, 1980, quote on p. 527.
11. Tom R. Burns *et al.* (eds.) *Work and Power, The Liberation of Work and the Control of Political Power*. Beverly Hills, CA: Sage, 1979.
12. William Graham Sumner, the famous sociologist-teacher of the budding capitalists at Yale presented this view in What Social Classes Owe Each Other (1883). His constant admonition to his enthralled young listeners was "Gentlemen, get capital!"
13. In Veblen's day their main virtue was to engage in "conspicuous consumption" a la *The Great Gatsby*. Thorstein Veblen, *Theory of the Leisure Class*. 1899.
14. Lise Vogel, *Marxism and the Oppression of Women: Toward a Unitary Theory*. New Brunswick, New Jersey: Rutgers University Press, 1984. "It is the provision by men of means of subsistence to women during the child-bearing period, and not the sex division of labor itself, that forms the material basis for women's subordination in class society." See also Norma Stoltz Chinchilla, Enlightened Leninism, revolutionary feminism, *Guardian*, (New York) 23, March 21, 1984.
15. Maurice Dobb, The Essence of Capitalism," pp. 57–60 in R. C. Edwards *et al.* (eds.), *The Capitalist System*. Englewood Cliffs, New Jersey: Prentice-Hall, 1972, quote from p. 59.
16. R. Appellbaum, Marx's Theory of the Rate of Falling Profit, *American Sociological Review 43*:67–80, 1978.
17. Until the point shortly before widespread open class warfare and revolution, the ruling bourgeoisie is much more self conscious as a class and thus is more directing of state and cultural influences affecting and protecting its interests. However, this hegemony is never complete nor completely unified. There are internal ruling class conflicts as well as the main struggle between the working class and the bourgeoisie. Through the process of struggle and capitalist development working class consciousness grows.
18. Karl Marx, *Capital*, Vol. 1, (first published in 1867) quoted in R. C. Edwards, *et al.* (eds.), *The Capitalist System*. op cit. (note 15):65–66.
19. Ibid, quote from p. 66 of Edwards op. cit. (note 15); some delitions (. . .) added.
20. Frances Fox Piven and Richard A. Cloward, *Poor People's Movements, Why They Succeed, How They Fail*. New York: Random House, 1977.
21. This is true even though the NHS is not perfect and shares the dominance of bourgeois influence with the rest of U.K. society. Vicente Navarro, *Class Struggle, the State and Medicine*. Oxford: Martin Robertson, 1978.

22. I will come back to considering the distinction between a "social wage" or "social expenses" when considering the role of the state and the work of Mandel and O'Connor below. In brief, the social wage is "externalized" as far as the individual capitalist is concerned, while he pays the "money wage" directly. In a total accounting both can be considered part of variable capital which is the element in the factors of production over which the capitalist has most control and flexibility in his pursuit of the highest rate of profit (depending on how organized and resistive labor is).
23. Harold L. Wilensky, *The Welfare State and Equality*. Berkeley: University of California Press, 1975.
24. *The Cost of Social Security, 1964-65*. Geneva: International Labour Organization, 1972.
25. Vicente Navarro, Social Class, Political Power, and the State and Their Implications in Medicine, *International Journal of Health Services 7*:255-292, 1977, quote from p. 286.
26. I. Wallerstein, *The Modern World-System: Capitalist Agriculture and the Origins of European World-Economy in the Sixteenth Century*. New York: Academic Press, 1974; I Wallerstein, *The Modern World-System II: Mercantilism and the Consolidation of the European World-Economy, 1600-1750*. New York: Academic Press, 1980; A. G. Frank, The development of underdevelopment, pp. 94–104 in *The Political Economy of Development and Underdevelopment*, Wilber (ed.), New York: Random House, 1973; A. G. Frank, *Dependent Accumulation and Underdevelopment*. Monthly Review Press, 1979; J. O'Brien, with assistance from Salah Eddin el Shazali, *The Political Economy of Development and Underdevelopment, An Introduction*. Development Studies and Research Centre, University of Khartoum, Sudan, 1979; F. Greene, *The Enemy: What Every American Should Know About Imperialism*. New York: Random House, 1970.
27. V. Navarro, *Medicine Under Capitalism*. New York: Prodist, 1976; V. Navarro, The economic and political determinants of human (including health) rights, *International Journal of Health Services 8*:145-168, 1978; E. R. Brown, *Rockefeller Medicine Men*. Berkeley: University of California Press, 1979; E. R. Brown, Exporting Medical Education: professionalism, modernization, and imperialism, *Social Science and Medicine 13A*:585-595, 1979; H. Waitzkin, A Marxist view of medical care, *Annals of Internal Medicine 89*:264-278, 1978; H. Waitzkin, How capitalism cares for our coronaries, in *The Doctor-Patient Relationship in the Changing Health Scene: An International Perspective*. E. B. Gallagher (ed.), pp. 317-332; DHEW Pub. No. (NIH) 78-183. Fogarty Center, NIH, Bethesda, MD, 1978.
28. A. G. Frank, *Dependent Accumulation and Underdevelopment*. New York Monthly Review Press, pp. 10–11, 1979.
29. S. Jonas and M. Dixon, Proletarinization and class alliances in the Americas, *Synthesis 3*:1-13, 1979.
30. R. J. Barnet and R. E. Muller, *Global Reach: The Power of the Multinational Corporations*. New York: Simon and Schuster, 1974. See also

L. Turner, *Multinational Companies and the Third World*. New York: Hill and Wang, 1973.

31. H. Magdoff, *The Age of Imperialism*, New York Monthly Review Press, 1969, p. 198 as cited in J. O'Brien with assistance from Salah Eddin el Shazali, op. cit. (note 26.)
32. V. Navarro, The economic and political determinants of human (including health) rights, *International Journal of Health Services 8*:145–168, 1978.
33. S. Amin, *Neo-Colonialism in West Africa*, Harmondworth, England: Penguin, 1973.
34. Ray H. Elling, The Capitalist World-System and International Health, *International Journal of Health Services 11*:21–51, 1981. Also Ray H. Elling, Industrialization and Occupational Health in Underdeveloped Countries, *International Journal of Health Services 7*:209–235, 1977.
35. The term "metanational" is used and the dimensions and implications of these increasingly mammoth entities are provocatively examined, including their incursion into socialist states within the world-system, in John Borrego, Metanational Capitalist Accumulation, Reintegration of Socialist States, pp. 111–143 in Christopher K. Chase-Dunn (ed.), *Socialist States in the World-System*. Beverly Hills: Sage, 1982.
36. Michael Burawoy, Between the Labor Process and the State: The Changing Face of Factory Regimes under Advanced Capitalism, *American Sociological Review 48*:587–605, October 1983.
37. V. I. Lenin, *The State and Revolution*. Peking: Foreign Languages Press, 1976; emphasis supplied on p. 9. However, it should be noted that there is disagreement among Marxist scholars as to the role of the state. Some see it less as an "executive committee of the ruling class" and more of a mediator of the class struggle acting generally (but not exclusively) in favor of the bourgeoisie. Thus from this latter point of view, action to influence state action on behalf of workers' interests is seen as worthwhile. The enervating notion that nothing can be changed until the revolution takes place is rejected in this latter view. Eric Olin Wright, *Class, Crisis and the State*. London: New Left Books/Verso, 1978. Also Fred Block, The Ruling Class Does not Rule: Notes on the Marxist Theory of the State, *Socialist Revolution 7*:6–28, 1977.
38. Here Mandel cites Marx. *Grundrisse*, p. 533 for the formula of "the general conditions of production," and Engels, *Anti-Duehring*, p. 386 on "the general external conditions of the capitalist mode of production" which must be protected "against the encroachments as well of the workers as of individual capitalists." He also offers as earlier modes of surplus expropriation the great irrigation systems in the so-called Asiatic mode of production, and the transport of vast supplies of corn to Rome and other large cities of late antiquity. In modern times, one could suggest the postal service, an increasingly state supported rail transport system, which, in the U.S.A. seems to concentrate on transport of industrial products, leaving the passengers to Detroit and the automobile manufacturers, etc. It is interesting to speculate that the current and possibly growing strength of European, especially German, capitalism as well as Japanese capitalism, in world-system

competition may be due in large part to the frank embrace between monopoly capital and the state in these systems whereby the general external conditions of production—railroads, etc.—are well supported and maintained. By contrast, large parts of U.S.A. capitalism suffer from a strong dose of reactionary suspicion of anything governmental more appropriate to the *laissez faire* age of capitalist development.

39. E. Mandel, The state in the age of late capitalism, pp. 474–499 in *Late Capitalism*. London: Verso, 1975.
40. Akwasi Aidoo, "Social Class, the State and Medical Decolonization in Ghana: A Study in the Political Economy of Health Care," unpublished Ph.D. dissertation, Social Science (Sociology) and Health Care Program, University of Connecticut, 1985, quote from p. 112. The original essay is found in Antonio Gramsci, The modern prince, in *Selections from the Prison Notebooks*. New York: International Publishers, 1971 (original about 1923).
41. Mandel, op. cit. (note 39); L. Doyal with I. Pennell, *The Political Economy of Health*. London: Pluto Press, 1979, p. 475.
42. Marlene Dixon, Dual Power: The Rise of the Transnational Corporation and the Nation State: Conceptual Explanations to meet Popular Demand, *Contemporary Marxism* No. 5:129–146, Summer 1982.
43. Barnet and Muller, op. cit (note 30); L. Turner, op. cit. (note 30).
44. V. Navarro, op. cit. (note 32); J. E. Mulligan, The Dominican Republic: Police state protection for U.S. Corporations, *WIN 12*:8–12, 1976.
45. Holly Sklar (ed.), *Trilaterialism: The Trilateral Commission and Elite Planning for World Management*. Boston: South End Press, 1980.
46. R. H. Elling, Industrialization and occupational health in underdeveloped countries, *International Journal of Health Services* 7:209–235, 1977; M. Chossudovsky, Human rights, health, and capital accumulation in the third world, *International Journal of Health Services 9*:61–75, 1979.
47. J. O'Connor, *The Fiscal Crisis of the State*. New York: St. Martin's Press, 1978, p. 6.
48. A. Salpukas, Depressed industrial heartland stressing urgent need for help, *New York Times* 129, August 19, 1980: A1 and D9; second of five articles in a series on reviving industry.
49. On the increasing internationalization of capital as well as the increasing worldwide Taylorization of labor, see Jonas and Dixon, op. cit. (note 29).
50. Piven and Cloward, op. cit. (note 20).
51. It is important to recognize that some aspects of the health "services" also fulfill this social control function. An important work developing a conflict view of mental health services has been completed by Sylvia Kenig in "Limits on theory: a case study of the relationships of market and state to theory in the community mental health movement." Ph.D. dissertation. Program in Social Science (Sociology) and Health Services, University of Connecticut, 1981.
52. J. O'Connor, op. cit. (note 47):7.
53. E. Mandel, op. cit. (note 39):483.
54. Here Mandel gives this notation: "This fully corresponds to the logic of Marx's analysis of capital, which explicitly emphasizes that 'the highest

development of capital exists when the general conditions of the process of social production are paid out of deductions from the social revenue.'" *Grundrisse*, p. 532.

55. Mandel, op. cit. (note 39):483.
56. *Fortune* (Magazine) (1980):101.
57. W. E. Simon, *A Time for Truth*. New York: McGraw-Hill, 1978.
58. Evan Stark, The Epidemic as a Social Event, *International Journal of Health Services* 7:681–705, 1977.
59. Though recall that the alliance of workers and students came very close to toppling the DeGaulle government in France in 1968 (see the book, *The Merry Month of May*) and Lyndon Johnson declined to run a second time for the presidency in U.S.A. and Swedish miners went out on strike with broad support from organized labor in that country, so there were potentially major upheavals in many core capitalist nations.
60. Genovese, a student of Gramsci, as quoted in E. A. Krause, *Power and Illness: The Political Sociology of Health and Medical Care*. New York: Elsevier, 1977, p. 258.
61. June Jordan, The Language of the Powerful Imprisons the Powerless, *The Hartford Courant*, May 9, 1984: B11. This sensitive, perceptive poet offered the example of women on a panel at a women's liberation conference trying to cite the deeds and names of black women who had contributed so much to the struggle in U.S.A. and hesitating because they couldn't think of the names! Finally, people in the audience contributed: "Sojourner Truth!" "Harriet Tubman!" But the deepest feelings were aroused by the observation that two black women at a university, who may be lonely, as they are very few in number, may even adopt the prejudices of the wider society and hesitate to greet each other in passing on campus. The plea was "Let me know you see me; let me know you exist."
62. This and related ideas, several of which I have drawn upon liberally in this chapter, are developed further in Vicente Navarro, Work, Ideology, and Science: The Case of Medicine, *International Journal of Health Services 10*: 523–550, 1980.

CHAPTER 6
An Ideal Model and Strategy for Comparing Cases

INTRODUCTION

This chapter presents a method within a method. The overall method of the study can be described as that of "in-depth" contrasting case studies. "In-depth" is in quotes because time and resources obviously placed limits on the depth and detail of fieldwork and the study as a whole. This method was discussed in relation to the field of cross-national studies of health systems (CNSHS) in Chapter 2; and the Appendix gives the detail of the fieldwork design and data gathering techniques. The strategy for choosing the six "contrasting cases" (study countries) was suggested in Table 1 in Chapter 1 and related discussion.

But having chosen cases to be compared and settled upon an approach and means of gathering information, one still needs a scheme for highlighting the information to be used in contrasting or comparing the cases.

The ideal typical method I followed was first developed by Max Weber in his cross-national studies of bureaucracy.[1] There has been considerable ferment in sociology around Weber's intent in using this method. Was it a neutral, value-free device for abstracting from social reality to report its variations in relation to varying conditions–a kind of natural history of social forms seen through the glasses provided by the model? Or did the model carry both theoretical and value-laden meaning, that is, were the dimensions of bureaucracy not only significant for abstracting the essential elements of complex formal organizations,[2] but were these elements seen as desirable for achieving more efficient, effective human organization? In deciding upon an answer, it seems important to recognize that the giant shadow of Karl Marx hung over the life and times and work of Weber and other earlier sociologists such as Durkheim and Toennies. Their work can be interpreted as the efforts of generally conservative intellectuals[3] to find ways of shoring up capitalist forms of production organization which Marx saw as essentially caught in class conflict and eventually failing and being taken over by the working class along with state power. Thus, I am convinced that the method was used not only to describe and analyze complex formal organizations, but was also an attempt to bring about social change in a desired direction—greater productivity with increased efficiency and lowered alienation of workers (and other "office holders") who could look forward to advancement according to merit.

Weber himself spoke of the method as a "mental experiment" whose ideal elements would not likely all be found anywhere in reality in their entirety. But conditions associated with their greater approximation might be studied and compared with conditions associated with the lesser occurrence or achievement of the ideal elements so as to foster policy changes and other rational social action in the direction of conditions which appeared to foster the ideal elements.

I believe the bad repute into which imperfect, oppressive bureaucracy has fallen since Weber did his work (around the turn of the Century) have clouded his intent to find a more ideal and desirable form of organization. Such titles as "Up the Organization;" "The Organization Man;" and even Orwell's "1984" remind us of this general distaste for bureaucracy gone wrong. Weber's conservative orientations led him to assume that top-down rational-based (as distinct from traditional or charismatic) authority was the generally to be desired base of human organization. We can see how his ideal elements of bureaucracy were indeed ideal, that is, to be desired in his mind:[4]

- *Appointment by merit*: staff in the organization are engaged and promoted on the basis of certification of adequate formal training designed to give them competence in the tasks they have to perform and/or success in previous work experience of the same type. No one is hired on the basis of "who you know" or "who you are" by family name, sponsorship, etc. No one is hired who is unqualified or unable to do his job. Those who can not or do not measure up are to be demoted or fired.
- *Specialization*: The complexity of functions and tasks to be performed by the whole organization requires a rational, specialized division of labor. Thus the individual must perform only a limited set of tasks. This will lead also to expertise in this special work. No person is to be all things to all clients or customers.
- *Written rules*: Written codification of jobs and expectations allows for merit-based coordination of a large number of people performing a wide range of specialized tasks. These rules also allow for merit-based advancement and recognition. These rules are set by the policy makers (the board or its equivalent) and carried out by management which has a wide overview of what is needed to coordinate the whole organization. These top office holders have both the skill and power (the authority of the office) to carry out these key management functions. Situations which are not covered by rules are decided by the management and may lead to the establishment of new rules.
- *Impersonal relationships*: Personal biases must be avoided. Nepotism and other forms of favoritism are likely to undercut objective, competent performance of tasks and handling of materials (money, production goods) and clients or

customers. Therefore relationships should be oriented to organizational concerns and be friendly without love/hate connotations. Organizational goals must always be kept in mind and always take precedence over personal goals and ambitions. Thus the organization has an existence and continuity independent of any individual or set of individuals at any level. The organization is thus protected against collapse due to personnel turnover.

- *Separation of policy and management*: To help assure perspective as well as centralization and coordination, policy making is limited to a select group at the top of the organization. The working staff and lower and middle levels of management do not make policy. Policy makers are the Board of Directors and/or the Chief Executive Officer. Policies are created in relation to changing knowledge, resources, and other conditions and radiate down through the specialized, graded hierarchy of the organization for effective day to day operations.

In his comparisons involving India and China as well as examples from Persia and elsewhere, Weber found certain political and cultural conditions, such as widespread nepotism, which did not allow the realization of these elements as easily as in the more "rational" West.

The modern industry and local production unit or factory as well as other complex formal organizations such as hospitals are essentially designed after this model. The effects of "scientific management" or "Taylorism" and the assembly line, as well as subsequent waves of organizational philosophies, such as the "human relations" school of thought, have all had an influence. But the essential elements can be found in Weber's model.

I have cited the elements of Weber's model at some length (though in no way as completely as he presents them or as completely as they have been examined in the subsequent organizational literature) both to support the position that he intended it as a desirable, value-laden model and to allow explicit contrast with the Model I will suggest for improved OSH and OSH-PHC. On the one hand, I want to make it clear that I am following what I believe to be the Weberian tradition of value-oriented sociology and I am using his general method. On the other hand, I am standing Weber on his head, so-to-speak, for the realities of organizational life and secreting of information from workers in the whole complex of works-inspectorate agencies (OSHA-type organizations) primary care providers and others seems to me to call for a much greater degree of democracy and worker control. It is this element and its implications which I emphasize in the following ideal-typical model.

A MODEL FOR IMPROVED OSH AND OSH-PHC

I have just provided a general introduction to this model as an ideal-typical statement of desirable elements to be striven for in efforts to improve OSH conditions and surveillance and the relations of workers' health to PHC. Here some more specific introductory comments and observations are in order as well as a general rationale for the model per se, before presenting and discussing the elements.

An ideal OSH and OSH-PHC system would include a near infinite number of "elements" or units of concern, depending upon how one decided to categorize and discuss them. For example, each standard for each of thousands of hazardous chemicals and work conditions might be discussed as an "element." To be as ideal or protective of workers as possible (without attention for the moment to the practical politics of labor-management struggles and the role of scientific research and measurement and such strategies of resistance to worker-oriented change as "cost-benefit analysis"–see Chapter 3) I might say, for example, as regards benzene, that the standard should be "no measurable amount" or "no more than one part per million (1PPm) of air, for a continuous eight hour work day," rather than the 100 PPm which had been the standard in the U.S.A. until 1971 when OSHA adopted 10 PPm. Subsequently, increased research evidence of its relationship to aplastic anemia and leukemia and pressure from organized labor forced a move by OSHA to 5 PPm, with court battles still underway concerning a level of 1 PPm.[5]

But obviously such a detailed model would be more like an encyclopedia,[6] too unwieldy to serve as an analytic device for comparing OSH systems in different nations. Thus the model suggested here is limited to six relatively general organizational elements: *policy; sponsorship and control; education; organization; information*; and *financing*.

With these elements achieved as indicated below, one could expect more detailed, specific components of a system–such as highly protective specific standards vigorously enforced–to fall in place or go along with the elements identified and discussed here. This would be so because, hopefully (at least this is my intent) these elements are conceptually connected. That is, they link back to the basic structure of the political economy which will have to be radically transformed in most instances for these elements to be realized, and they work forward, so to speak, to affect more detailed aspects of the OSH and OSH-PHC system.

If we think back on the framework suggested in Chapter 5, it will be clear that most if not all OSH problems stem from the inequitous exploitation of relatively powerless workers engaged in producing value which is expropriated for the private use of a ruling elite. This exploitation of the worker and expropriation of the surplus value he or she produces is carried on with the object of realizing the highest rate of profit possible in order to keep up with or even grow in relation to competing capitalist enterprises in the world-system. Usually it is cheaper and easier (especially in conditions of high unemployment) to replace a worn out worker than to invest in equipment or engineering or material changes which would keep the worker healthy. How else can we explain the fact that workers, more often minority workers, are still dying from cancer due to exposure to hydrocarbons over the coke ovens of steel mills–the same substances discovered by Percival Pott to be causing scrotal cancer in chimney sweeps 200 years ago?[7] How else can we explain the fact that hundreds of thousands of workers were exposed to asbestos with many dying from asbestosis, lung cancer, and mesothelioma, with many more yet to die, without their ever having been told that there was a hazard, even though this hazard was well known to management and recorded in the scientific and clinical literature? As Epstein summarizes this history:

> While the dangers of asbestos have been well recognized for over five decades, the industry was able to largely ignore or suppress these until studies in the 1950s by Doll in England and by Selikoff in the U.S. (sponsored by the asbestos insulators union) established the relationship between asbestos and asbestosis and cancers at various sites. Since then the industry has employed scientific expertise to advocate its position in professional journals and to successfully resist and prevent effective government regulations. With an estimated 8 to 11 million workers having been exposed since World War II, the toll of asbestos occupational disease and cancer is now reaching epidemic proportions. Workers have begun to exercise legal initiatives, including third-party and medical malpractice suits, resulting in multimillion dollar awards. . . .
>
> As the realization is dawning that asbestos is too expensive in terms of disease and death to continue using, the industry is beginning to develop the strategies of relocating in lesser developed countries and promoting the use of fiberglass as an asbestos substitute in the United States. . . .[8]

Health and safety on the job have been given as little priority at national policy level as possible (considering that differing limits have been placed on this indifference according to the events—sometimes a disaster such as the Gauley Bridge tunnel—and the level of struggle and public outcry) in favor of investing in further profitable production, rather than investing in preventive health and safety measures. Others who have held the power and authority have knowingly assigned workers to hazardous situations on the job, have kept information from workers, have sponsored and controlled the development and flow of information, and have inordinately influenced standards and enforcement by government agencies. Education of workers, as well as primary care providers and even OSH specialists has been inadequate. Organization of OSH and PHC services have been inadequate and have kept the OSH and PHC services disjointed, not to say, antithetical. Financing arrangements have not motivated people concerned with OSH and PHC to take essential preventive measures as regards widespread and serious OSH hazards. It is with a view to turning these inimical conditions around that the following model is offered.

At the *policy* level, the defined course of a nation should be to give the highest possible priority to the protection of workers' health. However inclusive or exclusive one's definition of "productive forces" which lie at the base of a society's existence and way of life,[9] surely the core of such forces is made up of the working men and women who produce all valued products, most of which we unfortunately take for granted in our everyday lives–our chairs, the clothes on our backs, etc. Such a policy would of course require an effective worker-oriented power structure, state apparatus and supporting culture.[10] It is not likely to occur in situations of capitalist production with high unemployment where workers are seen as disposable objects more replaceable than the machines they work with.

This element of the model is complex, as are the others. It therefore requires some further discussion before it can be properly appreciated. For example, it is one thing to find authoritative written statements giving great importance to a healthy and safe workplace as in U.S.A.'s OSHA law:

> Sec. 2 . . . The Congress declares it to be its purpose and policy, through the exercise of its powers to regulate commerce among the several States and with foreign nations and to provide for the general welfare, to assure so far as possible every working man and woman in the Nation safe and healthful working conditions and to preserve our human resources– . . .
>
> Sec. 5 (a) Each employer – (1) shall furnish to each of his employees employment and a place of employment which are free from recognized hazards that are causing or are likely to cause death or serious physical harm to his employees; . . .[11]

This is a fairly strong statement. A statement from another of our study countries may be even stronger in recognizing psychological and emotional hazards as well as those which "are causing or are likely to cause death or serious physical harm. . . ." But it is a reasonably strong statement. Yet it is another thing to watch the waxing and waning of actual pursuit of this policy, to the point where, under the Reagan regime, OSHA has been greatly diluted and, some would say "unravelled."[12]

Thus assessing or measuring this dimension (as is true for the others) will be a more qualitative than quantitative procedure,[13] requiring an examination of authoritative written and oral assertions, but also an assessment of actual pursuit of a policy as evidenced primarily in government actions or those stimulated or sponsored by government (since we are concerned with a nation's policy and not that of, say, a particular union which may give OSH a very high priority when the government does not). In this assessment, there is some likelihood that there will be overlap with other dimensions. But this is not bad, for the several dimensions are seen in any case as interwoven, forming a system, rather than as entirely independent, discrete elements.

If the policy direction just indicated is to be implemented in fact, then *sponsorship and control* of the OSH services and supporting elements should be in the hands of the workers themselves–the peoples whose lives and well being are at stake. The sad history of hired experts of management and government keeping the known hazards of asbestos from the workers would have no place in a proper OSH system.[14] The spectacle of the pot (Johns Mansville Corporation) suing the kettle (the U.S.A. government) only underlines the depravity of the situation.[15]

If there is one element more central than any other in the model, worker control and sponsorship is it. Yet it, like the other elements, is complex and requires careful assessment.[16] These subtleties can not all be discussed here. Many will come out in the presentations and discussions of each case (Chapters 7-12). But to illustrate, whereas the OSH physician in U.S.A. is, with few exceptions, an employee of and beholden to management,[17] in the G.D.R. the OSH physician, as is true of all physicians, is an employee of the Ministry of Health. Thus, since the G.D.R. bills itself as "a workers' state" (though this will require detailed discussion when we come to this case–Chapter 9) one would say that workers have a greater degree of control and sponsorship of OSH physician's and their activities in G.D.R. than in U.S.A. This is undoubtedly true, yet for a full assessment and understanding of this element in G.D.R. one has to observe that, although the OSH physician reports to the district physician or regional physician representing the Ministry of Health, he or she also reports to the management and the union in the workplace, and the management supplies the material means for the work. Thus, as the Chief Physician in a large steel mill put it, "You have to get along with the management." This is only common sense.

But it shows that there are always multiple influences and subtle, yet possibly powerful ones at that, and it is a matter of assessing the place of workers and their representatives among these influences which will allow an understanding of a case as regards this dimension.

Although it is obvious, it is important to note that worker control and sponsorship is important throughout the system, not just with regard to the services of OSH physicians. The general lack of expertise available to workers and their representatives, including sound laboratory and epidemiologic studies, as well as information on OSH hazards generally, has been and continues to be one of the major failings of most OSH systems. Also, and I will make special note of this in the concluding chapters, the lack of worker control or influence over the PHC services, as compared with their sometimes greater voice as regards OSH services, has to be seen as a major failing of all the systems studied.

Education to recognize and act to prevent and avoid hazards on the job would ideally be widespread but would be especially emphasized for workers and for a variety of experts. Special attention will be given to OSH education for workers. And, once again, there are many subtleties. If the G.D.R. places great emphasis on this aspect of this element, but one of its main tools in this regard is the "work protection control book" (Arbeitsschutzkontrollbuch) in which each employee must sign off that he or she has been instructed with regard to some hazard in the workplace, a judgment must be made as to whether this is live, effective education or whether, it is formalistic, wooden and ineffective. However ineffective it may be judged, it will probably rate higher than no education at all or a "blame the victim" approach to worker education.[18]

Beyond the workers, if we limit ourselves here for illustrative purposes to physicians, three kinds of preparation are called for.

First, and possibly most important, undergraduate medical education and postgraduate and in-practice education for PHC providers should (1) alert the student and community physicians to the depth and range of problems which might come out of the workplace; (2) prepare them to always take a thorough work history whenever any suspicion of a work-related problem is presented to them; (3) prepare them to follow up with other workers who might be similarly exposed, including understanding epidemiologic surveillance, and the work inspection and information reporting apparatus; and (4) become thoroughly acquainted with the plants and other work situations in their area of service. Obviously, this last (point 4) would be best served in a general system in which PHC providers had defined geographic areas to cover. This is an example of one dimension of the model, education, intertwining with another, organization.

As we examine different countries we will see considerable variation as regards undergraduate and other medical education. One country has OSH questions on its national boards but no required hours of OSH instruction. Another has both. Others neither. Again, there are subtleties as regards the impact and meaning of these things.

Second, there should be expert preparation for in-plant physicians, certainly more than the three-week course, but not necessarily four years of special training. Perhaps something like two years of didactic and practical in-plant experience under teaching supervision is called for. Clinical, preventive, and epidemiologic surveillance techniques and perspectives must be taught. Some people prepared at this level might work for unions, depending on the organization of OSH services.

Finally, the teacher-researcher would have some four years of specialty preparation and be expected to contribute new OSH knowledge and pass it along in a variety of ways. In addition to serving in university teaching positions, these experts might fill key positions in the OSH inspectorate and other policy level positions.

Organizational issues overlap with other dimensions, particularly sponsorship and control. Whatever the specifics of the arrangements, the key objectives are: (1) the experts must serve the workers' interests; it is the workers' health and safety which must be protected; and (2) integral connections should be provided between workplaces and PHC providers such that these latter can properly serve the sick person and follow up to protect other workers.

In addition to the more familiar problems of how to organize OSH service in large plants,[19] there are the special problems of dispersed work forces. Sometimes it may be a very large firm or workforce, but one which is widely dispersed on different small to medium-sized construction sites. Surveillance and follow-up for the OSH problems of agricultural workers are another special concern of this sort, as are the problems of providing OSH services for small plants.

As noted earlier and as will be elaborated later on, if a nation's health services system is fully regionalized,[20] its peripheral PHC units will be geographically oriented. In this case, even if the education of practitioners in OSH matters is inadequate, there should be a tendency for them to get to know the workplaces and the associated hazards in their service area. This organizational element can be quite important in forging adequate PHC-OSH links, but is not in itself sufficient to assure proper identification and follow-up (both individually and epidemiologically) of OSH cases treated in the PHC setting.

Information. Communication from PHC provider to the works inspectorate and to plant physician for work group surveillance and epidemiologic follow-up; links from PHC provider to plant physician for case follow-up; national work-hazard identification surveys and follow-up; and special studies would ideally all be developed in a proper OSH-PHC information system.

If geographic coverage by the PHC services, as discussed above, were combined with computerized records, epidemiologic surveillance could be built in. Whenever a disease showed up with greater than expected incidence in a particular plant, or in relation to a type or similar types of jobs in different plants, or in relation to certain substances or work practices, there could be a red flag setting off proper follow-up.[21]

Also with regard to this dimension, there may be gaps between formal and actual requirements. Most systems formally require PHC providers to report work-related disease to the works inspectorate (or to the Ministry of Health, etc.) however, there may be many reasons including poor training, a rushed examination in the doctor's busy practice, few orientations favorable to working class patients, and lack of financial incentive which interfere with the realization of this formal requirement.

Financing must allow for OSH policy to be pursued and the organizational goals to be fulfilled and must protect the experts' motivation to serve the workers.

If we take the situation of private office practice in U.S.A., what incentives are there to identify and follow through on a work-related disease—both for the individual patient and for a possibly large group of workers similarly exposed in the same plant or other plants? If the case involves testifying on a worker compensation case, it can be very time consuming.[22] Similarly, calling the plant physician (if there is one; which there usually is not in U.S.A.) or even visiting the plant to help assure that the worker who has been sick is properly reintroduced to the job, or to a new more supportable job, can be very time consuming. And time is money in the private office practice. If physicians in general are on salary in a nation's health service system, some greater or at least comparatively greater motivation to diagnose and follow-up work-related disease can be expected, though there may be other problems.

Beyond the payment of individual PHC providers and others, there is the overall adequacy and way of financing the OSH services. If money comes from the employer (ignoring for the moment that it actually comes from the workers who produce all value, but may lack control over the disposition of what they produce) there may be one set of motivations, whereas if the funding comes from the National Health Service or the Ministry of Health it may carry with it another set of motivations.

A STRATEGY FOR COMPARING STUDY COUNTRIES

Although as we have just seen, the interwoven elements of the model will require largely qualitative assessments to achieve understanding of them in each system, it will be possible to give an overall ranking on each dimension for each country in comparison with the other countries. Since there is no established zero point, scale, or calibration of levels on these dimensions, these rankings will only make sense when considered in the context of the discussion of each case. Thus, if I judge Sweden to be higher than any of the other countries on worker education (though it is not perfect) it will be important for the reader to know that some 110,000 union health and safety representatives have been given forty hours or more of OSH training, and there are small study circles in most workplaces and a monthly magazine, *The Working Environment*, is distributed free to hundreds of thousands of workers, thanks to funding from *The Work Environment Fund*.

The model will probably be most useful as a summation device for each case. That is, I intend to offer a general description of the OSH system and provisions for PHC in each county after having discussed the political economy and history and current situation as regards the strength of the workers' movement. Then each country chapter will close using the above model as a way of summarizing the OSH and OSH-PHC situation.

Although there may be some tenuous aspects to generalizing on the basis of the model, in a summary chapter I will use it as an heuristic device for making cross-national observations on OSH and OSH-PHC relations.

NOTES AND REFERENCES

1. Max Weber, *The Methodology of the Social Sciences.* Edward A. Shils and Henry A. Finch (trans. and eds.). New York: The Free Press, 1962.
2. The literature in this field is too voluminous to cite and review here, nor is this my purpose. The interested reader may pursue the field through such general works as James March, (ed.), *Handbook of Organizations.* Chicago: Rand McNally, 1965; J. Eugene Haas and Thomas F. Drabek, *Complex Organizations: A Sociological Perspective.* New York: Macmillan, 1973. In recent years, the field has grown from primary concern with internal organizational composition to recognition of the determining influence of the political economy and other aspects of the organizational environment. For this emphasis see J. Kenneth Benson, (ed.), *Organizational Analysis, Critique and Innovation.* Beverly Hills: Sage, 1977; W. L. Zwerman, *New Perspectives in Organizational Theory: An Empirical Reconsideration of the Classical and Marxian Analyses.* Minneapolis: Greenwood, 1970; M. B. Brinkerhoff and Phillip R. Kunz, (eds.), *Complex Organizations and their Environments.* Dubuque: Brown, 1972.
3. Weber's politics were frankly conservative. Durkheim and Toennies had openly expressed socialist leanings, but on a spectrum they might best be described as social democrats, not prepared to see state power and control of the means of production taken over by the working class.
4. The elements of an ideal-typical bureaucracy as presented here are abstracted and paraphrased from a variety of sources, but most importantly Max Weber, *The Theory of Social and Economic Organization,* A. H. Henderson and Talcott Parsons (trans.), Urbana, IL: The Free Press, 1947.
5. This research work is briefly reviewed in Richard R. Monson, *Occupational Epidemiology.* Boca Raton, FL: CRC Press, 1980, pp. 201–202. He concludes, "If a standard of 1 PPm results in no major economic disruption, it clearly is indicated." He goes on to ask ". . . why has it taken so long to make this determination? Benzene has been used in industry since at least 1900 and thousands of workers have been exposed to levels that likely were several thousand PPm [shoemakers and rotogravure workers are cited, but I would add many hundreds in the petro-chemical industry and thousands of persons pumping gasoline]. Since the role of benzene in the etiology of aplastic anemia has been known since at least the 1920s, one might expect that workers exposed to benzene would be monitored for other adverse effects, including leukemia." Monson goes on to answer his own question in terms of the lack of data: "Unfortunately, common sense arguments such as this must play a relatively weak role in the setting of standards. While it might be expected that exposure to benzene in the early part of this century was very high, no data exist." I would go further and say that no data exist because it was not in capital's interest to find or reveal such information and the workers lacked awareness and control. See Chapter 4. Also, see The Official Body Count, Chapter 2 in Daniel Berman, *Death on the Job, Occupational Health and Safety Struggles in the United States.* New York: Monthly Review Press, 1978.

6. *The Encyclopedia of Occupational Safety and Health* published by the International Labour Office in Geneva is a very useful and important reference source of such an extent that it could be used as a beginning basis for such a detailed model, though all the suggested levels in the ILO work might not be as ideal as workers and their representatives would have them.
7. Joseph K. Wagoner, Occupational Carcinogenesis: the two hundred years since Percival Pott, pp. 1–3 in *Occupational Carcinogenesis*, special issue of the *Annals of the New York Academy of Science 271*:1–3, 1976.
8. The Asbestos 'Pentagon Papers', pp. 89–102 in Samuel S. Epstein, *The Politics of Cancer*, revised and expanded edition, New York: Doubleday, Anchor Press, 1979, quote from p. 102. On the involvement of core-nation firms such as Johns-Manville in asbestos production in semi-peripheral and peripheral nations, see Barry Castleman, Double standards: Asbestos in India, *New Scientist*, pp. 522–523, 26 February, 1981.
9. Some ultra-rightist forces seek to obfuscate or defuse the inherent struggle between capital and labor in capitalist societies by defining the bosses and workers as on the same team against "nonproductive elements" such as welfare recipients, the elderly, and who knows who else. How far will this fascist thinking go? Perhaps to eugenic purifying of "the race," to keep other needy persons from diminishing the rate of profit in the face of increasing competition from ever larger aglomerations of capital worldwide. See William Simon, *A Time for Truth*. New York: McGraw-Hill, 1978.
10. Here the Gramscian idea of a hegemony is important. The media, even most scholars, and all information-culture-producing elements are aligned in a functional systemic way (not necessarily in a conscious conspiratorial fashion) to define in a dynamic changing mode a given set of values, beliefs, and directions as the right ones. For an application of this idea to medical, health problems, particularly in U.S.A. see Elliot Krause, *Power and Illness, The Political Sociology of Health and Medical Care*. New York: Elsevier, 1977. As applied to working people and their representatives in the U.S.A., unions and others must continually struggle against devaluing, disparaging images and propaganda. The usual public and media utterance about unions is to immediately cite corruption in the Teamsters Union as typical, while never mentioning the numerous white collar crimes of bank executives whose bonding fees these days are astronomical because of past misdeeds of their kinds, and never mentioning that no official of the UAW at any level has ever been indicted. For Gramsci's development of this idea see, Antonio Gramsci, The Modern Prince, in *Selections from the Prison Notebooks*. New York: International Publishers, 1971.
11. Public Law 91-596 91st Congress, S2193 December 29, 1970 (the OSHA law).
12. Philip J. Simon, with a Foreword by Ralph Nader, *Reagan in the Workplace: Unraveling the Health and Safety Net*. Washington, D.C.: Center for the Study of Responsive Law, 1983.
13. For a good statement of qualitative methods in sociology see Claire Selltiz, et al. *Research Methods in Social Relations*. New York: Rinehart, and Winston, 1976. Also Matthew B. Miles and A. Michael Huberman, *Qualitative Data Analysis, A Sourcebook of New Methods*. Beverly Hills: Sage, 1984.

14. Epstein, op. cit. (note 8).
15. Cris Oppenheimer, Asbestos Firm Files Suit Against U.S. *The Hartford Courant* (July 20, 1983): A1 and A10.
16. Many of the subtleties involved are covered in Tom R. Burns, Lars Erik Karlsson and Veljko Rus, (eds.), *Work and Power, The Liberation of Work and the Control of Political Power.* Beverly Hills: Sage, 1979. Also see David H. Wegman, Leslie Boden and Charles Levenstein, Health Hazard Surveillance by Industrial Workers, *American Journal of Public Health 65*: 26–30, 1975.
17. This factor along with poor training often leads U.S.A. workers to mistrust the company doctor and to refer to him as "the company quack." Sometimes the sense of responsibility of the physician to the company is so strong that it leads him (seldom a woman) to view the company, rather than the worker as his patient. The case of Dr. Grant who was responsible for much of the human misery and death in the Tyler Texas asbestos plant was cited in Chapter 4. See Paul Brodeur, *Expendable Americans.* New York: Viking Press, 1974, p. 245. See also, The Captive Specialists, pp. 94–104 in Daniel Berman, *Death on the Job.* New York: Monthly Review Press, 1978. Also see Peter Orris, et al., Activities of an Employer Independent Occupational Medicine Clinic, Cook County Hospital, 1979–1981, *American Journal of Public Health 72*:1165–1167, 1982.
18. On the "safety first" campaign and ideology blaming "the careless worker" in U.S.A., see Daniel Berman, *Death on the Job*, op. cit. (note 17).
19. Allen H. Storm, Occupational Health and Safety Programs in the Workplace, pp. 81–90 in Barry S. Levy and David H. Wegman, (eds.), *Occupational Health, Recognizing and Preventing Work-Related Disease.* Boston: Little, Brown and Co., 1983. Also Henry Forbush Howe, Organization and Operation of an Occupational Health Program, revised edition—Part 1, *Journal of Occupational Medicine 17*:360–400, 1975.
20. I have identified ten interwoven elements of regionalization as another ideal-typical model for comparing overall health service systems of nations. See Ray H. Elling, *Cross-National Study of Health Systems, Political Economies and Health Care.* New Brunswick, New Jersey: Transaction Books, 1980, esp. pp. 100–101.
21. An important inventory of symptoms and diseases which should raise a red flag as possibly being work-related is given in David Rutstein, et al., Sentinel Health Events (Occupational): A Basis for Physician Recognition and Public Health Surveillance, *American Journal of Public Health 73*:1054–1062, 1983. See also, Mark R. Cullen and James M. Robins, Diseases of the Workplace: The Clinician's Role, *Connecticut Medicine 45*:363–369, June 1981; also Robert McLellen, Investigating an Occupational Illness: Rationale and Method, 1981, 17 pp (reproduced).
22. Peter S. Barth with Allan Hunt, *Workers' Compensation and Work-Related Illnesses and Diseases.* Cambridge: MIT Press, 1980.

CHAPTER 7
Sweden

The table and discussion in Chapter 3 offered some general comparative information on the six study countries. Sweden is the second smallest in population with 8.3 million people in 1981. It is comparatively large in area, the fourth largest country in Europe, with 174,000 square miles of surface area (450,000 square kilometers). Of this, only 10 percent is farmland.[1] Half this area is covered with forest, one of the major natural resources of the country. Many lakes dot the relatively even countryside. A chain of mountains in the northwest rises to almost 7,000 feet at some peaks (Kebnekaise, 6942 feet). The jagged coast of this long country (977 miles or 1,574 kilometers North to South) stretches for hundreds of miles along the Baltic Sea and is lined with thousands of islands.

For its far northern latitude, Sweden has a comparatively mild climate due to the warming influence of the Gulf Stream. Stockholm, the capital, is at almost the same latitude as southern Greenland, yet it is warm and usually sunny in summer and has average temperatures just below freezing with moderate snowfall in winter. Nevertheless, the northern half of the country can experience bitter cold and winters there can be severe. For this reason, some 90 percent of the people live in the southern half of the country. The major cities are here: Stockholm, the capital, with 1,384,000 population; Göteborg with 694,000; Malmö with 453,000.

Swedes have lived in the territory of present-day Sweden for at least 5,000 years–longer than nearly any other European people. Although, as a nation, Sweden has remained a kingship until the present day, it formed the earliest parliament on the European continent in 1435–the Riksdag, which included some representation from all classes, though of course it was overwhelmingly dominated by the feudal lords. Swedes of the Viking times, and under King Charles XII (reign from 1697-1718) when Swedish armies fought their way into Russia and Poland as far south as the Ukraine, might have been characterized as warlike. But in modern times, Sweden has stayed out of Europe's wars for more than 100 years; thus avoiding the terrible loss of life and capital and some of the disruption which struck the other study countries, particularly in WW II.

The mechanization of farming and the limited demand for manpower in industry created many problems of subsistence during the latter half of the 19th century. Poor harvests and famines resulted in the first emigration wave, between 1868 and 1873, when 3.2 percent of the country's population emigrated. The widespread agricultural crisis in Western Europe and the relatively high level of prosperity in the U.S.A. resulted in the emigration of 540,000 Swedes (11.5 percent of the population) between 1880 and 1893. After this, emigration declined and between 1930 and 1976 was outweighed by immigration.

The country has a low birth rate. Since WW II, net immigration of some 600,000 from neighboring countries—Finland, Denmark, Norway, Germany—as well as from elsewhere in the world has accounted for more than half the population growth. In northern Sweden, the Lapps, an ethnic and linguistic minority, have a tradition of reindeer herding and have migrated freely into Norway and Finland and back, though the advent of the snowmobile and other changes has tended to bring a more sedentary life and some recognition of nation-state boundaries.

In addition to extensive pine forests, natural resources include water power, iron ore, uranium, and other minerals. Products of the sea have also been important historically. The country lacks significant oil and coal deposits.

THE POLITICAL ECONOMY AND WORKERS' MOVEMENT

Cheap water power was a major factor in Sweden's industrial development, and today around 13 percent of the country's energy supply comes from its hydroelectric plants, many of them on the main northern rivers. About 75 percent of energy consumed in Sweden comes from imported oil and coal. Seven nuclear reactors currently provide 27 percent of Sweden's electrical energy. A referendum in March 1980 determined an expansion of the nuclear program to a total of twelve reactors.

The largest iron reserves are in the far north. The high-phosphorus ore containing 60–70 percent iron is mainly exported. Low-phosphorus ore found in central Sweden was formerly integral with the country's special steel industries, which still have a high level of skilled technology and research.

Sweden's vast forests of spruce, pines and other softwoods supply a highly developed sawmill, pulp, paper and finished wood product industry. Despite high domestic consumption, Sweden exports 40 percent of its forest products.

Sweden is a constitutional monarchy with a parliamentary form of government. The King, Carl Gustaf XVI, since 1973, has only had ceremonial functions as Head of State. Parliament consists of one chamber whose members are directly elected for simultaneous three-year terms according to a proportional representation system. Sweden has had universal suffrage since 1921 and the voting age is now 18. The Social Democratic party held power alone or in coalitions from 1932 until 1976, when the non-Socialist parties–Center, Liberal and Conservative–won a majority of the seats in Parliament and formed a coalition cabinet.

In the September 1979 elections, the Center got 18.1 percent of the votes, Liberals 10.6 percent, Conservatives 20.3 percent, Social Democrats 43.2 percent and the Communists 5.6 percent. The non-Socialist parties stayed on in power with 175 parliament seats (174 for the Socialist parties). A new coalition cabinet was formed with the Center party leader as Prime Minister. In May 1981 the Conservative ministers left the cabinet, and the Center and Liberal parties formed a minority government under the same Prime Minister.[2]

Thus during the fieldwork period of this study (Fall 1981) the non-socialist parties had been in power for five years, but a swing back to the workers' party, the Social Democrats, was underway. The Social Democratic Party subsequently regained power and is in power now. Thus for more than forty-five years, much of it continuous time, a party representing workers' interests has been in power in Sweden. This has been a reformist, rather than a revolutionary movement toward socialism. This point is elaborated below.

While deep, continuous cultural roots enrich and strengthen the identity of this people and lend it great cohesiveness, these characteristics could not erase class conflict. As in other capitalist nations, there have been long and sometimes sharp struggles between workers and owners. Industrial capitalism came later to Sweden than to the U.K. and some other countries. In 1870, Sweden had just 9 percent of its labor force in manufacturing and crafts, while the U.K. already had 43 percent in industry. When industrial capitalism developed toward the end of the 19th century, it did not differ basically from capitalism in other countries. "The working people were exploited with the same ruthlessness. The country was marked by mass poverty and severe social conflicts."[3]

The lateness of industrial capital's development in Sweden, along with another special characteristic, its high concentration from the beginning,[4] help explain the development of an especially powerful and well organized working class. It is this power of the working class which has led to a dynamic, but general understanding, with capital, or "social peace" as some observers have termed it.[5]

It is perfectly true that for certain periods, sometimes long periods, Sweden has had fewer strikes and labor strife than most other industrialized countries.[6] But such labor peace has not held sway in all periods. In fact, until the working class, spearheaded by organized labor, consolidated its political interests and achieved an important say in government, when the Swedish Social Democratic Party was elected to power in 1932,[7] labor unrest had been higher and strikes and disputes outnumbered those in any other European country and the U.S.A. Again when the Social Democrats lost power in 1976, labor unrest heated up culminating in the especially bitter miner's strike of 1980 which drew sympathy from much wider labor circles.

> Contrary to prevalent "social peace" and "deference" myths, class struggle in Sweden has been, until recently, very intense. The industrial scene has been frequently characterized by great labor unrest. Actually, until the Swedish Social Democratic Party became the governing party in 1932, labor disputes in Sweden outnumbered those in any other European country and in the United States. The gaining since then of important parcels of political power (i.e., control of the government for forty-four years) gave the Swedish working class an enormous power. Capital was thus forced to put long- over short-term interests and general (class) over specific (enterprise) interests. This is the explanation for capital's collaboration. As soon as the working class weakened, that collaboration declined most dramatically. After the Social Democratic defeat in 1976, employer compliance declined substantially and employer defiance of state interventions to protect workers' safety and health increased significantly. Social peace was replaced, once again, by labor unrest, a situation that characterized Swedish industrial relations until the 1982 election that returned the Social Democratic Party to power.[8]

Access to power through its political arm, the Social Democratic Party, has not always kept labor satisfied. In 1969-1970 wildcat strikes (not controlled by the union or the party) started in the northern mines and spread to affect the sympathies and mood of the entire labor force. This unrest fit within the climate of the times also evident in other Western countries—the anti-Vietnam protests in the U.S.A. and elsewhere; the "Merry Month of May" in France in 1968 when an alliance of students and workers almost toppled the deGaulle government; "Hot Autumn" in Italy in 1968; and the 1968 strikes in Spain. For whatever reasons these events of the late 1960s took place,[9] they had a radicalizing effect and in all our study countries a spate of important social legislation was passed, including important OSH laws. Sweden was no exception and I will come back to these developments. The point here is that underlying the power of the Swedish workers' movement through its party and unions, there has been a wide *class* consciousness and solidarity and an awareness, sometimes brought to the fore through wildcat strikes and other unrest, that a revolutionary, more violent path to socialism is always an option to pursue should the reformist, peaceful path begin to falter. Let us examine the fundamental strategies of the Swedish workers' movement in attempting to establish a socialist way of life.

Three key strategies of the Swedish labor movement have been evident throughout the long struggle toward this aim: a close tie has been maintained between the party and the unions; the movement has been a reformist, rather than revolutionary one; and an effort has been made to enhance the workers' movement as a class struggle, keeping it mass-based as a living popular movement. Thus the spirit has been democratic and issues have been broad, those of life in general—education, health, culture, living conditions, participation—not just the narrow concerns of the "business unions" in U.S.A. which have largely focused on wages, hours and certain aspects of working conditions.

Already at its first congress in 1889, the Social Democratic Labour Party, although formed on a clear Marxist basis, passed a resolution which "rejected the idea of a revolution by violence, unless the rulers of the country themselves provoked it as a 'desperate measure of self-help.' "[10] The aim was to take over the state and thereby control the means of production and the distribution of the surplus which was to be a social and not a private surplus.

The strategy for taking state power distinguished the Swedish workers' movement, from the so-called Erfurt Programme of the German social democrats which otherwise served as a model. Hjalmar Branting, the first Chairman, stressed the importance of the role of trade unions. The distinctive feature was "the heavy emphasis on the close link between the party and the trade union movement."[11]

> In summary, socialism was going to be reached through legislation, upon a mandate of an electorally expressed majority, as the will of universal suffrage. The aim was clear. As indicated by the first program of the Social Democratic Party (1897), "Social Democracy differs from other parties, in that it aspires to completely transform the economic organization of bourgeois society and bring about the social liberation of the working class." Socialists were going to abolish exploitation, destroy the division of society into classes, remove all economic and political inequalities, end the wastefulness of capitalist production, and eradicate all sources of injustice and prejudice. Rationality, justice, and freedom were the guiding goals of the social democratic movement.[12]

While great successes were realized, especially after the Social Democrats took power in 1932, the growth of the working class as a majority never materialized as predicted and the party's hegemony was never complete (even though working class solidarity has been strong in Sweden as a matter of deliberate strategy—the second major point of strategy which is examined below), thus the workers' party was always in power as a minority party. As a consequence, always forced to make alliances in order to wield power, the strategy of reformist socialism was gradually transformed into one of social reform within capitalism.

In brief, reformist socialism, as opposed to insurrectional socialism (identified with the communist movement), was the gradual socialist transformation of society via the electoral process. The goal was indeed socialism. And the socialization (or collectivization or nationalization, terms used with great ambiguity in the Party's documents) of the means of production was considered to be the principal method of realizing socialist goals and hence the first task to be accomplished by social democrats after the conquest of power. Socialization or nationalization was the manner by which socialist revolution would be realized. Reformist socialism was not to be the alternative to revolutionary socialism but, rather, was considered to be the best way to reach it.

Some of the assumptions in the strategic scenario of the socialist movement proved to be wrong, however. One was that the working class, the proletariat, did not become the majority class. The proportion of workers in the electorate grew from 28.9 percent in 1908 to 40.4 percent in 1952.[13] In the interwar period, the expected majority never materialized. Moreover, the labor movement never attained hegemony in civil society. Consequently, not all workers voted socialist. Because of these facts, the Social Democratic Party never became a governing majority party in that period but, rather, a minority party which needed the establishment of class alliances and political coalitions to be able to govern. These alliances imposed the postponement of the long-term aim of socialization. While waiting to become a majority party, the most important tasks for the Social Democratic Party were: (1) to prove to the general electorate that it was a party fit to govern, i.e., a governmental party; and (2) to defend the short-term interests of the working class. . . .

In the world of production and at the workplace, the state should merely regulate the labor exchange and mitigate the damage created at the workplace. The property, ownership, and possession over the means of production and over the labor process should and would continue to be in capitalist hands. In its forty-four years of government, the Social Democratic Party has nationalized only the drug stores (a very minor element in the distributional network of medicaments) and very few other economic units. This focus on distributional policies implied that the state would support and facilitate the privately owned production in order to increase the product to be distributed. As Wigforss indicated,[14] "Because Social Democracy works for a more equal and more just distribution of property and incomes, it must never forget that one must produce before one has something to distribute." Without nationalization of the means of production, the only way to increase productivity was to encourage the profitability of private enterprise. Indeed, within a capitalist system, the primary condition for the distribution of a product is the maintenance of higher profits and of a strong process of capital accumulation. Thus, the possibility for carrying out the social democratic project depends on the profitability of the private sector and on the willingness of the capitalist to cooperate. Higher profits and rationalization are a condition for that distribution to

> take place. Thus, the Social Democratic Party became the party of capitalist rationalization, cooperating with the capitalist class in restructuring the Swedish economy to make it more internationally competitive.
>
> In this new scenario, socialism became a program of social reform within a capitalist system. From being a reformist socialist party, the Social Democratic Party became a social reform party. . . . A major element in the new strategy was the social contract between labor and capital. According to that contract, capital would continue to own the means of production and would control the major elements of the labor process. Labor would bargain for a more equitable distribution of the social product. To that effect, labor would collaborate with capital to increase the productivity of the privately owned capitalist system. Thus, labor would help capital to rationalize the capitalist system. The implications and consequences of this rationalization for the working conditions of the Swedish working class have been many. . . .[15]

In spite of this overall "compromise with capitalism," the second fundamental strategy, working *class* formation, has kept the workers' movement comparatively strong.

With the lateness and concentration of industrial capital's development came the workers' counter ability to organize primarily in the industrial sphere and also in a relatively concentrated manner. Already in 1908, 46 percent of the central labor organization's (LO–Landsorganizationen i Sverige) membership were members of local industrial unions as distinct from craft or mixed local unions. By 1933, 67 percent were in industrial unions and by 1975 it was 85 percent with 5 percent in craft and 10 percent in mixed locals.[16]

When LO succeeded in organizing a nationwide strike for universal suffrage in 1902 (note that this was a class issue on the path toward control of state power, and not simply a workplace issue of the kind which preoccupied the attention and efforts of organized labor in U.S.A., when the broader Debbsian socialist forces finally lost out to Gomperism) the oligopolistic owners were shocked and formed the central confederation of employers (SAF–Svenska Arbetsgivareföreningen). This association waged an aggressive struggle against organized labor until the Social Democrats took power in 1932 and the first basic agreement was reached between SAF and LO at Saltsjobaden in 1938. Pushed by the union-busting efforts of SAF (the central confederation of employers) LO responded by developing greater centralized power of its own, at the same time following through on its primary principle of creating broad awareness, support and solidarity among the working class.

In addition to widespread organizing, two specific strategies were key. First, progressive centralization and coordination of collective bargaining was achieved to cover all workers in the same branch of industry, whether in large or small locals, so that they could not be played off against one another. In short, there could be no givebacks from United Auto Worker locals of Chrysler with which to beat down the workers in the UAW locals of GM and Ford. Second, wage solidarity based on job requirements, rather than productivity or profitability of capital, kept the unifying condition of equal pay for equal work across the economy, regardless of the workers' characteristics (men versus women is a key point divisive of working class solidarity with regard to this matter in the U.S.A. still today).[17]

> The intention is to establish a more equal distribution of earnings, and through this, to retain a homogeneous and united working class movement. Under this policy, there may be some scope for local decision-making. The system of payment is often decided at the local level, but the deviations allowed in a particular branch or company are limited. The goal is class solidarity, uniting labor and avoiding the divisions within the working class that may strengthen the class of employers in its continuous conflict with labor. This has been defined by a major theorist of the Swedish labor movement as an expression "of the socialist movement's primal ideology of equality, which has given the Swedish labor movement much of its strength, vitality and autonomy."[18]

By contrast, much of collective bargaining in U.S.A. is carried out at the very local level and the focus ("business" versus "uplift" unionism)[19] is often on getting the best wage and hours deal possible for one's own members, regardless of what else is happening at the workplace (OSH hazards, production speedups and stress or whatever) regardless of what is happening to one's brothers and sisters down the street in the same kind of (but a competing) shop and regardless of what is happening in terms of health, education, welfare, culture, and politics broadly in the society.[20]

From its position of class-based strength, the Swedish workers' movement has been able to pass labor laws which are conducive to wide success in organizing. Locals are organized on a closed shop basis, that is, if there is a union, and there almost always is, then all non-management workers in the shop belong. They are also organized in a single union. If it is a steel mill, for example, the draftsmen and secretaries will all belong to the steel workers' union.[21]

One can argue to what extent the aim has been accomplished. Clearly, the "take over" of the state and of the means of production have never been completed. Four decades of Social Democratic rule ended in 1976 with the election of a conservative coalition government, though the Social Democrats are back in power today. And 95 percent of the economy is in private hands. In recent years, as the economy turned downward, along with most other capitalist economies (in 1981 Sweden showed a 1.1 percent decline in gross domestic product, while the Nordic nations on average showed -0.5 percent) unemployment rose to 2 percent in 1980 and probably reached 2.7 percent in 1981.[22]

Whatever the society's shortcomings and the weaknesses of the workers' movement in Sweden, it has to be said that much has been accomplished across all of social life and the workers' movement must be evaluated as very strong indeed in comparison to what one finds in the U.S.A. and F.R.G., with the other study countries somewhere in between. On the first point, the level of living, a great deal could be said. Comments on health and health services will be made in the next section. In general, it is probably adequate for present purposes to say that Sweden, with $11,340 Gross Domestic Product per capita in 1980, and income comparatively equally distributed (since the taxes which on average are more than 50 percent of income, are steeply graduated) enjoys one of the highest levels of living in the world and has one of the most complete sets of social welfare provisions.[23]

On the second point, the strength of the workers' movement, perhaps it would be enough, in addition to the background just presented on its class-based strategy, to cite current estimates of the proportion of workers organized into labor unions—from 95 to 98 percent of blue-collar and some 75 percent of white-collar workers! The confederation of blue-collar workers, the LO (Landsorganizationen i Sverige) represents some 2 million workers (2,140,814 in 1981) organized in twenty-five national unions and some 1,800 local and regional unions which in turn are made up from some 10,000 trade union clubs in the workplaces. Since each union is formally autonomous, the confederation can only recommend that agreements of the confederation be accepted.[24] Nevertheless, because of the need to counter the oligopolistic moves of capital, the combined influence of the confederation and internal politics between unions and the confederation has increasingly led to solidarity and coordinated support for the agreements which have been reached. The arrangement for white-collar and professional workers is somewhat different, but generally the PTK (Privattjanstemannakartellen) reaches agreements with the SAF and recommends these to the twenty-one white-collar unions representing over 1 million workers (1,062,366 in 1981) and twenty-six professional associations representing some 225,000 workers. Third, this direct negotiation nationwide has entailed a comparatively small role for the government and law in determining relations of production. This point will be recalled when we take up the present situation of OSH as regards small plants.[25] Fourth, this concerting of Labor's power (not to say concentration, for this is not the same and does not appear to be the case) has enabled the working person, through their union representatives, to have a strong voice on the development of social welfare legislation, and all other areas of government concern including foreign policy.[26] Thus LO and PTK have representatives on every kind of commission and committee at all levels of government, and, as will be noted, the local levels are of great significance.

In conclusion of this brief presentation of the Swedish political-economic context, it remains to reinforce and elaborate the last point concerning the importance of the local level in the authority and governing structure of the country. The fourteen national ministries are rather small units mainly concerned with preparing new government bills for submission to the Riksdag.[27] Implementation and enforcement of laws is carried out by about eighty relatively independent central administrative agencies (ASV, Arbetarskyddsverket, The National Board of Worker Protection or Occupational Safety and Health is one of these) and twenty-four County Administrations. Each county has a popularly elected council which is empowered to levy an income tax which is mainly used to cover some 75 percent of hospital and medical care costs. There are some 280 municipal districts, each with its own popularly elected council. These bodies too collect an income tax mainly used to operate schools, child and old-age care, utilities, housing, and cultural and leisure activities.[28] Thus many of the most important decisions of everyday life are reached in county and municipal councils, even if these are taken within a framework of national laws. One could go on to discuss the many voluntary sport and cultural bodies at local level. In general, Sweden's authority structure can be characterized as comparatively decentralized yet concerted.[29]

THE HEALTH SYSTEM

Voluntary health insurance operated by local societies existed in Sweden from the mid to late 1800s. It was made compulsory for certain worker groups in 1931. In 1962 compulsory health insurance coverage became universal. Thus all Swedish residents are protected by national health insurance which covers hospital and medical care and sickness pay.[30] If a person is ill, or must stay home to care for a sick child, he or she receives 90 percent of lost income, up to a ceiling of $22,000 annual income (Skr 120,700).[31] While this arrangement is very generous and goes a long way toward making needed care available on an equal basis to those least able to pay and to those most in need, there are certain self pay or deterrent pay items the patient must cover. For example, a visit to a public salaried physician employed at a hospital or other public outpatient facility costs the patient $4.50 (Skr 25) paid to the County Council. A visit to a private physician costs the patient $5.50 (Skr 30) the rest of relatively standard, set fees being paid to this kind of practitioner by the insurance. Patients pay no more than $7.25 (Skr 40) for any one course of drugs. Thus, for example, a year's supply of "The Pill" for contraception would cost this amount. Patients pay more for dental care—50 percent of any amount in a single treatment period up to $550 (Skr 2500). But dental care is provided free through age sixteen (and in some locales through age seventeen, eighteen, or nineteen) by the national dental service.[32]

In general, the system is publicly financed and controlled. Only about 5 percent of physicians are in private practice. These receive 17 percent of visits to physicians, while 30 percent are to public physicians employed in district centers, and 53 percent are to public physicians in hospitals. Patient payments provide just 4 percent of total health system costs while 60 percent come from County Council taxes; 15 percent from general State subsidies; 9 percent from the national health insurance system; and the remaining 11 percent from State grants including those for medical education, research and psychiatry. While one might classify the Swedish health system as a National Health Insurance (NHI) system, on a chart classifying nineteen industrialized systems according to degree of government financing and control, Maxwell places it immediately after the U.K. which follows the U.S.S.R.[33] Thus, as with the other two, it should probably be seen as a National Health Service (NHS) system.[34]

While there is a high degree of public financing and control involved in any NHS, in the Swedish case it is important to underline the high degree of regionalization and local control within a national framework.[35] The national framework, as in other areas of public life is provided by two agencies–one the legislative development body, the Ministry of Health and Social Affairs which is part of the cabinet; and the other, a relatively independent administrative arm, the National Board of Health and Welfare. But the real functioning structure of health services delivery is the regionalized one. There are seven regions each serving on average something over 1 million population.

Ideally, the most important level of this structure is the primary care level, or point of entry into the system. Unfortunately, this is somewhat fractionated, being made up of the three kinds of points of first contact cited above (hospitals, district physicians, and private physicians) plus industrial medical services (now covering some 50 percent of Swedish workers–more on this in the next section) school health services, and certain elder care institutions and programs. Serious efforts are now underway to reorient these regional structures toward greater support to a more rationally organized form of primary care as there has been too much emphasis on hospital care.[36] There may be room in these efforts to reconsider the place of OSH services in relation to PHC. I will come back to this point more than once. While there would seem to be many grounds for moving toward a locally-based public organization for PHC (other types are hospital-based public PHC; regulated private PHC; and unregulated private PHC),[37] one important strength of the present system is its orientation toward and responsibility for defined local geographic areas ranging in population from about 5,000 to 50,000. Overall responsibility for this level and much of its financing is in the hands of the 275 municipalities which themselves range in population from about 5,000 to 700,000. As noted, this strength is qualified by the uncoordinated plural points of entry to care.

The second, most fully developed level of the regional systems is the district and county hospitals, run and financed almost entirely by the County Councils of which there are twenty-three plus three which are not part of a council area–Göteborg, Malmö, and the island of Gotland. This general picture of major County responsibility for the hospital and medical care system has held for the last 100 years. The district hospitals serve from 60,000–90,000 inhabitants and have at least four specialties—internal medicine, surgery, radiology, and anesthesiology. The one central hospital in each county (a couple have more than one central hospital) serve between 200,000 and 300,000 inhabitants and offer from fifteen to twenty specialty services.

The regional hospitals receive the most complex cases on referral and have all specialties. If the Örebro hospital is eliminated as a regional unit, as recommended in a government study, there will be six regional hospitals. These are in Umeå, Uppsala, Stockholm, Linköping, Göteborg, Lund-Malmö. These hospitals are generally regulated and financed by agreements among the counties, though on the day I visited Uppsala (September 17, 1981) there was a massive protest meeting by staff at all levels against announced central government budget cuts—this being the only Swedish hospital still supported centrally. The cuts were being proposed by the Conservative government of the time under the rationale that PHC needed more emphasis. But the hundreds of protesters complained that there was no evidence that the government was giving more money to PHC; thus, the move was simply a disguised cut, but, a cut nevertheless.

The weight or thrust of this system lies at the County, and increasingly the local level; thus there is much room for local involvement and determination. But, since it is a NHS, there is a back and forth between the center and the local level. Some recent developments designed to place greater emphasis on and rationalize the periphery illustrate this point:

> The national government and the county councils have in these days made an agreement regarding the payment from the state to the county councils for the health care service. Earlier the county councils were payed a certain amount for each patient they treated (a sort of fee-for-service system between the state and the county councils) in addition to the money the county councils get through taxation. The new system says that the county councils will be payed only on the basis of the number of individuals living in each county (capitation fee). To this is coupled a regulation of the private doctors. Till now there has been no regulation of the establishment of private practitioners. The new thing is that in order to start a private practice you have to come to an agreement with the county council, and the number of private practitioners must not increase in the big cities (where there is a surplus of physicians). The establishment of private practice in the sparsely populated areas (where there is a shortage of physicians) is free.[38]

Much more could be written about the Swedish health system. It should be mentioned that the level of health by all indicators is very high,[39] though most would agree that this has more to do with the level of living than with the health services *per se*. The oft-cited high suicide and alcoholism rates in fact are not as high as in some other countries.[40] This does not mean these are not problem areas. It should also be mentioned that the health services themselves are as expensive as any in the world. In part this relates to the heavy concentration on hospital services and to the aging of the Swedish population,[41] though the cost issue is much more complex. Today the health services take over 9 percent of the GNP, about Skr 30 billion, as compared with some 3 percent in 1960. Thus efforts to further rationalize the system include this concern, though an increased PHC thrust may actually cost more.[42] But the Swedes do not spend much on military defense in comparison with many other countries,[43] and their commitment to social welfare has been strong, so such costs for health care may not be of deep concern (though I noted above the cut-back atmosphere and resistance in one nationally-supported regional hospital). But my major purpose here is not to lay out the Swedish health system in detail. My hope is to present enough of its general structure, and character to serve as a background for understanding the OSH-PHC relationship. In discussing this relationship in the following subsections, other aspects of the health system will be brought out—for example the preparation of general physicians and occupational physicians.

THE OSH SYSTEM AND SERVICES

The History of Swedish Occupational Safety and Health (OSH) Services

The first OSH legislation in Sweden was drafted nearly one hundred years ago. It was aimed primarily at preventing accidents on the job. Today there is legislation which covers health conditions and services and ergonomic and psycho-social aspects of work environments. But as we will see, legislation has usually played only a confirmatory role, the main developments being those reached in negotiations between "parties on the labor market."

At the turn of the century and before—the 1880s and 1890s—there was little of the present national health system. True, the County Councils were already supporting their hospitals. But much of Swedish industry was in isolated areas—the mines, lumber industry and paper mills. The biggest ones among these industries began to hire physicians to treat accidents. These then began to offer care to workers and their families in the community, since sometimes they were the only medical-health personnel in the locale. Several aspects of this early experience have been carried forward into today's forms of organization. First, large plants continue to have their own OSH services, while medium and small plants are covered through combined services (public or private OSH service centers reaching several plants) or are not covered at all. Second, OSH services have continued to stress medical care, that is, personal health services as distinct from (if not opposed to) preventive care. Third, the large-plant industrial medical services may still serve as a reference point, at least in isolated areas, though this last aspect has declined as the NHS has become more comprehensive and reached all of Sweden. Still another, a fourth, characteristic implied by these early arrangements has carried forward, that is, that the OSH services are voluntary and private. To this day there is no obligation by law to provide OSH services. This aspect has been under reconsideration through the so-called, Wessman Commission (after Yvonne Wessman, Executive Secretary of the Företagshälsovårdsutredningen). In early 1984 this Commission issued its report. There had been many delays and a split vote, with labor representatives and the Head of ASV favoring organization of OSH services by law. But a majority was against making OSH services a requirement of law as has been done in Finland, so there will be little change and no government action, at least not in terms of new legislation. In any case, OSH services have been greatly expanded and extended under collective bargaining, though there is still a serious problem (as in most other countries) of coverage for small workplaces. Finally, a fifth characteristic, paternalism, still lingers in certain ways, but has largely been done away with under the increasing pressure from organized labor to have a determining voice in conditions and relations of production, including OSH services and the hiring and firing of OSH physicians and other personnel.

The bargaining between labor market parties reaches back to "the December Compromise" of 1906 between the SAF and LO. This was a victory for labor in that it recognized their right to organize. But many of the gains were lost in the General Strike of 1909 when several unions had to disband. Nevertheless, before the general strike, relations between LO and SAF were almost always conflictual. Subsequently, these two major powers in the labor market more or less stopped trying to erase each other and more often carried on negotiations.[44] The first Social Democratic government assumed power in 1932 and trade unionists now felt they could depend upon the government for supportive measures, for example, against unemployment. Following the strikes and struggles of the Great Depression years, the first basic agreement between SAF and LO was signed in 1938 in Saltsjobaden (named after the seaside resort near Stockholm where the document was signed). Both labor and management were anxious to ward off legislation which might have restricted their powers to bargain directly and freely. This agreement outlined a negotiating procedure to deal with disputes in-so-far as possible without open conflict and it prohibited a number of direct actions on both the parts of management and the unions.

> However, this agreement says very little about other working conditions, which were left for the employer alone to decide. Here the trade unions took a defensive position: they reserved the right to react to obvious disamenities such as safety hazards, but they did not lay claim to active participation in shaping work practices, technical equipment, organization and environment.[45]

While no specific measures were adopted in the 1938 agreement, the Employer's Confederation (SAF) from that time on engaged a medical advisor. In 1950 the noted authority, Professor Sven Forssman was appointed to this position. It was not until 1964 that LO employed medical expertise of its own in the person of Dr. Erik Bolinder.

The controlled economy of WW II involved great strains toward increased efficiency and production. Incentive pay schemes and speed ups coupled with greater mechanization and rationalization were accepted reluctantly and were often seen as repugnant.[46] A number of organizational forms were adopted at this juncture to enable continued cooperation between employers and trade unions. Joint committees were one such mechanism. These enabled employees to keep informed about planned changes in technical equipment, environment and organization. These committees did not change the basic decision power of employers confirmed in the 1906 "compromise." This is (from labor's point of view the infamous) paragraph 32 of the statutes of SAF which says it is the right of the employer to employ and to dismiss workers and to conduct and to divide the work.[47]

As part of this same WW II atmosphere, in 1942, SAF and LO concluded their first agreement explicitly on occupational safety, including formation of the Joint Industrial Safety Council. Today, this Council consists of a board composed of representatives from LO (blue-collar workers) PTK (white-collar workers) and SAF (owners and managers). This board acts through a secretariat financed by the labor market parties to train union safety delegates and company supervisors, conduct some studies, and otherwise provide information on working environment matters. Today these activities are carried out within the framework provided by the Working Environment Act of 1978 which revised once again the Workers' Protection Act first passed in 1949. As is traditional in Sweden, the impetus to revise the law in 1978 came from the very important Working Environment Agreement of 1976 negotiated between SAF on the one hand and LO and PTK on the other hand. In this agreement, it was determined, among other things, that employees should have a majority on the works safety committee and that it should be the central body for environmental matters in the work place, including medical and technical (engineering) concerns, though employers retained a veto or final approval of decisions involving monetary outlays.[48]

To reach this present stage, to which the next subsection is devoted, various steps were taken following the WW II period. As mentioned, the Workers Protection Act was passed in 1949.

In 1954, the SAF adopted some "Guiding Principles for Company Health Services." This document identifies twelve duties of the company medical officer including:

> He shall perform his work as company medical officer in accordance with "scientific and proven experience." He shall seek to maintain a good state of health among the employees and bring them back to good health as quickly and efficiently as possible after injury or sickness, paying regard to socio-medical considerations and the

> nature and proper interests of the company. He shall seek to work in a spirit of cooperation with other doctors treating the employees to the benefit of the sick or injured person.[49]

From the timing of this statement and reading between the lines of the document itself, one can notice some concern for competition, or at least uncertainty, as regards relations of company services to the developing general health services for which the national health insurance law was to take effect on January 1, 1955. The document even states:

> This medical care should be provided by the company medical officer without a fee from the company, being paid by the patients themselves, if no other agreement is reached. It is useful if all details concerning the state of health of the employees are centered on one doctor through his treating them, although the right to a "free choice of doctor," which is also included in the health insurance scheme, offers significant psychological advantages. If the company's medical care is of a high standard, it can be expected from experience that even given the free choice of doctor more than 90 percent of the company's employees will use the company doctor, if only on account of the short waiting times, ready availability, etc. The overview of the health of the employees and the other experience that is acquired in this way by the company doctor is particularly valuable.

In 1976, a new agreement on local industrial safety activities was reached and guidelines for company health services were drawn up. These appear to have drawn heavily on the 1954 document of SAF, perhaps reflecting SAF's dominance as regards medical OSH expertise for some years before as noted.

In 1974 the three major parties (LO, PTK, and SAF) signed an agreement covering insurance for occupational sickness, injuries and accidents. Full compensation was assured after ninety days, regardless of income level. Recall that 90 percent of salaries up to a limit (Skr 145,000 per year in 1984) is provided for non-work related sickness under the general sickness insurance. If a person's pay is more than this, he or she will still receive only 90 percent of the limit. After the ninety days, if the person's disability is accepted as work-related, 100 percent of the normal pay is received, regardless of its level.

This is a far cry from the situation in Connecticut, one of the more liberal states in U.S.A. There is usually some provision for continuance of ordinary pay for a period of non-work-related sick leave. But most often these sick days must be accumulated through work—so many work days for every allowed sick day up to some limit, often thirty days. If there is no union (usually the case, since only some 15 percent of workers are organized) there may be no sick leave provision at all or it will vary widely from one place of employment to another. Non-union employers exposed to being unionized sometimes offer a better package of benefits than a union shop in a competing or comparable labor market as a way of fighting against a union being voted in. Whatever the case as regards pay during ordinary sickness, in general it in no way approaches the uniform 90 percent up to about $20,000 (variable in dollars according to exchange rates) which the workers' movement has achieved in Sweden. Compensation for work-related disability in Connecticut is limited to two-thirds of pay up to the average weekly wage which in 1984 was set at $340.[50] Disregarding variation from time to time due to exchange rate, in the first ninety days the work-disabled Swedish worker will receive up to 90 percent of about $20,000 or $18,000. The most a Connecticut worker could receive would be two thirds of $17,680 (52 x $340) that is, $11,785. Furthermore, even at this meager level there is great suspicion of disabled workers on the parts of companies; most cases are fought hard; and a campaign is underway to cut these costs back, especially since compensation for work-related disease is being increasingly recognized, claimed and awarded whereas previously the system mainly compensated injuries due to accidents, fires, explosions, etc.[51] After ninety days, the Swedish worker receives 100 percent of his or her income if the condition is accepted as work-related, regardless of income level!

The major effect of the 1976 Working Environment Agreement[52] and the 1978 Working Environment Act was to give employees greater decision making rights about their own work environment and it brought the white-collar workers (PTK) in as a more formal partner. This greater worker control was certainly necessary, for, among other reasons, there had developed some of the same lack of worker trust in company medical services as one finds in the U.S.A.[53] and other countries.[54] Also, it came as part of a general push for greater democratization of the workplace, continuing labor's attack on the notorious paragraph 32 cited above. The most important of the steps in this general direction was the Act on Employee Participation in Decision-Making which came into force in 1977. It contains fundamental rules on the right of association and negotiation, right to information (which relates to the workers' right to know about hazards under the Work Environment Act) and the unions' right to veto of certain matters (which relates to the employee members of the plant safety committee having right of approval of plant physicians to be hired, etc.). This movement continues, though it may be slowed by more difficult economic times. But at its 1981 Congress LO called for joint ownership control of companies through Employee Investment Funds.[55] With the Social Democrats coming back into power in 1982, this call is becoming a reality and nothing has frightened the bourgeoisie more in recent years:

> On a Tuesday morning in October, about 75,000 people gathered here for one of the biggest political demonstrations in Sweden's history. They were an unlikely crowd for such a rally, mainly with business and banking backgrounds.
>
> The target of their anger was a mouthful to pronounce lontagarfonder, considered the hottest issue to hit Sweden in many years.

> It is a government plan for "wage-earner funds" that are intended to give workers through their unions, a share in the ownership of companies that are privately held.
>
> Sweden's affluent business community has struggled against the plan for a decade arguing that it is the leading wedge of state control of industry.
>
> That view is rejected as nonsense by spokesmen for Prime Minister Olof Palme's Social Democratic government, who see creation of the funds as a limited step that will encourage workers to restrain their wage demands and will encourage investment to protect jobs . . .
>
> As perhaps the world's leading welfare state, Sweden is also the most heavily taxed, with nearly 70 percent of the gross national product devoted to education, health and social security.
>
> But the country is a model of successful capitalism, with virtually all business in private hands, including such internationally respected names as Volvo and Electrolux.
>
> The wage-earner funds would alter the balance of power between management and labor by the end of the decade. They are to be funded from a new tax on excess corporate profits and a small payroll levy—one of the major sources of employer dismay.
>
> By the time they reach full size in 1990, according to Edin, the five funds around the country—receiving about $50 million each annually—will own 5–6 percent of the country's industrial and commercial shareholdings.
>
> That would make the union-controlled funds more than twice as big as Sweden's next largest institutional holder, Ab Skandia, an insurance giant.
>
> While this would make the interests of labor in the nation's enterprises a considerable force to be reckoned with, the legislation specifically prevents the unions from owning more than 49 percent of any company.
>
> Moreover, the funds' earnings are earmarked for use in the expensive Swedish pension system so that they cannot be recycled.
>
> Leaders of the Swedish employers' federation and the trade union organization are agreed that the dispute has undermined the traditionally cooperative relationship between them, enshrined in thirty-five years of national wage talks, which broke down this year.
>
> "There isn't a single member of the bourgeoisie in Sweden who likes these funds," an economist said.[56]

Throughout this history, the OSH laws which often followed and confirmed the agreements negotiated between parties in the labor market also brought certain state apparatus to facilitate the agreements and provide a work inspectorate (like OSHA in the U.S.A.) and a research component (like NIOSH in the U.S.A.). These developments are summarized in the following taken from a National Board brochure.[57]

> *From the Occupation Hazards Act to the Work Environment Act* – 1889 can be taken as a starting point in tracing the development of work environment questions in Sweden. That year the

> Parliament passed the Occupational Hazards Act. Three labor inspectors were appointed the same year. In 1912 new legislation was passed: the Workers' Protection Act. 1949 saw the foundation of the National Board of Occupational Safety and Health, at the same time as a new Workers' Protection Act came into force. This Act remained operative until 1978, though several amendments were passed in the meantime. In 1966 The National Institute of Occupational Health was established. The closing years of the 1960s saw an awakening of great interest in matters concerning the occupational environment that led to the appointment, in 1970, of the Work Environment Commission. In 1972 the National Institute of Occupational Health merged with the National Board of Occupational Safety and Health. In 1974, following proposals by the Work Environment Commission, amendments were made to the Workers' Protection Act. Among other things, safety delegates were given increased power and the National Board of Occupational Safety and Health and the Labour Inspectorate were reorganized. In 1978 the Work Environment Act entered into force. The final report submitted to the Government by the Work Environment Commission in 1976 included the draft text of a completely new enactment—outline legislation to be supplemented by regulations issued by the National Board of Occupational Safety and Health.

One of the not well known aspects of this history is the part played by dissident workers, students, physicians, lawyers, scientists and others. For some years before 1976, a socialist-oriented organization, SARB, joined these several interests in various actions for improved OSH.[58] The group published a book on hazards and recommended standards[59] and worked with local groups of workers on chemical and other hazards. According to one knowledgeable interviewee, they had started to get into questions of plant closings, worker take over and alternative production before they disbanded. Perhaps they had less longevity as an organization than similar groups in Denmark and Finland and the COSH groups in the U.S.A. because organized labor was not open to them and was itself quite strong and successful in achieving OSH gains by following the cooperative, reformist spirit of the Saltsjobaden agreement—enough to achieve the quite advanced agreement of 1976 and law of 1978.

Whether the Saltsjobaden spirit can prevail and continue to serve adequately to gain improvements in OSH and other spheres of concern as the crisis of capitalism continues and heightens remains to be seen. Plant shutdowns continued to rise markedly between 1960 and 1970,[60] and the supposition is that this will continue as production shifts within Sweden and the world-system.[61] Following the massive strikes and lock outs of May 1980 when some 800,000 workers were out across the whole LO/SAF sector over unsatisfactory wage hikes in relation to inflation, some people said the Swedish experiment with "freedom under responsibility" (the Saltsjobaden spirit) had ended.[62] To me it seems too early to say this, but it also seems problematic how long this spirit can last. Thus in future we may see groups such as SARB coming back into the picture if organized labor can not win significant battles against the alignment of employers.

The Current OSH System and Services

The concept, work environment, covered in the 1976 Agreement and the 1978 law is a good deal broader than traditional OSH. It is at least as broad and perhaps more authentic than the concept of "humanisation of the work world" (Humanisierung der Arbeitswelt) which the F.R.G. has adopted as a policy direction.[63] This is mentioned here for comparative purposes. It will be taken up again in the F.R.G. case (Chapter 10). This idea means fitting work conditions to the requirements of men and women, not the other way around. But little is said in F.R.G. about democratizing the workplace and psychosocial factors at work as these are affected by control over the work process (and not the worker's home life, etc.). Thus in Sweden one finds very serious attention being given not only to the more familiar accident hazards, to chemicals and physical health hazards generally, and to the more recently identified field of ergonomics,[64] but also to worker participation, self and work-group determination at work, effects of piece work, shift work, incentive systems, and psychosocial factors related to physical and mental illness. Some very significant research has been done along these lines, for example, linking lesser job decision latitude (worker control) to coronary diseases.[65]

This breadth of concern for conditions at work related to health, and one might say well being, appears to be reflected among leading figures in all spheres of interest whom I contacted through interviews. Thus one can find that the chief safety engineer in an outlying OSH Center serving several small industries is especially concerned with ergonomics questions, though he is not ignoring slippery floors and fire and explosion hazards. The trade unions' programs give explicit attention to psychosocial concerns as a part of the push for democratization as discussed above.

The head of the Medical and Social Unit in the Supervision Division of ASV (the National Board of Occupational Safety and Health) has staff in occupational medical sociology and psychology and is himself a psychiatrist with well rounded concerns for all sorts of work hazards. The safety representative of the Employers Confederation with whom I spoke also showed an awareness of ergonomic and psychosocial factors, though the conception here may be more individually than collectively oriented. Also, there is concern over a steadily rising absenteeism from work. But as Gardell, the most widely recognized researcher and theorist in this field puts it:

> Preventive measures must be directed towards different sections of society and working life. It is no good taking measures solely or principally at individual or group level, as many still imagine. It is

> believed it will help if one includes psychologists in firms' health departments. This may be justified, but obviously the structural aspects of the problem cannot be solved at that level.[66]

It would be fascinating to further inquire into this broad ennobling conception which is evident on the Swedish OSH scene and the concrete studies and steps which have been taken to pursue it,[67] but I must come back to laying out the realities of the present OSH system and the OSH services within it before passing on to consider links with PHC.

The base of the system is formed by the 110,000 or so safety delegates elected through their union structures within large plants and some 1,400, so called, regional safety delegates to cover employees in smaller plants. These more than 110,000 safety delegates have each received at least forty hours of instruction organized through the Joint Industrial Safety Council. As the reader will recall, this Council is controlled by a board made up of representatives from SAF, LO and PTK. While the employer remains legally and financially responsible for the work environment, considerable was done by the 1976 Agreement and 1978 Act to place workers on an equal footing in the decision process.

For example, the employee-elected safety delegates are empowered not only to interrupt work that poses immediate danger to the employee, but also to stop a production process which involves long-range hazards, until an inspector from one of the twenty district offices of ASV arrives. According to informants, this power has not been used often. But it has been used and it is there as a strong motivating force to aid in cleaning up the workplace. These delegates are elected for three-year terms. There must be one at every place of work with five or more employees. He or she takes part in planning for new equipment and processes as well as overseeing current operations. These duties are done during paid work time and it is up to the delegate to decide how much time is needed for these duties. Also, any employee has the right to leave their own work to confer with the safety delegate.

Also every place of work with fifty or more employees must have a joint labor-management safety committee on which a majority are employees. Since almost all Swedish workplaces of this size are organized, this means that employees function on this committee not as individuals exposed to management favors or retribution, rather committee members function as representatives of an organized workforce. In some instances there may be poor representation, especially where there is no union, a weak union, there is special competitive pressure, perhaps from a similar company in another country also producing for the world market, or the individual union representative does not show solidarity with his or her union brothers and sisters. Other members include the employer's representative, and occupational health and safety staff. This committee draws up guidelines for the firm's work environment and supervision and is also responsible for the occupational health.

This means among other things that the workers usually have a majority voice over the hiring of plant physicians and safety engineers who are paid by the employer. Those interested in an announced position submit their applications to ASV and the three or four best ones available at the time are forwarded to the company. Minimum training standards, including short courses given by ASV have been spelled out and are detailed below. The medical and technical parts of the service often still function too independently of each other and there is often too much medical dominance. But the 1976 Agreement gave explicit attention to correcting this.[68] Also, the OSH service director for the plant must present his annual report and plans for the next year to the committee. Since one half the funding for the service is paid by the National Health Service for acute care services (which many argue orients these services too much toward care rather than prevention) and the safety committee must approve the plan before payment is received, this is a powerful voice.[69] However, this voice is one of general guidance and direction and not one meant to influence detailed case decision-making. In fact it is expected and explicitly agreed that the physician and engineer and their staffs are expected to act freely according to scientific and professional standards. If there is disagreement on the committee, outside experts can be called in, usually from the Labour Inspectorate of ASV.

The Labour Inspectorate is the field arm of ASV, also referred to as the National Occupational Safety and Health Administration. It covers the field operations of both the Board (which has research, training, information, and standard issuing responsibilities) and the Inspectorate *per se*. The National Board or OSH Administration is an independent agency (that is, funded by government, but is not subject to political appointments as with Cabinet changes) with SAF, LO and PTK represented as well as members of parliament and others on its governing board. There have been no women on this board. There is a staff of some 1200 and an annual budget of some Skr 160 million. Limits have been set on some 120 chemical substances and the list is regularly revised. Unions recently demanded more standards and more stringent standards.[70] In general, Swedish standards are more protective of workers than those in U.S.A. For example, the noise standard is 85 db while it is more than twice as high in U.S.A. at 90 db (since there is a doubling of sound at about every 3 db).[71]

Many more things are spelled out in the 1976 Agreement regarding the OSH services, including their responsibility to cooperate with public health authorities and general health services and hospitals.

A recent case of an occupational physician being fired illustrates some of the complexities of the specialty OSH services and the complex of forces surrounding them.[72] When asked about it, another expert occupational physician said this person overstepped his boundaries and started acting like the Labour Inspectorate. "He said to the owner, 'You must do this and this to clean up this factory and properly protect the workers.' So the owner said, 'You go to hell!' "

In this case, the union members on the safety committee went along with the management in the firing. Some dissident union and professional people (former SARB members) said this is because the union in this plant is threatened by present economic circumstances and has been convinced by management that they face stiff competition from a similar industry in Germany and that the old non-choice between "your job or your life" must once again be made, for additional expenditures on OSH could lead to the plant closing. One side says the physician involved has a history of being emotionally unstable. The other side says he is known for his outspoken ways and political commitment. No one said he is poorly trained in OSH or incompetent technically.

Except for a very small number of university hospital-based experts in OSH (these are located only in a few regional centers–two in Stockholm, one in Lund, one in Örebro, and one in Uppsala) there is nothing like a certified OSH specialty in medicine in Sweden. The Wessman Commission which has recently studied OSH services (Företagshälsovårdsutredningen) has suggested two kinds of specialities–one for all regional and district hospitals to be primarily clinical and diagnostic (as is now the case with the few hospital-based specialists–though they only consult with other specialists and do not control beds) the other to be active in workplaces and be more preventive oriented. This latter group of OSH physicians would be based in large plants or in centers serving several medium to small plants. If such a recommendation implied more university involvement as some think it should, it would have to be taken up by the Commission on Professional Education (Oversynsutredningen) of the Board of Higher Education which issued a report in November 1982 stressing the importance of social sciences in medical education.[73]

To qualify in one or the other way as an OSH physician as things stand now, following the European kind of high school (perhaps equivalent to two years of college in the U.S.A.) a person studies five and one-half years before taking the final qualifying exam to be a physician. Then two years on internship and four years of study and practice lead to specialization in internal medicine, surgery, etc. There is no specialty in OSH. There are the few hospital-based experts plus a small number of noted authorities with long experience like Drs. Forssman, Bolinder and others who work for various central bodies, the National Board, etc. In 1959 a course was established for the few company doctors who could take it in periods over a year. This development has been elaborated and become more systematic under ASV's guidance since the 1976 Agreement:

> The Occupational Health Advisory Committee has declared that only personnel meeting the formal requirements which it has recommended should be employed in occupational health services. The Committee's recommendations refer to basic training, conditions of service and specialist training for occupational health physicians, nurses and physiotherapists. The Advisory Committee has also been

anxious to issue recommendations concerning the qualifications of safety and hygiene engineers. A number of efforts have been made to this end, but so far without success.

2. *The training of occupational health service personnel*
Sweden has specialized occupational health training for occupational health physicians, safety and hygiene engineers, occupational health nurses and occupational physiotherapists. These categories begin by training as doctors, nurses, physiotherapists or engineers in the normal way at university, college or (in the case of certain engineers) upper secondary school, after which they can undergo specialized training for occupational health services. Specialized training of this kind is arranged by the National Board of Occupational Safety and Health, which is the central authority responsible for occupational safety and health in Sweden. The Board includes a Department concerned with training and research, and this Department is responsible for specialized occupational safety and health training. . . .

2.1 *The training of occupational health physicians*
The Occupational Health Advisory Committee has recommended that holders of occupational medical appointments (i.e., occupational health physicians)

- should be registered as physicians in Sweden (which requires five and one-half years' training and one year's practice at the university),
- should be qualified general practitioners or else hold specialist qualifications in general internal medicine "or some other adequate field" (four to five years' practice and training) and
- should have completed a course of specialized training organized by the National Board of Occupational Safety and Health.

The occupational health physician training organized by the National Board of Occupational Safety and Health comprises roughly nine weeks' theoretical instruction. Physicians not already employed in occupational health services on commencement of their training also have to complete six months' practice of this kind.

The training is spread out over a period of ten or eleven months, and the week-by-week content of the course is as follows.

1. The aims of occupational health services, their organization and their position in society and enterprise. Legislation, insurance and collective agreements.
2. Occupational medicine.
3. Mental and social aspects of the occupational environment.
4. Occupational physiology.
5. Occupational hygiene. Medical statistics.
6. Epidemiology, occupational medicine.
7. Ergonomics. Socio-medical aspects of the occupational environment.
8. Field trips.
9. Additions to the above programmes, e.g., reports on field trips. Specialized sectoral questions or problem spheres.

About ninety physicians are admitted to this training every year, and over 900 have been trained so far. The great majority undergo the training after obtaining occupational health service appointments.[74]

By 1984 there were some 1500 Swedish physicians, many with ten to fifteen years in general practice, or another specialty, who have taken the basic course offered by the National Board. I believe this is an eight to twelve hour course spread over one year. If I have this correct, the in-plant medical "specialists" in OSH have had less training in their special field than the 110,000 union safety delegates who have each had forty hours. In any case, it is not a great amount of training. But by comparison with the OSH physician training situation in U.S.A. it is a remarkable achievement.

By 1984, NSV has also given basic OSH training to some 3000 OSH nurses, about 1500 OSH engineers, and to some 400 OSH physiotherapists. There are also quite a number of psychologists and ergonomicists as well as managerial and clerical personnel in the OSH services in Sweden. The recent Wessmann report covers all these types.

Many other aspects of the Swedish OSH system could be discussed, but mention can only be made here of a few. The Swedish National Product Control Board tests substances found in work environments (including imported machines, chemicals, etc.). The Work Environment Fund awards some Skr 400 million annually for research and development projects–for example, the Mora project, some aspects of which will be discussed–and for training and information. There is also a National Work Environment Council for public employees; this is similar to the Joint Industrial Safety Council. Finally, considerable is done to educate workers. Beyond the 110,000 safety delegates who have received a forty-hour course, there are some 300,000 other workers who have participated in "study circles" and have completed a basic course in work environment since 1974 as offered by the Joint Industrial Safety Council.[75] The plan is to go on to advanced and supplementary courses. In addition, the magazine, *Arbetsmiljö*, is mailed free every week to the home addresses of the 110,000 safety delegates and others.[76]

OSH-PHC LINKS

Focusing more specifically on the OSH services and their relations to PHC what are some of the strengths and weaknesses?

First, the in-plant services are in place only for larger plants, those with 1000 or more workers. And multiple plant OSH centers reach mainly those between 200 and 1000 employees and not all of these are covered. Thus only about half of Swedish workers are covered by OSH services. This varies greatly by type of industry. One knowledgeable informant estimated that 90 percent of paper workers and 80 percent of metal workers are covered.

With a special interest in coverage of small plants, I visited the Mora project, a special effort on the part of a free-standing district OSH center to enlist and serve small plants in a wide region. Some aspects of this project are clearly successful. For example, the medical and technical OSH service offered to those plants which have enlisted seems quite complete and good. However the sense of worker participation and control seems minimal as the system of regional safety delegates mentioned above seems too loose. Also there is a very interesting and apparently successful involvement of general practitioners once a month in a plant visit followed by an educational session. Thus workers not covered by OSH services (or those who are, but who choose to go to another care source) have a chance of having any work-related problem recognized.[77] But the overall objective of enlisting the large majority of plants seems elusive. In a special study, an area was chosen for intensive contact and selling of the idea. Of ninety-six firms chosen for such attention, only four or five signed up. This was about the same as in a comparable area which did not get such intensive educational and sales efforts.[78]

I heard of a hospital-based OSH policlinic in Skövde which was reputedly more successful than any other OSH center in enlisting small plants for its services. I was unable to get there and only succeeded in interviewing the Director by phone. Subsequently, a close collaborator sent the following information:

> Your third question was about the situation in Skövde and how they had succeeded to reach the small scale companies in their integrated OSH-PHC work. I spoke to the chief in Skövde, but he could not give the answer you wanted, because they do not know the number of the total small companies in the region (i.e., they have no denominator if you want to calculate a relative figure). If you look at working places with less than ten employed, they have thirty-nine companies, 397 farmers (who will be excluded after the end of this year—then a special OSH service will start for them, 'Lantbrukshälen') twenty-two administrations belonging to the church, and five administrations belonging to the government. In the table below I show you the total situation for the Skaraborg-Skövde OSH-PHC so you can calculate some figures which perhaps are compatible with the Mora figures.[79]

Table 5. Small Employers Receiving Services from the Skaraborg-Skövde OSH Policlinic

	Number of employed				
Type of working place	*<10*	*11-20*	*21-50*	*51-100*	*>100*
Companies	39	28	32	17	16
Farmers	397	2	3	—	—
Ecclesiastical administration	22	5	1	1	—
Public administration	5	5	3	2	3
Local government	—	—	—	—	11

If something like the Mora Project or that of Skövde could be spread by negotiation to cover all workers, LO and PTK would support it. The Wessmann Commission on OSH Services has not recommended that it be done by law as it has been in Finland. Even though the unions wanted a legal solution, they were ambivalent, since this would tend to take away their bargaining power both locally and nationally in this sphere of concern by placing the decision-making in the hands of government.

A special project on OSH for farmers run out of Stockholm but with decentralized project areas appears to have had real success in reaching and involving farmers and looks like its relations to PHC should be more widely adopted.[80] But the problems remain. One GP treated a farmer wrongly eight times before he learned it was a case of farmer's lung.

There don't appear to be any national studies or data which would show where workers go for their first help outside of family and friends when they first feel sick. The physician visit figures cited previously plus one limited but very interesting study of the patient flow problem as between factory and district medical services in Luleå,[81] suggest that the proportion of workers first seeking help at district doctors and hospital services is high. Theoretically, there is a system for proper follow-up of work-related cases. Any known reportable work disease and any disease suspected of being work-related should be reported by the PHC provider to the Labour Inspectorate which in turn should follow through in the plant, and with the OSH services at that work site, if there are any. But it doesn't work well.

One of the major troubles is that most physicians lack working class backgrounds, are poorly trained and/or are uninterested in OSH problems. Most informants agreed that this is especially true of older physicians in Sweden. They see a change for the better in younger physicians and medical students.

Medical schools offer a varying amount of OSH education. Estimates range from as little as seven required hours to as many as forty in one or the other of the six medical schools. A student I asked to inquire for me at the Karolinska Medical School in Stockholm had a difficult time getting a clear answer, even after speaking with responsible professors. The answer turned out to be "two to three weeks," but this is an estimate including some other aspects of social medicine and an uncertain amount of OSH material worked in with other subjects such as dermatology. There was no course specifically titled or directed primarily toward OSH. Union people with whom I spoke wanted to see much more than forty hours of OSH education included in medical education. Besides the hours of education there are also possibilities of including OSH questions on the qualifying exam. The recent Commission for Higher Education (Oversynsutredningen) has made a serious proposal on this matter in its November 1982 report. Among other problems involved in making a change, as one LO informant serving on the Commission put it, "One can't just keep adding. If hours are added in this area, some must be taken away in another area. This leads one into the thick of faculty politics." Other problems include the need to provide significant faculty and other resources in this sphere of concern if increased OSH hours were to be effectively taught. In spite of the difficulties, the Commission recommended greater stress on the social sciences in medical education as its main point. It also gave special proposals regarding courses in family medicine (ten weeks) long term care and gerontology (three to four weeks) occupational medicine (two to three weeks) and social content of health care (no time specified). With regard to OSH it is a helpful but rather weak statement: "The education in occupational medicine in elementary education of the physicians ought to be systematically integrated with education in, first of all, allergology, family medicine, audiology, dermatology, pharmacology, clinical psychiatry and social medicine." At this point (p. 75 of the report) according to my informant, there is a note suggesting that OSH material can also be worked into biochemistry, lung medicine, medical rehabilitation, neurology, ophthalmology, oncology, orthopedics, and pathology. The informant writes:

> My comment: there ought to be implemented a special course in occupational medicine. This should take up the definition of the subject, short history, and the special features of the subject. The student shall learn how to take an occupational history. Visits to work sites and work at an occupational health center is also important.

These are proposals. It remains to be seen if the government will fashion a bill carrying with it a mandate and the resources to implement the suggested changes.

In the meantime, the Wessman Commission on OSH Services has arranged with the medical schools and the Medical Society that a one week post graduate course in OSH be offered as one which can be selected among six which must be taken for the specialty certificate in General Medicine and possibly certain other fields such as General Internal Medicine—but I am unsure on this point. This one week medical-school based course would complement the one offered by ASV for certification as an in-plant OSH services specialist.

There is also the question of physicians out in practice. The Mora Project cited above suggests a way to involve such physicians, especially those most likely to be offering primary care, but that is a project carried out with seventeen general practitioners in one locale. It is unclear what measures, if any, can be taken to make practicing physicians across the country more aware, concerned and knowledgeable about OSH.

In addition to problems of education for present and future PHC personnel, there are policy, organizational, financial and sponsorship or control concerns affecting the OSH-PHC connection.

At the policy level, while it is clear as previously described that a major new push is developing in the National Health Service to emphasize and support PHC, it is only very recently that this thinking has recognized the OSH sphere of concern as something to be included in the PHC thrust.[82] However, the Wessman Commission on OSH services has recommended the establishment of OSH clinics in hospitals as a back up and support to OSH services in large plants and possibly as centers for OSH services for small plants.

Organizationally speaking, the multiplicity of PHC forms, about which there is concern and movement to provide greater rationality, makes it difficult at this stage to spell out in any detail the connections, responsibilities, points of information flow, etc. as between the OSH and PHC services. In a way one could say that the County Council health services are competing with themselves—at least as between the hospital outpatient services and the district doctors (then there are the private practitioners who also act as PHC providers). Both the hospitals and the district doctors are under the County Councils; yet they compete with each other. The district doctors have demanded that a note of admission from them should be obtained before a patient can go to the hospital's specialty outpatient services.[83] But this is seldom observed, though for certain specialties such as ophthalmologists and orthopedists it may be observed, especially in the north of Sweden where there are shortages in these and certain other specialties.

Financially, since OSH services are half paid by employer contributions (presumably more for the preventive efforts) and half paid by the general health insurance (presumably mostly for acute and other clinical care) there should be better possibilities of sharing between the OSH and PHC services.

In terms of sponsorship and control, the present system of OSH services which is almost entirely separate from the NHS is as strong and good as it is (though there are still many weaknesses, some of which have been noted, for example, coverage for small workplaces) because Labor has been strong and has fought to develop this service in cooperation with management and the government. And, as noted, workers have gained a significant degree of control over their OSH services. In part because of this control old fears of "the company doc" have begun to disappear and workers may be increasingly willing to bring their problems to the OSH services covering their plant; however, as also noted there are little data on this. Were some kind of merger to take place with the general health services, there is a fear on labor's part that they would lose control and old suspicions of medical services oriented toward owners' interests would be rearoused.[84]

A recent important development is the issuance of a report, "Primary Health Care's Cooperation with Worker Health Services" by the Health Care and Social Services Planning and Rationalization Institute in Stockholm.[85] This report carries with it a policy-level impetus for further attention to the PHC-OSH problem as the development of the report involved key representatives from all geographic areas and many industrial, union and academic centers of the country. It goes into nearly every facet of the problem recognizing that there are severe shortcomings in the present arrangement of things. Perhaps its special emphasis is on the ways in which PHC providers could be helpful to the OSH services to properly rehabilitate the sick or disabled worker once he or she has returned to work. But other aspects such as the need for more training of PHC providers in OSH and better recognition and reporting of OSH conditions by PHC providers are also discussed. This is a welcome, even if very recent development.

QUESTIONS, OBSERVATIONS AND RECOMMENDATIONS

One of the subtlest yet deepest concerns one has about the Swedish system is the extent to which corrections of hazards may be postponed or worked at slowly when urgency in terms of workers' lives may be called for. Behind the cloak of relative harmony between Labor and Management—the reformist approach adopted very early and confirmed in the 1938 Agreement of Saltsjobaden—lurks the danger that six months, a year, or more may be used to correct a ventilation or other problem when it should be done immediately in the interest of workers' health. This concern heightens when the economy slows and both Labor and Management start cooperating "to meet the competition," "to save jobs," "to increase productivity," etc. I talked with a family therapist (psychologist) who sees many workers. Some have an alcohol or drinking problem or other psychosocio-medical problems. By his experience these often originate or are caused by pressures in the workplace.

> And the unions are not always strong. In one company recently they accepted to work twelve hours a day because the employer says there is a crisis and to get out of it we have to sacrifice. Sacrifice what? Your health? Your benefits? The Swedish worker gained an eight hour day in 1900. So what is this?

This problem is by no means limited to Sweden as we will see. In various forms and variations it is all too universal—sometimes crude and obvious, sometimes subtle but also threatening. Also, it would be wrong, perhaps especially in the case of Sweden, to overdraw this problem. Clearly, the Swedish worker through his and her union organization has achieved much in OSH for workers in other countries to envy. Still, the concern over the old non-choice—your job or your life—is there.

There is some reason for concern about union solidarity across sex, age and other lines—in spite of the impressive overall strength of unions in Sweden. There were demonstrations against foreign workers in 1980 as times got a bit tougher. Walking by the gate of a factory of some 800 workers in Norrköpping one day, I spoke with a woman just leaving work to pick her child up from school to take the child to the doctor. After asking about factory medical services (to which few workers would go she said as the doctor is there only one half day a week) I asked how many women worked there. "About 600." "Is there a good union?" "Not very." "Are there any women among union officers?" "No." In the bit of experience I had visiting Union headquarters offices, I saw few women in other than secretarial positions. Again, this kind of problem is not unique to Sweden by any means. At its 1981 Congress, LO gave explicit recognition to these concerns and issued a general call for renewed grass roots democratization of the union.

Within the realm of specialty services for OSH (connections to PHC are taken up below) it seems important for the unions to gain access at all levels to their own truly independent and competent experts. Even at top levels, though here I am on very sensitive ground and admit that I have only a "sense of the situation" and little firm data, there seems to be a need for greater autonomy and critical approaches on the part of union technical and medical experts *vis-a-vis* management experts. Part of the problem is that these men grew up together, so to speak, developing a virgin field and doing so with outstanding success and accomplishment. The struggle then was to get OSH services accepted into companies at all. Even now, some say there is a need to further integrate these services into the factories so they can speak with accepted authority, while building needed bridges to PHC at the same time. But as one knowledgeable SAF informant put it, after I had observed that there seems to have been a history of agreement on almost everything between management and union experts "Yes, sometimes before a public meeting we have asked each other what points we can disagree on so we don't always look like we are too much in agreement."

While OSH special services for large and increasingly for medium size factories seem well on the way to establishment, there is clearly a need to find better ways to cover the small to medium size factory and other workplaces. This is a complicated matter. For example, on finance and content, if the average cost of OSH service is some $100–$200 per worker per year, what should the owner of a dress shop or even a small laundry be told they will receive? What is *health* service? Should there be hazard inspections, psychological examinations and counseling? Annual physicals? The Mora Project can offer a model for what to do and how to organize it, but the full enlistment of factories and shops may require some kind of legislation (especially for nonunion shops). A better possibility may be a framework of legislation within which there would be negotiation; this would be more in line with Swedish tradition.

The manpower problem appears serious—even for large plants. There probably should be two kinds of specialties developed for OSH within medicine,[86] as seems to be happening in some other countries as we will see. The one could be the more research-teaching, clinically oriented, medical school and/or hospital based-person. The other would give preventive and treatment services and be based in workplaces or in centers with other OSH specialists covering a number of workplaces. These OSH service specialists should understand and participate in epidemiologic studies of OSH problems, perhaps carried out by the medical-school-based OSH specialists, under joint union-management OSH committee supervision. Perhaps the universities should play a greater role in the preparation of both these OSH medical specialties and the research without losing the involvement of ASV and work settings.

As regards linkages between OSH and PHC;

- medical and nursing education in OSH should be considerably increased and improved, and required of all students,
- postgraduate OSH education should be offered and perhaps required, especially for PHC providers, but possibly other specialists as well,
- in this last connection, a national medical sociological survey might be carried out to determine where workers first go for professional help when they first feel sick. This should be done to reflect men and women, different ages, different industries, rural as well as urban areas, types of complaints, availability of OSH and PHC services, and perhaps other factors. This information might have several uses, including redesign of PHC services, but would be particularly useful for designing a strategy (high priority points) for postgraduate OSH education,
- the planning for restructuring of PHC services which is now going on should more fully take into account needs for (1) better referral and follow-up connections to OSH services in large and medium plants (2) participating in the OSH coverage of small plants and (3) inclusion of the voices of organized labor in the design and development of PHC-OSH services.

SWEDEN AND THE MODEL— A BRIEF SUMMARY[87]

Policy

Clearly rates high among the cases studied. The long history of labor-dominated government and strength of the workers' movement has turned the class struggle in the direction of the workers' interests. The struggle is not over, and not all battles are won. But the protection of workers on the job clearly has a very high priority. The recent report on PHC-OSH cooperation shows growing policy-level concern with this aspect as well.

Sponsorship and Control

Workers have a strong voice as regards all aspects of specialty OSH protection. For example, union representatives make up a majority of each plant's health and safety committee (required by law) and have the effective hiring and firing power as regards physicians and other OSH personnel. The Law of Codetermination puts unions on a par with management in determining production goals and how they will be carried out. Thus there is a minimum of nonsense about "trade secrets" and the worker's right to know about the materials he or she is exposed to.

Although there might be mechanisms, since labor representatives are on nearly every board and commission in the country, labor has not seemed to have exercised a voice in the training and provision of PHC. Probably this reflects a conceptual gap, rather than a power gap. I hope this study can help to correct this situation.

Education

This country is very strong on worker education. In addition to the 110,000 health and safety representatives trained with forty or more hours in hazard recognition and other aspects of OSH, there are "study circles" covering most other workers and many publications and other materials are provided.

Education of physicians is more problematic. There are commissions now investigating this area. At undergraduate level, a medical student at Karolinska could not find out exactly how many hours there are on OSH topics. It appeared there were less than six, depending on how some material on lung and skin diseases would be defined. There is great variation among the other schools; generally it is on the low side. A Commission in Higher Education has recommended improvements but the recommendations are not strong.

For practitioners, there is a promising project in Mora in which GPs are invited by physicians in a regional OSH policlinic intended to cover medium to small plants to make a plant tour once a month, return for supper together, and then discuss the problems patients might present from such a work site and what to do for treatment and follow-up prevention. This, however, is not a country wide effort.

The training of specialists for OSH is also not well-defined. There are very fine experts in the field. But the kind of two-level model suggested above seems to be something realizable in the future, while most in-plant physicians have a variety of short courses. However, the National Board for OSH has minimum standards and is involved in the filling of positions, recommending three candidates from which the plant health and safety committee finally selects one.

Organization

The arrangements for OSH specialty services, especially in large plants is well developed. There would be many positive things to say in this regard. The small plant problem is not solved. The Mora Project mentioned earlier was specifically set up to enlist and serve small plants in a defined region. But even in a part of the region in which the staff of the OSH policlinic devoted special efforts with intensive use of the media, visits to plants, etc. only 4 percent of plants signed up.

The organizational arrangements for links to PHC providers are worked out on paper but there are many problems in reality. The PHC provider is supposed to report any suspected work-related illness to the work inspectorate for follow-up and continued handling by in-plant OSH personnel. But PHC providers are generally not trained to recognize OSH problems and only some of them are organized to cover defined geographic areas—district physicians (the other main PHC input points are hospitals and private physicians). In many medium-sized and small plants there is no OSH service to follow-up on any reported problem, though the health and safety committee can play an important role.

Information

Ideally, a reporting system such as just suggested, from PHC provider, to inspectorate, to plant services, should exist in fact for proper treatment of individual workers and for epidemiologic and preventive follow-up purposes. At least two other facets of information development, flow and use are important.

First there should be an easy working relationship between in-plant physician and the PHC provider who may have picked up a case. Possible pitfalls include fear of one stealing the other's patients for monetary and other reasons, the workers' trust or lack of trust in one or the other of the physicians, and misunderstandings of what information should be communicated to protect the worker and his or her job. I believe the situation in Sweden is strong in this regard, because in-plant physicians, being under worker control, are more trusted than in previous generations. But the PHC providers remain a weak link.

Second, there should be a national workplace hazard identification and follow-up program for every worker. No such program has been launched in Sweden. In G.D.R. such a program is being actively pursued.

Finally, special studies have been extensively carried out and have been given widespread practical impact in Sweden. These have been conducted not only on a wide range of physical and chemical hazards, but on ergonomic questions, and very extensively on stress questions. One important prospective study showed that in a national sample of male workers, those with very little self determination at work (they said "no" to each of four questions such as "Can you leave your job once a day for ten minutes when you need to make a personal phone call?") were at much higher risk of having a coronary.[88]

Financing

Half of the money for OSH services in Sweden is paid by employers, half by the national health insurance. Presumably, the first half is intended to support preventive efforts and the second curative or treatment services. There is some dispute as to whether this 50-50 apportionment of effort and financing is the best. One would expect that this amount of funding from the general health services would make the PHC-OSH links more integral than they are. I see the continuing fractionation of the PHC services between hospitals, private practitioner, and district physicians as problematic. Study commissions are at work trying to devise a way to rationalize the PHC level of care. If the total job were put in the hands of district level public physicians, geographically assigned as in the G.D.R. and U.K. a better possibility for proper PHC-OSH linkages would be created.

In general summary, the OSH situation in Sweden is comparatively quite strong. Perhaps it is the best in many respects among our study countries, though there are important weaknesses or missing pieces on which the G.D.R. makes a better showing (while being less adequate than Sweden in authentic worker involvement and control, so it is not a toss up, rather these are two strong, but different cases). As regards the OSH-PHC link there are many problems. The situation leaves much to be desired as is true for the other study countries. My overall working hypothesis that a strong workers' movement will have achieved through its struggles a favorable OSH system as well as a favorable OSH-PHC relationship seems to be only half correct. The OSH situation is quite good in Sweden. The OSH-PHC situation is not.

NOTES AND REFERENCES

1. This and other points made in this introductory section are largely taken from "General Facts on Sweden" from the series, "Fact Sheets on Sweden" Stockholm: The Swedish Institute, July 1981. Other issues of this series which have been helpful are "Sweden," June, 1980; and "The Swedish Population," August, 1980.
2. The above information is taken directly from the reference in Note 1.
3. "Swedish Labour Movement." Stockholm: AIC (International Centre of the Swedish Labour Movement) nd (approx. 1978), quote from p. 4.
4. I am indebted for many details in this chapter and for confirmation of my own similar analysis to Vicente Navarro's important article: The Determinants of Social Policy. A Case Study: Regulating Health and Safety in the Workplace in Sweden, *International Journal of Health Services 13*: 517–561, 1983. On the concentration of capital, Navarro states: "From the very beginning, it was oligopolistic. Sweden was a small country, and the market for its industries was primarily abroad. Its industry was based on export-oriented manufacturing and was highly concentrated. This concentration of property enabled the owners of those industries to cooperate on their labor policies, making them a powerful group to resist." (p. 525). While Navarro gives no specific documentation of this point it is covered in Ulf Himmelstrand, Sweden: Paradise in Trouble in I. Howe (ed.), *Beyond the Welfare State*. New York: Schochen Books, 1982; Also W. Korpi, *The Working Class in Welfare Capitalism*. London: Routledge and Kegan Paul, 1978. Just how giant the capital concentrations are in this small country and how intra-bourgeoisie conflicts are resolved is suggested by Stephen D.

Moore, Volvo, Wallenberg Call Truce in Sweden, *The Wall Street Journal* (May 1, 1984), p. 38.

5. Kelman applies the notion of social peace to the relative lack of antagonistic relations between Swedish capital and labor in the domain of OSH regulations and enforcement (lack, that is, when compared with the highly adversarial situation in U.S.A.). To the extent that he explains the social peace, he appeals to shared values within an aura of respect for authority (the regulating agencies and government in general) said to stem from the days of the absolute kingship. Steven Kelman, *Regulating America, Regulating Sweden: A Comparative Study of Occupational Safety and Health Policy.* Cambridge, MIT Press, 1981. What he ignores are the sometimes bitter struggles and eventual strength of the workers' movement which has simply forced capital to come to terms. On the vital importance of the Swedish workers' movement achieving the strength to "exchange" (or I would say transform) the political structure, see Walter Korpi and Michael Shalev, Strikes, Industrial Relations and Class Conflict in Capitalist Societies *The British Journal of Sociology 30*:164–187, June, 1979.
6. In recent decades there have been fewer labor disputes in Sweden than in most countries, though as the economy turned down in the most recent years, disputes have grown and some have involved demonstrations against foreign workers. See Lennart Forseback, *Industrial Relations and Employment in Sweden.* Stockholm: The Swedish Institute, 1980, esp. the Table on p. 65. Also, "Labor Relations in Sweden," Stockholm: The Swedish Institute, Fact Sheets, August 1981. The comparative "labor peace" in the last fifty years in Sweden has been possible precisely because the workers' movement has achieved the strength to force the kind of "political exchange" which Korpi and Shalev describe, op. cit. (note 5).
7. However, not as a majority party. In fact the Social Democrats have always been in power as a minority party forced to make alliances and thus gradually forced to concentrate on short-term gains for workers rather than long-term goals of socialism such as placing the means of production and distribution of surplus under control of the workers.
8. Navarro, op. cit. (note 4): 524–525.
9. Navarro, op. cit. (note 4): 536ff. suggests that this widespread unrest occurred in response to capital's long-term policy of managerial and technical rationalization or Taylorism which entailed mechanization of the work process and deskilling, not to say, dehumanization, of labor. This is appealing as an historically rooted explanation of the widespread, cross-national labor and citizen unrest of the late 1960s to early 1970s and probably helps explain the fact that important OSH laws were passed in this period in every Western study country (a different set of explanations might be needed for the G.D.R. which also saw important OSH developments in this period—though one might argue that a certain mimicking or "keeping up with the Joneses" goes on between G.D.R. and F.R.G.). Still, the gross inhumanity and injustice of the Vietnam War carried out first by the French, then by U.S.A. as the dominant core nation in the capitalist world-system, probably

also played a role along with other factors in each country (the civil rights movement, the environmental movement, the women's movement in U.S.A.) in creating this climate of revolt and change.

10. "The Swedish Labour Movement," op. cit. (note 3): 4.
11. Ibid, p. 5.
12. Navarro, op. cit. (note 4): 531.
13. Navarro cites A. Przeworski, Social Democracy as a Historical Phenomenon, *New Left Review 122*:45, 1980.
14. T. A. Tilton, A Swedish Road to Socialism: Ernest Wigforss and the Ideological Foundations of Swedish Social Democracy, *American Political Science Review 73*:516, 1979.
15. Navarro, op. cit. (note 4): 532, 533 and 536.
16. W. Korpi, *The Working Class in Welfare Capitalism*. London: Routledge and Kegan Paul, 1978, is cited by Navarro, op. cit., Table 2, p. 526.
17. L. O. "Lönepolitik", Stockholm, 1971.
18. Navarro, op. cit., p. 527. He takes the concluding quote from R. Meidner, "Samordning och solidarisk lönepolitik under tre decennier" (Coordination and the solidarity wage policy over three decades). In *Tvärsnitt* (Cross Section). Stockholm: Prisma, 1973.
19. This distinction is discussed in relation to narrow versus broader union-sponsored medical care insurance plans in Joseph W. Garbarino, *Health Plans and Collective Bargaining*. Berkeley and Los Angeles: University of California Press, 1960.
20. In fact, the weakness of labor's situation in U.S.A. does not come only from its internal fractionation but from the strength of the capitalist class which has state power generally on its side. Thus it would be very difficult to organize a national general strike since the Taft-Hartley Act passed in 1947 outlaws even local sympathy strikes in which, let us say, the steel and auto workers and teamsters would go out to support the Greyhound bus drivers in their recent, less than successful strike. States have also passed "right to work laws" which abolish closed shop organizing.
21. In Connecticut, the 12,000 plus workers at the Electric Boat (a nuclear submarine Division of General Dynamics) are organized into some ten different unions—pipefitters, boiler makers, electricians, carpenters, etc. Most are kept from undercutting each other in key bargaining issues by belonging to an umbrella bargaining organization, the Metal Trades Council (MTC). But the 2,000 member Marine Draftsmen's Association is separate. It has been out on strike for nearly eighteen months. The company is seeking to break the union by hiring scabs and not bargaining in good faith and there is very little the MTC can do to be supportive.
22. Jan Kristiansen, Sweden: Trying to Agree on a Remedy, *International Herald Tribune*, p. 75, 7 October 1981. The expectation for 1982 was -0.3 percent, while for the Nordic countries together it was a 1.0 percent increase. But compare these unemployment figures with 12 percent in the U.K. and 9 percent in U.S.A. at that time. The recession high-point for the U.S.A. was 10.8 percent unemployment with 12 million workers out of a job!

23. For example, see the following Fact Sheets published by the Swedish Institute: General Facts on Sweden, Stockholm, July 1981; "Social Insurance in Sweden," March 1981; "Old-Age Care in Sweden," May 1981; "Child Care Programs in Sweden," February 1980; "Die Grundschule und die Gymnasialschule in Schweden," December 1979; "Environment Protection in Sweden," April, 1981. Also, "The Cost and Financing of the Social Services in Sweden in 1979," Stockholm: National Central Bureau of Statistics. Also, *The World Almanac and Book of Facts, 1980*, "Sweden" New York: Newspaper Enterprise Assn., pp. 580–581.
24. On the other hand, "An LO affiliated union cannot declare a strike without the approval of the LO executive board, if the stoppage involves more than 3 percent of its membership." "Labor Relations in Sweden," Stockholm: The Swedish Institute, Fact Sheets on Sweden, August 1981, p. 2.
25. Specifically, the question is whether legislation should be adopted to require OSH services coverage for small plants as has been done in Finland. The "Wessman Commission" has only very recently issued its report (late 1983) and urged OSH services coverage for all workers in the next ten years, but did not call for legislation, only voluntary measures.
26. International Solidarity, Neutrality and Disarmament, pp. 10–11 in *The Swedish Labour Movement*, op. cit. (note 3).
27. In 1981, the 349 members included ninety-two women. By party the total was divided as follows: 154 Social Democrats; 73 Conservatives; 64 Centrists; 38 Liberals; and 20 Communists. "General Facts on Sweden, 1981" op. cit. (note 1): 24.
28. "General Facts on Sweden" op. cit. (note 1): 1.
29. The other logically possible types (within a dynamic historical space and not as a set of rigid 2x2 categories) are decentralized fractionated, centralized concerted, and centralized fractionated. Not all these types are represented among the study countries although there is some variation as will be noted. I have elsewhere suggested the decentralized concerted system as most likely to be supportive of a fully regionalized, PHC-oriented health system. R. H. Elling, *Cross-National Study of Health Systems, Political Economies and Health*. New Brunswick, New Jersey: Transaction Books, 1980. Also B. Kleczowski, R. H. Elling, and D. Smith, *Health System Support to Primary Health Care*, Geneva: WHO Public Health Papers, no. 80, 1984.
30. Hospitalization was primarily covered through general taxation supplied by the local county councils which each controlled its own hospital network. Only small charges were made to the insurance funds for hospital care. Hospitals have traditionally swallowed the largest share of health funds in the overall system. In recent years attempts have been started to give more emphasis to ambulatory care and PHC.
31. The figure in Swedish Kronar (Skr) for 1984 is 145,500 per year or 395 a day (Correspondence from Professor Urban Janlert). At today's exchange rate (8.066 Skr to the U.S.A.$) this works out to $18,040.
32. These figures are taken from "The Health Care Delivery System in Sweden" Stockholm: The Swedish Institute, Fact Sheets on Sweden, January 1980. Other sources used in this section include "Social Insurance in Sweden"

op. cit.; P. Owe Peterson and K. Ichimura, "Facing the 1980's, A Review of the Swedish Health and Medical Care System" Uppsala: University Hospital, 1980; "Primary Health Care Today, Some International Comparisons," HS90; "Primary Care and Care for the Elderly, Highlights from Reports Published in Swedish," Stockholm: Swedish Institute for the Planning and Rationalization of the Health and Social Services (SPRI) 1979; "The Cost and Financing of the Social Services in Sweden in 1979," op. cit. (note 23).

33. Robert Maxwell, *Health Care, The Growing Dilemma: Needs Versus Resources Western Europe, The U.S. and The U.S.S.R.* New York: McKinsey and Co., 1974.
34. The NHI-NHS distinction is not a hard and fast one. Usually physicians contract to give services or receive fees for each service in an NHI. In an NHS physicians tend to be on salary from the government and hospitals are publicly controlled. For a discussion of the main moves in the NHS direction see Mack Carder and Bendix Klingeberg, Towards a Salaried Medical Profession: How 'Swedish' was the Seven Crowns Reform? pp. 143–172 in Arnold J. Heidenheimer and Nils Elvander, (eds.), *The Shaping of the Swedish Health System*. New York: St. Martin's Press, 1980.
35. Ragnar Berfenstam and Ray H. Elling "Regional Planning in Sweden: A Social and Medical Problem," in D. E. Askey (ed.), *Health: A Major Issue*, (special issue of *Scandinavian Review*)*63*:40–52, September 1975.
36. Aside from a general qualitative assessment of this over-orientation toward specialty hospital care involving such subtle but nevertheless powerful matters as the medical prestige-reward-motivational structure and public understandings of "health" care, various quantitative indicators have been used: the figure cited earlier, 53 percent of visits to doctors take place in hospitals; another is the bed to population ration, 17 per 1,000 which is quite a lot higher than in most other countries. Gunnar Wenstrom, "Notes for a Talk on The Future of Swedish Health Services" Stockholm, unpublished, 1981; "Health Policy in Practice . . ." op. cit. (note 32) "Starting Points and Guidelines HS90, The Swedish Health Services in the 1990s." Per Olof Brogren and Ulf Nicolansson, "Primary Health Care," Göteborg: The Nordic School of Public Health, June 1951, unpublished draft; Berfanstam and Elling op. cit. (note 35). Also, Jan-Erik Spek, "Why Is the System So Costly? Problems of Policy and Management at National and Regional Levels" pp. 181–207 in Heidenheimer and Elvander op. cit. (note 34).
37. "Primary Health Care Today, Some International Comparisons," Stockholm: The Swedish Ministry of Health and Social Affairs, HS90, 1981.
38. Personal communication from Professor Urban Janlert, April 19, 1984.
39. R. F. Tomasson, The Mortality of Swedish and U.S. White Males: A Comparison of Experience, 1969–71, *American Journal of Public Health 66*: 968–974, 1976; A. S. Häro, "Health Facts and Figures" in D. E. Askey, (ed.), *Health: A Major Issue*, op. cit. (note 35).
40. "Denmark, West Germany and Switzerland have higher suicide rates than the Swedish one of 20 per 100,000 inhabitants" and "Only 2 percent of the population over fifteen years of age put away more than 6.3 ounces of

pure alcohol a day. That compares with 9 percent of the French, 7.4 percent of Italians and 4.8 percent of West Germans." Mats Hallvarsson, Figuring Out the Swedes, *SAS Magazine*, pp. 60–63, Fall 1981. Also, "A comparison of thirty-four countries shows Sweden in the 30th place" (in alcohol consumption). Alcohol and Drug Abuse in Sweden, Stockholm: The Swedish Institute, Fact Sheets, January, 1981, p. 2.

41. "Primary Care and Care for the Elderly," op. cit. (note 32).
42. Kleczkowski, Elling, and Smith, op. cit. (note 29).
43. Ruth Sivard, *World Military and Social Expenditures*. New York: Institute for World Order, 1983. In 1980 Sweden spent 3.3 percent of its GNP on the military; U.S.A. spent 5.6 percent; the U.K. 5.2 percent; the G.D.R. 5.0 percent; F.R.G. 3.2 percent; and Finland just 1.6 percent (from Table II of Sivard).
44. For some further detail and a general interpretation of the strength of the Swedish workers' movement, see the opening section of this chapter. Since this is not a history of labor in Sweden, many important steps are skipped in this brief presentation. For more detail, see "The Swedish Labour Movement" op. cit. (note 3). Lennart Forseback, *Industrial Relations and Employment in Sweden*. Stockholm: The Swedish Institute, 1980. Also Korpi, op. cit. (note 4) and Korpi and Shalev, op. cit. (note 5).
45. *The human work environment; Swedish experiences, trends, and future problems*, A contribution to the UN conference on the human environment. Stockholm: Ministry for Foreign Affairs and Ministry of Agriculture, 1971 (Report of working party chaired by Professor Sven Forssman).
46. Navarro, op. cit. (note 4) analyzes two kinds of capitalist rationalization—technical (Taylorization) and managerial—and Labor's responses as central to the dialectic of class struggle in Sweden.
47. A fair amount of labor history in Sweden can be seen as a struggle to dismantle or demolish this paragraph piece by piece. "The Swedish Labour Movement" op. cit. (note 3) esp. cartoon p. 14.
48. Navarro, op. cit. (note 4) cites this fact of employer power of the purse as an important weakness of the OSH system in Sweden. I agree, but sophisticated locals have learned some tricks such as having an agreed, prebudgeted amount to work with. Also, the worker-dominated joint OSH committee must approve the plant physician's and engineer's OSH report and plan for the coming year which are submitted to the NHS. Without this approval, the employer is not reimbursed 50 percent of the OSH services cost by the NHS.
49. Swedish Employers Confederation, "Guiding Principles for Company Health Services." Stockholm: SAF, 1954, pp. 5–6 (unpublished).
50. "The ABC's of Workers' Compensation" 7th edition. Hamden: Connecticut State AFL-CIO, June, 1983.
51. Carey W. English, Cutting Costs, Abuses in Disability Insurance, *U.S. News and World Report*, pp. 80–81, May 28, 1984. "In the decade between 1972 and 1982, premium costs for insurance soared from 4 billion dollars to more than 15 billion dollars. Nationally, typical employers now pay $2.43 for every $100 of payroll to cover workers against occupational hazards,

according to the Council (National Council on Compensation Insurance)." The emphasis of the article is entirely on getting workers back to work by various pressures, including keeping disability pay low. There is not one word about prevention or keeping hazards out of the workplace in the first place. For the labor view of shortcomings of the U.S.A. compensation system, see American Labor Education Center, "Workers' Comp: Making the Employer Pay," *American Labor* No. 16. For example: Nearly 50 percent of workers' compensation money does not go to workers at all. It goes to insurance companies, lawyers, doctors, and state governments instead." Lawyers fees are said to range between 25 and 40 percent of any award. Unfortunately, this statement says nothing about prevention of disability either. For a scholarly general treatment of the field, see Peter S. Barth and H. Allan Hunt, *Workers' Compensation and Work-Related Illnesses and Diseases*. Cambridge: MIT Press, 1980.

52. "Arbetsmiljöavtalet" Stockholm: SAF, LO, PTK, 1976. The English version lacks the agenda and protocol and cartoons included in the Swedish version, but provides the complete text and discussion: "Working Environment Agreement." Stockholm: SAF, LO, PTK, 1979.
53. Susan Mazzochi, Training occupational physicians, suppose they gave a profession and nobody came? *Health PAC Bulletin* (New York) No. 75:7–10, 19–24 March/April, 1977.
54. In the U.K. organized labor seems particularly suspicious of physicians and refuses to look to the NHS for answers in the OSH field as the NHS to this day, the largest employer in Europe, has failed to establish OSH services for its own employees in spite of hospitals not being particularly healthy places to work.
55. LO Tidningen Nr 39. Årgang 61 24 September 1981; Also "Swedish Trade Union Confederation." Stockholm: LO, 1978, p. 23.
56. Washington Post, Ownership Program For Swedish Unions Angers Businessmen, *The Hartford Courant*, p. A27, February 17, 1984.
57. "National Swedish Board of Occupational Safety and Health." Stockholm: The Board, 1980, p. 5.
58. "Vad hände med SARB?" *Motpol* Tidskrift Fur Sjukvårdsdebatt, Special issue on work medicine Nr 8 (1980) and Nr 1 (1981): 8.
59. *SARB, Arbetsmiljö På Vems Villkor? En bok Om Risker I Arbetet.*
60. *The human work environment* . . . op. cit. (note 45): Figure 1, p. 13.
61. "The Trade Union Movement and the Multinationals." Stockholm: LO, 1976.
62. Lennart Forseback, "Postscript" pp. 133–134 in *Industrial Relations and Employment in Sweden*, op. cit. (note 6).
63. Hans-Ulrich Deppe, Von 'Humanisierung der Arbeitswelt' kann noch keine Rede sein, *Frankfurther Rundschau* Nr. 229, p. 14, 3. Oktober, 1981.
64. "Ergonomiskt åtgärdsprogram." Stockholm: LO, no date (probably 1979) (reproduced).
65. Robert Karasek, et al., "Job Decision Latitude, Job Demands, and Cardiovascular Disease: A Prospective Study of Swedish Men," *American Journal of Public Health 71*:694–705, July 1981.

66. Bertil Gardell, Production Techniques and Working Conditions, *Current Sweden* (published by The Swedish Institute, Stockholm) No. 256 1–16, quote from p. 15, August 1980. An important statement. See also Bertil Gardell and B. Gustavsen, Work Environment Research and Social Change: Current Developments in Scandinavia, *Journal of Occupational Behavior 1*: 3–17, 1980 and Bertil Gardell, Scandinavian Research on Stress in Working Life, *International Journal of Health Services 12*:31–41, 1982. Important work by other researchers in this sphere of concern which I believe will become the key battle ground between the working class and the owners in the future includes M. Magnusson and C. Nilsson, *Att Arbeta på Obekvän Arbetstid* (To Work at Inconvenient Working Hours). Stockholm: Prisma, 1979; T. Akerstedt, Altered Sleep/Wake Patterns and Circadian Rhythms, *Acta Physiologica Scandinavia, Suppl.* p. 469, 1979; Arthur Kornhauser, *Mental Health of the Industrial Worker.* New York: Wiley, 1965.

    On absenteeism it was noted in 1977: "Job absenteeism has increased steadily and has doubled in approximately ten years. The last two years show a certain tendency towards levelling off. This increase applies to both long- and short-term absenteeism." Kaj Elgstrand, "Training and Education in Occupational Safety and Health in Sweden." Stockholm: National Board of OSH, November 1977 (reproduced paper No. 13) quote from p. 9.
67. For example, the Volvo experiment where the union and management negotiated production quotas for work teams, then the teams were free within some limits to arrange their time and approach to the work (including exchange of jobs) as they saw fit. Also, "Towards Democracy at the Workplace, New Legislation on the Joint Regulation of Working Life," Stockholm: The Ministry of Labour, International Secretariat, March 1977.
68. "The goal of company health services is to work towards the optimal adaptation of the work to human conditions. This requires systematic technical and medical efforts. The technical side of company health services is led by a specially-trained safety engineer and the medical side by a specially-trained doctor. These two functions are independent of each other and should preferably be placed under common management. The safety engineer and company doctor are both to be regarded as advisory impartial experts. It is up to the company in consultations with its employees to test and implement proposed measures. The company health services experts are responsible for measuring and evaluating environmental factors and for submitting proposals for measures to be taken." From "Working Environment Agreement." Stockholm: SAF, LO, PTK, 1979, Section "The Functions of Company Health Services," pp. 39–51; quote from first paragraph under "Remarks" p. 40.
69. In 1983 there was discussion about the way OSH services are funded. The government sought to cancel the 50 percent support from the NHS. This would have placed the burden on the industries entirely and presumably would have erased the false distinction between prevention and treatment. But it would have meant the end of many independent OSH policlinics such as the one I visited in Mora which serves a number of medium to small workplaces. Thus the proposal was withdrawn and no change was made.

70. "Occupational Health Services in Sweden" Fact Sheets on Sweden, Stockholm: The Swedish Institute, 1980, p. 2. Also, Criteria Group for Occupational Standards, ASV (National Board of Occupational Safety and Health) Solna, Sweden S-171. Reprinted in *Arbete och Hälsa* No. 21, pp. 1-62, 1981.
71. Other comparisons of the more protective situation in Sweden were given in Table 2 in Chapter 3 taken from Navarro, op. cit. (note 4) who gives great attention to documenting this point, since Kelman, op. cit. (note 5) had discounted the more protective OSH situation achieved by the workers' movement in Sweden.
72. "Omstridd läkare väljer rättegång." *Dagens Nyheter 3*, 27 Augusti, 1981.
73. This is not just an incidental matter. In the same context of concern, this Commission also took up various aspects of social medicine, including OSH. I will come back to this report in the next section on OSH-PHC links. The point here is to note how this political-economic context with a strong workers' movement influences all aspects of medicine in a broader social direction–not only OSH. LO had its voice on this commission in the person of a physician on their staff. It is likely to be some time before organized labor in U.S.A. has a direct voice in medical education.
74. Yvonne Wessman, "Training of Occupational Health Personnel in Sweden." Stockholm: The Commission on OSH Services, 1976. This statement also covers training specifications for OSH nurses, and OSH physiotherapists. There are also observations on OSH safety and hygiene engineers, but as noted these specifications have not been generally agreed. It is interesting that Industrial Hygiene has not been as clearly distinguished from safety and hygiene engineering in Sweden and some of the other study countries as it has in U.S.A. See also, Kaj Elgstrand, "Training and Education in Occupational Safety and Health in Sweden." Stockholm: National Board of OSH, November 1977 (reproduced).
75. As reported by I. Soderstrom in a paper presented to the ILO/WHO meeting in Sandefjord, Norway, 16-19 August 1981. Published in the Conference Proceedings, *Education for Occupational Safety and Health*. Geneva: ILO, 1982.
76. Examples are published each year for international interest in English: *Working Environment, 1980, 1981, 1982, 1983, 1984*.
77. Arne Bjernulf, et al., "Samarbete företagshälsovård–distriktsläkare ger bättre underlag for riktade insätser," Mora-projektet II *Läkartidningen 78*: 3211-3212, 1981.
78. Arne Bjernulf, et al. "Småföretag har behov av företagshälsovård men frivillighet get inte heltäckning" Mora-projektet I *Läkartidningen 78*:3210, 1981.
79. My thanks to Professor Urban Janlert, Karolinska Institute, Social Medicine Department, in Luleå for his detailed letter of December 14, 1982.
80. "Välkommen till Lantbrukshälsan." Stockholm: Lantbrukshälsan AB, pamphlet, no date (about 1980).
81. Finn Diderichsen, "Vårdkonsumption och Patientflöde inom Företagens och Landstingets Sjukvård i Luleå." Karolinska Institutet Socialmedicinska Institutionen Forskningsenheten i Luleå. *Socialmedicinsk tidskrifts* Nr 43, Stockholm, 1977.

82. The County Council statement gives clear recognition to the working environment as one in which "environment-oriented preventive measures" can be taken. See "Health Policy in Practice . . ." op. cit. (note 32): 11. Also, a very recent Spri report explicitly on OSH-PHC relations is significant (see note 85 and associated text).
83. This would help give the Swedish system one of the main strengths of the Danish system. See Erik Holst and Marsden Wagner, "Primary Care is the Cornerstone," pp. 30–39 in Donald E. Askey (ed.), *Health Care in Scandinavia* (special issue of *Scandinavian Rev.*) (September, 1975).
84. Other considerations: "One motive for not turning this sector over to the county councils has been the desire to guarantee that resources will be allocated to preventive health measures, and not to curative medicine in the hospitals—which has been a tendency in county council administration. Another reason is that "occupational health services are something one does not willingly give up. Having ready access (usually free of charge) to a doctor on the premises has clearly been a valued fringe benefit." Uncas Serner, "Swedish Health Legislation: Milestones in Reorganization since 1945," pp. 99–116 in J. Heidenheimer and Nils Elvander (eds.), *The Shaping of the Swedish Health System*. New York: St. Martin's Press, 1980, quote from p. 115.
85. "Primärvårdens somverkan med företagshälsovåarden." Spri report 146 Stockholm: Sjukvårdens och socialvårdens planerings-och rationaliseringsinstitut, 1983. ("Primary Health Care's Cooperation with Worker Health Services"). My special thanks to Professor Urban Janlert for sending a taped translation of this report along with the report itself.
86. I am less well informed on the safety engineering side, so do not make observations on this. However, there appear to be needs in this direction too. Serious consideration should be given to developing the Industrial Hygienist in Sweden as this kind of person can often link protection against chemical hazards and clinical and epidemiologic concerns of OSH physicians and there is a need to bring the medical and technical closer together.
87. The reader is reminded that the rationale and content of the dimensions of this model are provided in Chapter 6.
88. Robert Karasek, et al., op. cit. (note 65).

CHAPTER 8
Finland

THE LAND AND THE PEOPLE

With only 4.8 million people in 1981, Finland is the smallest of our study countries in terms of population.[1] Its area of 337,000 square kilometers (130,000 square miles) makes it second in geographic size to Sweden among our European study countries (the reader will find other direct comparisons of study countries in Table 3 and associated discussion in Chapter 3). The density is sixteen inhabitants per square kilometer, but settlement density is quite varied as between north and south, so this average figure does not give an adequate picture. In fact, more than three-fifths of the population live in the six southernmost provinces with just over one-quarter of the country's surface area. Thus Uuisimaa province surrounding the capital city, Helsinki, located on the southernmost shore along the Gulf of Finland (the eastern arm of the Baltic Sea) had a density of 93.3 residents per square kilometer in 1964 compared with only 7.4 in Oulu province in the center-north, and just 2.3 inhabitants per square kilometer in Lapland, the nothernmost one-quarter of the country.[2] The density is further affected by the fact that Finland has tens of thousands of lakes which cover 9 percent of its area.[3] Some 57 percent of the country is covered by forests; 8 percent is cultivated, and cities, towns, roads, railroads, and other uses take up 26 percent of the area. Some 38 percent of the people still live in the countryside, with 62 percent in urban areas.

Finland is the northernmost country in the world. Yet, like Sweden and the rest of Scandinavia, it is warmed by the Gulf Stream and prevailing winds from the milder southwest and west bring warmth and moisture to the peninsula. Thus there are no glaciers, no permanent snow, and no permafrost. Average temperatures in the coldest month of February range from a surprisingly high plus 12 to minus 5 degrees Fahrenheit in the far north and from plus 19 degrees to minus 28 degrees Fahrenheit in the South. In July, the corresponding figures are 52 to 57 degrees in the north and 59 to 63 degrees in the south (all in

Fahrenheit). Thus the climate has easily supported a modern agricultural and industrial society, whereas areas in Greenland, the north of Canada, Alaska, etc. at the same latitude have not been so developed. It is not as long a country as Sweden but it stretches some 1100 kilometers (almost 700 miles) from the Norwegian border far above the arctic Circle to the Gulf of Finland. Helsinki, the capital, lies on the Gulf, just a bit more northerly than Stockholm which is located not quite 400 kilometers (250 miles) west-southwest across the Baltic Sea. Finland has some 4,600 kilometers (2,850 miles) of coastline stretching southward along the Gulf of Bothnia (the northern arm of the Baltic Sea) to the Gulf of Finland (the eastern arm of the Baltic) and from there eastward to the border with the U.S.S.R.

The country's culture as well as climate is affected by this maritime location. For example, during my visit in the Fall of 1981, I enjoyed immensely to see the fishing boats in Helsinki harbor, in from the islands to market their wares of home preserved anchovies, herring and other sea products one final time before winter might set in. It was a festive time when householders from the capital and further afield stocked up on these delicious things.

Finland shares borders with Norway in the far north, Sweden in the northwest, and with the Soviet Union along its whole eastern length. This relationship to its neighbors, particularly Russia and Sweden has deeply affected its history. It is a country which has always seemed to balance between East and West.

Finland was part of the kingdom of Sweden from 1154 to 1809, when it was united to Russia as a Grand Duchy. Finland declared itself independent following the Bolshevik Revolution in Russia in 1917.

The Finnish people's origins are more varied and controversial than is true for the Swedes (Chapter 7). Indeed, the very names by which Finns call themselves and their land, *Suomalaiset* and *Suomi*, "have not been satisfactorily explained."[4]

The earliest settlements date back to approximately 8,000 years B.C., and, curiously enough, were found by Finnish archaeologists in the far north, from the coastal areas of the Arctic Ocean (now part of Norway) to the present Finnish border area. This may be explained by the melting of the continental ice sheet of the last Ice Age. While most of the Finnish peninsula remained covered by the ice sheet and great bodies of water which later receded into the tens of thousands of lakes of present-day Finland,

> . . . a tundra, overgrown with dwarf birch, bordered the glacial margin, both in the north and in the south. There, wild reindeer, arctic fur-bearing animals, and—in the coastal waters—fish, offered primitive hunters and fishermen a chance to eke out a livelihood.[5]

The earliest relics of human habitation in the south date from about 7,200 B.C.

About 3000 B.C. people with a new Stone Age culture, called the Comb-Ceramic culture, spread throughout Finland. The name stems from the comblike decorations on the tall earthenware vessels of the period. These people still lived primarily from a hunting and fishing economy. The only domesticated animal was the dog. Living on open sandy beaches near waters which abounded with fish, these were relatively self-sufficient people, though trade did occur to the east and southeast. Flint was brought in from Russia and amber from East of Prussia. The assumption is that these people probably had a different origin from those in the rest of Scandinavia because Stone Age man was poor in navigational means and traveled mostly overland.

By contrast, the people who subsequently came upon the scene, termed the Battle Ax people after one of their prime weapons, were excellent sailors. They migrated to Finland from the Baltic region about 1800 B.C. and by 1600 B.C. had established themselves as the dominant cultural group. They apparently cultivated some crops but concentrated on raising cattle. The trade orientations now were to Sweden, the sea and the west generally.

The so-called Kiukainen Culture which evolved (the name is taken from the site of archaeological finds) was a hybrid of the Comb-Ceramic and Battle Ax cultures. While oriented to the west, it maintained contacts with the peoples in the east—almost the exact situation of modern Finland, though there is no necessary direct causal link between the pre-Bronze Age situation and today, perhaps only a coincidental parallel.

The Bronze Age in Finland was commercial in character as there was little or no local production of the metal.

> Scholars have disagreed as to whether this culture was maintained in Finland by a foreign, Scandinavian population or whether the shift from the Stone Age to the Bronze Age represented a cultural loan. Evidently, however, the coastal region had been settled by a small ruling class, consisting primarily of merchants, while the original native stock remained the same. Along the coast, the latter gradually adopted the use of bronze, whereas the inhabitants of the interior continued to hunt and fish in the old way with implements of bone and stone.[6]

This geographic and developmental division, supported and reinforced by a class structure, also carried forward, albeit through various transformations: from the spread of the Roman Empire as far north as the southern part of Finland; through the period of Swedish hegemony during the Feudal Period; to Russian dominance in the 1800s; the rise of the Finnish nation after 1900; independence in December 1917; the Civil War between "Reds" and "Whites" in 1918; to present-day capitalist Finland.

This history and class divisions are tangled up with disputed scholarship concerning the Finnish language and racial origins of the people. In an effort to change images of a relatively isolated people, and to establish that his people had wide connections with the world, M. A. Castren traveled widely in the east to Mongolia and south to Hungary and elsewhere looking for linguistic connections of the Finnish language.[7] His conclusions, reported in 1849, were that this language, and therefore this people (not a necessary connection accepted by modern students of language, race, and ethnicity) were related to one-seventh of the people of the globe—the Turco-Tataric, Mongol, and Manchu-Tungus language groups. But popularly this got used to create class-based, racist divisions. Castren's patriotic intents were turned by the dominant, ruling class Swede-Finns into impressions of "Mongol" origins and a supposed inherent inferiority of the Finns based on the cultural hegemony of the time which cast suspicion and fear and lower value on "the yellow hordes" and peoples from the East.

In summary of the origins of the Finnish people:

> It is a well-known fact that the Finnish language differs greatly from the other languages of Europe. Only Estonian is closely enough related to Finnish to enable a layman to discern a common linguistic base. Hungarian is often mentioned as a relative of Finnish. Actually the relationship is very remote, being about as close as that between English and Persian. . . .
>
> In so far as philological and related evidence throws any clear light upon the problem, it points to present-day European Russia as a possible original 'home area' of the Finns. But even this conclusion has been challenged by scholars whose claim to distinction is no less impressive than that of the defenders of this locus as the area of original dispersal.[8]

In short, we don't know, nor does it matter. What matters is not the reality of "racial" origins, but the ways in which perceptions of such origins have been built up and used to defend class interests.

THE POLITICAL ECONOMY

The development of national consciousness from 1840 on was stirred, among other things by the widespread but abortive European revolutions of 1848. This consciousness was also stirred by Elias Lohnrot's long epic poem, *Kalevala*, first published in 1835, in which his welding together of many folk poems suggested that the Finns were not mere forest dwellers but a spiritually gifted people.

But clearly what happened, as the Finnish nation was becoming conscious of itself from the 1840s on, was a disguised class-struggle carried out in terms of the "Finnish cause" (the working and peasant classes) and the "Swedish cause" (the bourgeoisie—to a considerable extent indigenous of mixed Finnish-Swedish origins and to some extent imposed, of primarily Swedish origins).

> The opponents of the Finnish cause frequently contended that the Finnish language was "undeveloped" and therefore could not be used as an official language until some time in the future. This idea had been put forth years before champions of the "Swedish cause" appeared after the 1850s. . . .[9]

Of course, the Swede-Finns were not entirely homogeneous as a class, even though they made up the large bulk of the ruling class.

> The Swede-Finn language and nationalist movement remained largely an upper-and-middle-class effort until the 1890s; the Swedish-speaking common folk played no part in it, and the leadership of the movement made no real effort to seek their support. It was not until 1896 that Swede-Finn leaders at long last came to the realization that the "Swedish cause" could no longer be defended without enlisting for it the full support of the Swede-Finn common man. The soundness of this conclusion was more than adequately proven ten years later when the new parliament bill and franchise law, promulgated on July 20, 1906, fundamentally changed the national legislature.[10] The introduction of the single-chamber Parliament of 200 members and the extension of the right to vote to all adult women as well as men made it clear that, unless the Swede-Finn political leaders obtained the support for their cause of the Swedish-speaking farmers, fishermen, laborers, and other common folk, they could not expect to reach even the unpretentious representation that a full utilization of their numerical strength—at that time, nearly 13 percent of the total population—might ensure. . . .
>
> As a result, the Swedish People's Party was launched in May 1906. Its program provided a platform on which the Swede-Finn intellectual, farmer, and businessman could stand shoulder to shoulder with the Swede-Finn butcher, baker, and candlestick maker.[11]

Once again the working and peasant classes were divided within themselves and from each other along racist lines—one of the oldest in the book of capitalist tricks. Today, just 6.4 percent of the population are primarily Swedish-speaking, while 93.5 percent are primarily Finnish speaking. Yet Swedish remains one of the two official languages and everything from street signs to government documents are printed in both Swedish and Finnish. In 1979, the Swedish People's Party held just 5 percent of the seats in Parliament (10 of 200) yet the influence of this small segment, especially its business, banking, and industrial circles remains strong.

In the meantime, the nationalist struggle for independence from Russia also served to disguise class interests as Finland's industrialization took hold even later than that in Sweden. Nationalist strivings by the bourgeoisie for greater autonomy were sometimes conducted on the basis of absolute loyalty to the Tsar and directly with the Tsar, without regard for relations between the Russian and Finnish peoples. But Bismarck frightened many European rulers, including the Tsar, and the imperialist grip was tightened to ensure military security of border defenses. The February Manifesto of 1899 called for thorough Russification. This astounded the Finns. Resistance of many sorts developed with the parliamentary development of 1906 as one result when the Tsar's control had been generally weakened through the 1904–1905 war with Japan as well as the internal revolts of 1905. This parliamentary development changed little, however, (see note 10) and it was not until the Soviet Revolution of October, 1917 that Finland got her real chance for independence. Independence was declared on December 6, 1917 when the new Soviet state was weak and plagued by the prospect of invading armies from capitalist countries whose rulers feared the establishment of a socialist workers' state.

The class formations and struggles which motored the Soviet Revolution were also present in Finland, but less developed. In the early months of 1918, when General Carl von Mannerheim, who had served in the Russo-Japanese War, was declared Supreme Commander and Regent, the Civil War (or as some call it, Class War) broke out between "Reds" and "Whites." It was short but bitter, ending in victory for the Whites and exile of the Red leaders to the U.S.S.R. There they formed the Finnish Communist Party in exile.

> But a party in exile soon loses contact with the political realities of the home country. For a while after its foundation the Finnish Communist Party tried to initiate a new revolutionary movement in Finland, but its banning from all parliamentary, cooperative and trades-union activities had the effect of reducing its popularity. In 1920 an undercover Communist Party was formed which tried to combine illegal underground activity with parliamentary action in conformity with the policy then laid down by the Comintern. This Socialist Working People's Party achieved marked success in the elections in 1922, when it gained 14.8 percent of the votes on its first appearance at the hustings, but it was banned from parliamentary activity in 1923 and all its deputies were imprisoned despite parliamentary immunity. . . .
>
> As right-wing radicalism gained ground in Europe a similar process was at work in Finland. At the beginning of the thirties all Communist activity was made illegal by the "Communist laws." The Party went underground, where it was to remain for a long time. In subsequent elections up to the outbreak of the Second World War the Communists transferred their support more and more to the Social-democrats. . . .
>
> When Finland signed the armistice with Russia in September 1944 the Finnish Communist Party was legalized and it resumed activity at once, using another name for tactical reasons. In order to gain a wider base it joined with dissident Left-Wing Socialist groups to form the Finnish People's Democratic Union. This Union was the first well-organized attempt to put into effect in post-war Europe the new Communist Party line, and Yrjö Leino, one of the most

> prominent leaders of the Union, was the first Communist minister in a non-Communist country in Europe. Under close Communist control, though without the outward appearance of Communism, the Union has represented Left-Wing radicals in the Finnish Parliament since the war.[12]

The Communist movement is somewhat divided between the east of the country where relatively naive "Backwoods Communism" is said to prevail as a social protest against poverty, historically prevalent unemployment, and inadequate living circumstances generally, and the southwest where "Industrial Communism" is said to prevail as a self conscious ideologically sophisticated political movement built along Marxist-Leninist lines.[13] Whatever the composition, the People's Democratic Union (later called People's Democratic League) has played an important role since the end of WW II. Communists have been in the governing coalition from 1945 to 1948, 1966–1971, and from 1975 until the present. In 1966, for example, the Union joined the Social-democrats and Centre Party (previously the Agrarians) to form a "National Front" government of Center and Left elements.[14] The People's Democratic Union had three ministers and the second minister for Social Services.

Thus Finland has not been without its working class representatives who sometimes exercised real, if not dominant, power. Yet, as we have seen, these representatives and those they represented have also experienced vicious repression over long periods of time. In addition, the cultural hegemony of obedience to higher authority (including the bourgeoisie's treatment of the Tsar as "our Grand Duke") seems never to have been thrown off. So there is more evidence of paternalism in Finland than in Sweden. In general, the workers' movement has not achieved the same strength it has in Sweden.

This is true in spite of a class structure which favors the working class numerically. Although there are different, non-Marxist ways of defining and measuring social class, as one such investigator put it in 1952, "Finnish Society is in structure decisively a society of workers and farmers, of physical laborers."[15] This is reflected in Table 6 taken from Rauhala.[16] Independent owner farmers are not included in these figures. Although there are changes suggesting an increasing middle class to accompany the development of more high technology industry and communications, Finnish society is still two-thirds worker-based. Also, one can expect in Finnish society the same kind of proletarinization of important segments of the so-called, middle class as Braverman has observed for other core capitalist nations where corporate structures have enlarged and computer and other high-tech-dependent work has become routinized and has displaced many workers into service industries or unemployment.[17]

The, so called, "upper" class, although only three percent, seems inflated in Table 6 by the definition used. These are people whose characteristic feature is "a feeling of social security, and independence based on economic independence or a highly developed sense of awareness."[18] From agriculture it includes large

Table 6. Social Stratification of Finnish Society, 1940, 1950, 1960

	Thousands			Percent		
Class	*1940*	*1950*	*1960*	*1940*	*1950*	*1960*
Workers	921	854	881	62	55	51
Farmers	356	311	280	24	20	16
Middle Class	163	342	518	11	22	30
Upper Class	45	46	52	3	3	3
Total	1,485	1,553	1,731	100	100	100

Source: Urho, Rauhala, "Yhteiskuntaluokka," *Yhteiskuntatieteiden käsikirja*. Helsinki: Otava, II: 1006-1011, 1964.

landowners; from industry and business, the independent industrialists and managers and well-to-do merchants; from service work, the highest functionaries, especially the self-employed who have a university education. In the framework suggested in Chapter 5, the ruling class is much smaller, those who own and control the means of production, probably much less than one percent.

In terms of voting power, the classes are aligned with parties in such a way that in 1945 workers' parties (Social Democrats and People's Democratic Union) received 48.6 percent of the popular vote, the Agrarian Party 21.3 percent and others 28.6 percent. In 1951, the corresponding figures were 48.1, 23.2 and 27.9 percent.[19] The governing coalition has usually included the Communist Party. In 1981 the Parliament was made up as follows: (1975 in parenthesis): Social Democratic Party 26.0 percent, seats 52 (54), People's Democratic League including the Communist Party 17.5 percent, seats 35 (40), Centre Party 18.0 percent, seats 36 (39), Conservatives 23.5 percent, seats 47 (34), Swedish People's Party 5 percent, seats 10 (10), Christian League 4.5 percent, seats 9 (9), Rural Party 3.5 percent, seats 2 (2), Liberal People's Party 2.0 percent, seats 4 (9), Others 0.0 percent, seats 0 (2).[20]

In important ways, this alignment of forces relates to the modern version of the age-old geo-political role Finland has played of balancing between East and West. Today, Finland is generally oriented toward Western capitalism, but Soviet influence is strong; thus Finland has played a balancing role. This balancing has sometimes offered advantages, at least economically. Finland has had free-trade agreements with the EEC and the U.S.S.R. since 1974, and similar agreements exist with most of the Comecon countries. In 1979, 16.2 percent of exports were to Comecon countries. While much of the capitalist West has experienced an economic downturn, ever since the oil crisis of 1974, and the rest of Scandinavia has experienced declining rates of growth since 1977, and even a net decline in 1981, Finland's eonomy has held up. It has been especially strong (comparatively speaking) in the last few years when other Scandinavian economies have been weakest. In 1979 it had a 7.2 percent increase in gross domestic product (GDP) while Sweden's was 3.9 percent. In 1980, the figures were 4.9 percent versus 1.4 percent, and in 1981, 2.0 percent compared with 1.1 percent. Many people explain this by Finland's stepping in to sell high technology to the U.S.S.R. when President Carter imposed economic sanctions on the U.S.S.R. over the invasion of Afghanistan. Possibly the current sanctions imposed by President Reagan because of Soviet pressure on Poland will have similar positive results for the Finnish economy, though before this happened, there was a sense of growing uncertainty, if not pessimism.[21] However, with the beginning of a recovery in 1983 in U.S.A. and other core capitalist nations, Finland is seen as oriented once again to the West primarily.[22]

While Finland's overall economy has been strong in recent years, its employment policy and social welfare measures are not as strong or complete as those of Sweden. In *Helsinki Guide* for Autumn 1981, the head of the Association of Advertising agencies (possibly with an eye on Western capital) lists 10 points about the country. Point 7: "Finland is not very socialized (sic) and the factories are not owned in large numbers by the State." Also, in spite of a strong economy, unemployment was at 6.6 percent in 1979 and averaged over 5 percent in 1981. These figures compare with Sweden's figure of around 2 percent. It is also the case that even though Finland spends a comparatively high proportion of GNP on social services, this does not match Sweden's expenditures, and the disparities in household income are greater in Finland than in Sweden. In 1962, the last year for which data from the World Bank were available for both countries (1963 for Sweden) Finland's poorest 20 percent of households had just 2.7 percent of income while the richest 5 percent had 20.9 percent. In Sweden the comparable figures were 4.6 percent and 17.4 percent.[23]

In 1980, inflation measured by the consumer price index reached 13.7 percent (but in 1978-79 it had been only 8 percent and hopes were to hold it to 10 percent in 1981). Real wages have risen every year since 1968, except for 1977 and 1978. The per capita GDP in 1980 was $8,900 (Finnmarks 39,100).

There is plenty of money and many government programs reflect this, that is, research and other programs (including the very important National Institute of Occupational Health about which more is written below) have not been cut as they have in U.S.A. and U.K. But the unemployment rate makes Trade Union leaders and rank and file cautious in their demands for improved OSH as well as improved wage and other benefits. There is a history behind this timidity which distinguishes Finnish labor from Swedish labor.

While Sweden was late in entering the age of industrial capitalism, Finland was even later. At the end of WW II, about half of the population was still agrarian. Today 13.8 percent of workers are still in agriculture, forestry or fishing, though some of the forestry is fairly heavily capitalized. This means that Finnish labor does not have the depth of experience in organizing that it does in Sweden. Even though the first Finnish union, Bookworkers, was formed in 1894 and the first Confederation of Unions (SAK) was formed in 1907, during the Great Depression, the semi-fascist Lapua movement succeeded in having the government dissolve SAK and its affiliated unions. Unions had a hard time until 1944. Collective agreements between Labor and Management were very rare up to WW II. "During the Winter War of 1940 the labor-market organizations finally acknowledged each other's existence and established the bases for collective bargaining in the General Agreements of 1944 and 1946. The terms of the 1946 agreement are still largely valid today.[24] The flow of collective bargaining is shown in Figure 1.[25]

This process is similar to what one finds in Sweden between SAF (The Employers Confederation) and LO (Blue Collar Confederation) and PTK (White Collar Confederation). But there are at least two important differences. In Sweden, the forces, especially union forces, are more concerted (there are fewer central organizations) and the central organizations, well backed up by careful internal political development over the years, speak with a more authoritative voice, even if component national unions and their locals must approve agreements reached at central level. Whereas there are two trade union confederations in Sweden, there are four in Finland. The largest of these, SAK (Central Organization of Finnish Trade Unions) was formed in 1970.[26] It has twenty-eight member unions and over 1 million members. The biggest unions in this group are the Metal Workers (163,000 members)[27] Municipal Workers' and Employees' Union (144,000) Joint Organization of Civil Servants and Workers (115,000) Union of Commercial Workers (100,000) and the Building Workers' Union (90,000). The TKV (Confederation of Salaried Employees) was founded in 1922. It has nineteen member unions with some 325,000 members, the majority of whom are employed by State or local authorities. TVK has recently won a reputation of being especially strong in representing the interests of women workers. The Akava (Central Organization of Professional Associations) was founded in 1950 and has 165,000 members in forty-seven unions. Teachers and Engineers are big in this Confederation. The STTK (Confederation of Technical Employees' Organizations) was founded in 1946. It is made up of thirteen unions with some 120,000 members. Technicians are big in this group. Clearly there is potential for competition and even conflict within and between some of the above combinations.[28]

The employers too are more divided. Whether this leads to weakness may be arguable; perhaps it offers less of a concentrated target and is therefore harder to influence. There are seven confederations! (This compares with Sweden's one in the private sector and one for public employers). The largest and oldest of these is the STK (Finnish Employers Confederation) formed in 1907 with 4,800 member companies employing some 590,000 blue-and-white collar workers.

Thus, while Finland's workers' movement must be rated as strong (it has about 80 percent of both blue-collar and white-collar workers organized and has periodically had a labor-oriented government) it does not have the same

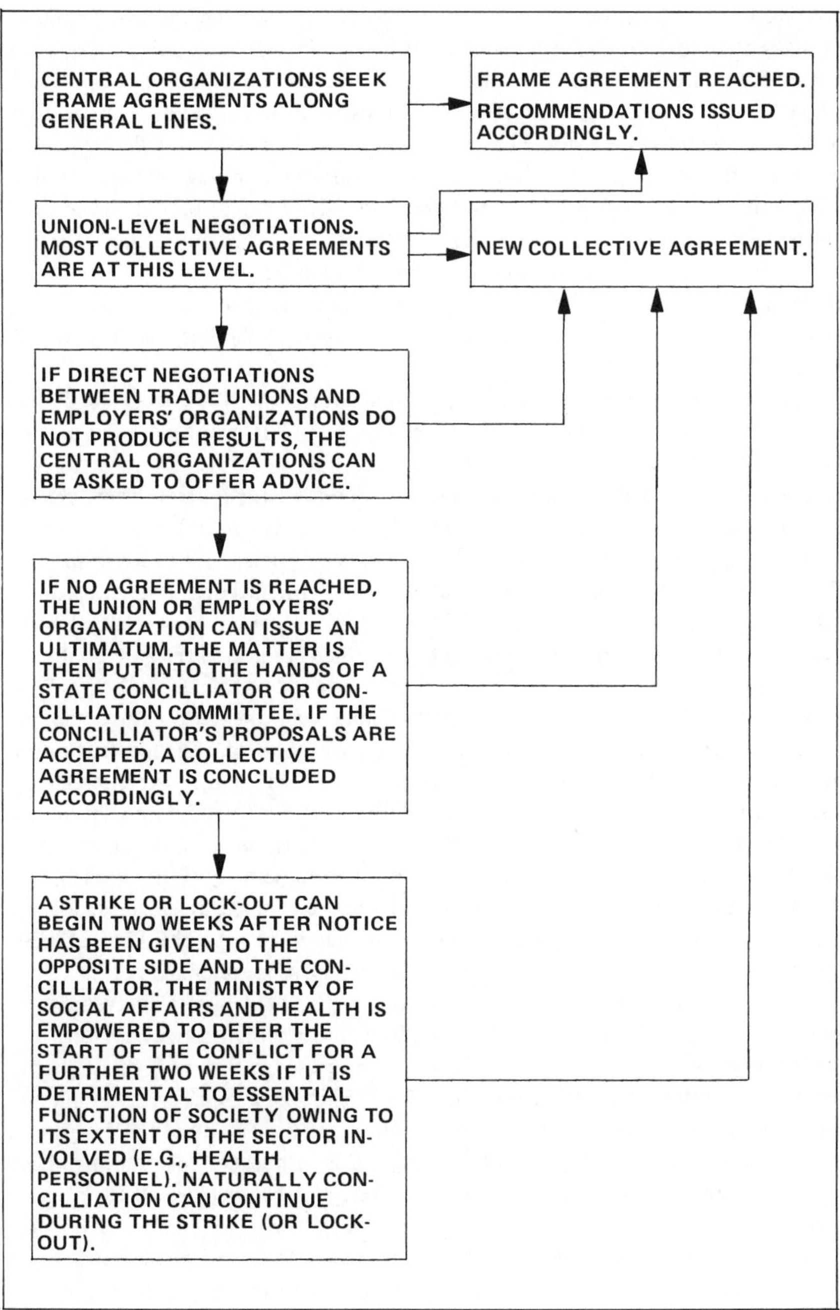

Figure 1. Collective bargaining in Finland—a flow chart.

continuity in power or organizational strength as does the Swedish workers' movement. Thus, for example, people just stared in wonder when, in a meeting with OSH and other officials of the Metal Workers Union, I said LO could seek to influence medical school curricula as regards PHC and OSH because they have representatives on the boards of every kind of school. Whereas it is *de riguer* in Sweden to include LO and PTK on all major government commissions and in major policy bodies and decisions, this kind of participation is not always open to Finnish workers' representatives. Thus at central level there is less influence of organized labor in Finland than in Sweden. At local level, there may be almost as much, with great variations from factory to factory and locale to locale, especially, as noted earlier, where the heritage of paternalism is strong.

For example, I visited a large paper mill with some 4000 workers. It was formed in 1910 on the base of an earlier saw mill. This giant works in a town of 25,000 has been the dominant element for many generations. It was organized, but clearly had a long history of paternalistic control. There one of the most disturbing experiences of my trip occurred. I found myself standing outside one of the main gates with the company physician and a physician from the regional office of the National Institute of Occupational Health at 10:55 AM, yet we could not go in. More startling, some forty to fifty human beings–workers–who had ridden their bikes or had walked to the gate could not come out. They were locked in! They could not get out until the bell rang at 11:00 AM signifying the official shift change and the automatic lock was released. I can not always read faces, but the looks of resigned frustration seemed unmistakable. As we finally went in, the guard on duty exercised extensive security precautions while several TV monitors for surveillance were within his view. Lunch for the three of us was in a private dining room at an exclusive club for management level personnel. I can not complain about the food. It was delicious. But the elitism, paternalism and hierarchism of the place made me very uneasy. I believe this overall condition related also to the kind of OSH services found at this plant. I'll mention this further subsequently.

There is a subtler and less certain aspect to the workers' movement in Finland. One gets the impression that it has achieved practically everything that has been achieved in Sweden and other countries–on paper.[29] But it is my feeling after interviews with some union leadership persons, as well as rank and file, that many of these things are less realized in reality and seem less rooted and less well-known as clear rights. Perhaps this is because many things have been adopted by well-intended leaders but have not been struggled for at grass roots levels.

In part at least, this seems to be so because the state has played a more directive role on behalf of the interests of capital and the ruling bourgeoisie than in Sweden where the government was more or less continuously under Labor's control from 1932 on. Thus there have been vigorous strikes and serious instances of labor unrest and strife in Finland. In 1956 a general strike resulted in seven million workdays lost. But state controls have been instituted:

> Since the occurrence of widespread strikes around the turn of the century, work stoppages caused by mass labor disputes have occurred regularly, mostly during periods of economic well-being and political unrest when workers have the most to gain and disputes are most easily intensified. . . .
>
> Because labor tranquility is important from the viewpoint of the entire national economy and society [read capitalist owners and ruling bourgeoisie], the state has undertaken steps to prevent and reduce labor confrontations. . . . In accordance with a 1962 law on mediation of labor disputes, there is a full-time national mediator and five part-time district mediators, in addition to which the ministry for social affairs can designate temporary mediators or mediation boards. . . .
>
> Freedom of action by the disputants is limited in two respects. There can be no work stoppage (either strike or lockout) unless two weeks notice has been given to the opposing side and to the mediator of labor disputes. If it is felt that an intended work stoppage, or its extension, will affect society's vital activity or will significantly endanger the general interest, the ministry for social affairs can, in order to provide sufficient time for mediation, forbid a work stoppage or its extension for up to fourteen days from the moment that the existence of a labor conflict is announced.[30]

Thus the eye tooth strength of the workers' movement—the right to strike—has been partially pulled.

Finland has no long history of something like the County Councils playing as large a role as they do in Sweden. The Constitution provides for local self-governing Communes able to levy an income tax. The present 464 Communes headed by elected councils play an important role in education, health, welfare, public works, and land utilization and planning. But because of the lack of a long history of local determination and, instead, a history of obedience to central authority mentioned above, and other factors, I offer the overall assessment that Finland has an authority and governing structure which is more centralized than Sweden's. Also, because of the splintering of political parties and other groupings, I would say Finland is less concerted. This does not mean that it is highly centralized and very fractionated. As we will see, the local Communes play an important role in financing and control of health services.

THE HEALTH SYSTEM

The national health insurance and services offer very complete coverage—for those who can not afford private services. Thus the Finnish system is essentially dualistic. Even the ambulance services, so important in a country with many rural based residents and great distances, are only 40 percent publicly run—by local communes: the other 60 percent are privately run, with costs to patients partly compensated by insurance.

On a visit to a municipal health center in Helsinki one morning where I was to speak with a general practitioner about OSH, I saw some forty to fifty patients waiting to be seen. In general, the clinic seemed busy and crowded, even though it was said to have an appointment system. The doctor with whom I spoke informed me that she and six other doctors and three nurse practitioners are assigned to care for an area of 80,000 population. "So the government in fact expects as many people as possible to go to private doctors. They encourage this by giving minimum support to this center." The charge for a visit to a private doctor is "about 7 pounds sterling," she informed me—a little over $15.00. Only some 5 percent of Finnish doctors are exclusively in private practice, but many hospital doctors on full time salaries (one estimate is 30 percent) also run private practices in the afternoon and evening (health center doctors who are paid for thirty-six hours get fee-for-service for overtime and may also run small private practices). When I asked if this system led to sharp class differences between patients seen in the different PHC settings, she said, "Yes; it is a sad thing. So there is a class difference also between the doctors. Most of the private doctors are specialists. We are general doctors." She went on to note that the big cities probably feel this split in the system most acutely, for under the new PHC policy[31] municipal health centers in medium and small towns are being well enough supported to draw physicians to them.[32]

The statistical picture suggests a more inter-mixed public/private system nationally. Though persons in the lowest fifth according to income made a higher proportion of their total physician visits to municipal or public health centers in 1968 (about 58 percent or 52 of some 90 visits per 100 adults during 100 days—the measure used in the study) those in the highest fifth according to income also used the public physicians but to a lesser extent (about 47 percent or 38 visits out of 81 per 100 adults during 100 days). The poorer group used private physicians less (these services are also public in the sense that 60 percent of the fees are reimbursed by the National Health Insurance). Thus the lowest income group made some 18 of 90 physician visits to private doctors (20 percent) while the highest income group saw private physicians about 21 times out of 81 visits (25 percent).[33] To further complicate the picture, some employers cover employees' expenses for visits to a private physician as a voluntary (or collectively bargained) addition to the OSH services which are now mandatory for all workplaces.

The sickness insurance pays on average only 45 percent of lost income from work (recall the Swedish figure of 90 percent). The upper limit is $8.70 a day (Fmk 38) with an additional amount for dependents. Payment begins only after eight days from onset of an illness and can run for fifty weeks. Care in public health centers and hospitals is essentially free with only small patient fees. The

National Health Insurance for private care covers 60 percent of fees as long as these are not above a set tariff. Compensation for medicines is 50 percent. Funding of this system comes from 1.5 percent of gross wages of employees and 2 percent from employers.

While private, market-oriented care plays a significant role, especially in some settings, "the health centre system has grown very quickly in the last seven years. Nowadays about 75 percent of primary health contacts take place at communal health centres; the rest are handled by private practitioners and subsidized by sickness insurance."[34] These centers are now the main providers of PHC and are conceived under the Primary Health Care Act of 1972 as the source of comprehensive PHC for the entire population in the future. This should be offered in an integrated manner with equality of distribution of resources according to the act and free services for the individual.

There is a significant degree of community involvement in these centers as they are largely financed and controlled by the Communes within the national guidelines of the PHC Act. Thus the minimum population to be served is defined as 10,000–15,000 and in no case less than 8,000. In no case should there be fewer than three physicians to allow free time, proper coverage and colleagueship, though this is not always enforced. A number of the some 470 Communes have "federated" in order to meet these size limits. There are now some 213 public PHC centers, 100 of which serve a single Commune; the rest serve two or more Communes. In addition to physicians, there should be other personnel to assure that a number of services are provided in a functional whole with the personnel functioning as a team. While many of these concepts sound like the latest buzz words when they are compared with the realities, the thinking and actual movement which has gone into the Finnish PHC system is quite impressive.[35]

Without saying exactly what agency (the health center or others?) will do each set of tasks, the act makes the Communes responsible for a broad range of PHC services:

- health promotion and disease prevention, including maternity and child health care, health education for the whole population and special guidance for the aged, advice on family planning, and screenings
- primary medical services for the population, including most common x-ray, laboratory and rehabilitation services, home nursing and also local hospital services especially for those in need of long-term care
- school and student health services
- dental health services
- ambulance services
- occupational health
- participation in the training of health personnel[36]

For our purposes here, the inclusion of occupational health in this list is especially noteworthy. We will come back in the next section to remarking on the striking difference between this conception and the organization of OSH services in Sweden.

While the conception of the health system in Finland is in many ways admirable, there are problems with its realization. We have already noted the fractionation caused by a dualistic private/public system. Also, while there is a conception of geographic orientation of the system—e.g., for the PHC centers and their Communes—and of hierarchy of referral and regionalization, the hospital catchment areas and referral responsibilities are not well articulated with each other, with the PHC centers, or with the central hospitals. In part, this is because the system of central or regional hospitals to back up local Commune-run hospitals was only completed in 1979.[37] Part of the confusion stems from the lack of well-defined intermediate administrative units and authority. There are twelve regional or provincial offices of the National Board of Health, but there are twenty-one central or regional hospitals.[38] Also, like the Swedish system, it is too hospital focused. At the end of 1978 there were 15.5 beds per 1000 inhabitants.

> By the start of the 1970s the country had reached a state where 90 percent of all health expenditures went on hospital treatment, with specialist medicine taking the lion's share, leaving a mere 10 percent for primary health care.[39]

The heavy financial burden for local hospitals and hospital care is shared between local and central governments, the central subsidy running from 39 percent to 70 percent depending on the prosperity of the Commune, the average being 50 percent. More or less this same system works for the central hospitals with Communes "federated" to pick up their share of one of the twenty-one central hospitals. The same formula applies to health center financing if expenditures fit within plans previously submitted to the national authorities. This gives local authorities a strong incentive and central authorities a strong tool to reorient the system toward PHC. Whether this will lead to lower overall costs seems highly questionable.[40] At the time of the PHC act in 1972, overall health care expenditures had been rising at twice the rate of the GNP. As is true for other industrialized societies, this relates to higher morbidity and medical care costs associated with a rising age structure. In 1950, 6.6 percent of the population were sixty-five and older, by 1975, 10.8 percent were in this age group. The total health care cost figures in relation to GNP for recent years is not known. Public expenditures per person on health in 1980 amounted to $410 while GNP was $10,333 or 4 percent.[41] These were only public monies and, as already noted, private funds play a big role in health care. But in 1975, for the first time, capital expenditures in health center facilities (about $50 million) exceeded investments in hospitals (about $35 million). Also, salaries for health center physicians have been raised to equal and sometimes surpass those paid to hospital physicians.

Much more could be written about the Finnish health system. But my purpose is to focus on the OSH-PHC relationship. Thus in conclusion of this section only a set of general observations are offered on the overall situation.

Finland is relatively healthy. Its infant and maternal mortality rates are among the lowest in the world. But 19.5 percent of people forty-five to sixty-five were on disability pensions in Finland in 1974, compared with 10 percent in Sweden, 10.8 percent in Norway, and 10.6 percent in Denmark. An exceptionally high mortality of working-age men, especially from coronary heart disease (CHD) has led to the saying, "We have the healthiest bodies and the sickest men." The situation was serious enough to call for the launching of a massive way-of-life, health education, health surveillance project in North Karelia. Intermediate results have now been reported and suggest a variety of successes,[42] though there are arguments as to simultaneous but unstudied influences on the adjacent comparison population which experienced similar declines in CHD as well as arguments over other methods questions.[43] Perhaps, in addition to pressure from organized labor and other groups, it is especially because of such deficits in the health of the working age population, that recent major steps have been taken in Finland to improve OSH.

THE OSH SYSTEM AND SERVICES

The problems of work disease, injury, and death are widespread and serious in Finland as in other study countries. Curiously enough, but not too unexpectedly, considering the vagaries of OSH information laid out in Chapter 4, the reported problems have become more numerous, if not more serious, as the awareness and concern accompanying the legislative OSH spurt of the 1970s have begun to take effect in perception and reporting.[44]

Detailed calculations indicated that "at least a third of the Finnish labor force still works in an environment which involves daily exposure to agents or factors containing health risk."[45] "Rough approximations" of the occurrence of a range of risk factors (not all) are given in Table 7.[46] This table only serves to suggest some of the dimensions of the problem (it seems a bit absurd to even offer an estimate of 9 percent for those exposed to "psychic strain" at work when the real figure must be many times this, depending on what definition and consequences one is willing to consider).[47] In 1981, the rate of reported accidents causing at least three days' absence from work was 620 per 10,000 employed workers. The accident rate has been generally increasing but has also fluctuated strongly according to economic conditions. In times of economic downturn, unemployment is high, so fewer people are exposed for fewer hours; thus there is less turnover of personnel, thereby lessening the risks from newness to and unfamiliarity on the job. Also, in depressed economies the production pace is usually slower, allowing more time for production planning and adequate time

Table 7. Approximations of the Numbers of Workers Exposed to Various Occupational Risks in Finland and the Number of Registered Consequences Per Year[45]

Exposure	*Population at risk*	*Percent of total labor force*	*Number of registered occupational diseases or accidents in 1981*
Noise	300,000	14	1,856
Chemicals	1,000,000	45	1,307
Monotonous work	200,000	9	1,396
Psychic strain	200,000	9	–
Accidents	1,300,000	59	115,125

for production maintenance, etc. In the late 1970s unemployment was high and especially so for young people. In 1977-1978 it reached 27 percent in the age group fifteen to nineteen and 13 percent in the group twenty to twenty-four.

> Young men (age group 15-24) have an accident frequency rate which is around 140-170 accidents per 10^6 manhours. It is twice the accident frequency rate of men over thirty. Due to unemployment, approximately 20-25,000 occupational accidents did not occur in 1978.[48]

The incidence of reported occupational diseases was 23.7 per 10,000. This represented a continuation of the increase during the 1970s and in 1980. About a quarter of these diseases were skin problems due to various chemical exposures. About half were due to "physical factors" (presumably back problems and other ergonomic problems). Others were lung and other diseases related to chemical and biologic exposures and to undefined work hazards. Fatal accidents have declined in recent years. "However, the success in preventing traffic accidents [those to and from work, also counted as work-related] was better than that of preventing workplace accidents."[49] Losses caused by work accidents and diseases were estimated to approximate 4.9 billion Finnmarks or (according to a recent exchange rate) some $853 million.[50]

Unlike Sweden where nearly the whole OSH system has developed through the struggles of the workers' movement and direct negotiations with the employers' confederation, Finland's OSH system has developed much more through legislation and central expert guidance, often with an eye on what has happened in Sweden and elsewhere. This is only a relative statement, for there is obviously a history of labor's struggles over working hours, wages and conditions[51] and many OSH elements have grown out of these, some unique to Finland and some common.

Among the common elements in OSH history is the development of medical services for large paper mills and other large mills and factories as these often were in isolated areas where workers and their families had no other medical services. Even today this history may have a bearing on the problem under study, the OSH-PHC relationship. In the large paper mill I visited, the chief plant physician said his clinic continues in many instances to treat family members of workers and yet there are easy relations with other physicians in the town and hospital. "After all, the company once owned the hospital," he said.

Thus, as in Sweden and other countries, the large mills, plants, factories, mines and other undertakings are those with the most coverage with special OSH services. Whether these services are adequate and whether workers have and exercise an authentic voice in determining these services will be considered.[52]

The Industrial Inspection Act passed in 1927 was mainly aimed at general work conditions and safety matters and remained essentially unchanged until the 1970s.

The first Labour Protection Act regulating OSH was passed in 1958 and is still in force, though importantly modified through the law of 1978, about which more below. In 1972 the "Administration of Labour Protection Act" and in 1973 the "Act on the Supervision of Labour Protection" were passed. These replaced the 1927 law and made labour protection one of the responsibilities of the Ministry of Social Affairs and Health. A Special Labour Protection Advisory Board with representatives of the most significant labour market organizations was set up to deal with questions of principle, unification and cooperation in the field of labour protection.[53] The overall structure in relation to health and welfare is given in Figure 2.[54]

Actually the situation is more complicated than this. Work at Sea comes under the Safety in Maritime Work Act. The Mining Act contains safety provisions for miners in addition to those found in the Labour Protection Act of 1958 as revised in 1972 and 1973. Provisions affecting other worker populations are found in the Employment Contracts Act, Young Workers' Protection Act, the Domestic Workers Act, and other labor laws. But the principle provisions are in the 1958 Act as revised. This is supplemented by Government orders, orders issued by the Ministry of Social Affairs and Health, and instructions from the National Board of Labour Protection. These last concern:

- machines
- economic sectors (e.g., construction)
- particular hazards (e.g., noise)
- dangerous chemicals (e.g., asbestos and other carcinogens)

These are binding on employers, importers, subcontractors, etc.[55] There is also an Occupational Disease Act which gives compensation to workers for work-related disease under the same terms as laid down in the Industrial Injuries Act. These insurances are required of all employers with more than twelve employees

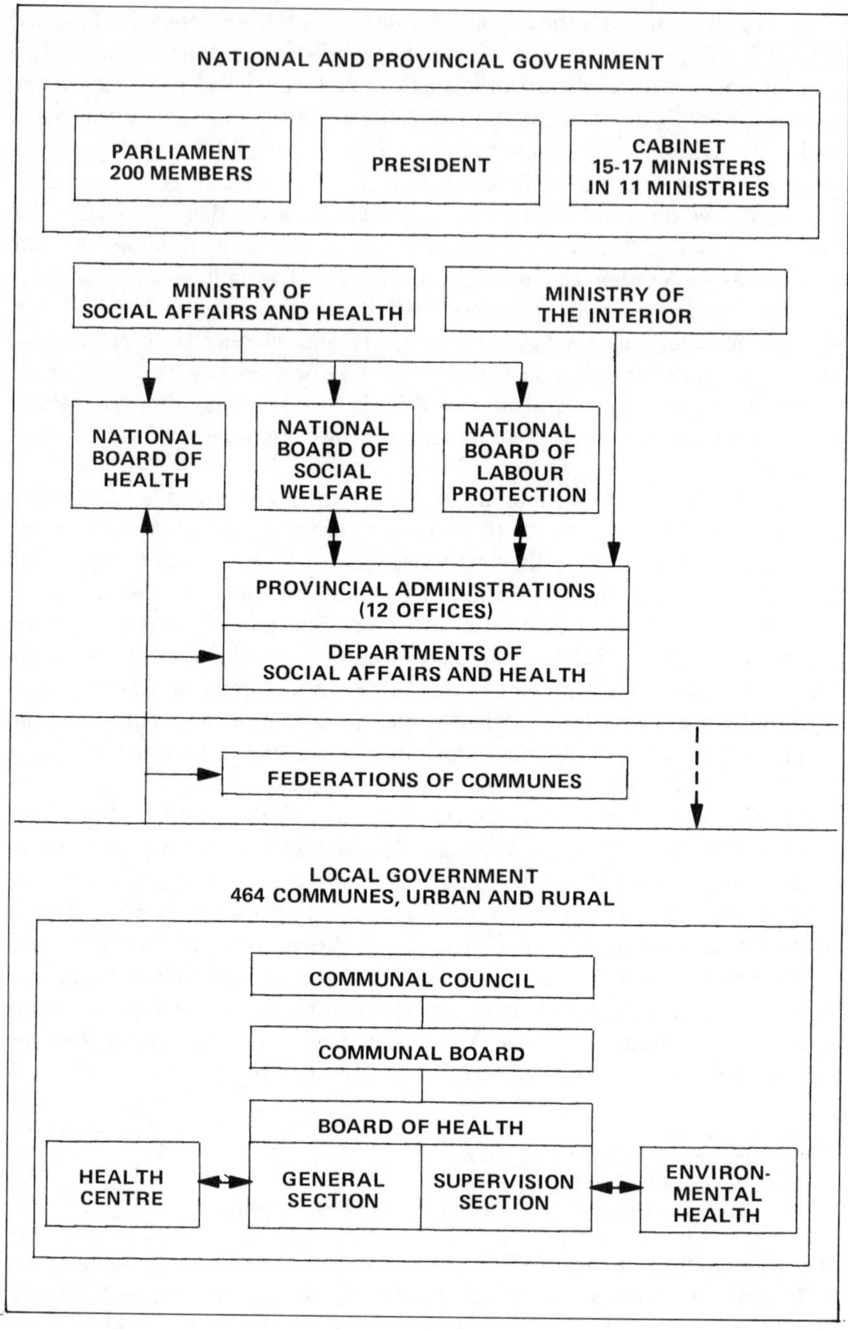

Figure 2. Health administration in Finland.

and 60-80 percent of pay after taxation is allowed. If incapacity is permanent, compensation can reach 90 percent of pay under a limit (recall that in Sweden it is 100 percent after the first ninety days if proven work-related, whether the disability is permanent or not, with no limit on salary level).

> All these developments have led to a rather complex system of labour law and norms in Finland. At the lowest level, the employee and his employer can agree on individual terms of employment and working conditions. Secondly, labour-market organizations sign collective and general agreements which also regulate terms of employment between employers and employees. And finally there are laws regulating both individual terms of employment and collective relations between organizations.[56]

Still one more very important and relatively unique component must be mentioned—the Institute of Occupational Health. It deserves more space than can be given here, since it is such a central reference point for the whole OSH system and has just become a reference/training center for all of Scandinavia with secure funding from the Scandinavian Conference of Ministers. In 1978 the Institute became a semi-autonomous government body. Its Board of Commissioners of thirty people and Board of Directors of thirteen include representatives of the major labor market parties as well as government and scientific medical and technical persons. In the 1970s: 1) it issued more than 2,000 research and investigation reports; 2) ran a network of national and regional clinics in which 50,000 persons were examined for work-related diseases; 3) carried out close to 250,000 industrial hygiene measurements; and 4) distributed considerable educational material and gave 200,000 person-course days of labor protection education to workers and professionals.[57] The publication list is truly impressive as is the description of current research.[58] In fact, the overall work is so impressive, one wonders in what ways the National Board in Tampere (the equivalent of OSHA in U.S.A.) may be a duplication, even if the Tampere Board has more statutory authority (though the Director General of the National Board is Chairperson of the Institute Board, which may aid coordination).

In summary of these many new protective laws and related changes, we can note that in the late 1960s through the 1970s, as in other study countries, there was a veritable rash of measures as seen in Table 8.

To understand the real extent and depth of this complex machinery in offering actual protection to workers, we have to go back briefly to recall the situation as regards strength of the workers' movement. As noted earlier, the proportion of Finnish workers who are organized is very high—perhaps 80 percent overall. Some branches are virtually 100 percent organized—engine drivers, teachers and nurses. Some 95 percent of metal and engineering industry workers are unionized as is true for seamen, and workers in paper and printing.

Table 8. Some of the Organizational Measures Taken During the Past Ten Years to Improve Occupational Health and Safety Conditions in Finland

Year	*Action*
1969	Agreement between parties of the labor market with regard to occupational safety at workplaces. Result: Center of Occupational Safety and branch committees for occupational health and safety work established.
1973	Acts passed concerning occupational health and safety control at workplaces. Result: a. National Board of Labor Protection established and the factory inspectorate revised. b. Became necessary to establish a safety committee in companies with more than twenty employees and select a safety representative of the workers in companies with more than ten employees.
1974	Labor Protection Laboratory established in the State Technical Research Center.
1974	Professorship in safety engineering established at the Tampere University of Technology and improved possibilities to teach in this field arranged at other universities.
1978	Acts passed concerning the Institute of Occupational Health, with no immediate consequence—the Institute had previously been strengthened, reorganized and its resources improved (1973 on).
1970–1980	Many other new activities started and activities rearranged within various organizations.

Source: Jorma Saari, op. cit. (note 48).

But some sectors are well below 60 percent and it varies for public-sector and white-collar workers. Also, there was little organized strength at all until after WW II and the present level was only reached after rapid growth in the 1970s.

Thus there has been and there still is much room for an activist labor/medical/technical group (a COSH group in U.S.A. terms) to function on a voluntary basis—the Socialist Health Front. Its twenty local departments spread across the whole country. There are some 200 organizational or collective members, the majority are union locals representing thousands of workers in different sectors; and there are some 3000 personal members. The Front's interests and actions extend to "health care, social policy, working and living environment, as anchored in the interests of the working people."[59] The Front publishes a magazine issued five times a year–*Terveysrintama*. It has also issued books, student texts, pamphlets and posters. One of its most exciting efforts with important implications for the future is its critical analysis of management-sponsored "work humanization." This is seen simply as another more sophisticated or subtle sociological approach to *intensification of labor* which can (in addition to "humanization") be analyzed in terms of the technical (production flow, machines, etc.) the organization of work (standardization, piece work, etc.) physiological (ergonomics) and psychological (incentive systems).[60] A great deal of the action has been to improve OSH. For example, since the legal status and completeness and stringency of standards has been an issue, the group developed its own book of standards with input from workers and medical/technical people. In one case, workers in a foundry examined the book in their union's "study circle" and decided that the heat standard applied to them–i.e., to take a fifteen minute break every hour when the work room temperature is over 100°F. When they put this into practice, the employer complained and asked the Institute of Occupational Health (interestingly, not the National Board) to send out investigators. Two Institute people found it so hot in the work area that they too took fifteen minute breaks every hour, so the story goes, so the workers won. The group has also offered an annual national weekend seminar as well as other local courses taken by many workers.

Curiously enough, some especially knowledgeable government and industry OSH experts denied to me the existence of any such group when I described the COSH groups in the U.S.A., the Aktionsgruppen Arbejdere Akademikere in Denmark, and the once active SARB in Sweden. In semi-joking reference to the Danish group, one of these authorities said, "Denmark is rotten. Finland is more built on consensus." Some other informed people admitted knowing of the group and said it had had a larger impact in the early 1970s than it has had recently. This, they said, was because the Social Democrats withdrew and formed a similar, but not very active group of their own. In any case, it seems to have been at least a noteworthy gadfly (the denials of some officials notwithstanding) which helped foster the important legislative measures of the 1970s.

One of the most important of these measures and the one of most direct concern to this study, is the Occupational Health Care Act of 1978.[61] This act continues an agreement on OSH services reached in 1971 between major labor market parties but extends it by law to all work places including farms and small shops. Part of the continuity is that while employers are responsible to provide the care, 60 percent of costs will be reimbursed by the general Sickness Insurance Fund for both preventive and curative care, though the hope is that prevention will be emphasized. Historically, the expectations of workers for clinical care and the preparation and orientations of most plant physicians are not likely to lend much support to this hope. Nevertheless, the reimbursement system could be more encouraging of preventive care than that in Sweden (where 50 percent is reimbursed exclusively for personal care) though the system will in no way approach the exclusive preventive orientation of the OSH system in F.R.G. (which is largely market-determined and perhaps less satisfactory in some other respects, as we will see in Chapter 10). In order to cover all workplaces, and in keeping with the dualistic public/private character of the overall health system, two outside-plant forms of service are provided for, since smaller workplaces can not afford their own in-house services–the municipal OSH center which may be an offshoot of the PHC centers discussed earlier; and the private OSH consulting surveillance service. I visited examples of all three and will briefly discuss some strengths and weaknesses of each. But before doing so, I want to come back once more to considering labor's position in relation to this act.

Obviously, in small and other shops where labor is not organized, the workers will not have a strong voice. Nevertheless, the act requires that information be given to workers on hazardous materials and conditions "and any guidance they may need in order to avoid such hazards" (para. 3.(2)). They must also receive exams after a physician is recruited if they are engaged in work involving any special hazards or if there is reason to suspect such hazards (para. 2(4) and (5)). There are also provisions for workplace visits to search out hazards, to conduct job-placement exams, and for follow-up and rehabilitation as well as proper referral of injured or ill workers. Also, before an employer undertakes any of these services or alters them once in operation, "the matter shall be discussed by the labour protection committee . . . or, where there is no such committee, with the labour protection representative" (para. 4). The catch here is that even though the labor protection committee has a majority of workers (for example, the largest sized committee in a big plant will have twelve members–six blue-collar, three white-collar and three from the owners with the medical and technical people sitting in as advisors) they may not all be in the same union, and they do not have veto power over the hiring of a physician, as they do in Sweden. In many respects, the committee is only advisory.

But one advisory function which gives the committee real power is that it must see and sign off on the OSH services plan for the coming year before the Sickness Insurance fund will pay the 60 percent reimbursement. If the committee says no, the Sickness Insurance asks the Institute of Occupational Health. If they say no too, the money is not paid. There have only been twenty or so such cases and in most the Institute backed the employer, because, according to the Director, the OSH service plans were OK, the problem was the workers expected a more treatment-oriented program. Other signs of comparative weakness of workers include the fact that health and safety delegates cannot legally stop the production process in the face of a long-term or even an immediate hazard as they can in Sweden. Of course, where there is a well organized local, there are ways of "tagging" a machine or other hazard to take it out of service until the hazard is removed. But here we are discussing formal powers and in general one must say they are less for labor in Finland than in Sweden. Obviously, the problems of authentic labor participation, not to mention control, are more difficult in the case of the two types of out-of-plant services, especially the private one.

Extensive training has been organized through the unions and the Institute and there is a useful book covering all aspects of labor protection for this purpose.[62] The reported training has been about as extensive as in Sweden:

> The activities at the local level are carried out by municipal health authorities, municipal labor inspectors, and plant-level safety commissions, safety delegates and occupational health physicians and nurses, who all are present in the local organizations because of legislation. The total number of personnel in the Finnish occupational health and safety system comprises more than 100,000 persons (about 5 percent of the labor force) who have received specific training for occupational health and safety functions (1 + 3 weeks). About 3,600 of the 100,000 are full-time occupational health and safety experts.[63]

But one well informed person said the system of safety delegates does not work as well as in Sweden. He said this is because of Sweden's tradition of worker/management cooperation. In his view, Finland is more adversarial, like the U.S.A. But I think labor is simply stronger in Sweden and management goes along, so it looks like cooperation.[64]

In order to provide the physicians and other manpower necessary for the new coverage required by this law, a major push has been mounted. There are now two medical specialties for OSH (technically, only one, as one is a subcategory of the other).

The older one, of which there are very few in number, can be called "clinical industrial medicine" and is largely based in hospital teaching centers. The new one can be called "occupational health" and will function in one of the three forms of plant-related OSH services named above. The Institute of Occupational Health will play a major role in the preparation of these specialists. It will be a big job. While the act says 100 percent of workers must be covered by 1983, only a little over half were covered in 1981. In 1981 there were some 1000 physicians involved to some degree in OSH; also 1200 nurses and 800 others. Many more are needed and only 300–500 of the present OSH physicians have a minimum of training in OSH, especially on the preventive aspects. The real hope (perhaps optimistic) is by 1990 to have all places covered with specialists who have had at least forty to sixty hours of training and who are full time OSH people who regard their work as a life career. All in all, thus far, the implementation and training mandated or implied by the 1978 law seem to have gone very well indeed:

At the moment more than 75 percent of the active labor force is provided with such services. The areas not covered are small industries and the self-employed, e.g., farmers. The National Occupational Health Service Program, which was partly given in the form of a new Occupational Health Service Act, stipulates the provision of preventive occupational health services for all employees independent of the branch of industry, size of the company, or of whether one is an employee or self-employed. Theoretically, then, coverage is 100 percent. The implementation of the act has taken place in five separate steps according to an order of priority. The last step began 1 August 1982, and the entire act will be in Force by 31 March 1983. Numerous problems, concerning the content and methods of occupational health services for farmers and the self-employed, have been met in the stages of implementation. Therefore two major research programs have been started so that more information can be gathered on the most effective methods for providing services in agriculture and to small enterprises.

In spite of the difficulties already mentioned the enforcement of the new act has proceeded well. At the end of the third stage of implementation, 89 percent of the companies controlled by labor inspection were covered by occupational health services, and 96 percent of the labor force working in the branches of industry of stage 3 were provided with services . . .

Specific training curricula for occupational health physicians and occupational health nurses, have been prepared, and a new medical specialty in occupational health services has been established. The old specialty in occupational medicine, which has existed in Finland for twenty years, was simultaneously transformed to a sub-specialty for occupational health services. After full implementation of the program the Finnish occupational health services will employ 1,000 physicians and 2,500 nurses, which makes one physician per 1,800 workers and one nurse per 740 workers (ratios which have already been achieved for 75 percent of the active labor force). This situation implies that 500 new occupational health physicians and about 1,000 nurses should be trained during the next few years, as some of the presently working occupational health personnel also needs training. Since the general primary care services will be given by the municipal health center without charge, the capacity of occupational health physicians can be effectively directed toward preventive work.[65]

One of the curious aspects of the structure is that the National Board of Health (run by physicians) is responsible for assuring the quality of the medical part of OSH services, the Institute does the research and training, and the National Board of Labour Protection is responsible for enforcement of standards and the rest. It is unclear what kinds of conflict and cooperation or lack of enforcement this leads to, but it shows that medical people like to have medical and not other kinds of people supervising their work.

Now some observations on the three types of OSH service.

In the large, paternalistic sawmill, papermill, and plywood factory I visited, the chief doctor is well trained, has a sizable staff and very complete program as required in the law. This program was pretty much in place long before the 1978 law. It involved pre-employment and placement exams, special hazard exams, rehabilitation and follow-up, plant visits and hazard assessment, some worker education programs, etc. Two rather curious things came out related to the role of the workers and unions. The pattern is for the several nurses to each spend a full day a week out in "her plant." When I asked what they do out there, the doctor said, "they do hazard assessment, take measurements like hearing, etc." I said, "Hearing measurements out in the noisy plants?" He answered, "Well, you know, just to show the workers you're doing something. What they are really doing is public relations." When I asked if the Labour Protection Committee has to approve his plan (apparently here there is a committee for each of a number of plants) he said, "Oh yes. Every January is nothing but meetings. But they are very brief. I say here is what we've done—number of exams, treatments, etc. and here is what we're going to do. And they say 'Yes, OK'." I asked, "There's no discussion?" He said "Hardly ever." Another unfortunate thing is that about 1000 workers—mainly women, interestingly enough—were heavily exposed to epoxy in the plywood factory for a couple of years in 1968. There were many instances of acute reactions so they instituted more controlled use of the substance. But there has never been any systematic follow-up of the 1968-70 cohort for long-term effects. When I asked, the answer was "Well, I've had ideas about a lot of things that could be done, but there's never enough time." We had time for a very posh and pleasant two-hour lunch in the management's private club. Unfortunately, sociologically speaking, we have to consider the possibility that the whole OSH service has more of a "PR" function than a real, preventive function.

On a visit to a medium-sized communal OSH center covering 3,600 workers in 137 workplaces I found its staff well prepared and fairly complete (the need for industrial hygiene services is not well figured out, but this may be a national problem). Also, the conception of its work seemed fairly comprehensive and there was an impressive record and information system, including a computer linkage which would also show worker visits to PHC centers (if the information was sent back by the PHC Center or private practitioner). But there was little evidence of worker involvement and few efforts at worker education. Also, if there are 300 employers in the town and one large one has its own OSH services and the private OSH center (see below) covers forty-three workplaces, that still leaves 119 employers without OSH coverage in this area. The building industry has been hardest to get involved. The space in the Center's office is small, so much of the examining work is done at the plants. Often the space there is also inadequate. They have no mobile van, so sometimes they do a hearing test, for example, in their car. In spite of the relative lack of worker control and involvement, one strength is that the staff seemed worker-oriented and concerned and they are able to maintain this position, because they are on salary from the municipality and independent of direct management control (though managements must agree to sign up for and help pay for their services).

At the private OSH center covering forty-three employers with some 800 employees, I spoke with the nurse who really runs the OSH program. Her putative boss is a part-time physician with only a small amount of OSH training. This physician was not available, so the young Associate Director of this private group practice—with forty part-time physicians covering everything in addition to OSH—met with us. He is a hypertension specialist and had to defer to the older, more experienced nurse for answers as to the OSH program (though in male-dominant fashion he expected all questions to come to him). Although he finished medical school only 3½ years ago in Helsinki he said he had "only a small course, I think a few hours" in OSH which included a visit to the Institute of Occupational Health. The nurse had no OSH in school (she finished in 1963). But she became a public health nurse and had a two-month OSH course at the Institute of Occupational Health. She said there is still nothing on OSH in basic nursing education. The OSH staff is the part-time doctor, one chief nurse, three other nurses and two office workers. One of the motivations for this private group to do OSH is that employers often agree to contract for all care with them when they sign up for OSH. This is true for about two-thirds of the workers covered. "Why do employers come here and not to the municipal center?," I asked. The Associate Director answered "First they want privacy and immediate access. Second, we have so many specialists and there is immediate consultation available. Also, I think they aren't afraid of the costs, since the government pays 60 percent." After the employer signs up, the workers are called in one by one for exams. There seemed to be less certainty about health hazard evaluations and environmental measures in the plants. It would be nice to know what "privacy" for employers really means.

There is an impressive action research project to offer OSH services to farmers in the Kuopio region and evaluate results.[66]

The present national budget for occupational health services for all sorts of workplaces is about 30 million Finnmarks, 250 Finnmarks per year per worker, or about $43.[67]

OSH-PHC LINKS

During my visit with the general practitioner in the municipal health center in Helsinki—before I had indicated my interests in OSH, though the National Board of Health people may have given some tip when they arranged the appointment—I asked about a hypothetical forty-five-year-old patient coming with an asthma. I asked what questions she would ask, "Well, if he has a cat or a dog at home. Or if his parents had asthma. Or if he is sensitive to something else, like something he eats," she replied. "What if he works where there is a lot of metal dust?" I asked. "Oh, yes I'd ask about the work too." Six years out of medical school, this otherwise committed, and, it seemed to me, able physician, couldn't exactly remember any OSH course; she thought she remembered that some OSH material was worked into the teaching of things like skin and lung diseases. She has had no OSH training since. Even though her main interest is in maternal and child health, this general physician was rather critical and unhappy with the OSH situation. She does see workers and is a committed person. To get anything done in the workplace, she has to refer the patient to the Institute for a specialist's word. "The employer won't do anything on my word," she said. "Will the employer change when the Institute asks?" I enquired. "Yes, maybe just the job this worker does." "And if there are fifty others with the same exposure?" I persisted. "I'm afraid they will most of them just have to get sick before anything is really done to change the conditions."

Others may not take such a dim view. Presumably, when OSH diseases are reported to the National Board, the inspectorate should follow up and see who else is similarly exposed and demand corrections. Whether it really works this way, beyond the reporting, I don't know. One is not as worried about this in large plants where there are unions and where the unions are strong. For example, in one large factory organized by the Metal Workers, twenty or so painters refused to use epoxy paint. The National Board sent its health and safety expert and negotiating team and convinced the employer to substitute a larger molecule paint and install better ventilation. But in medium to small-sized plants it seems uncertain. Fortunately, there is a serious effort to evaluate the effects of the law in smaller plants.[68]

I asked several times of different sources if there are figures on where workers go when they first seek help for perceived symptoms outside the family. Some people suggested that close estimates could be made from available data on source of care indicated in Sickness Insurance reports, but these might not distinguish first point of contact for each illness episode. Such information would be useful in planning a strategy for postgraduate medical education for PHC practitioners most likely to see workers outside of the OSH services. Because some 40 percent of workers are not yet covered by special OSH services and because many of those covered would go elsewhere when they first feel ill, the proportion going to a non-OSH, PHC provider when they first feel ill must be quite large—at least 50 percent.

Because the job of preparing enough physicians and other personnel for OSH coverage is so mammoth, little is being done to provide a minimum of OSH education for non-OSH PHC providers; yet, clearly, this is a need.

Theoretically, this need to reach PHC providers with OSH information would not be so great if the geographically-oriented publicly provided and regulated PHC centers were solely responsible for PHC. The geographic responsibility could lead providers to get to know the factories and other workshops and the unions and workers and their concerns. But, as we have seen, this direction for the future is far from realized as hospital care is still so important and the fractionation of the system by the often part-time private PHC services is unfortunate. I did not hear of any project like the one in Mora, Sweden, in which OSH specialists take PHC providers on plant tours once a month and then discuss what kind of work diseases patients from those plants might bring to the PHC services.

There is also a need for improved and increased OSH education in undergraduate education for nurses and doctors. For physicians, the hours seem to vary considerably from one school to another. In Tampere where the OSH component and the whole Department of Public Health is strong,[69] there are more than thirty required hours, including field work in factory or other work settings. In other schools the hours and staffing seems to be much less adequate (though I do not have complete information on all schools).

SOME OBSERVATIONS, QUESTIONS AND RECOMMENDATIONS

Above all else, if workers' health is to be better protected in Finland, organized labor must get itself together in more solidary, better informed, more effective ways at all levels. This may be difficult when there are threats of plant closings, job losses and actual plant closings. Still, the workers' movement has achieved much in a short history and much of the needed legal and other infrastructure is in place. Certainly the job security provisions are far better than those, say, in U.S.A. And the law and moves to cover all workers with OSH services is remarkable when we realize that there is no standard or requirement for OSH medical and related services in U.S.A. or Sweden where the Wessman Commission turned the idea down as a legal requirement. But new demands should be raised to enable workers to play a more authentic role in determining OSH standards, actual conditions at work and OSH services, for what they have may be admirable in many ways, but the workers do not control it.

Workers should find out what it means when the plant medical director sends his nurses out into the plant "for public relations." They should find out what it means when a private OSH service director says that employers sign up with his service "because it gives more privacy." For Labor to firmly reject the old non-choice choice of "your job or your life" in these and other ways, it will, among other things need reliable medical/legal experts of its own. In this connection, even if only for the gadfly value it would have, organized labor might reinvolve itself, informally at least, with the Socialist Health Front.

On the particular problem of special concern to this study, the OSH-PHC connection, organized labor should insist on an authentic role in determining the form and function of PHC services and their organizational and financial and functional relations to the OSH services at commune level. The unions should also seek ways to influence postgraduate and undergraduate education of PHC providers in favor of an increased awareness, concern and understanding of OSH.

At the national health policy level, the current impressive and generally desirable push to rationalize and enhance PHC services should be carried out with a much more complete cognizance of OSH than now seems to be the case.

The need for increased and improved OSH education for PHC providers has been covered above. In this connection, especially with devising a strategy for postgraduate education in view, a national sample survey of where workers go for their first medical contact for each illness episode would be of help. This should reflect multiple factors as suggested for a similar proposed study in Sweden (see p. 174). Perhaps the Scandinavian countries could cooperate in such a study.

There may be a need to better define the OSH relations between the Institute, Board of Health, Board of Labor Protection and other national bodies. This may be the query of a naive outsider who does not easily grasp the complexity of relations and responsibilities. But as one knowledgeable informant put it, "It was a terrible mistake to put the National Board in Tampere. People are spending all their time traveling." The traveling complaint may be only symbolic of other concerns.

As to PHC, some of the local organizational, financial and sponsorship problems have been alluded to above. The idea is to have competent PHC providers who: 1) are well informed about OSH in their area; 2) have the guidance of organized workers to support their efforts so that their services will be trusted by the workers; 3) are not misdirected by financial considerations; and 4) have organizational and informational means of seeing that occupational cases can be properly followed up in terms of rehabilitation for the individual worker and preventive efforts for other workers who may be similarly exposed. Without this, the specialty OSH services now being so vigorously developed in Finland will not achieve their full potential.

FINLAND AND THE MODEL

In Chapter 6 the dimensions of an OSH and OSH-PHC model were presented and discussed. Here a brief summary of the Finnish case is offered in terms of the model's dimensions.

Policy

Clearly OSH is of major concern in Finland. The 1978 law to provide OSH medical and related services to cover all workers is unique among study countries and appears well along in implementation. The National Institute for

OSH is another especially strong component of this system. It is less clear that the standards are as strong as in Sweden and there is even some uncertainty as to their legal status and enforceability. Unfortunately, awareness of the OSH-PHC problem seems minimal; in any case, it is evident that little has been done to forge effective links between the workplace services and the general health services. Other strengths and weaknesses covered in this chapter suggest that this case falls toward the high end of our range for six study countries as regards clear national policy directed toward improving OSH. Although money is not a measure of all things, perhaps the overall costs of OSH protection research in Finland as compared with the same costs in Sweden are suggestive of the relative position of these two countries on this dimension:

> More than twenty organizations carry out occupational health and safety research in Finland, and the total sum of funds used for occupational health and safety research in 1980 was about FIM 50 million. This was 2 percent of the total research budget in the country and about 1 percent of the losses caused by labor accidents and occupational diseases. The Finnish funds for occupational health and safety research were about 17 percent of the respective funds in Sweden and about one-third of the sum allocated for occupational health and safety research in Sweden per active worker in industry.[70]

If we reason that Sweden's total GNP was $116 billion in 1980, while Finland's was $49.4 billion, then if Finland spent one/third on OSH research per worker of that which Sweden spent, it was doing about 78 percent as well (.333 spent on OSH research divided by .426 as the ratio of Finland's GNP to that of Sweden). I have no illusion that these matters can be so exactly quantified; yet the figure suggests about the range of comparative effort I would have guessed.

Sponsorship and Control

By comparison with Sweden, this is the weakest dimension of the Finnish model. The whole system has a heritage of paternalism and for long periods vicious repression of socialist workers' forces, even fascist dominance in some periods. Thus workers in general and their leading representatives, the unions, have not had time or learned to demand and exercise a desirable degree of control. Thus even though the OSH system is impressive, especially in its coverage of workers with OSH medical and related services, union health and safety reps do not have the legal right to stop the production process in the face of a perceived hazard as is true in Sweden; nor do the joint union-management OSH committees afford a dominant voice to the worker; thus workers and their representatives do not have the control over hiring and firing of OSH physicians and other service personnel as they do in Sweden. Finally, while it would be inconceivable to set up major study or advisory commissions in Sweden without at least equal representation for LO and other workers' representatives, such groups in Finland are often defined in terms of expertise and may or may not include workers' representatives.

If Finland is less strong on this dimension than Sweden, it is well above U.S.A. How it stands in relation to F.R.G. and the U.K. is more complex and less certain, and must await the presentation of these cases before discussion. The G.D.R. is a special case we take up in the next chapter.

My summary impression is that whereas Sweden's special strength is action in the workplace itself under the control of a union-dominated health and safety committee, in Finland the action is in the National Institute for OSH, in the laws and in the agencies generally which bear upon the workplace but from outside the workplace.

Education

Finland has done as much as any other study country and more than most on this dimension. With some 96,000 workers trained in OSH (though there is a little uncertainty whether this figure also includes OSH agency personnel) considering the country's size (4.8 million people versus Sweden's 8.3) it has done as much to reach workers as its neighbor—at least in terms of numbers; however, the nature and interpretation of content and ideology must be considered when workers' sponsorship and control of educational programs is not assured. A great deal of the actual effect of this education of workers concerning OSH hazards and what to do about them depends on the local union's organization and ability to force changes through collective bargaining and other ways. The general impression I have is that while unionization is almost as widespread in Finland as in Sweden (perhaps 80 percent compared with 85 percent) neither the role of the state in terms of labor laws, nor the history of labor solidarity versus paternalism, allows the expectation that this extensive educational effort will be as effective in protecting workers as the almost equally massive effort in Sweden where some 110,000 health and safety reps have been trained plus hundreds of thousands more workers given some training through local union "study circles." But the picture shows a lot of variation within Finland. In some situations the union local is strong and militant. The example was given above of the local in a foundry which pressed for the establishment of a heat standard (15 minutes break every hour if the temperature is more than 100°) with the assistance of the Socialist Health Front.

Finland is stronger than any other study country in its efforts to prepare OSH specialists. Two levels of OSH physician specialists have been formally recognized and curricula adopted. The long recognized in-depth specialist is a clinician teacher, maybe also researcher, who tends to be hospital based and may take part in academic programs. The new specialty is workplace-based or OSH-policlinic-based (serving many medium to small plants) a preventive specialist who may also give treatment. This development is in response to the 1978 law, unique among our study countries, requiring OSH medical and related services coverage of every workplace. The training and production of the many new people needed for implementation is well underway. The National Institute for OSH has played a remarkable and important role in these developments as in all aspects of OSH in Finland. There is still a need to improve continuing education for those who have been in the field for some time.

The preparation of medical students and of PHC practitioners to know the dimensions of the OSH problem and be alert to picking up work-related disease and properly follow it up is as poor as in any other study country. The Medical School at the University of Tampere has an especially well-developed program and an impressive faculty group devoted to this sphere of concern, but other schools have little going on in OSH education by comparison. I did not learn of any special programs as in Mora, Sweden, to reach PHC practitioners concerning OSH.

Organization

The National Institute for OSH in Helsinki clearly provides the leading organizational focus not only for research and training, it's primary concerns, but for the OSH system in Finland. Its staff and work are very impressive. But there is some confusion in relations to the enforcement arm, that is, the National Board for Labour Protection, located in Tampere; and in relations to the National Board of Health in Helsinki, which has responsibility for supervising the medical aspects of the 1978 OSH Services Law. There is little evidence of authentic worker involvement in these institutions, to say nothing about worker control. Thus the work of some of them may be impressive in terms of scientific publications and numbers of people trained but what these things mean in terms of supporting working class versus bourgeois interests is very much open to question. Relations between PHC and the OSH services in large plants or other OSH services for medium to small plants—which may be either public municipal OSH policlinics or private OSH services—are confused at best. While referrals back and forth between OSH services and PHC or more specialized general services are supposed to carry adequate information and entail proper follow-up as regards job placement of a recuperating worker and other matters, this is seldom the case.

The split and yet intermixing of the general health services between public and private services does not allow a coordinated approach to geographically defined PHC areas. Thus the advantages which could be realized from PHC providers getting to know the workplaces and hazards in their area (aside from special OSH training—which is also lacking as noted above) are not realized. This inadequacy is probably less serious in rural and semi-rural areas as public, geographically assigned PHC services tend to predominate in these areas.

In short, while there are some special organizational strengths in the Finnish OSH system, particularly the National Institute, the general picture leaves much to be desired in terms of worker involvement and control and clarity of mission and relations.

Information

There is evidence of considerable special effort on this dimension. The work accident reporting system is quite complete and has yielded some very important insightful studies (see note 48). A growing awareness of work-related disease seems to be the primary explanation for increasing rates of these illnesses (note 44). But the summary table presented earlier is frankly based on rather crude estimates and there remain many problems with the OSH information system, including, of special interest here, the inability of PHC providers to report OSH problems because they have not been trained to recognize them and may not be specially motivated to report and follow-up cases they see.

The research productivity, that is, special studies of the National Institute, the Academic Department in Tampere, the Social Research Institute of the Social Insurance Institution, and other agencies is quite strong and impressive. The voice workers have in these studies and the access unions have to findings and control of the use of this form of information seems unsatisfactory, primarily because workers and their representative organizations lack their own expertise. However, the Socialist Health Front (a COSH-type group in U.S.A. terms) has played an important role in this regard.

Financing

Funds for OSH services and the OSH system in general in Finland are quite generous.

> The financial resources of various organizations are approximated at a total of FIM 540 million per year. This sum is 11 percent of all the annual losses caused by labor accidents and occupational diseases and 0.3 percent of the gross national product. The total allocation of funds to occupational health and safety activities in Finland is about FIM 245 per capita of the total active labor force per year, whereas losses caused by the consequences of risks are approximated to be FIM 2,300 per capita of the labor force per year.[71]

In U.S. dollars (according to a recent exchange rate) this works out to some $43 per active worker per year for all OSH activities (the figure may be closer to $50 if earlier exchange rates were used). No similar figure has been found for other study countries, so exact comparison is not possible; yet it is my impression that this is comparatively generous.

The flow and control of the financing is another matter. OSH services remain very much a responsibility and under the control of management. Thus, if the services in a large paper mill are sometimes for "public relations" (to give the workers the impression that something is being done for them) this is not the best situation. Similarly, if a number of small plants sign up with a private OSH service "because their managements want privacy," this is not the best either.

In general, and speaking in terms of comparison with other study countries, the OSH system in Finland is very strong. The very crucial aspect of worker involvement and control locally and nationally seems less strong than in Sweden. But as an expert-determined system it is very impressive. However, the OSH-PHC linkages are very weak and undeveloped.

NOTES AND REFERENCES

1. Information for this introductory section is taken from several sources: "Facts About Finland, 1981," Helsinki: Union Bank of Finland; Ruth L. Sivard, *World Military and Social Expenditures, 1983*, Washington, D.C.: World Priorities, 1983; and several historical works which are cited more specifically in following notes.
2. Jaakko Nousiainen, *The Finnish Political System*. John H. Hodgson (trans.), Cambridge: Harvard University Press, 1971, p. 4.
3. This figure comes from "Facts About Finland . . ." op. cit. (note 1). Another source says "Somewhat more than 11 percent of Finland consists of Lakes and accounts for the oft-quoted phrase 'the Land of the Thousand Lakes'." John W. Wuorinen, *A History of Finland*. New York: Columbia University Press and The American-Scandinavian Foundation, 1965, p. 4.
4. Eino Jutikkala with Kauko Pirinen, *A History of Finland*. Paul Sjöblom (trans.), New York: Frederick A. Praeger, 1962, p. 9.
5. Ibid, p. 4. Subsequent information on the origins of the Finnish people is also drawn primarily from this source, but others, including Wuorinen, op. cit. (note 3) have also been consulted.
6. Jutikkala, op. cit. p. 6.
7. "There is only *one* thing by which I have been deeply moved, and I can live only for it. Everything else is of secondary importance. I have decided to prove to the people of Finland that we are not a nation isolated from the world and world history, but that we are related to at least one-seventh of the people of the globe. If the cause of this nation is served thereby, all will be well. . . . Grammars are not my objectives, but without them I cannot reach my goal." From a letter dated October 18, 1844 to Johan Vilhelm Snellman II from M. A. Castren in the manuscript collection of the University of Helsinki as quoted by Wuorinen, op. cit. (note 3) p. 15.
8. Wuorinen, op. cit. (note 3): 11 and 16.
9. Ibid, pp. 164–165.
10. Finland was under rule of the Russian Tsar at this time, but under the pressure of nationalist strivings for autonomy and the weakening of Russian forces through the disaster of the Russian-Japanese War of 1904–1905 as

well as through the internal rebellions of the time, the Tsar agreed to an elected parliament and universal suffrage. "The franchise was universal on the principle of 'one man one vote,' the ballot was secret, representation was proportional, and for the first time in Europe women were given the vote. As a result the working class was brought into the political arena and the Social-Democrats gained 40 percent of the seats in the first Parliament." Mikko Juva, A Thousand Years of Finland, pp. 17–36 in Hillar Kallas and Sylvie Nickels (eds.), *Finland, Creation and Construction*. London: George Allen and Unwin, 1968, quote from p. 34. Juva goes on to note that this reform did not in the least change the Tsar's absolute power; once he had regained control of Russia, the grip tightened on Finland once more and laws passed by the Parliament were summarily rejected.

11. Wuorinen, op. cit. (note 3): 172–173.
12. Olavi Borg, "Communism–Finnish Style," pp. 67–72 in Kallas and Nickels, op. cit. (note 10) quote from pp. 69–70.
13. Ibid, pp. 71–72.
14. The Metal Workers' Union identifies only the Social Democratic and the People's Democratic League as "parties of the labour movement" and defines all the others as "bourgeois parties." "Finnish Metal Workers' Union," Helsinki: The Union, 1977 (pamphlet).
15. Heikki Waris, *Suomalaisen yhteiskunnan rakenne*, 2nd edition. Helsinki: Otava, 1952, p. 185, as translated by J. H. Hodgson and quoted in Nousiainen, op. cit. (note 2): 15.
16. As given in Nousiainen, op. cit. (note 2): table 7, p. 17.
17. Harry Braverman, *Labor and Monopoly Capital: The Degradation of Work in the Twentieth Century*. New York: Monthly Review Press, 1974.
18. Waris as quoted by Nousiainen, op. cit. (note 2): 17.
19. Nousiainen, op. cit. (note 2): 18.
20. "Facts about Finland, 1981" Helsinki: Union Bank of Finland; Also "Finland" p. 535 in *The World Almanac and Book of Facts, 1980*, New York: NEA, 1980.
21. Jan Kristiansen, Finland: Downturn This Year, *International Herald Tribune*. pp. 7 and 8, 7 October, 1981.
22. Stephen D. Moore, Finland Decides It Has to Look West for Growth, *The Wall Street Journal*, p. 38, June 26, 1984.

"In this year's first quarter, exports to the West by Finnish companies rose 32 percent from a year earlier. Trade with the Soviet Union–still Finland's largest single trading partner–dropped 20 percent.

"Last year, Finnish companies bought 147 foreign companies, raising the number of units abroad to more than 1,200. Direct investment by Finnish companies in the West rose 79 percent last year, to $261.1 million from $145.7 million in 1982.

"This rapid international expansion and efforts to raise foreign equity capital have received strong support from the socialist government. 'Companies have completely free hands,' says Finance Minister Ahti Pekkala. 'We look with favor on private companies raising capital abroad. They strengthen themselves, and strengthen the country too in the process.' . . .

"Finland isn't turning its back on the Soviet Union. Shortly after taking office, Finnish President Mauno Koivisto traveled to Moscow to renew the countries' longstanding cooperation and mutual defense pact ahead of schedule. Since then he has visited Moscow frequently, trying to forge relations as amicable as those built up during the twenty-six-year tenure of his centrist predecessor, Urho Kekkonen.

"For the Soviet Union, Finland is a prized source of Western products, particularly ships and heavy machinery. The Finno-Soviet bilateral trade system involves no payments in convertible foreign currency, which the Soviets demand even from their East bloc satellites.

"In turn, East trade is a buffer for the Finnish economy against business cycles of the West, which are unusually strong in a country so dependent on the forestry industry.

"While the relationship with the Soviets remains the cornerstone of Finnish politics, analysts say Moscow isn't likely to object to increased economic links between Finland and the West as long as they don't influence foreign policy. And they say that at any rate, in Finland's private sector at least, Soviet influence on decision-making is negligible. . . ."

23. Shail Jain, *Size Distribution of Income, A Compilation of Data*. Washington, D.C.: The World Bank, 1975.
24. Timo Kauppinen, "Industrial and Labour Relations in Finland," Helsinki: The Committee for Industrial and Labour Relations, 1981. The Winter War of 1940 resulted in the ceding of some border territory considered in Moscow to be crucial in the struggle against Nazi Germany. For this brief, but devastating war, Finland was actually an ally of the fascist forces.
25. Ibid.
26. "SAK Today" Helsinki: SAK, 1981.
27. "Finnish Metal Workers' Union" Helsinki, 1977.
28. This may occur along political as well as trade or class lines. For example, at the 1976 Congress of SAK, 64 percent of delegates were Social Democrats and 36 percent were Communist or People's League. "Finnish Metal Workers' Union" Helsinki, 1977, p. 10.
29. For example, the employment safeguards and employee participation agreements and laws are impressive. Kauppinen, op. cit. (note 24). Also "Labour Protection and Legislation" Helsinki: Government Printing Centre, 1977.
30. Nousiainen, op. cit. (note 2): 421–422.
31. "Primary Health Care in Finland." Helsinki: Ministry of Social Affairs and Health, National Board of Health, 1978: booklet of 29 pp.
32. Some evidence for what this informant is saying is given in the booklet describing PHC (note 31). On pp. 28–29 "a typical medium sized health centre in middle Finland" is described. For a population of 32,571, there are one chief medical officer, one chief dental officer, one nursing superintendent, one manager, nine doctors, ten dentists, twenty-eight public health nurses, thirty-two and one half clinical nurses and other therapists and technicians; forty-two auxiliary nursing personnel; and seventy-two other personnel. This adds to sixty-two per 10,000 population or one

health worker for every 162 people. The plan is for this health manpower to population ratio to rise to one per 143 in 1983.

33. Esko Kalimo, et al., "Need, Use and Expenses of Health Services in Finland 1964–1976," Publication A: 18 of the Social Insurance Institution, Helsinki, 1982. The figures are interpolated from a table on p. 181 of the report; thus are close but may not be exact. My thanks to Dr. Elianne Riska of the Research Institute for Social Security, The Social Insurance Institution of Finland, for bringing this information to my attention in personal correspondence.
34. "Primary Health Care . . ." op. cit. (note 31).
35. This seems especially noteworthy when one observes that the Finnish PHC Act preceeded the Declaration of Alma-Ata by six years: *Alma-Ata 1978, Primary Health Care*. Geneva: WHO/UNICEF, 1978.
36. "Primary Health Care in Finland" op. cit. (note 31): 19.
37. "Health Services in Finland," Helsinki: Ministry of Social Affairs and Health, National Board of Health, 1979. Most of the information in this section comes from this booklet and the one on PHC (note 31). Opinions and impressions come from interviews and observation.
38. "A special weakness in this country is the intermediate level of administration. The very multitude of names under which it goes in Finland—provinces, regions, districts, federations of communes—gives some idea of the lack of cohesion." Ibid, p. 32. Curiously, the power to approve local PHC plans has been decentralized to these uncertain intermediate units: "Primary Health Care . . ." op. cit. (note 31): 13. See also Nousiainen, op. cit. (note 2): Chapter X, pp. 312–344.
39. "Health Services . . ." op. cit. (note 37): 17. This seems like an unlikely set of figures if one takes into account that expenditures for public health measures, for OSH, school health and outpatient dental as well as medical services, and for drugs used out of hospital, and possibly for eyeglasses, hearing aids, etc., are usually counted as part of "all health expenditure." Nevertheless, the statement offers a clear, strong impression of the hospital-oriented thrust of the system.
40. Bogdan Kleczkowski, Ray Elling, and Duane Smith, *Health System Support to Primary Care*. Geneva: WHO Public Health Papers Series, no. 80, 1984.
41. Ruth L. Sivard, *World Military and Social Expenditures*, op. cit. (note 1): Table III, pp. 38–39.
42. Pekka Puska, et al., Changes in Coronary Risk Factors During Comprehensive Five-Year Community Program to Control Cardiovascular Diseases During 1972–1977 in North Karelia, *British Medical Journal* Vol. 2 for 1979: 1173–78; Alfred MacAlister, et al. Theory and Action for Health Promotion: Illustrations from the North Karelia Project, *American Journal Public Health 72*:43–50, January 1982.
43. Edward H. Wagner, The North Karelia Project: What it Tells Us About the Prevention of Cardiovascular Disease. *American Journal Public Health 72*: 51–53, January 1982. Also I wonder what attention was given to occupational stress and other CHD-related exposures at work.

44. The following figures and table are all taken from an overview of the OSH situation by the Head of the National Institute of Occupational Health in Helsinki: Jorma Rantanen, Occupational health and safety in Finland, *Scandinavian Journal of Working Environment and Health* 9:140–147, 1983.
45. Ibid, p. 142.
46. Ibid, p. 142.
47. For a more qualitative, subtle, telling estimate of the problem in modern capitalist society, see Studs Terkel, *Working*. New York: Avon, 1975; Also Braverman, op. cit. (note 17).
48. Jorma Saari, Long-term development of occupational accidents in Finland, *Scandinavian Journal of Working Environment and Health* 8:85–93, 1982. This is an excellent article including detailed information on other factors including the increasing shift to blue-collar work, the probable effects of automation, changes in statistical reporting and awareness as well as insurance, and the rash of protective changes adopted from 1970–1980.
49. Rantanen, op. cit. (note 44): 142.
50. Calculated from figures given by Rantanen, p. 141.
51. The general strike of 1956 was noted earlier. As regards hours, working hours per year per person in industry have declined from over 2400 in 1920 to under 1900 in 1980. "Labour Protection in Finland." Helsinki: Ministry of Social Affairs and Health, 1980.
52. On the concept of "authentic participation" see A. Etzioni, *The Active Society*. New York: The Free Press, 1968.
53. "Administration of Labour Protection in Finland," Tampere: National Board of Labour Protection, 1977. See also, Nousiainen, op. cit. (note 2): 422–424.
54. "Health Services in Finland," op. cit. (note 37): 13.
55. The term "binding" is unclear. I heard differing opinions as to whether there is any legally enforceable OSH standard in Finland, though most see the asbestos standard as a legal requirement. I'll mention this point again when I discuss an activist group of workers and health technical people—The Socialist Health Front.
56. "Labour Protection in Finland," op. cit. (note 51): 12. This fractionation is noted and a good overview of the system given in Jorma Rantanen, "Occupational Health and Safety in Finland," *Scandinavian Journal of Working Environment and Health* 9:140–147, 1983.
57. "Institute of Occupational Health Annual Report, 1979." Helsinki, 1980.
58. "Current Research Projects, 1981." Helsinki: Institute of Occupational Health.
59. "Socialist Health Front of Finland; Over Ten Years of Fight for Citizens' Health, Labor Protection, Social Security and Better Environment." Helsinki: The Front, 10/81, p. 2.
60. Maria-Lüsa Rantala and Antero Honkasalo, *Työsuojeln Työn Voimaperäisyys* 350. Espoo: TKY Otapaino, 1976. English summary pp. 127–128.
61. "Act Respecting Works Health Care" dated 29 September, 1978. Act No. 743. *Legislative Series, 1978-Fin 4*. Geneva: ILO, 1980, pp. 127–129.

62. Matti Ylokoski and Jorma Rantanen, *Työterveyshuolto etc.* Helsinki: Institute for Occupational Health, 1979 ("The Green Book").
63. Rantanen, op. cit. (note 44): 141.
64. See the discussion of Navarro's and Kelman's positions on this matter in Chapter 7. Also their exchange in *Journal of Health Politics, Policy and Law 9*:157-165, Spring 1984.
65. Rantanen, op. cit. (note 44): 144.
66. Ilkko Vohlonen, et al. "A Feasibility Study of Organizing Occupational Health Services for Farmers," Helsinki: Social Insurance Institution, 1982.
67. Rantanen, op. cit. (note 44): 145. The calculation into U.S.A. dollars is done at the recently reported rate of 5.75 Finnmarks to the dollar. It may be that the figure should be closer to $50 per worker as the dollar has been especially strong in 1984.
68. Elianne Riska, sociologist with the Social Research Section of the Social Insurance Institution is director. Her preliminary report to the Medical Sociology Research Committee meeting as part of the 10th World Congress of Sociology, organized by the International Sociological Association in Mexico City in August 1982, suggested great progress, but also some unanticipated difficulties such as reluctance of employers to sign up for OSH coverage because of various misunderstandings.
69. Some of the research on class and position in the production process in relation to stress and morbidity is truly ground breaking. For example, Vesa Pervi, "Occupational Status, Working Conditions and Morbidity Among Employees in the Engineering Industry." Publications of the Medical Faculty No. 12, University of Tampere 1977; also Seppo Aro, "Stress, morbidity, and health-related behavior, a five-year follow-up study among metal industry employees." *Scandinavian Journal of Social Medicine*, Supplement 25, 1981. This work strongly suggests that within the working class, there are unfortunate divisions with the less skilled and less well paid more often being exposed to the greatest heat, and/or to the greatest pressure from foremen or other stresses.
70. Rantanen, op. cit. (note 44): 145.
71. Ibid, p. 141.

CHAPTER 9
East Germany and German History

Historically speaking, the present-day thirty-five-year-old division of Germany into the German Democratic Republic (the G.D.R. or East Germany) and the Federal Republic of Germany (F.R.G. or West Germany) is a new division, but division is not new to Germany. In fact, the concept of a modern German nation state did not take hold until relatively late and was only confirmed in the successful outcome of the Franco-Prussian war of 1870–1871.

Because of this history it is not possible to discuss the geography and origins of the German people in the same way as we have for Sweden and Finland, since the boundaries have been changing ones.[1] Nevertheless, there has been a core, both geographic and political-economic, as well as, cultural continuity. Indeed, these qualities make the comparison offered in this and the next chapter between G.D.R. and F.R.G. especially interesting, for it is their common heritage which makes their political-economic divergence following WW II especially significant—a controlled experiment, so to speak, on the impact of political economy on OSH and OSH-PHC, with "other factors" (cultural background, etc.) controlled or "held constant" (though as we will see, changing cultural hegemony goes along with changing political economy).

I will briefly present here the geographic and historical core background common to the two Germanies, including labor history and health system. Then the post WW II background, labor organization, and present health systems and OSH and OSH-PHC situations will be discussed: for G.D.R. in the last part of this chapter; for F.R.G. in the next chapter.

THE COMMON HERITAGE

History—The Land, People and Rulers

Generally speaking, as is the case today, Germany has been bordered in the east by Poland and the Czechoslovak lands and peoples; in the south by Austria and Switzerland; in the west by France and the Benelux nations; and in the north by Denmark and the Baltic Sea. Thus it lies in the heart of Europe and, divided or not, has long played a key role in the dynamics of European history.

The land area itself is divided into two rather distinct regions–the flat northern plains and the hilly, sometimes mountainous south. The sandy northern coast is low and The North German Plain is broken by swamps and lakes, especially in the northeast. The elevation begins to rise in the *Mittelland*, an area of upland plateaus and low mountains separated by broad river valleys. This middle region contains the famous Bohemian Forest in the East and the Black Forest in the west. The Bavarian plateau rises, until near the southern border some of the Alpine peaks reach close to 10,000 feet. The Zugspitze, famous for its skiing and beauty, is the highest, reaching 9,721 ft. The Harz mountains in the southeast and the Erzgebirge along the Czechoslovak border are not as high, but also offer winter sports possibilities.

Germany is crossed and drained by several major European rivers which also contribute importantly to transportation and industry. Most notable is the Rhine River stretching from Lake Constance and the southern border with Switzerland up along the border with France through the western part of the country into the Netherlands where it empties into the North Sea. Along the present border of East Germany with Poland, the Oder River flows northward into the Baltic Sea. Other major rivers include the headwaters of the Danube flowing out of Bavaria through Austria and eventually to the Black Sea. The Main River is a tributary of the Rhine and flows through Frankfurt am Main (another sizable city of the same name is in the G.D.R., Frankfurt an der Oder). Other tributaries of the Rhine flowing through central and western Germany are the Elbe, the Ems, the Weser, and the Moselle which flows out of France.

The climate is temperate. The same ocean currents which warm Scandinavia warm the far north of the country, while the high Alps cool the south. Winter temperatures average a bit below 20°F in the south and about 30°F in the north, while summer temperatures are in the mid-60s throughout the country.

Though south Germany is largely agricultural, there are major industrial cities as well–Frankfurt am Main, Munich, Stuttgart, Pforzheim and Ulm, Nuremberg, Augsburg, Mannheim, Karlsruhe–all in present day West Germany (F.R.G.). The middle section of the country has perhaps been historically more industrialized and many of these urban concentrations are in present-day East Germany (G.D.R.)–Leipzig, Dresden, Karl Marx-Stadt (formerly Chemnitz) Halle and Erfurt. The heaviest urban industrial concentration of all has been and is in the western sector along the Rhine–Cologne, Duesseldorf, Essen, Duisburg, Dortmund and others. Berlin served historically as the major inland port and industrial and governing center of the country. It remains an important city today with the G.D.R.'s capital in East Berlin, while West Berlin is a province (or Land) of F.R.G. surrounded by the G.D.R. Thus it is a divided city, something of a shadow of its former greatness, though taken as a whole, both East and West Berlin, it is still the largest city. The North Sea ports of Hamburg and Bremerhaven in the West and the Baltic port of Rostock in the East are also important urban commercial-industrial centers.

With no natural boundaries besides the Alps in the South, the territory has been open to numerous migrations and invasions over the course of European history. Teutons, a Germanic people, were among the earliest settlers. The territory was divided among several tribes of Teutons–The Franks, the Saxons, and the Thuringians–led by elected kings. Around 200 AD other Germanic tribes arrived from the east, notably the Goths and Burgundians, followed by the Huns and the Vandals whose migratory, warlike character led to conquest but only brief rule over the other tribes.

The Roman Empire spread its influence across the Alps as far as the Baltic Sea, but never successfully conquered and ruled all the Germanic tribes at once.

Only the Franks succeeded to conquer and weld the several tribes into a single empire. They had settled in the Rhine valley in the late 300s AD and by 500, following their adoption of Christianity and receipt of support from the Popes, had become a greater power than the Roman Empire north of the Alps. In 732 they fought off a Muslim invasion of Europe. Their height was reached under Charlemagne when they controlled a vast territory ranging, south to north, from Central Italy to the Baltic Sea, and, west to east, from the Pyrenees mountains along the present border between Spain and France, to the Elbe River flowing through the center of present-day East Germany. Charlemagne became the main protector of the Popes and in 800 was crowned Emperor in Rome. This laid the political and religious foundations for the later Holy Roman Empire.

But disunity followed Charlemagne's death in 814. In 843 the Treaty of Verdun divided the Frankish territory in three parts. The eastern, predominantly Teuton portion, between the Rhine and the Elbe, went to Louis the German. This became the core of modern Germany.

The tribal divisions and areas of earlier times formed the bases in the developing feudal economy for dukes and princes who vied for wider control. By 900 one Teuton clan, the Saxons, emerged dominant and with their ruler, Henry the Fowler, extended their control to the Oder where they fought off attacks from the Magyar and Slavic peoples to the east and south.

A series of Saxon kings were crowned as Emperors of the Holy Roman Empire. By 1030 this Empire became comparatively prosperous on a feudal economic base with many thriving towns in vigorous trade with each other. Henry III who reigned from 1039 until his death in 1056 probably held more power than any other. But he was the last with such power for a long time. A gradual but deep disintegration took place, with the twenty-year "Great Interregnum" occurring from about 1250–1270 when no one ruler ruled. Instead, the local princes, dukes, bishops and the towns became the major centers of power.

Only in 1273 was another German prince crowned to head the Holy Roman Empire. This occurred because Pope Gregory X had become fearful of the growing power in France. This was Rudolf of the relatively minor House of Hapsburg in Austria. But from these beginnings grew the mighty Hapsburg Empire which dominated southern Germany and much of central Europe until modern times, though it never reached into Prussia in the north of Germany, the birthplace and core of modern Germany.

The weakening of the Holy Roman Empire engendered by the Lutheran reformation and Protestant-Catholic conflicts entangled with political-economic changes. The Peace of Augsburg of 1555 granted to each prince the right to determine the religion of his subjects. But in 1618 began a long series of conflicts called the Thirty Year's War which brought the final weakening and the rise of France. The German states were divided. Only the Hapsburg Empire survived to hold sway in the east, having fought off the Ottoman Turks in the 1600s.

In the latter half of the 1600s, Brandenburg, later Prussia, began to grow from a small duchy into a powerful state. The ruling House of Hohenzollern established an efficient centralized administration and expanded its territorial control through a nobility-officered army which became the central institution of the state. In 1740 Frederick the Great succeeded to the throne. He immediately invaded Silesia, a large wealthy Austrian province. This action precipitated fifteen years of war which engulfed most of Europe. In 1763, Prussia won Silesia, doubling its own population and natural resources thereby. Germany was then polarized between the Prussian north and Austrian south. Religious differences tended to correspond and persist over succeeding generations, the north becoming increasingly Protestant, the south Catholic.

Following Frederick's death in 1786, the system did not function as effectively and Prussia fell easily to Napoleon at the Battle of Jena in 1806. Austria too succumbed to Napoleon's forces and the Holy Roman Empire was dissolved. But the French occupation ended in 1813 after Austria joined forces with Prussia and Russia to push the French armies west of the Rhine.

The Congress of Vienna held in 1815 after Napoleon's defeat created the German Confederation, a lose grouping of thirty-eight independent states. This reactionary concoction dominating the southern German states was primarily an invention of famed Austrian foreign minister, Prince Klemens von Metternich who sought tight monarchical control over the forces of bourgeois liberalism which the French Revolution and Napoleonic era had awakened throughout Europe.

In Prussia too reactionary forces reasserted themselves over the moves which before 1815 had led to tax and administrative reforms and the abolition of serfdom. The risings of 1848 constituted a broad set of rebellions seeking to reestablish liberal bourgeois and worker and freedman interests against the aristocracy and resurgent feudal order. This movement included demands for a unified Germany. An 1848-49 attempt to establish a unified republic failed, with the Hapsburgs once more able to impose their rule. But in 1844 Prussia had established an administrative device, the customs union (Zollverein) among half the German states. On this basis unification moves gathered strength.

Modern Germany begins only after 1862 when Otto von Bismarck, the "iron chancellor," led Germany to European dominance. In 1866 in a brief Seven Weeks War, Prussian armies defeated the Austrians. Austria and several south German states joined with Hungary in the Austro-Hungarian Empire. In 1867 a German Constitution was adopted for a new confederation, but Prussia was clearly the controlling center.

Through his policy of "blood and iron" and efficient, centralized bureaucratic control, Bismarck quickly led the nation to become the major force in Europe. He imposed a humiliating settlement on the French, following the war of 1870-71. France ceded the industrially powerful areas of Alsace and Lorraine, paid a large indemnity, and supported a German army of occupation.

The south German states, except for Austria, were now included in the North German Confederation.

> The new Prussian state was a curious mixture of democratic and authoritarian institutions. Advanced social welfare laws existed side-by-side with legislation curbing the Roman Catholic Church and suppressing the socialists. Both the central government and the twenty-five states comprising the union had monarchial forms of government.[2]

Bismarck's rule ended in 1890 when the Kaiser, or emperor, William II, who opposed Bismarck's extreme authoritarianism and diplomatic techniques, decided to administer the system directly. But William followed a course of aggressive imperialism which antagonized Britain, France and Russia and led eventually to WW I. Defeat of Germany in WW I; the disasterous Weimar Republic, leading to the Nazis' rise to power and murder of millions of Jews and other people; and WW II are too well known to be detailed here in this background vignette. Aspects dealing with the class struggle and workers' movement are taken up in the next section on political economy.

While this background vignette has dealt primarily with the land and people and governance of Germany, one of the richest cultural inheritances of the world in art, music, philosophy, political-economic theory, and science was evolving in parallel. The 18th Century was especially rich. One need only mention names like Johann Sebastian Bach (1685-1750), Immanuel Kant (1724-1804), Johann Wolfgang von Goethe (1749-1832), Wolfgang Amadeus Mozart (1756-1791), Johann Christoph Friedrich von Schiller (1759-1809), Ludwig von Beethoven (1770-1827) and Georg Wilhelm Friedrich Hegel (1770-1831) among many others who could be named to make the point. In the 19th Century, the giants of political-economic theory, Karl Marx (1818-1883) and

Friedrich Engels (1820-1895) appeared. Their writings stirred the working classes of the time and are still generating new works and energies in the struggles of working people and supportive intellectuals the world over. In Science names like Max Karl Ernst Ludwig Planck (1858-1947) and in medicine and public health Rudolf Virchow (1821-1902) serve to index the breadth of Germany's cultural development. It is Virchow's famous aphorism, "Medicine is a social science and politics is medicine on a large scale," which gives the present work its overall direction.[3]

The Political Economy, Pre WW II

Under the North German Confederation of Bismarck:

> Each state enjoyed a large degree of autonomy, although the federal government, or *Reich*, was empowered to administer a common communications system, maintain an army, and conduct foreign affairs. In addition to the *Kaiser*, or emperor, the federal government had a legislature with two houses–the *Bundesrat*, where the states received representation, and the *Reichstag*, where the people were represented through a system of universal manhood suffrage.
>
> The economic growth of the new empire was astounding. In 1860, for example, German steel production did not even equal that of France; by 1900 it exceeded that of both England and France combined. By 1900, moreover, German naval power rivaled that of Great Britain, the traditional 'mistress of the sea.'[4]

More so than in most other countries in which industrial capitalism was taking hold, the old feudal and aristocratic order hung on in Germany. In fact, many in the old hierarchy simply shifted to become the new capitalist owners and ruling class. Thus there was less mobility for the peasantry turned workers than in other countries.[5] The social structure and its cultural hegemony emphasized authority and hierarchy even more than in other study countries.

But control by the "nobility" did not ameliorate the rawness of capitalism. Probably it exacerbated it. In any case, the squalor in towns and sub-subsistence wages and conditions at work were no less appalling in early capitalist development in Germany. Virchow's now famous report of 1848 on the typhus epidemic in Upper Silesia, which he had been investigating beginning in 1847, detailed some of the disgusting living and working conditions. He laid the blame at the feet of the government and ruling class. Naturally, no efforts were spared to suppress the report.[6] It was on this experience that he developed his view of society, politics and disease cited earlier.

In the face of these conditions, workers were not silent. In concert with the Revolution of 1848, Karl Marx and Friedrich Engels published *The Communist Manifesto*. Communists and anarchists began to establish their leadership with the working class. But the success of the counter-revolution in defeating the purposes of the national assembly which began meeting in Frankfurt am Main in May 1848 to establish a democratic republic of unified Germany (albeit under an elected Emperor), meant the failure of bourgeois democracy. Almost prophetically, Engels wrote, "Thus vanished the German Parliament, and with it the first and last creation of the Revolution. . . . Political liberalism, the rule of the bourgeoisie . . . is forever impossible in Germany."[7]

> If liberal democracy was not able to realize German unity, if the action of the people themselves was incapable of attaining this supreme goal of all Germans, then this meant that liberal democracy was a failure. And if unity could be achieved by might, by the action of the sword, by intrigue, diplomacy and war, then these instruments of political action justified themselves by their realization of a noble end. If the German bourgeoisie had fulfilled its historic task in 1848 as the French bourgeoisie a century before that, Bismarck would never have found the arena in which to play his role and he would have remained the Prussian Junker landowner that he was in 1847.[8]

The counter-revolution also brought a severe repression. Many fled and many were exiled. Herwegh, Heine, Marx, and Liebknecht remained exiles for the greater part of their lives. Marx with his family could barely eke out an existence writing articles for the *New York Tribune* while living in London and working at the British Museum on his three volume magnum opus, *Capital*. Also,

> A large number of German political refugees came to the United States. . . . Many of them fought in the Civil War in the battle against slavery and became prominent in the Republican party. Others became active in the American labor movement and helped build a socialist party in the United States. All these forces for democracy and social reform were lost for good to the German Reich.[9]

In spite of the failure of 1848, the *Communist Manifesto* became a basic document of the proletarian movement and when democracy was able to reassert itself as in the Weimar Republic, following the defeat of Germany in WW I, and in both of the present Germanies where one finds liberal democracy in F.R.G. and socialist-oriented democracy in G.D.R., the ideals and experience of 1848 have been looked to for guidance.

The anti-Socialist legislation of Bismarck (some of it importantly in the health field, which I will take up in the next section) seriously hampered organizational work, so that it was not until after 1890 that significant, but then rapid expansion of trade unions took place. At that time, the unions were divided into three types. The Socialists were known as Free Trade-Unions; those of religious sponsorship were Christian Trade-Unions; and those sponsored by progressive managements were called Hirsch-Duncker Unions after the two employers instrumental in first setting up this type. In spite of these divisions, they usually stood together in wage struggles and they had representatives in the Reichstag in the respective parties with which they were aligned. But they had no labor party of their own as was the case in Sweden. The strongest unions were in steel and metal production generally, mining, and in the construction industries. A number of great strikes were waged by Ruhr coal workers in 1872 and 1889, dock workers in Hamburg in 1897, textile workers in Saxony in 1903, and steel workers in 1905. Table 9 gives the growth of the three types of trade unions from 1891 to 1913. The Socialist unions were clearly dominant. The management-sponsored ones were smaller and less numerous and less accepting of women workers.

In the developing capitalist world-system, German imperialism began to play a major role and compete with British and French colonialism in Africa and elsewhere. Those who follow a Leninist line of reasoning would argue that this factor was key in keeping the western European working classes from adopting a consistent thoroughgoing revolutionary stance.

> . . . to Marxists who follow Lenin's discussion of imperialism, the success of capitalism in finding solutions for its structural tendency to crisis is attributable chiefly to its capacity to subjugate the less developed world. The fruits of conquest, manifested in the superprofits derived from the exploitation of raw materials and the superexploitation of labor through the export of capital are shared with a narrow stratum of the working class in the metropolitan countries. To be sure, according to Lenin, the mass of workers in advanced countries remained imprisoned by unrelieved exploitation of their own labor. In his view, the nonrevolutionary character of the Western working class as a whole had its roots in the domination by skilled workers and the reformist trade union and Socialist leaders over the entire labor movement. In turn, the success of imperialism provided the material underpinning for the development of reformism both as ideology and political practice.[10]

The authoritarian cultural hegemony, and organization of the ruling class and its stance toward organized labor played a major role in repressing the workers' movement. While some of the smaller industries came to appreciate some of the gains to be realized in labor peace through collective bargaining,[11] heavy industrialists like Stumm and Thyssen bitterly opposed any worker organization. Baron von Stumm, the "King of the Saar," declared that he would never allow any "artificial creations" like union contracts or constitutions to come between himself and his employees.[12]

In 1873 the iron and steel manufacturers founded the *Zentralverband Deutscher Industrieller*. This drew in owners across a wide spectrum from steel to hats and paper. This organization became quite influential in government

Table 9. Membership in German Trade Unions 1891–1913

	Free Trade-Unions			*Hirsch-Duncker Unions*			*Christian Trade-Unions*		
	No. of Associ-ations	*Membership*		*No. of Associ-ations*	*Membership*		*No. of Associ-ations*	*Membership*	
Year		*Total*	*Women*		*Total*	*Women*		*Total*	*Women*
1891	62	278,000	– –	18	66,000	– –	–	– –	– –
1896	51	329,000	15,000	17	72,000	– –	–	8,000	– –
1900	58	680,000	23,000	20	92,000	– –	23	79,000	– –
1905	64	1,345,000	74,000	20	117,000	– –	18	188,000	12,000
1910	53	2,017,000	162,000	23	123,000	– –	22	295,000	22,000
1913	49	2,574,000	230,000	23	107,000	6,000	25	343,000	28,000

Source: Koppel S. Pinson, *Modern Germany, Its History and Civilization*. New York: Macmillan, 1954, p. 247.

(even though the non-aristocratic, bourgeois owners were never accepted socially in higher ruling and military circles). This organization did not actively oppose Bismarck's social legislation which he shrewdly used as a major tool to undercut the Socialist workers' movement, "but they vehemently opposed anything that went beyond those measures."[13] In fighting child labor laws they declared, "It seems to be more reasonable to set children to work at pleasant jobs and let them make money, than to allow them to go idle and become wild." A measure forbidding night work for women was opposed as a violation of the principle of "liberty of the people to work whenever they want to." To remain impoverished, "uneducated, and lacking in intelligence" was seen as inevitable and natural for the worker. But a fall from grace for owners was a disaster. Thus to gain sympathy for owners who they said might become impoverished in the face of workers' demands, a pamphlet was issued in Dortmund which stated, "It is indeed sad to remain poor all one's life, but to fall from a condition of well-being to one of poverty and cares of livelihood is a hell on earth."

While the smaller industries separated in opposition to such intransigent views, and with shared interests of their own, as the *Bund der Industriellen*, in 1919 they reunited with the big ones as the *Reichsverband der deutschen Industrie*.

Addressing the 1905 meeting of the Verein fuer Sozialpolitik, which was on the subject of "Labor Relations in the Private Giant Industries," Max Weber complained that this German policy toward labor gives the impression "that it seeks not so much power itself as, above all, the appearance of power, the passionate aspiration for power." And this, he continued "is in the blood of our employers" (a not very sociological view I might observe). These owners, he said, rule over their industrial enterprises after the manner of *paterfamilias*, demanding stern obedience of those under them.[14]

Labor, by contrast with Capital, and by contrast with the Swedish umbrella organization, LO, never succeeded to form a unifying umbrella organization.

While I can not detail here the subsequent history of the workers' movement through WW I, the Weimar Republic, the Great Depression and rise of fascism, and WW II, it is important to underline that workers in the German Labour Front and others on the left offered steady and heroic resistance.

The Revolution of 1918-1919 came close to becoming the transformation of an advanced capitalist society which Marx had predicted, as distinct from the 1917 Soviet Revolution in a less developed country. A Congress of Workers' and Soldiers' Councils met in Berlin on December 19, 1918. The House of Hohenzollern was driven from power and for a time it looked like a worker-based system of Soviets would be established. But divisions on the left,[15] the influence of the still strong conservative officer corps, and the allied forces who had won WW I, and civil war in Bavaria during which the seeds of Nazi violence began to grow, led instead to the hardly unifying and never effective Weimar Republic. "The revolution had spent its force, and the next ten years would be a story of continuous struggle against the onset from the right and the final overthrow of the system established by the November revolution."[16] This background plus the disasters of rampant inflation, then deep depression, and widespread unemployment were perfect breeding grounds for Hitler's rise to power. But resistance continued,[17] even during the darkest days, in the face of capitalism gone mad under Naziism, when thousands were sent off to labor or extermination camps, along with six million Jews,[18] and millions of Poles, and while 20 million Soviet people lost their lives along with millions of allied soldiers in WW II. While these horrendous and still undigested events of German and human history deserve much more attention, the specific point here is that the German workers, in spite of often heroic efforts, faced a set of circumstances and opposition which prevented them from ever gaining anything like the strength of the workers' movement in Sweden and Finland.

However, the two Germanies of today are very different in this regard. I will come back to the workers' movements in each of the present-day Germanies. It may be enough to observe here that the Soviet-sponsored socialist hegemony in the East draws strength from opposition to Germany's disastrous and inhumane history of authoritarianism and fascism and provides a very strong, even if formalistic position for organized labor. In the capitalist-sponsored hegemony of the West, organized labor enjoys a but-little-stronger position than in U.S.A., where it is weakest of all among our study countries.

The Health System Pre WW II

Elsewhere I have offered a more detailed description and comparison of the health systems in G.D.R. and F.R.G.[19] Here I will use some of that material in a briefer description of the system inherited by both present-day Germanies. More will be said about each health system–in the following sections of this chapter on G.D.R. and in the following chapter on F.R.G.

Germany adopted the first compulsory national health insurance in the world, the "Sickness Insurance Act," passed on June 15, 1883. At an earlier date, Bismarck had perceptively said, "The social insecurity of the worker is the real cause of their being a peril to the state."[20] Thus when he had his chance, even though Liberals called him a socialist, he pushed through this and certain other social welfare measures to "cut the legs off" the socialist workers' movement.[21]

In many respects, the overall story covering what was one system and now are two is the gradual addition of types of coverage and kinds of groups covered. This is shown in Figure 3 covering the period up to WW II.[22] From a system initially covering only the most underprivileged 15 percent of the population, now, in the G.D.R., according to Article 35 of the 1968 Constitution, "every citizen of the G.D.R. has the right to the protection of his health and working capacity . . . Medical care, medicaments and other medical benefits are

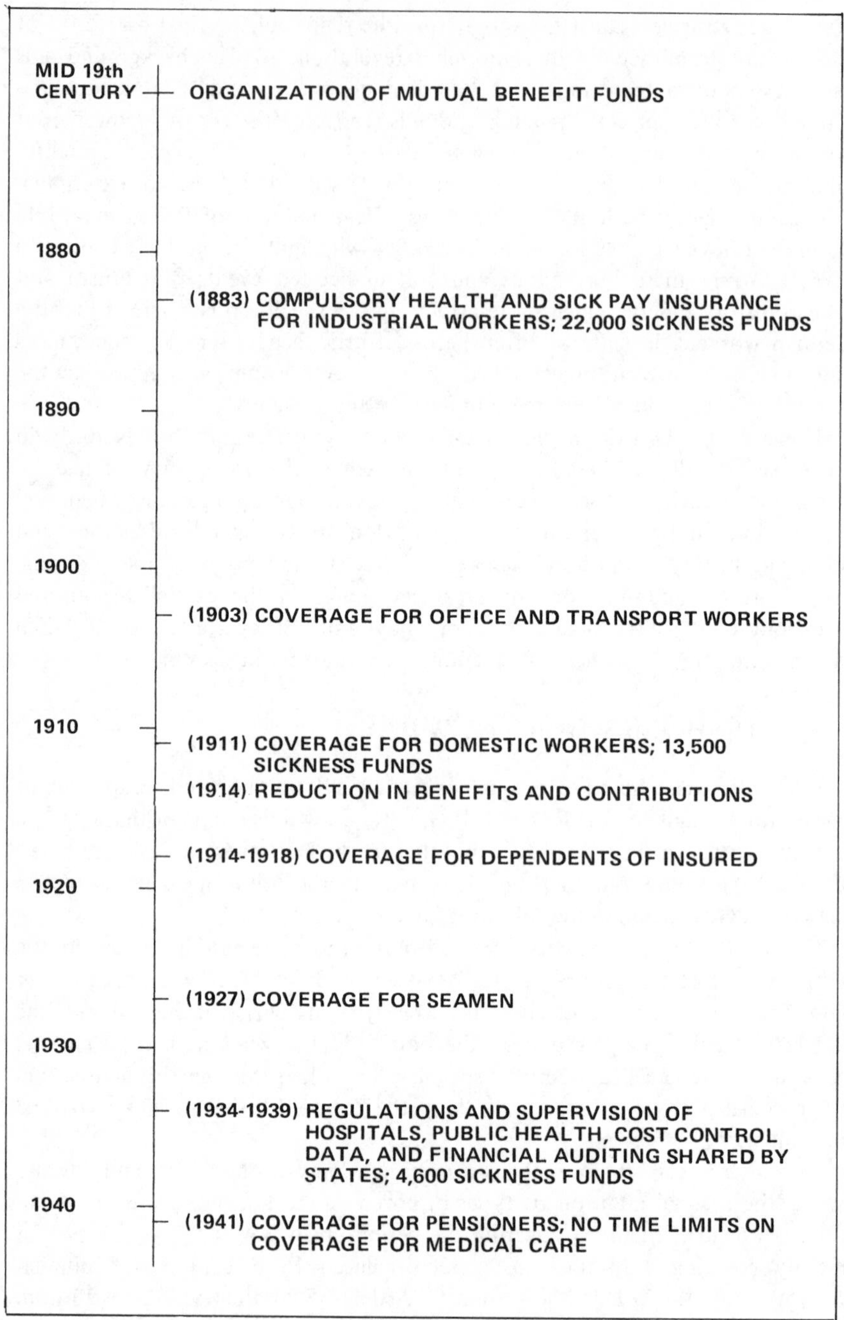

Figure 3. Progression of German National Health Insurance (1883–WW II).

granted free of charge . . . on the basis of a social insurance system."[23] The health insurance in the F.R.G. expanded until it now covers about 90 percent of the population, though there are certain deductibles and some items are only partially covered.[24]

Clearly these two countries present special opportunities for cross-national comparisons. With a common historical and cultural background, considerable "control" is gained over these important factors. While they do not exactly represent a laboratory experiment in political economic influences on health and health care, and in this study, provisions for OSH and OSH-PHC linkages, they come about as close to this as one can find in real life. Perhaps North and South Korea would offer another such pair of nations. In any case, the rationale or logic of choice for contrasting cases discussed briefly earlier (Chapter 2) is here altered. In these two cases, instead of holding resource levels approximately constant while selecting contrasting cases as regards health levels, one is holding background constant and looking for impacts of a single (although highly complex) cluster of hypothesized independent influences, namely, political-economic differences. With a sharp split in this regard in 1945, the two countries have since developed under sharply divergent political-economic forms (though recall that world-system influences reach into the G.D.R. too, even if not as much as in F.R.G.).

THE GERMAN DEMOCRATIC REPUBLIC (G.D.R.)

I take up the G.D.R. description first simply because I went there first and organization of my field notes, materials and order of recall leads me in this direction. I have fairly extensive field notes and because the Ministry of Health arranged such a full program of visits to OSH service centers in large factories, district OSH service centers serving many workplaces, and a rural OSH-PHC center, and to the National Institute in Berlin, I probably got a more complete view of the operating strengths and weaknesses of this system than I did of others included in this study. Thus my visit included interviews and discussions in Berlin, Seelow, Frankfurt an der Oder, Eisenhuettenstatt, Grossraeschen, Jena, Erfurt, and Karl-Marx Stadt. Though I could cope with German well enough to participate in the interviews and discussions and take reminder notes at the same time, it was a great help, for later clarification of my notes as regards unfamiliar terms and unclear points, as well as for travel arrangements, to have a senior staff member of the National Institute of Occupational Medicine (Zentral-institut fuer Arbeits-Medizin–ZAM) accompanying me throughout the business aspects of my visit. I would like to offer special thanks for this friendly and well-informed assistance. Although this guide or advisor ("Betreuer"–another translation is "nurse" which might be more appropriate) and I often chose to attend cultural events, etc., together, because I believe we enjoyed each other's company, it should be noted (because popular impressions in the West often suggest rigid restrictions and control over every aspect of life in the G.D.R.) that during many leisure hours I was completely free to walk, visit stores, and speak with ordinary workers and others. In the Appendix to this chapter I attach a lengthy commentary my guide made on an earlier draft to give the reader a clear idea of points of agreement and disagreement. Many of his points have been accepted. Some have not. The reader can tell by comparing my present text with the critique in the Appendix.

Most important for the existence of the present two German nations, was the defeat of Nazi Germany in WW II by the Allied Forces after horrendous loss of life and destruction on both sides. The Yalta Agreement between Churchill, Roosevelt, and Stalin provided for occupation of different sectors of the former German nation by the Four Powers (France, U.K., U.S.A., and U.S.S.R.) and ceding of certain territory–for example, that East of the Oder-Neisse line to Poland. The Soviet occupied zone, including the Soviet Part of the four-power-administered Greater Berlin was proclaimed the German Democratic Republic on October 7, 1949 and made fully sovereign in 1954, just before the U.S.S.R. ended the state of war with Germany in 1955 (the Western Allies had ended it in 1951).

But the common heritage runs deep and is very rich. One of the special pleasures I had on this visit was to enjoy a short concert of Johann Sebastian Bach's compositions played on instruments of his time in The Bach House in Eisenach. On a walk in Erfurt I came upon a house with a plaque saying Goethe had stayed there on his frequent visits from nearby Gotha. Of course, the contributions to political-economic theory and praxis of Marx, Engels and other internationalists such as Lenin are also well recognized in this part of Germany. Recently, the common heritage of the two Germanies has been rediscovered and reemphasized and new cooperative moves begun as one way of toning down the frightful level of East-West tension caused by the Reagan administration's introduction of new short-warning, potentially first-strike nuclear missiles into F.R.G. and other NATO nations.[25]

The Political Economy of G.D.R.

When West Germany (the Federal Republic of Germany–F.R.G.) entered the West European Defense Community (NATO) in 1952, the G.D.R. set aside a prohibited zone three miles deep along its 600 mile border with the F.R.G. It also established a separate phone system in Berlin. In August 1961, the G.D.R. built a fortified wall cutting Berlin in two after more than three million people had emigrated to the West. Between 1961 and 1978, over 50,000 more had fled. Also thousands of retired persons were allowed to leave and some 20,000 people held in jail were freed upon payment by the F.R.G. of

$250 million. These facts are cited in part because they have an important bearing upon OSH. This in the following way:

The present area of the G.D.R. was heavily involved in agriculture before WW II, though, as noted in the first section of this chapter it also had several industrial cities. At the end of the war, whatever industry there had been was completely destroyed or removed as war reparations. In the strong push to develop through industrialization there was a severe shortage of labor. The flight of thousands of what in the G.D.R. might be called bourgeois-oriented workers was very harmful to the country's development. There is no unemployment. There is still a severe shortage of workers. There are even concerns for the future since the present population of 16.8 million is not quite reproducing itself. Thus, in addition to the preventive and comprehensive care measures usually taken in a rather fully socialized general health care system (see below) special pains have been taken to develop an effective OSH system—perhaps the most effective of any of the study countries (though there remain serious points for critique and also concerning the OSH-PHC relationship as will be discussed).

Starting out in the Stalinist era, the G.D.R. was a highly centralized Communist state. Its authority structure remains rather highly centralized. I would say it is generally concerted—that is, dissident elements have either left or keep a very low profile. But after suffering severe economic problems under the central, master planning and decision-making approach, a "new economic system" was introduced in the mid 1960s.[26] This allows factories and other undertakings to sell on a market which is open within some broad planned guidelines and to make a profit as long as this is reinvested in the undertaking or used to provide bonuses to the workers. The companies are called VEB for "Volks Eigene Betriebe" or People's Own Undertakings in contrast to the private designations of Western capitalism—GmbH, "Gesellschaft mit beschraenkter Haftung" in West Germany; Inc. in U.S.A.; and Ltd. in the U.K. By the early 1970s, the economy was highly industrialized. Today it is something like eighth in the world in industrial production. Only some 11 percent of the labor force is involved in agriculture, and most of the farms are large collective undertakings with a high degree of mechanization, including areal spraying over immense fields which offer a striking contrast to the tiny private plots one sees after flying across the border to the West. Recently, some new "Combinations" (Kombinaten) have been encouraged, for example, in high technology areas. These are insulated from the domestic economy and function in as aggressive and entrepreneurial a fashion as any Capitalist enterprise to export goods, especially to hard currency countries. They are allowed to operate somewhat outside of the rules of the domestic economy. As mentioned earlier, the influences of the capitalist world-system are felt here too.

Since the mid 1970s, the G.D.R. has enjoyed the highest level of living of all communist nations. By 1974 the World Bank calculated that the G.D.R. had overtaken the U.K. in per capita income. "At $3,710 per person it was ahead of Britain's $3,590."[27] In 1976 the GNP was $70 billion; per capita income $4,000. By 1980, the GNP had risen to $7,226 per capita.[28] It is obvious to the visitor, that this income is fairly equitably distributed. One can not find gross signs of poverty as in many other countries, nor did I see signs of gross affluence–though the provision, especially for foreign visitors, in some of the luxury hotels at $75 a day seems a bit overdone for a society aiming for a classless society. But I did not receive any exact data on income differentials. Whether the chief engineer in a large steel or chemical works earns eight or ten or more times what the street cleaner earns I don't know.[29] Data reflecting the distribution of income in G.D.R. and F.R.G. for 1970 show the latter with greater disparities. Whereas in F.R.G. the top 10 percent of households received 29.1 percent of the income and the bottom 10 percent got only 2.2 percent; in G.D.R. the same percentages were 16.9 and 4.[30]

I looked also for regional disparities as one impression I had heard was that "only the Berlin stores are loaded with goods–to impress the foreign visitor." This was not true. One could get into a traffic jam in Erfurt or Karl-Marx Stadt almost as often as in Berlin where the number of personal cars as well as other motor vehicles is remarkable. I'm sure there is a larger selection in Berlin stores, but those in other cities had lots of goods. On occasion, because the market or planning mechanisms do not always work, some item is out of stock. In a Drogerie (like a U.S.A. Drugstore but without pharmaceuticals, since these are sold in special stores–Apotheke) I asked for paper handkerchiefs and was told they arrive only four times a year and are quickly sold out. This was not a serious matter. I was offered Zellstoff, soft paper-like sheets of about 30x50 cms used for all kinds of things–especially diapers. As long as one had the patience to tear them, they worked fine for my purpose, though they may not have looked as nice.

There has been a chronic problem with living quarters.

> . . . Massive investments were pumped into industrial projects in order to develop the G.D.R.'s economy. House-building suffered a comparative neglect. In the 1950s the G.D.R.'s per capita house-building production was only a fifth of the Federal Republic's. By 1960 it had only risen to 40 percent. By 1961 the G.D.R. had built just half a million homes since the war, an average of some 30,000 a year.
>
> The results of this neglect are visible to any visitor. Far too many of the old pre-war tenements, their brickwork unpainted for a generation, still stand in the center of East German cities. The G.D.R. has one of the oldest housing stocks in Europe. Their amenities are poor. In 1971 only 11 percent of all flats had central heating, 58 percent had no inside lavatory, and 61 percent no bath. Whereas in West Germany more than half the homes were built after the war, in 1971 four out of five in the G.D.R. were not. Half of

> the G.D.R.'s homes were more than fifty years old. East German flats are smaller than those in the Federal Republic. The commonest form of flat has only two rooms. But because East Germany's population size has remained almost static, the country at least has more space per person than other East European nations. Comparative figures for 1970 show that East Germans had 19.4 square metres, Poles 13.5, Czechs 11.6, and Soviet citizens 11.2. West Germans had 24 square metres each. . . .[31]

But the priorities are now swinging away from heavy industrialization and apartment buildings and other housing construction is evident everywhere.

> The rate of house building and the amount of investment in it have considerably increased since 1971. More money is going into repairs, many of which are done collectively. Local authorities organize 'Join-in' competitions between people in different neighborhoods to encourage them to make improvements. In Berlin they have now renovated some of the old tenements, which were shaped like square-cornered figures of eight. The central section where the flats were darkest has been torn down, so that the block is now a large hollow square with more light in the inward-facing rooms.
>
> The target for the five-year-plan period ending in 1975 was for 525,000 new or improved homes. In the next five-year-plan up till 1980 the target is an extra 750,000 flats as well as a 6 percent increase in the investment in each one, which will mean a larger number of three- and four-room flats. There is also to be a 50 percent increase in the budget for repairs. The 1971 party congress also decided to allow people to build their own homes. Partly in order to overcome the country's shortage of building labour, and partly in response to the demand for smaller and less uniform constructions, the party gave the go-ahead for do-it-yourself building. Nowadays in East German outer suburbs on summer evenings and at weekends alongside the car-washers and garden-waterers the new amateur house builders are hard at work. The man shoveling sand outside a half-finished house is probably the owner. He may have started with no capital of his own at all. According to a family's size a couple can get a credit of up to 80,000 Marks to build one of the forty-four different types of house-design supplied by the Building Ministry. The money covers materials and labor. The only problem is that the builder must find his mates. Often these new do-it-yourself homes are in small neighborhoods where several houses are going up. People from the same factory or office help each other. Factories have to assist in supplying bull-dozers and cement mixers to their employees at weekends. The local authority offers working drawings and plans, and can send plumbers and electricians if the family gets stuck. The scheme has become very popular and between 10 and 12 percent of the new houses in the G.D.R. are now built by owner-occupiers themselves. The average loan is 65,000 Marks, of which roughly 40,000 goes for materials. On this sum only 1 percent interest is paid. On the other 25,000

> there is an extra 4 percent. For a finished house with five rooms, kitchen, bathroom, cellar, and garage the monthly cost would be around 136 Marks a month. The amount is low, but it is officially not allowed to be set higher than the rent for a new state flat of the same size.
>
> Housing in the G.D.R. is cheap. This is the main consolation for the serious housing shortage. Rents for a new state flat were fixed on January 1,1967 and have not been increased since. In Berlin they only cost between 80 and 90 Pfennigs. For a three-roomed centrally heated flat this works out at roughly 123.25 Marks a month in Berlin, and 97.50 Marks outside.[32]

I had several bits of bad luck with the trains, at least I was told my experiences were unusual. But in general the material life is really very good. I have experience in both industrialized and in underdeveloped countries and would say there is little resemblance between the G.D.R. and an underdeveloped country. It also seems certain that the poorest in the G.D.R. are in no way as bad off as the poorest in U.S.A. and are probably better off than many in F.R.G. (see below) and many Brittons today. On the other hand, the life of the poorest in G.D.R. does not approach that of the same group in Sweden. So it is a middling to well off country in material terms and distribution of resources is comparatively equitable.

The sense I have with regard to authentic citizen involvement is more problematic. It is a terribly complex matter and I don't pretend to have the last word to say on this difficult, yet extremely important question. It is possibly the most important matter–also for workers' health and safety in its fullest sense. All of the problem does not lie with the centrist tendencies of the present governing structure. Some of it is a matter of cultural history. For example, on a visit in 1976 to the Neubrandenburg area for a meeting on chronic disease epidemiology, I went with a group to a rural policlinic. This was the area from which Bismarck originated, where the Junkers held sway for many generations. We saw farm men half bowing and lifting their caps to the doctors. When one of the visitors asked about the mechanisms for handling patient-complaints, one of the clinic staff said they have to actively encourage serious criticism from patients concerning human relations and other clinic matters or they'd never get any feedback because of this heritage of hierarchism. As Steele puts it:

> Another characteristic of the G.D.R. which is part of a long German tradition is the way authority runs from top to bottom. In its modern guise it is described as "democratic centralism" and "the leading role of the party," but it has the effect that lower-level cells within the party obey higher ones without question, and that social bodies outside the party take the lead from it. The phenomenon is remarkable when compared with Poland which tempers Communist rule with a tradition of anarchy and romanticism. The "good" side of the coin is German respect for the law. . . . The "bad" side of the

> coin is submissiveness to authority. People do not stand up to be counted. They lie down in order to be ignored. . . . Talking with German officials from the Free German Youth, the trade unions, or the SED [Sozialistische Einheitspartei Deutschlands–Socialist Unity Party of Germany], one notices how often the phrase begins "Our task is to . . ."[33]

No one should jump to the conclusion that the Communist Party of the G.D.R. (SED) is itself lacking in integrity and any degree of independence from the Soviet Communist Party. It is true that the presence of Soviet military might (carefully restricted to bases because of continuing resentment) puts the U.S.S.R. in important ways in overall control. But the inheritance of heroic struggle against fascism, even to the death for many comrades, and their German roots are factors too strong for there to be no independence and integrity within the SED itself.

> A month after becoming Chancellor Hitler used the pretext of the Reichstag fire to start mass arrests of Communists. Hundreds of members were rounded up. Others escaped abroad or went underground. . . . Scores of those arrested died later in prison or concentration camp. Many who went to the Soviet Union were killed during Stalin's purges. But those who survived became the core of East Germany's post-war elite. None of the new post-war Communist states had as relatively large a group of seasoned party cadres as the G.D.R.
>
> They were to become known as "activists of the first hour" and were held up in the new society as its leading heroes. Fritz Selbmann, a former KPD [Kommunistische Partei Deutschlands–German Communist Party] deputy in the Reichstag and after the war a deputy minister of industry described them in his book *The First Hour* as the people who in the days when Germany had suffered total military defeat and the complete collapse of its economic, political and social order after the darkest night in German history, took over the task of leading the German people out of the misery into which they had been thrown by the crimes of Nazism.[34]

There is the history of hierarchism and sexism, and so one finds considerable variation. In some PHC and OSH policlinics and centers I visited, women were in charge and there was broad open involvement of staff in our discussions. In one or two centers I visited one must say the old authoritarian school of thought and action was very evident.

Beyond such signs of democracy or its lack in terms of personal interaction, there is the possibly more important question of collective action. As I noted above, the factories etc., are called VEB (people's own undertakings) instead of Inc. or Ltd. But what this really means in practice is hard to say. If one accepts that the Communist Party is the leading power of the working class, it can be argued that indeed the working people are in control in the G.D.R. But if one looks for a significant degree of independent worker control in "their" own factories and "their" own health services, it is not to be found. I want to be careful not to overstate this point, for perhaps I was still excitedly impressed by having visited Poland for several days before coming to the G.D.R. I happened to be in Poland on Wednesday, October 28, 1981 when Solidarity brought the country to a standstill for one hour in a general strike. The excitement and fresh creative spirit was truly something enobling of the human spirit. The whole history of Solidarity is this way.[35] Now it is crushed or smouldering under military dictatorship. But early on in the visit to the G.D.R. and then several times subsequently I made known my intense interest in the part unions play in OSH. Yet whenever we visited an OSH center, nothing was said about the unions unless I asked. Union people did not accompany us on plant visits; they did not take part in greeting me. Certainly they were not in charge. On only one occasion was a union spokesperson present.[36] I believe even this one time was an accident (not necessarily a bad sign). This woman was the Chairperson of the Social Insurance Advisory Committee of the Union and one of her tasks was to come each day and check through sick slips with one of the clinic doctors. She happened to be there and was invited into the meeting (I'll say more on this when the OSH system is discussed, since this is an interesting and important function).

Again, I want to caution myself as well as the reader about overdrawing this impression of relative lack of authentic union participation and control. There is a formal structure which undoubtedly gives the unions a significant voice and probably there is some variation in the extent to which this voice is exercised formally and informally. It is also argued by some on the left in this country that independent unions are not needed in a Communist-Party-controlled state, since the state functions on behalf of workers. But in 1984, I am nervous about Big Brother in U.S.A. in G.D.R.–everywhere–and would like to see independent unions.

Because of the complexity, yet great importance of the matter of worker control through the state structure, the Communist Party (SED), the unions, and other organizations (for example the FDJ or Free German Youth and the DFD, Democratic League of Women) I will use both my own experience and Steele's observations from his chapter, "How Much 'People's Democracy'?"[37] to go into the structure and its functioning in more detail, particularly as regards the party and the unions. Local illustrations will be drawn from Eisenhuettenstadt an entirely new (that is post WW II) steel plant employing some 10,000 workers and its surrounding town of close to 50,000 people close to the border with Poland which Steele appears to have studied extensively, and which I was able to visit and study as regards OSH services and the hospital and general health services.

The Amended Constitution of 1974, articles One and Two, states "The German Democratic Republic is a state of workers and farmers. It is the political organization of working people in town and countryside under the leadership of the working class and its Marxist-Leninist party. . . . All political power in the G.D.R. is exercised by working people. . . ." While there are other parties representing farmers, religious groupings, and so on, and these take part in elections of representatives to a national assembly, the ruling party is the SED. In July 1984 the SED had 1,907,719 members and 46,411 candidates. Of these 56 percent were workers.[38] In 1981, there were 2.17 million members and candidates active in some 79,700 local district, regional and national party organizations. Of these members and candidates, 57.6 percent were workers, 22.1 percent intelligenzia, and 5 percent farmers. One third were women.[39]

As in other nations controlled by communist parties, there is a parallel structure of governance and control at all levels. In the Eisenhuettenstadt works, for example, the SED party organization (the Betriebsparteiorganisation or BPO) consists of some twenty people who meet once every two weeks with the director who is himself a member of the party (often but not always the case). The BPO thinks of itself as the political economic conscience of the plant. The director must answer to his fellow party members who, taken together, work in most departments of the plant, though some are in the town's governing structure. The BPO cannot take a direct part in economic decisions, but they function as a control-inspection, and reflection group *vis-à-vis* the manager-director (the system requires only one person to be in charge and ultimately answerable in any factory, works, Kombinat, or other enterprise).

> The director himself is a member of the BPO and has to answer to his fellow party members at its meetings. In addition, the party secretary in the factory can sit in on all the director's meetings. Through its links "upwards" with the national party apparatus and "downwards" with ordinary workers via the various party cells in every branch of the factory, the enterprise party leadership has access to a wealth of information which may not be as readily available to the director through his own channels. It can bring pressure on him for changes. The BPO is in charge of the factory newspaper which gives it control of the most powerful means of communication within the factory. Even without that, the party's presence at every level of the factory makes it a strong, independent force which the director cannot afford to oppose.[40]

Another important general guidance structure is the Permanent Production Council. This brings together leading members of the trade unions, the FDJ, DFD, and key members of the technical and economic staff of the works. This group has purely an advisory role. But, since 1971, the director must report back to this council on his responses to their suggestions and explain if he rejects any of them. The closing of a department, expansion of another, introduction of new machinery and materials (thus helping to assure the workers' right to know) as well as changes in scheduling of shifts are among the matters within their purview. In Eisenhuettenstadt Kombinat Ort (the formal title of this steel works) the council has twenty-five members and meets about ten times a year. In branches of the plant there are related committees which usually meet twice a month.

> The trade union is the largest organisation in the factory with a nearly 100 percent membership. In the early days of the G.D.R. its role was mainly to organise production drives and help in fulfilling and over-fulfilling the plan. The plan, known as the collective contract, is drawn up jointly by the management and trade unions. In the period of post-war austerity, low pay, and high work norms, the trade union's image for most workers was not favorable. Increasingly the trade unions have been given more responsibility. They have become less like Western trade unions. They are in charge of all social security funds, from sick pay to the allocation of holiday places in trade union hotels. The trade union is in charge of the factory's hospitals and clinics. Under the collective contract the factory is committed to spend money on crèches, kindergartens, new sports facilities, leisure centers and housing. The trade union supervises industrial safety, and ensures that the factory management observes the rules. On this the G.D.R.'s record is particularly good. The West German Government's "Report on the State of the Nation" in 1971 admitted that industrial accidents hit 8.8 percent of the work force in the Federal Republic but only 4.1 percent in the G.D.R. The report commented: "The low accident quota in the G.D.R. is mainly the result of the system of labor safety which is apparently more efficient and more intensively controlled. . . . A more comprehensive catalogue of labor safety regulations, coupled with factory instructions, the general factory-based health service, accident research and the trade unions' strong functions in control and participation have produced a system which is superior to that of the Federal Republic." By 1974 the proportion of the work force affected by accidents had been reduced to 3.5 percent. . . .
>
> The picture that emerges is of a trade union and factory structure very different from the West. In the West a worker expects from his factory good wages, decent working conditions, and a clean environment. The trade unions' job (where unions are recognized by employers) is to negotiate and fight for them. In the G.D.R. a factory has a more central role. A worker expects it to provide kindergartens, holiday facilities, help with getting a flat, and a range of other services for which private agencies or at most local authorities would be responsible in the West. In the G.D.R. workers demand more from their factories. In turn more is demanded from them by the factories.
>
> The trade unions are different too. In the G.D.R. they are based on Lenin's model and have a radically changed role from their counterparts in a capitalist system. Instead of a basically negative and defensive approach, trade unions become "schools of socialism" as part of the process of building a new society. Their three-fold task is to minimize tendencies towards excessive bureaucracy in economic planning, to ensure that the social regulations and the

> labor code are observed in the factories and to protect their members' immediate interests. There is a rich field for potential conflict here. The trade unions are expected simultaneously to help with achieving production targets and also to defend members' interests. In looking after the latter, trade unions have certain rights. No one may be dismissed without trade union approval. Inevitably this works less well in practice. An official of the central council of the FDGB, the central trade union organization, conceded in an interview in Berlin that "sometimes the unions are only consulted afterwards. This must be improved." Unions deal with workers' complaints about unjust punishments or unfair treatment in the payment of bonuses. In a case at Eisenhuettenstadt a worker who had been absent for two days one September, and was mildly reprimanded at the time, was told at the end of the year that he could not share in the factory's bonus. He complained to the union that he had had no warning of this. They won the case for him.
>
> Generally in disputes unions seem to act more readily for an individual worker than for workers as a group. The original union function of combining workers' strength and solidarity together has given way to more of a welfare role. There is no right to strike. Occasionally workers act on a large scale. At the beginning of 1971, according to the Berlin official mentioned above, massive criticism by workers at a Berlin factory forced the replacement of the director. This incident, and my interview, took place soon after the riots in nearby Poland which caused the overthrow of the Gromulka regime. Could this happen in the G.D.R.? "We react earlier than the Poles do," said the official. "The Poles did not listen to the complaints. They took no notice. We have a system of regular reports from the factories."[41]

During my most recent visit, it was a very pleasant experience to see the Ministry of Health driver included at the same table when we were served lunch–a rather fancy one in the Town Hall in Frankfurt an der Oder,[42] courtesy of the chief district doctor's representative. It never occurred that the driver initiated any conversation. But he seemed to respond with ease when addressed and the conversation seemed open and easy in his presence.

It is also clear that women have achieved considerable equality. The heads of most of the OSH clinics I visited were men, but one in an electrical works was a woman and the head of one general PHC policlinic we visited was also a woman. There is a very high proportion of women in medicine–I think something like 65 percent. In general:

> Women have won a more equal place in society than ever before in Germany or in West Germany now. In 1974, 84 percent of women of working age had jobs, one of the highest ratios in the world.[43] The G.D.R.'s working mothers have more pre-school facilities for children than any of their Eastern or Western neighbours. Three-quarters of all pre-school children have places in crêches or kindergartens.[44]

While these achievements of women must be attributed in part to the overall shortages of labor power, just as in the realm of protection of workers in general, the achievements are none-the-less real for their political-economic origins. In fact, equality of the sexes before the law and in employment, education, and other spheres, is fundamental to the G.D.R. Constitution. A cultural hegemony has been established to back up the political-economic needs and opportunities.

So the system is very different from the capitalist one of free trade unions functioning in an adversarial relationship to management and seeking through a degree of political action (which varies enormously in its organization and effectiveness among our study countries) to set up laws and enforcement favorable to workers' interests, while Capital buys its own political influence to counter labor's efforts. But if it is different, does the G.D.R.'s system yield less net worker influence and control or more? I would say more, much more than in all other study countries, except Sweden which has its own very strong workers' movement, achieved through a long history of Social Democratic party struggle and rule. Still, there are limitations:

> It does not add up to workers' control, but does it give workers any significant participation? Several limitations have already been touched on. One is the leading role of the party which takes precedence over the power of the trade unions. The second is the principle of one-man management. Whatever the trade union recommends, the director and department heads have the last word. Then there is the general principle of democratic centralism which runs throughout the country's political structure and affects trade unions like all other organizations. The simple definition is that whatever the top level of an organization decides, lower ranks have to obey. Although in practice things are more complicated and the central council of the trade union has to listen to some extent to the rank and file, it has the right and the power in the last analysis to prevent any dissidence or opposition. Finally there is the role of the central planners and the VEBs which effectively circumscribe the actions of trade unions. In their discussion on the collective contract neither the factory manager nor the unions are able to decide what to produce, but only how to produce it . . .
>
> To talk of a "new class" is wrong. Decision-making, it is true, is in the hands of a relatively small elite. But the elite is not closed. It is based on a combination of education (mainly the younger members) and loyalty to the system. The members of this elite work hard, and their position is risky. The history of the party purges is enough to show that the higher you are, the farther you fall. There is no hereditary right to rule. Factory managers probably feel this the most. The legitimation of private ownership no longer exists. Directors can only justify themselves by their results. The party and the workers expect an enormous amount from them. The earlier part of this chapter described the wide role factories play in East German society. The man at the top of the pyramid has a proportionately

> harder job. He must be a specialist in his field (two-thirds of G.D.R. managers are economists, the rest engineers). He must understand Marxism-Leninism. He must arouse, initiate, and keep the trust of his staff, and the various social organizations in the factory from the party to the trade unions. He must be ready to work cooperatively but also take risks.[45]

There are many things to praise in the G.D.R. Its level of economic development and level of living. Its involvement of both women and men. Its very comprehensive social welfare measures. Its preservation and enhancement of culture, sports, and other spheres. But, in the most general terms, I believe Navarro's analysis and critique of the Soviet Union applies also to the G.D.R.[46] It is a system of state capitalism and not a worker's state, and too high a proportion of decisions are reached by centrally-oriented, bureaucrats, technocrats and experts. It needs to undergo much more thorough democratization and politicization before it can be called a socialist state. A great deal of control remains in the hands of an educated elite, however well intentioned these cadre are *vis-à-vis* the ordinary worker. For example, when I told a group meeting of health officials that an ordinary worker had told me that he had never heard there is anything wrong with asbestos and I recommended more general public information on the subject, the elitist, spoonfeeding approach came out. The reply was: "We don't want to alarm the people about things we can't change right away." Why not encourage a fully informed people to decide for themselves how they will live—or die?

The Health Services System of G.D.R.

The G.D.R. has a centrally planned, yet regionalized health services system. There are fifteen districts (Bezirke) with relatively complete health systems though some elements may serve more than one region, and some elements are only available nationally. Although there are public health and educational components at regional level, the main medical care instrument at this level is the regional hospital with all specialties—some also have a clinical diagnostic OSH specialist, though the National Occupational Health Institute in Berlin is the main referral center.

Below the regions are districts (Kreise). Here one finds both the district hospital and a number of multiple-specialty policlinics which offer the services of general practitioners as point of first contact. All staff at regional and district facilities are full time salaried people.

Although progress has been made in decreasing the private sector, in 1973 there were still eighty-two hospitals in the hands of religious communities and sixteen privately owned.[47] The number of consultations in private general practice and the number of such practices themselves are not published,[48] but the number is probably quite small. As Winter puts it:

> While freestanding office practice remains the dominant mode in the F.R.G., patients in the G.D.R. do not have to pay for care in policlinics and other outpatient facilities. They do have to pay in private practices. On this basis alone, private practice has lost its attraction in the G.D.R.[49]

During participation in the Third National Symposium on Non-Communicable Diseases Epidemiology, April 21-24, 1976, in Neubrandenburg, I was given the opportunity along with several other visitors to visit a rural policlinic. General practitioners or specially trained nurses offer first-line care to people not in the immediate vicinity of the policlinic. The policlinic had twelve GPs and a number of specialists in pediatrics, internal medicine, obstetrics/gynecology, surgery, orthopedics, and certain other specialties. There had been openings for two GPs and for a psychiatrist for some time. The policlinic offered, consonant with the general medical culture in both Germanies, a range of physical therapies including electric, sound wave, and water therapies. Special programs for childhood immunization, follow-up of alcoholics and of tuberculosis patients were integrated into the policlinic.

The visitors had a special interest, as noted earlier, in citizen control and complaints and learned that the sharp class differentiations of the past were hard to erase. The chief doctor of the district said she has to investigate every complaint, and the people had learned to voice themselves about matters of decent treatment and human dignity. One physician had to be dismissed for his arrogant attitude. She said people had not yet learned to inquire into what they viewed as the true preserve of physicians—the more technical medical matters. There was a moment of humor at this point when one of the staff observed that the school teacher seemed to be the one most ready to complain. This might have been a personal idiosyncrasy or, again, a class matter, the school teacher feeling more equal to the physicians than did the peasants and workers. Obviously, there is still much to be done before a fully democratized and politicized socialist society is achieved in the G.D.R. Still, it was also clear that someone else is in charge as compared with the days of the Junkers. The availability of this quite well endowed medical and public health facility in a primarily agricultural area seemed to me to offer a striking contrast with the medical care vacuum in most of eastern Connecticut—a rural section of the second wealthiest state in one of the richest countries in the world.

The policlinic makes referrals to a district hospital which in turn can refer special problems to the regional hospital or to the Charité in Berlin.

Inservice education, most of it focused around special problem areas such as alcoholism or improved nutrition, was explained to us as something which occurs regularly—in the staff room of the policlinic, in the examining rooms, and at special seminars at district or other levels. No mention was made of in-service OSH education. But at the time of this earlier visit, I did not ask specially about it. I will say more about this in a following section on OSH-PHC relations.

After closely examining the care and prevention systems for tuberculosis in a state of the F.R.G. and a district of the G.D.R., Empkie concluded:

> The main overall differences between the two systems appear to be that one (G.D.R.) is a centrally planned and executed program integrated at the periphery with other health services resulting in greater application of some preventive measures, while the other (F.R.G.) has a central advisory body (the DZK in Hamburg), but essentially consists of eleven provincial–Laender–programs separate from other health services at the periphery.
>
> It is tempting to attribute the differences in programs to the overall political structures in the two countries *per se*, but certainly other factors must be considered. For example, rather than political structure it may be the different interests of the power groups in the two countries which account for the dissimilar programs. In the G.D.R. where the labor pool is very small and the need for increasing industrial production very great, the economic necessity to protect the health of the workers is enormous. In the F.R.G. where full employment is not a necessity this may not be the case.[50]

In both countries, the health systems are taking up a larger and larger proportion of the social product. In the F.R.G., this appears in part to be due to general inflation and in part as a frightening matter internal to the medical care system over which there is little control. In the early 1970s, health care costs in the F.R.G. expanded at an unprecedented rate of almost 20 percent per annum.[51] In the G.D.R., by contrast, an increasing portion of the national budget has also gone for health and social welfare, but this has been a small rise as a percent of GNP, since GNP has been rising rapidly and the health care system is comparatively regionalized.[52] In any case, it is reported to be a planned and accepted direction of the system, though undoubtedly there are pressures to economize in G.D.R. too.

The official West German Report on the Situation of the Nation, 1974, (*Materialien Zum Bericht Zur Lage der Nation, 1974*) characterizes the overall organization of the two systems this way (p. 486):

> Health care which demands a further development of the health systems in both States has, from the same beginnings, evolved divergent concepts of health politics (Gesundheitspolitik). In the F.R.G. the self-government and responsibilities of the social insurances have been strengthened. The activity of the government apparatus is essentially confined to legal and financial conditions. The responsibility of the individual citizen is given a higher place. The health policy direction can be said to occur in a system which is open to different and to some extent divergent experiences and goals.
>
> In the G.D.R., the desire to build up the health protection of the people has led step by step to a central direction and planning of the

> health system. The Ministry of Health is the highest authority in a hierarchical system of health care and from this direction medical research is also thematically and organizationally connected.

As to the health status, both G.D.R. and F.R.G. share the relatively fortunate position of other industrialized countries. Yet by a number of measures, the G.D.R. seems slightly better off healthwise. This is true in spite of G.D.R. having had a lower health status in the years immediately following WW II, and in spite of having fewer overall resources per capita and in spite of spending less on health than does the F.R.G.[53]

Infant mortality has declined greatly in both countries since the war. In 1946 the rate was 131 in the G.D.R. zone and 97 in the F.R.G. zone. By 1972, the rate in G.D.R. had declined to 17.7 while F.R.G. had 22.8. In 1977, G.D.R. was down to 13.1 while F.R.G. was down to 17.4. By 1980 it had declined further in the G.D.R. to 12.1 and 13.5 in F.R.G.[54]

Childhood mortality per 1,000 population age one to four years was equal in the two countries in 1968-69. Since 1971, there have been no deaths due to measles reported in the G.D.R., while in 1971 alone, eighty-four children died of measles in the F.R.G. The maternal mortality rate was 16.5 per 10,000 births in F.R.G. in 1973, while it was 5.05 in the G.D.R. in 1971. By 1980, this figure had declined further to 2.3 in G.D.R.[55]

Empkie's study of tuberculosis control in both countries (cited above, note 50) showed the G.D.R. to have been more successful. The official F.R.G. report *Materialien Zum Bericht* etc., for 1974 gives new cases of all forms of tuberculosis per 100,000 for 1971 as 56.9 for G.D.R. and 74.0 for F.R.G. (Table 197, p. 480). Whereas infectious diseases and tuberculosis ranked second for men and third for women as leading causes of death in Germany at the height of the Great Depression in 1933, these were no longer among the eight leading causes of death in 1969. Although tuberculosis deaths rose just after the war, in the F.R.G. the death rate from tuberculosis (deaths from this cause per 100,000 inhabitants) fell to 17.3 by 1958 and 8.3 by 1970. The corresponding figures for the G.D.R. were 20.9 and 4.2 (1971).

This decline in infectious disease signals the "success" of both countries in moving firmly into the pattern of what the Germans call "civilization diseases" (Zivilisationskrankheiten). As is true of nearly every "developed" country (Japan is an interesting exception as regards coronary disease) diseases of the circulatory system, cancer, stroke, accidents, alcoholism, and nervous disorders, are the major killers and debilitators. With these diseases for which knowledge and clear control measures are often lacking, there is little difference between the two countries. One West German author attempts to use this fact to argue against the impression of his countrymen that the health system of the G.D.R. is better than that of their own country. But he chooses to ignore the achievements in infant, childhood, and maternal mortality, work-related deaths and morbidity, and the much lower accident and violent death rates in the G.D.R.

One less favorable situation in G.D.R. is the extensive air pollution due perhaps mainly to the fact that its major natural resource is a vast deposit of brown coal. People there seem to have got used to it. Connecticut, where I come from, has very bad air, but it is more from Ohio Valley SO2 and local automobile exhaust and (to my nose) isn't as noticeable. On a couple of occasions I had difficulty sleeping as the smell from burning brown coal was so disturbing. Of course the Ruhr and Rhein Rivers in F.R.G. are notorious sewers (though improvements have been made in recent years) so I can't really weigh the pollution comparatively.

Space does not permit to extend consideration of health conditions and status in the two Germanies as far as I would like. But I would be remiss in this report not to take note of the special attention devoted to occupational health and safety in the G.D.R. As a consequence there are remarkable differences.

> In the Federal Republic of Germany in 1972, 112.0 occupational accidents occurred per 1,000 employees (the total number of occupational accidents was 2,458,234, and there were 30,000 cases of occupational disease), while the rate in the same year in the German Democratic Republic was 36.8 per 1,000 employees.[56]

The OSH System and Services in G.D.R.

Herein I will first lay out the general legal and formal structural arrangements of the OSH system in G.D.R. relying here on two other fairly recent informal summaries sponsored by WHO/EURO,[57] and on official G.D.R. statements.[58] Then I will give some content as to the reported working of the system as this could be assessed from visits and interviews in a number of different kinds of OSH centers in different parts of the country. The next subsection will focus on OSH-PHC links. A comparison of the realities with the model presented earlier will be offered as the concluding section of this chapter. I will reserve direct comparisons between the G.D.R. and F.R.G. as regards their OSH and OSH-PHC situations for the end of the next chapter.

One of the unique and most important points about this OSH system, as compared with those of other study countries, is that it falls under the Ministry of Health and is defined as an integral part of the state-provided, regionalized health services described above. A second and related key point is that all OSH physicians and other personnel are on salary from the Ministry of Health, making them comparatively independent of local management and union direction in the undertakings they serve (though payment may come from lower levels of the health system—see the comment on this point in the Appendix to this chapter). These personnel are responsible for a full OSH program of both preventive and treatment activities. The shape of each local OSH program and number of personnel is set by centrally issued regulations recognizing number of workers in a plant and the existence of special risks.

Special statutory-based OSH programs, some of them demonstrations for evaluation and possible wider adoption, have been set up for certain industries. The one in the construction industry seems especially noteworthy and is already well established. Very interesting special efforts are underway also in agriculture, chemical production, public transport, and certain other industries.

In addition to in-plant OSH services serving a single large undertaking (which in some instances may have several plants located in different parts of the country with an OSH service headquarters in the headquarters plant and relatively autonomous OSH services in each of the subsidiary plants) there are OSH centers serving a number of undertakings, and, for the purposes of this report, there are some especially interesting examples of OSH services integrated with local policlinics.

OSH service programs of all sorts are backed up by fifteen regional or County (Bezirk's) Labor Safety Inspectorates. These agencies are under control of the Ministry of Health and function in the fifteen G.D.R. Counties as state supervisory organs attached to the County Executive's office. Personnel from the Inspectorates are free to come into a plant on their own but are more often called in to assist in terms of expertise, hazard measurement capabilities, and other resources not at hand in the local OSH service. Legally, the OSH services are only advisory to the management which has the ultimate responsibility for worker health and safety. However, there are ways in which the OSH services can bring legal force to bear on a management to correct workplace hazards. The Chief Physician in charge of the Labor Safety Inspectorate, after consultation with the Chief Physician of the Bezirk (County or region) and the Bezirk Executive, can give legally enforceable directions to factory managers on OSH matters.

A National Institute for Occupational Medicine (ZAM–Zentral Institut fuer Arbeitsmedizin) in Berlin is similar to the Institute in Helsinki in that it plays a key role in research, standard setting, referral treatment, preparation of personnel, and transmission of information. However, it is not independent like the Helsinki Institute. The Berlin institute is directly under the authority of the Ministry of Health. Local OSH services and Labor Safety Inspectorates frequently call on the National Institute for assistance. It is the center of the Occupational Medicine Research Association and coordinates all OSH research in the G.D.R., though more and more encouragement is being given to regional research initiatives.

As in other countries, a number of other governmental agencies play a role in relation to some aspect of OSH, be it insurance for industrial accidents and disease or other. In G.D.R., the *State Secretariat of Labor and Wages* drafts principles and legislation concerning labor safety and supports implementation in all sectors of the economy. The *National Board of Technical Supervision* is a central state organ concerned with preventing accidents and related damage. *The Supreme Mining Authority* sees to OSH issues in mining. *The National Board of Construction Supervision* watches OSH and other legal provisions in the building trades. There is also a *National Board of Nuclear Safety and Radiation Protection*.

Some comments were offered above on the situation of unions and more will be said below. Formally speaking, the Confederation of Free German Trade Unions (Freier Deutscher Gewerkschaftsbund–FDGB) has oversight control of OSH on behalf of its members as provided for under law and by decisions taken by its National Executive.[59] At local level, oversight is exercised by union shop committees and Bezirk or county Executives of the FDGB. In exercising this responsibility, the FDGB depends on the shop-wide labor safety commission (this represents management, OSH services–medical and technical–and the union), first aid people in each department, and the County Labor Safety Inspectorate. But managers are the ones legally responsible for OSH.

This structure rests on a fairly complex set of general and special labor protection laws which have evolved over the thirty years of the G.D.R.'s existence. Not all can be listed or detailed here.

The Labor Code of June 16, 1977 details the workers' basic rights as guaranteed by the Constitution. These include the right and duty to work, co-determination and participation in deciding the affairs of state and daily life (the national formal structure for this is an elected People's House of Representatives and several nominally independent parties, as well as the parallel party and effective governing and participatory structure discussed on pages 242–253), pay according to quality of work, education, leisure, and cultural and recreational opportunities, protection of health and capacity of work, care in old age and in case of disability, as well as security of income in case of illness or accidents.

The Labor Safety Ordinance, and Act of September 22, 1962, concerning the preservation and promotion of workers' health at work (or on the way to and from work) makes the managers of production and service undertakings legally responsible for OSH.

The ordinance on Industrial Health and Hygiene Inspection of June 30, 1977 provides for the powers and duties of the Labor Safety Inspectorate as discussed above. Included is supervision over the OSH services covering each workplace, implementation of safety and hygiene standards, and health education at work. In this last connection, for example, management must assure that workers have been instructed on general and special health hazards of their jobs. The OSH services may give the instruction. In any case, the worker must sign off in a sort of personal instruction diary (Arbeitsschutzkontrollbuch) certifying that he or she has been told of such and such a hazard. If something does happen, these documents become important for assessing blame and determining compensation.

One of the earliest OSH statutes, that of June 23, 1955, titled "Improvement of the Workers' Working and Living Conditions and of Trade Union Rights" in the 7th Regulative Statute provides for periodic physical exams to help early detection and avoid further damage if a work-related illness has started and to provide the worker with a job suitable to his or her health status in the case of rehabilitation needs or in general. A very important new law which has just come into effect will merge the periodic exam responsibility with a national systematic work hazard identification program. This will be discussed in more detail below.

An Ordinance on Compulsory Registration of Industrial Diseases was passed November 14, 1957 and is supplemented by the List of Industrial Diseases of September 18, 1968 and subsequent revisions.[60] This covers social support for those affected and, very important for this study, requires every physician to report any industrial disease or set of symptoms suspected of representing a work-related disease. How well or completely this is done may be a big question, but the legal provision is important.

Labor Safety Regulation Number 5 of August 9, 1973–Labor Safety for Women and Young People–protects working women, expectant mothers, nursing mothers at work, and young people against doing physically strenuous or health endangering jobs. In most respects this regulation simply adds emphasis as it repeats what is provided for elsewhere, but pulls it out for special attention in relation to these categories of workers.

The last few years have seen additional important legal developments. One was announced in February 1980 after agreement between the Ministry of Health and other central organs and the Central Board of FDGB. It provides for regional Work Hygiene Centers and more local related Advisory Stations.[61] These Advisory Centers are to operate under guidelines set out in agreements between the Ministry of Health and Bezirk and Kreis (district) Councils. These agreements will spell out relations between these OSH Advisory Centers or Stations and the local health system facilities (for example, policlinics, OSH Services, and hospitals) Labor Safety Inspectorate, the County or District Council, the ZAM in Berlin, and OSH scientific research programs in the local area. In general, it would appear that some of the functions of ZAM will be thus decentralized and or made more locally available. Almost any facility involved in OSH can be designated such a Center or Station. A well developed policlinic in a large factory can be so designated. It would then take on educational, planning and reporting functions *vis-à-vis* other undertakings and OSH personnel in the area.

An even more important development is the work hazard survey and follow-up program–Ordinance to Prevent, Report and Certify Occupational Diseases–announced in different parts in 1981.[62] This set of ordinances was built upon a large pilot experience involving monitoring of workplaces with a total of some 600,000 workers and examination of some 400,000 workers. In brief (since more will be said about what it means in relation to some particular places I visited) it means that every workplace must fill out and return a questionnaire identifying and measuring its jobs and any hazards involved including a rating of these hazards from urgently hazardous to harmless. Periodic and follow-up exams as noted earlier will be keyed to the kind of hazards. The information should also be used to set hazards aside, not just allow workers to be exposed and then protect them as much as possible after exposure. It is also intended to foster further scientific research into OSH problems.

In 1980 some 5.3 million workers were covered with OSH services. This was about 63 percent of workers of whom 22 percent were in small to medium-sized workplaces. Certain standards are employed to determine the staffing of OSH services. These are approximately as follows:

- In industrial process undertakings (large agricultural, forestry and food processing undertakings as well as factories), with allowances for special hazards, intensive work conditions, etc., there should be one OSH physician per 2,000-2,500 workers.
- In all other workplaces there should be one OSH physician per 2,500-3,000 workers.
- There should be one industrial hygiene (IH) specialist per 8,000-10,000 workers (though the concept of IH may call for a chemist or physicist and not an IH person as known in U.S.A.).
- One technical specialist (or safety inspector) per 5,000-10,000 workers.
- Two to three therapeutic and support personnel per physician.

Some 800,000 workers are under follow-up care for specific exposures (for example, workers using air drills who are in danger of developing "white fingers" etc. have to be examined every six months), for rehabilitation and introduction to a new job following exposure and illness on a previous job, or for reasons of personal susceptibility (tendency to asthma, etc.).

About four million prophylactic exams (including lab work) are carried out each year.[63] If an OSH physician finds that a worker is not able to do a certain job for health reasons, the worker can not continue, but the undertaking must offer another job fitting the person's abilities and health. If training and qualifications are needed to provide for this transfer, the undertaking must provide for these.

The OSH manpower picture included in 1980 some 2,130 full-time and 1,460 part-time doctors.[64] In general, works doctors have entered from general practice or some other clinical medical specialty, such specialties generally requiring five years' training and experience for full qualification. To fully qualify as an OSH doctor, they must take courses over a two and one-half year period. These are given by ZAM. There is also a five year program for training primarily clinical OSH specialists. These people are most likely to be located in or in relation to university medical colleges, all of which have chairs for industrial hygiene and medicine.

Plant physicians work with some 9,160 associated OSH personnel—nurses, technicians, and others.

There are some 420 industrial hygienists, though this term or profession is not known as such, rather, these tend to be people well trained, in chemistry or physics working in the occupational hazard field.

The Bezirk Labor Safety Inspectorates employ some 250 non-medical university and college graduates, for example, in chemistry, and 250 industrial safety and hygiene inspectors. Qualification for this inspection work is acquired through a three-year course at a college. After several years experience, these personnel can take course work to become industrial hygiene engineers.

Some Observations from On Site Visits. The ZAM has the strength of coordinating epidemiologic research for the whole nation; thus findings can be interrelated and lead to a coordinated research program. Unfortunately, there seems to be little effort devoted to questions of OSH in relation to stress, worker control and self-determination—questions which have taken on such importance in Sweden. The sizeable number of psychologists involved in OSH research are primarily conducting man/machine and productivity studies, rather than work democracy and stress alleviation studies. (Though see the rebuttal to this point in the Appendix.) Some work deals with the high sick absences from a psychological perspective—what it means for a worker to declare him or herself sick, etc. One general impression (I don't know if research-based) is that younger workers often take a nonchalant approach to asbestos and other carcinogens because the long latency period of cancer keeps them from taking the hazard seriously. On the other hand, I have already mentioned the worker who said he had never heard of the asbestos hazard. It seems that a formalism pervades the worker education efforts (signing off on hazard diaries, warning posters, etc.). Really creative approaches which would give workers a real sense of responsibility and control over their own health and safety on the job appear to be few and far between.

When a plant is found to be operating over a prescribed hazard limit, the Ministry of Health must give permission for it to continue operating. This is done with plans approved which reflect corrective actions in the short term (needed scientific research, etc.). The union must agree to these plans. While it is good to recognize hazards in this way and to have clear deadlines for specific corrective actions, it is far from clear that the workers whose whole lives and health are at stake are equal partners, to say nothing of being the final arbiters of this decision. In one case, a Chief Labor Safety Inspectorate doctor for a region or county (Bezirk) was asked about a large factory we had driven by. "Yes, that is our problem glass factory." "You mean it is hazardous?" "Yes, it has a number of different rather serious hazards and has had them for years." Why don't you close it down or reengineer it to get rid of the hazards?" "Well, that takes a great investment and the country needs the production." Presumably the union had to agree to each new plan laying out postponements of corrections.

Then there is the matter of hazard pay—e.g., x-ray technicians receive a special higher pay as compared with other technical workers with similar qualifications. The temptation must be there sometimes to pay for taking exposure risks, rather than investing capital in risk removal. (However, it may be released time instead of pay—see the comment on this point in the Appendix to this chapter.)

It is a firm principle of G.D.R. health policy (Gesundheitspolitik) that patients are free to choose where they will seek professional care. Thus one estimate is that where there is an especially well liked factory doctor, 90 percent of workers will first go to him or her. Where there is a well liked general doctor (Gemeindearzt) 90 percent will go there. There do not seem to be clear data on where workers do go when they first seek aid outside family and friends. The point of first contact with the health services has important implications for the OSH-PHC problem in focus here.

An interesting opportunity for worker or prospective worker education comes about when young people are examined in their late teens and given job counseling in relation to any health problems. For example, someone with eye problems is not encouraged to take up fine optics.

While the Ministry of Health pays the OSH Service physician and other personnel, thus making them independent of management and unions, the management of a large steel mill pays (perhaps this is general?) about 30 RDM per worker for the physical facilities and materials for the OSH services. As the Chief doctor in this case put it, "They pay for this chair I sit on, but they have nothing to say about what I do with it." This seemed a little overstated, for in another large plant the chief works doctor said he tries hard to get along with the management. "After all, the General Director provides the means for us to do our work" (anyone using a Marxist perspective based on the labor theory of value would know it is the workers who produce the value which provides the facilities he and his staff use—but see the comment on this in the Appendix).

The union in each undertaking (also in OSH policlinics) elects a trusted person as leader. This person is defined as a partner of the Director. He or she is said to have power, but not as a single person, rather as a representative of the workers. Meetings are always supposed to include this person (Gewerkschaftsleiter). But I met with many Directors of undertakings (mostly OSH services) and in only one was the union represented.

When a worker is certified as being disabled from work, he or she gets full pay. This decision is made by the union, though medical advice may be sought. When not disabled from work, only the sick money is paid. The definition of "work-related" is seen as too broad by some OSH doctors. One told of a worker certified as having a work-related injury after he had quite a bit to drink and got injured on the way home from a work-sponsored picnic. In part this judgment was made, so the story went, because the worker had met a woman at the affair and the two went off on their own, rather than taking the factory-provided bus.

The works management was held responsible because they had not told those attending that they had to take the bus. Some workers see the OSH doctor and other personnel as primarily concerned with productivity and cutting absenteeism. In spite of the independent pay and sponsorship of OSH through the Ministry of Health, there seems to be some climate of suspicion built up between workers and OSH service personnel as to sickness absence and work-related disability. This may place too heavy a burden of social control on the OSH services. Because of the extensive preventive efforts and independent sponsorship, however, I don't believe this mistrust reaches the levels it does between workers and "company quacks" in the West. But, surprisingly enough, from the perspective of independent unions in other countries, the control function is shared with the unions. As mentioned earlier, on the one occasion when a union representative was present in my meetings, it was because as head of the union's social welfare committee, she had come to check through the day's sick slips with one of the OSH physicians. If a worker has been out too often or too long, the union will send a representative to the home to see what is going on and of course to offer to help.

In one very large works with some 10,000 workers, the Chief OSH physician discussed the sickness absence problem at length and reported some of the studies they have done using an impressive computerized information system in which each sickness absence is recorded (this system has also been used for a variety of epidemiologic studies related to diagnosis–not just for the epidemiology of sickness absence in general). It was said that absences are higher just before vacation periods, on "blue Mondays," and they also vary according to the worker collective (department or work team). The collective running the gas-powered power plant in this works had only a 2 percent sickness rate. A number of collectives are at 10 percent or higher, with the overall average at 6 percent which is lower than the 8 percent or more national figure of three years ago and earlier. The power plant collective has well-educated, skilled workers who are highly dependent on one another or interdependent in their division of labor. The high absence collectives have heavier, noisier, dirtier work, the workers are less well-educated and do less skilled work and can more easily replace one another.

In relation to the problem of sickness absence, but clearly also in relation to the overall situation of the political economy and the general shortage of workers in G.D.R., it came out on a number of occasions, sometimes in relation to the use of alcohol, and other reasons for slack performance, that it is very hard to fire or otherwise remove a worker in the G.D.R. An example was given in one OSH service of a worker who says he is sick and hasn't been to work for five months, though physicians can't confirm that he has anything wrong with him, yet the undertaking can't fire him. This is no doubt an extreme example. But the philosophy behind the state policy involved is that the individual would have a hard time protecting himself from a large undertaking, so the laws must be strong on behalf of the individual worker. This is in striking contrast with Sweden, where as little as possible is done through laws and individuals protect themselves through collective action in strong independent unions.

Other Com Con countries (the Eastern counterpart to the European Economic Community countries) take a less lenient approach, for example, on the matter of worker compensation. In G.D.R., if the ZAM certifies that the disability is work-related, there is 100 percent compensation. A high paid crane operator who is certified as having a work-related disability and must be moved to another job gets the same high rate of pay. In Czechoslovakia, a commission judges each case and may say 25 percent of a work-related illness is the worker's fault, with corresponding cuts in compensation. This approach has been considered and rejected in G.D.R. because there is a shortage of workers and much is done—by the state, not through collective bargaining—to protect the worker's interest.

Much more could be written about the OSH system proper. For example:

- It is my impression that hazard exposure standards are as strict as those in the U.S.S.R., some of the strictest in the world—at least on paper.
- There is a very impressive central policlinic and program for construction workers with decentralized temporary health stations set up at each construction site of a size to warrant it. With about 75,000 construction workers, the central clinic has some 350 personnel, seventy-two of them physicians, 200 nurses and medtechs, etc.
- Even though there is a fair amount of OSH included in undergraduate medical education, attracting physicians into the field is not easy. OSH stands low along with surgery (what a contrast to U.S.A. where surgery is a real money making specialty and is quite popular). Still there is great variation. One school had twenty-five students enter OSH while another had none.
- The system of first aid with thousands of workers trained by the German Red Cross to cover each work area is truly impressive. OSH personnel usually take part in this training so preventive information as well as purely first aid and emergency information can be passed along.
- Again, as a humanitarian matter and as a concern related to shortage of manpower, special attention is being paid to occupational rehabilitation with the setting up of sheltered jobs and workshops.
- Although more could be done to make all workers and the public aware of the asbestos danger, a national registry and follow-up program has been established to follow-up for life the 6,000 some workers known to have been exposed.
- A widespread problem is relative separation between engineering and medical aspects of OSH.

OSH-PHC Links in G.D.R.

There is a clear consciousness among OSH and Ministry of Health leaders in the G.D.R. of these links and an awareness that they need special attention. Immediately this is more than can be said for most other study countries. In undergraduate medical education, there is a nationally stipulated curriculum including a certain number of OSH hours. One statement on this said there are thirty-two hours in the eighth semester with twenty additional hours taking the students into practical work in various undertakings. But there is more variation than I expected in a fairly centralized system. For example, one university medical school under the influence of a new OSH professor has cut out the practical factory visit hours. This school has two hours of general OSH per week in the eighth semester; and in the fifth year two hours per week are given to an interdisciplinary treatment of "productive capacity" (Leistungsfaehigkeit). As a confirmation of the training, all students have to prove their understanding of industrial medicine and hygiene as part of their general public health exam.

Since this level of training is relatively recent (I believe since 1975) physicians now out in practice tend to be less well trained in OSH. I don't believe that other specializations, say for general practice, or in internal medicine, include any OSH training. However, several of the OSH service programs I visited said they offer training programs for GPs and other physicians in the area. Nevertheless, problems remain, as illustrated by the following case.

In a general community policlinic in Karl Marx Stadt, A GP of about thirty-five years of age said she had almost nothing on OSH in medical school. This, she said did not reflect on the G.D.R., since she had had her medical training in Czechoslovakia. However, she took her specialty training to become a GP in G.D.R. and again there was almost nothing on OSH. "The subject was only alluded to. There should be much more," she said. Since the head of an OSH Service for a large combine (Kombinat) in the same city had told us that morning that he regularly offers seminars, discussion evenings, etc. on OSH to GPs and other physicians in the surrounding district, I asked her if she ever received an invitation to such a program. "No," was her answer. As to organizational arrangements and her connections to OSH, she said it is not good because she often knows patients best and yet she can not get into the workplace to have an influence in arranging the proper job, etc. Only specialists in OSH serving a given plant can do this. Since I had heard a great deal about worker education (signing off in work-hazard instruction diaries, the training of first aid workers, etc.) I asked if the workers she sees seem to understand what hazards they face on the job. Her answer: "The average worker is not well informed. He doesn't really know what dangers there are on his job—chemicals, heavy lifting, etc." This was an interesting case. She was obviously a physician who cares about workers and their problems; yet she had not had the opportunity to become properly informed. How many are there who do not even care about workers and their problems? When I asked what work-related health problems she sees most, she said, "back problems" ("Wirbelsaeule" which translates directly to "Spinal Column," but I think she included strained back muscles, etc. in her response).

If PHC personnel were responsive to workers' work-related health problems and were properly trained, the law provides for reporting of suspected OSH disease. Also, the health system of the G.D.R. has certain other strengths which should make the OSH-PHC connection an effective one. The geographic assignment of responsibility should lead policlinic personnel to get to know the industries in their area and the possible work hazards. However, more could be done to instruct them in this—along the lines of the Mora Project in Sweden.

The system of returning sick slips to the factory or other undertaking should allow the responsible OSH personnel to follow-up. However, in the large factory with some 10,000 workers mentioned earlier, the Chief OSH doctor said sick slips come in from some 500 different doctors and these slips are first used by the accounting office for several days (for pay reasons, etc.) before they are punched into the computer. At this point, he and his staff can do various analyses, but this is usually long-after-the-fact grouped data examined for trends over time. There is little case-related follow-up back into the workplace to see what may be making the worker sick—or simply alienated.

Two especially interesting policlinics were visited. These will be specially mentioned, even if briefly, as they suggest something close to a model for organizing the PHC-OSH connection.

One of these is a demonstration project involving a work inspectorate and a policlinic serving an agricultural area with a great deal of hot house cultivation involving a number of different chemical fertilizers and insecticides. There is a special staff of physicians, two chemists, one engineer, an OSH nurse, a graduate farmer, and assistants to conduct studies and health hazard surveillance. It acts as a district (Kreis) arm of the Regional Labor Safety Inspectorate. Among its studies is one on the time it takes for insecticide spray to settle in a hot house (glass house) so that it will be safe for workers to enter. Since this inspection-research unit is administratively separate from the general policlinic in the building in which it is housed, and since the policlinic only serves one-fourth of the district while the OSH unit serves the whole district, the arrangement may not be perfect as far as OSH-PHC connections are concerned. Yet it seemed clear from taking a tour with the young doctor heading the policlinic that the presence of the OSH unit has made him and his staff aware and pretty well informed concerning possible OSH diseases among the patients. A quarter of the GPs have a subspecialty in work medicine obtained in four weeks of courses spread over three years. It is planned that all GPs in the clinic will obtain this training.

Another policlinic came even closer to suggesting a model. This integrated PHC-OSH unit serves a semi-urban district with some 10,000 workers—Grossraeschen. Both families and workers are served. The larger undertakings are involved in soft coal strip mining, baking of bricks, glass making, etc. There are 115 undertakings, including one large one with some 6,500 workers, and fifty-eight small ones with fewer than ten workers. The clinic-OSH center has 120 personnel. One physician is in charge of the specialized OSH program. This involved, among other things, but very importantly, conducting and following up on the health hazard questionnaire and measurement survey of every undertaking in the district (mentioned earlier as part of what a 1981 law requires for all workplaces with over fifty workers). This information is transmitted to a card system for each workplace. Then one of the GP staff—all of whom have taken and who continue to take some OSH instruction each year—is responsible for a given work site or set of work sites (this can be part of a large plant). This physician may not do all the periodic exams but is the coordinator of the OSH program for his or her workplace(s). While the PHC personnel emphasize treatment and follow-up with a quite full awareness of and sophistication about OSH, the OSH physician functions primarily in a preventive way. To a considerable degree this setup broadens the interests of PHC personnel and makes OSH personnel aware of clinical OSH interests and limitations so that targeted in-service instruction can be offered. Some clinic hours are held inside the larger plants. Sick slips go first to the nurse serving a given workplace. Then these are gone over once a week with the physician responsible for that plant. If new projects or plants or even new processes within old plants are being considered, these have to have the advice of the relevant GP and OSH staff. There may still be problems. The clinical interests appear to dominate; thus the follow-up of workers who have been exposed may receive more attention than preventing hazards in the first place. Also, there is no union advisory board for this clinic or other evidence of strong labor input from the workplaces being served. But as a model of expert-based PHC-OSH integration, it was the strongest example I saw in any of the study countries.

Selected Observations, Questions, and Recommendations for the G.D.R.

There is no way to "recommend" that organized labor be "given" a more authentic role to play in this generally impressive, but expert-based OSH system. Power of this sort is not given. It will have to be taken. I saw no signs of a Solidarity-type movement as in Poland. This is too bad. Part of the background of solidarity includes a deep concern on the parts of workers over the hazardous work conditions they faced, even though the standards and measures for protection looked very good—on paper. I don't believe that work conditions are especially hazardous in G.D.R. The historical concern for protecting the short supply of labor, the complete absence of unemployment, the fairly elaborate OSH apparatus, as well as the figures which are available (those comparing G.D.R. and F.R.G. as given by Steele and Winter were cited earlier) argue against it. But there is a heavy emphasis on formal, legal things and one wonders how much is in the heads and hearts and actions of the workers themselves and their union representatives and those responsible for protecting their health. One of my deepest concerns in this regard is the extensive effort devoted to periodic exams of workers known to have been exposed to one hazard or another. Such assiduous follow-up is a good thing and missing or much less fully developed in other study countries. But what priority is given to redesigning a process, closing a hazardous plant, etc.—that is, fundamentally getting rid of work hazards? The schedules adopted for such measures—short term (1 year), medium term (3 years), etc.—may be agreed to by very unequal partners and certainly involve loss of health and life while these measures are being taken. Did those who are to lose their health, or their lives, give their full and well-informed consent?

While an admirable degree of attention is given undergraduate medical (and nursing?) education, there is still room for improvement. There is considerable room for improvement in the specialty preparation of GPs and other specialists. Little OSH training is now given at this level. A national sample survey of where workers first go for medical help when they seek help outside of family and friends might be conducted. This might take a number of factors into account as suggested under this same point in relation to Sweden. Such a survey could help in designing a strategy for postgraduate OSH education of PHC providers, that is, those physicians and other PHC providers found in the survey to be the most likely points of first contact would get priority attention as regards continuing education in OSH.

Since there is a need for more OSH physicians and it is as yet a relatively low prestige specialty choice (though some say it is improving) and since there is great variation in the number who enter this specialty from one medical school to another, a social psychological/sociological study might be conducted to determine what factors attract young physicians to and repel them from the OSH specialty.

Finally, the experience of the integrated PHC-OSH policlinic in Grossraeschen described above might be examined in more detail and, if found as positive as I think it is, this model might be considered for wider adoption. However, I would urge that a fundamental change be made in the control structure to put in charge an elected board of union representatives from the several plants served. However, if this kind of board or something like it (an advisory rather than directing board) is first set up by experts "for the workers' interests" it may not be effective. The workers of the G.D.R. may have to struggle for this sort of thing themselves if it is to be a vehicle for authentic worker participation and control.

The Model and the G.D.R.

In a number of respects the G.D.R. comes closest to achieving the several dimensions of the OSH and OSH-PHC model presented in Chapter 6.

Policy Dimension. A very high priority is given to OSH in deeds as well as words. To illustrate, the law of 1981 being implemented as I visited the country will bring about a thorough carefully analyzed work hazard survey for every worker and workplace in all plants with over fifty workers. The survey will be followed-up to make plans to remove hazards where possible, give periodic exams to workers already exposed to one hazard or another, and follow-up cases of disease to limit disability. Such a program would not now be possible to even put seriously before the fundamentally capitalist-controlled bodies politic of our other study countries.

There is still a problem as there is in other study countries of recapitalization to remove hazards. It is hard to judge the extent of this problem. It is not worse and because of policy differences it is probably not as serious as in other study countries (though capital resources are fewer, so one has to declare uncertainty in this matter). More than in any other study country there is an awareness and concern for the OSH-PHC connections. Some efforts have been undertaken. But much improvement is needed.

Sponsorship and Control. The G.D.R. gives a very strong place to worker control through the mechanism of the Communist Party. The unions function in a different way than in capitalist societies. They are in the picture, but as a cooperative, rather than an adversarial force. These complex arrangements were discussed at length above under the section on the political economy of the G.D.R. (p. 242) and can not be repeated here. In summary, there is considerable overall worker control through the party and to some extent through the unions and other plant structures such as the Permanent Production Council and the Plant Safety Committee. But there is also an inherited part of the cultural hegemony which lends great power to central authorities and experts. In short, the workers' movement is strong in the G.D.R., but it is organized in such a form as to need much further democratization, politicization and spontaneity—at all levels, but especially at local plant and community levels.

The sponsorship and payment of OSH personnel through the Ministry of Health, rather than by plant managements (even though these managements are legally responsible and accountable for OSH) allows a degree of independence and direct concern for workers' health which is important. There is no evidence of worker control over PHC services, something which would tend to bring about closer more effective relations between OSH and PHC. In general, the OSH system in the G.D.R. is well oriented toward protection of workers, but might best be characterized as primarily expert-determined technocracy ideologically oriented toward workers through overall party control.

Organizational Dimension. Organization is especially fully realized in the G.D.R. as compared with other study countries. Of first importance is the overall health system itself, a comprehensive NHS, rather fully regionalized, with geographically oriented PHC services. Thereby, even though there are other shortcomings (as in OSH education for PHC personnel) PHC providers tend to get to know the workplaces and hazards and likely work-related diseases in their catchment area.

OSH services in the G.D.R. are provided by Ministry of Health personnel within a regionalized structure pursuing a full range of responsibilities from identification and removal of hazards to ongoing treatment for work-related disease or disability, with work placement and follow-up according to the worker's needs and abilities as well as the firm's production needs and opportunities. At local level, these services are organized in a variety of ways. In large plants an OSH policlinic will provide the focus. In certain industries, for example, construction, central and regional OSH clinics provide backup to on-site satelite services. Regionally based inspectors keep check on workplace exposures and services.

In many ways, Grossraeschen, a community policlinic, provided a veritable model of well integrated OSH-PHC services.

In summary of the organizational dimension, the G.D.R. has a more comprehensive, coherent OSH service system with many variations, functioning within a regionalized national health service system. It shows some promising approaches to the OSH-PHC problem, but by no means has widely instituted clear ways of providing optimum connections between the general health services and workers' health and safety problems.

Educational Dimension. This reflects both major achievements and important shortcomings. Most important, worker education efforts of a certain sort are widespread, and through the Permanent Production Councils there should be no problem of workers' rights to know what chemicals and other hazards they may be facing. The ubiquitous system of Red Cross-trained first aid people for every so many workers also carries some OSH education with it, though this element could be enhanced. Every worker has a "work protection diary" in which he or she must sign off that they have had certain instructions. But this may be more formalistic and oriented toward assignment of responsibility in case of an accident than it is really effective, self-changing education.

More is done in undergraduate medical education to bring about an awareness of OSH than in any other study country. Not much seems to be done with practicing GPs and other PHC personnel as regards OSH. The development of OSH specialists is further advanced than in any other study country.

Financing Dimensions. This dimension did not strike me as particularly problematic in the G.D.R. The payment for OSH physicians and other OSH personnel comes through the Ministry of Health, while the material means for OSH services are provided by plant managements. In general, this seems to work well.

Information. In the G.D.R. the regionally-based health services provide an excellent epidemiologic information system for a wide range of health problems of special interest including workers' health problems. In addition, the recently passed Workplace Hazard Surveillance Act is providing an information base for all workplaces on number and type of hazards, workers exposed, and corrective and follow-up measures taken. However, there are still inadequacies in the preparation, awareness, and concern of PHC providers to identify and report OSH problems.

The Central Institute for Work Medicine (ZAM) in Berlin carries out extensive researches in the full range of work-related disease and has produced many publications. As a part of the Ministry of Health, ZAM is able to link its findings to policy and action. This work is being spread to regional and district centers under a recent law.

APPENDIX

Critical comments of an OSH expert in the G.D.R. (free translation from the German).

. . . I am impressed with your concern for the workers. Through our discussion you are aware that we share many thoughts. We can only agree with your closing statement that the workers of the DDR should be encouraged to further health protection in their interests. In this direction we made much progress in the past decades, some of which you give positive mention in your report. If we favor the involvement of the workers why should we exclude the health system? Your conclusion that the health system in the DDR is "expert based" is most likely due to the fact that your stay was too short to allow you an accurate observation and assessment.

Professor Hellberg from the WHO visited the DDR recently and in a discussion he made special mention of the fact that the entire system has provisions to allow the workers to take an active part in all phases of legislation. This is a change from a largely spontaneous community participation to solve problems.

It is most likely unusual for you to understand a system which has built-in provisions to give the worker joint decision-making in state and economy. It seems evident that your comments in the report were given from the point of view of the social structure of your own country. It would have been desirable if the historical development and social structure of the DDR from our point of view, the understanding of the workers, would have been given special mention, as the DDR is the only socialist country in your report.

I know that as a sociologist you are familiar with classical Marxism. Changes in class structure, education open to members of all classes etc., were all initiated by the workers themselves. But even in this social system experts in various fields are necessary. Today's experts come for a large part from workers' families or farmers. This does not mean however, that we no longer have a difference of opinions! Attitudes of members of the health system *vis-à-vis* workers which you mention are rooted largely in societal causes.

Now to comments on specific points in your report:

*p. 1 – The international workers' movement had a great part in fostering changes in worker's health protection. In socialist countries the workers do not need to fight authorities; instead the workers together with *their* experts find solutions for their problems.

p. 74 – I hope that the printed material and documents mailed to you will allow you a better insight.

p. 75/76 – Your discussion on the history of the DDR needs corrections but by now you should have the necessary material at your disposal. Concerning the question of reproduction of the population it is essential to incorporate the effects of sociopolitical measures in recent years on behalf of women and mothers (see Health Care in the G.D.R., p. 34/35).

p. 77 – For clarification on the role of "Kombinate" I enclose a copy of the legal regulations. Importance and number of Kombinate has increased substantially. Although independent, they do *not* work outside the economic system: they are an important basis of it.

p. 79 – Remnants of historical influence on doctor/patient relationships can still be encountered. The example of Neubrandenburg is certainly not applicable to the industrial section of OSH.

p. 80/81 – Had we known more about your special interests we certainly would have made arrangements so that you could have gained a better insight into the practical cooperation and joint decision-making of workers in their workplaces. Although there is a difference of definition of democracy, this should not assume that there exists no democracy in our system. The various institutions you visited had been informed that an American professor, preparing a study for WHO would visit, in order to learn more about the relationship between OSH and PHC in various countries. This was regarded as your main professional interest. I doubt, given the same situation, other countries would have arranged for you to meet union leaders. Can one really cite this as a good indicator of the joint decision-making of unions?

I cannot agree with Navarro. I am sure you are aware that we are dealing with politically-based values which in our country will find justified counter argument.

*Page numbers refer to an earlier draft report done for WHO/EURO. A few points have been changed to take these comments into account. Others have not, reflecting differences of views, opinions, or assessments of the situation. My special thanks to the expert informant who provided these comments.

p. 82 – For data on National Income and Expenditures for social and health care see "Health Care in the G.D.R." p. 41.

p. 91 – It is not the Ministry of Health which pays the occupational medical and other occupational health and safety employees; the payment comes from the various institutions within the health system (the monies come from the federal budget).

p. 92 – "Labour Safety Inspectorates" seems not accurate for our Work Hygiene Inspections. Labour safety is not part of the concern of the health system. The inspections are responsible for guidance and control of industrial hygiene and occupational health in the workplaces. Footnote 25: The management of a plant is responsible for keeping conditions in the workplace in accord with the guidelines of labour health and safety laws. Through the FDGB it is the workers who work for and control labour safety (FDGB–Association of Free Democratic Unions).

p. 94 – Safety Inspections of the district do not work directly with the Health Protection Commission of the workplace but when necessary the two work in close cooperation. The aim of worker protection instruction is not merely to alert the workers to possible hazards, but also instruct the workers in correct behavior so as to foster health and safety. All workers must be kept well informed on all important issues.

p. 96 – With respect to the recent (1981) workplace hazard identification and follow-up law, it is very important to spell out the analysis of exposures and body burden over time as the basis for occupational medical care: the assessment of the workplace is not based on questionnaires but instead, especially with health relevant factors, is based on measurements. Until now such measurements were frequently carried through by the OSH services. In the future, according to the law for these measurements, the firms themselves are responsible. OSH is responsible for expert advice and control, as well as the schooling of the measurement personnel.

p. 97 – Prophylactic examinations. This figure includes data on activities other than those of the doctor, such as laboratory tests, etc.

p. 98 – The courses for further specialization have different goals. The first leads to industrial physician, the second to specialist in industrial hygiene. In Health Care in the G.D.R. (p. 31) the figure for specialists in industrial hygiene is 900.

The Academy for Medical Education of the DDR is in charge of the higher education of physicians and nonmedical graduate students in medical institutions.

p. 98/99 – Afraid our conversations did not give you a good insight into my own specialty or you could not have reached the conclusion that psychologists in ZAM are mainly concerned with ergonomic problems. I personally have done extensive research on stress, etc. for many years. In cooperation with OSH services we researched 600 professions and occupations with regard to stress level and we interviewed nearly 10,000 and held group discussions with employees in different workplaces. The result of the study led to the inclusion of individuals with great psychological stress into the system of occupational medical care and the development of measures for psychoprophylaxis.

p. 99 – I do not know glass factories outside the DDR, but know from reading that no solution has been found to reduce heat, dust, and physical work.

p. 100 – The DDR pay system does not allow for extra pay for work which is hazardous. We try to minimize the hazards as much as possible and in some instances reduce the time of work but not the income.

The Marxian perspective: In a system where the people own the factories, management and OSH services are partners. The case cited by you points this out clearly.

p. 101 – The decision whether a worker's inability to work stems from a work-related condition is not made by the doctor but by the Union. If necessary, medical opinion is sought.

p. 102 – Acknowledgement of an occupational illness does not rest with ZAM. ZAM merely gives medical opinion.

p. 104/105 – Of course we have still many improvements to make in the relationship between OSH and PHC. However, we cannot base our opinion on the impressions of one GP. Every GP can telephone the OSH service and receive information on conditions of the workplace of the patient, etc. The GP in a highly industrialized area needs more than personal professional knowledge and initiative. He/she must obtain knowledge about the diverse work situations of the patients through OSH. In more rural areas we have many GPs who can obtain a very good overview of industrial and other work places in their territory and many function as the plant physician.

p. 108/109 – The responsibility for reduction of health hazards does not rest with OSH but management. The firms have legislative directives and management has to answer to the workers.

NOTES AND REFERENCES

1. A number of general sources have been used. An especially valuable one has been George Kren, Germany in *The Volume Library*, Nashville, Tennessee: The Southwestern Company, 1973, pp. 2157–63. One reflecting a view of Germany and its people and history before the present formal division is William Bridgewater and Elizabeth J. Sherwood (eds.), *The Columbia Encyclopedia*, 2nd edition, New York: Columbia University Press, 1950, pp. 768–769. Others include Henry M. Pachter, *Modern Germany, A Social, Cultural, and Political History*, Boulder, Colorado: Westview Press, 1978; Koppel S. Pinson, *Modern Germany, Its History and Civilization*, New York: Macmillan, 1954; and Erich Kahler, *The Germans*, Robert Kimber and Rita Kimber (eds.), Princeton, Princeton University Press, 1974.
2. George Kren, op. cit. p. 2161.
3. As quoted in George Rosen, *From Medical Police to Social Medicine: Essays in the History of Health Care*. New York: Neale Watson, 1974, p. 65. These words or ones close to them may have actually been first stated by Salomon Neumann a predecessor and contemporary of Virchow's, but they are usually attributed to Virchow who saw the body social and its pathologies in a somewhat mechanistic way derived from his view of and major contributions to cellular pathology. In fact, a quote correctly attributed to Virchow is: "The physicians are the natural attorneys of the poor, and the social problems should largely be solved by them," from "Virchow and the Medical Reform Movement" pp. 126–136 in Knut Ringen, *The Development of Health Policy: Norway, England and Germany*. Dr. P. H. Thesis, the Johns Hopkins School of Hygiene and Public Health, Baltimore, 1977. But, it should be noted, and not incidentally, that Virchow was deeply affected by the concepts of social class and class exploitation and struggle as seen in the work of Friedrich Engels on the ravages of early capitalism on the lives of the working class in England. See "The Social Origins of Illness: A Neglected History," pp. 65–85 in Howard Waitzkin, *The Second Sickness, Contradictions of Capitalist Health Care*. New York: The Free Press, 1983.
4. Kren, op. cit. p. 2161.
5. Stanley Aronowitz, *False Promises, The Shaping of the American Working Class Consciousness*. New York: McGraw-Hill, 1973, p. 138: "In contrast [to U.S.A.], Germany, the one European country embarking on the high road of economic development, was mired in feudal conditions typical of Western Europe. Even as European capitalism destroyed the old mode of production by force and violence, it had been obliged to accommodate itself to the old hierarchies and aristocracies. In fact, members of the old nobility in many European countries transformed themselves into new capitalists rather than give way to elements of the subordinate social classes of the old regime. In this sense, there was very little social mobility for the peasants in the transformation from feudalism to capitalism." See also Barrington Moore, Jr., *Injustice: The Social Bases of Obedience and Revolt*. White Plains, N.Y.: M. E. Sharpe, 1978.

6. Manfred Pflanz, Social Structure and Health: Methodological and Substantive Problems without Solutions in Part III of Magdalena Sokolowska, et al. *Health, Medicine, Society*. Dordrecht, The Netherlands and Boston: D. Reidel; Warsaw: PWN – Polish Scientific Publishers, 1976.
7. Friedrich Engels, *Germany: Revolution and Counter-Revolution*. London, 1933, p. 57 as quoted in Koppel S. Pinson, *Modern Germany*. op. cit. (note 1): 106.
8. Pinson, op. cit. (note 1): 106.
9. Ibid, p. 107.
10. Aronowitz, op. cit. (note 5): 139–140. But see Albert Szymanski, *The Logic of Imperialism*. New York: Praeger, 1981, especially Chapter 14 on the "Aristocracy" of Labor. He finds that the higher profits from investments in the periphery and semiperiphery do not trickle down to the working class but go for military repression and higher rewards for the ruling class.
11. Against the outraged cries of most other owners, Ernst Abbe, head of the Zeiss works in Jena, carried out one of the first experiments in worker democracy by turning over his entire plant to the workers. Pinson op. cit. (note 1): 248.
12. Pinson, p. 248.
13. Ibid, p. 248. The following quotes in this section are also from this source.
14. *Schriften des Vereins fuer Sozialpolitik 116*:214–215, 1906, as cited by Pinson, op. cit. (note 1): 249.
15. Pinson, op. cit. (note 1) details this history in an excellent chapter, The Revolution of 1918–1919, pp. 350–391.
16. Ibid, p. 410. A particularly perfidious bit of the division on the left was the sellout by Friedrich Ebert, leader of the SPD, following his election at the Circus Busch meeting to head the Republic which had been declared upon the Kaiser's abdication, November 9, 1918. Ebert and other of his right wing socialist friends were elected with the cooperation of the Revolutionary Shop Stewards (whose radicalism Ebert feared). Almost immediately after the Cricus Busch meeting, Ebert struck a deal with General Groener, Ludendorff's successor, Hindenburg's right-hand man. In return for the army's support, the government was to fight Bolshevism and support the officer corps in controlling the army. "Ebert's secret agreement with the General Staff was a stab in the back, not only for a socialist revolution, which for reasons to be set out in due course I do not think was a realistic possibility anyway, but even for a liberal revolution." Barrington Moore, Jr., op. cit. (note 5): 295.
17. Peter Hoffman, *The History of the German Resistance, 1933–1945*. Cambridge: MIT Press, 1977.
18. Arthur D. Morse, *While Six Million Died, A Chronicle of American Apathy*. New York: Random House, 1968.
19. Ray H. Elling, *Cross National Study of Health Systems*. New Brunswick, NJ: Transaction, 1980, especially Chapter 6, "Contrasting Case Studies of Selected Developed Countries."

20. As quoted by Henry E. Sigerist, From Bismarck to Beveridge: Developments and Trends in Social Security Legislation, *Bulletin of the History of Medicine 13*(April, 1943):365-388: quote and source p. 376.
21. Ibid., p. 380.
22. Gordon K. MacLeod, "The Consumer Under National Health Insurance in Germany," (Paper presented to the Program in CNSHS, Department of Community Medicine, University of Connecticut, Farmington, CT, May 17, 1974).
23. As quoted by Kaser in Chapter 5, "The German Democratic Republic," pp. 147-165 of Michael Kaser, *Health Care in the Soviet Union and Eastern Europe*. London: Croom Helm, 1976, quote from p. 147.
24. Manfred Pflanz, German Health Insurance: The Evolution and Current Problems of the Pioneer System, *International Journal of Health Services 1*: 315-330, 1971.
25. Henry Krisch, Common German Heritage Links G.D.R. with Federal Republic, *The Hartford Courant* (May 9, 1984): B11; Bonn Approves Major Credit for East Germany, *The Hartford Courant* (July 26, 1984): A14.
26. A valuable source for this section has been Jonathan Steele, *Inside East Germany; The State that Came in From the Cold*. New York: Urizen Books, 1977. See especially the chapters "The Soviet Connection Causes Problems," "Shaping an Economic Miracle," "How Much 'People's Democracy'?, and "Social Services from Cradle to Grave."
27. Ibid, p. 7. Steele goes on to observe that the psychological impact of this development was profound. "Remember the Cold War Clichés? Life there is 'drab'. The people's 'lot' is hard. The standard of living is not everything and there are many aspects of life in the G.D.R. which anyone will want to criticize. But it is no longer possible to argue that the system is both politically unacceptable and anyway economically inefficient."
28. Ruth Sivard, *World Military and Social Expenditures, 1983*. Washington, D.C.: World Priorities, 1983. There may be some disagreement as to whether the G.D.R. is ahead of the U.K. in personal income. Sivard reports the U.K. at $9,213 GNP per capita for 1980. As noted above (see note 27 and text) Steele cites a World Bank report from 1974 stating that the G.D.R. had passed the U.K. In any case, it is a relatively well-to-do nation. The U.S.S.R., by contrast, had only $4,564 GNP per capita in 1980.
29. Though these may not be good examples, for a job which in the West has low prestige and pay may be seen in G.D.R. as socially necessary and valuable. It is said, for example, that garbage collectors in Moscow are very highly paid.
30. Shail Jain, *Size Distribution of Income: A Compilation of Data*. Washington, D.C.: The World Bank, 1975, table 27, p. 41 and table 28, p. 42. Steele, op. cit. (note 26) pp. 145-147 gives some figures for 1970 and 1971 obtained from the Deutsches Institut fuer Wirtschaftsforschung in West Berlin. There was about a three to one ratio between top earners, that is independent people such as craftspersons and shopkeepers, who got an average 2,025 Marks per month and employees who received 720 Marks monthly. Members of cooperatives were in between with 958 Marks. In

terms of net household incomes, the bottom 20 percent got 10.4 percent, while the top 20 percent got 30.7 percent. In F.R.G. at the same time the figures were: bottom 20 percent, 8.3 percent; top 20 percent, 39.9 percent. In terms of particular jobs and functions: "These statistics can be fleshed out with some random examples from 1971. An ordinary university lecturer grossed between 900 and 1,600 Marks depending on whether he had a Ph.D. A doctor started at 700. A factory manager ranged between 1,200 and 2,500. The top party official in a county (Bezirk) earned between 1,200 and 1,500 and a full professor got up to 3,400 Marks. These figures suggest that in spite of the attempt to blur the status differences between manual and mental labor, the German tradition of paying senior academics exceedingly well lives on. None of these figures measures the value of undeclared perks. As in the West, certain groups have access to official cars, flats, and houses. Some party officials can use special shops: they get private rooms in hospitals. A relatively small number of people at the very top, probably not more than a hundred, are in the luxury bracket. Below this level there is not much wealth. Unlike the situation in the West there is no possibility for a citizen to own land on any scale. Each household is allowed two plots, one for their main house and one for a country cottage. No one can own shares, or have a foreign bank account. Any visitor to the G.D.R. can see that there are no major residential differences in East German cities and no wealthy suburbs. Standards are more uniform. In the countryside the large estates have been nationalized and their grand houses are in the hands of institutions, hospitals and trade unions." (Steele, p. 147).

31. Steele, op. cit. (note 26): 184–185.
32. Ibid., pp. 186–187.
33. Ibid., pp. 9–10. However, it should be noted that the workers' riots of June 1953 against the shortage of consumable goods were powerful persuaders convincing the SED leadership of the necessity to give workers a greater immediate return for their work.
34. Ibid., pp. 18–19. Much more could be said about this than I can take the space here. Steele op. cit. (note 26) goes into it in his chapter "Splits and Wounds on the German Left." One of the particularly devastating experiences followed Stalin's completely unpredicted non-aggression pact with Hitler in 1939 following which he turned over to Hitler 500 German and Austrian Communists, including several Jews. Margarete Buber Neumann, *Kriegsschauplaetze der Weltrevolution: Ein Bericht aus der Praxis der Komintern, 1919–1943*. Stuttgart: Seewald Verlag, 1967, p. 484. This history is not well advertised in the G.D.R. today. It is suppressed. But, needless to say, among party members who know of it, it only enhances their tendency to keep a respectful distance from their Soviet comrades.
35. Denis MacShane, *Solidarity; Poland's Independent Trade Union*. Nottingham, U.K.: Spokesman, 1981.
36. It can be argued that there were always union people present, since almost everyone belongs to a union. But none were union officials or speaking for the union. On one occasion, the local party representative in an OSH clinic, a woman physician, attended the meeting and spoke frankly and firmly but

did not dominate the meeting in any way, leaving it under the general guidance of the clinic director.

37. Steele, op. cit. (note 26): 127–149.
38. Ibid., p. 142.
39. *Die DDR stellt sich vor*. Berlin: Panorama DDR, May 1981, p. 76. There is no indication of what categories the others were in who would make these percentages add to 100; possibly the military make up 15.3 percent of the party. The exact wording in German: "Ihr [the SED party] gehoeren 2.17 Millionen Mitglieder und Kandidaten an, die in rund 79,700 Grund- und Abteilungsparteiorganisationen taetig sind. 57.6 Prozent der Mitglieder und Kandidaten sind Arbeiter, 22.1 Prozent gehoeren zur Intelligenz and 5 Prozent zur Klasse der Genossenschaftsbauern. Ein Drittel sind Frauen."
40. Steele, op. cit. (note 26): 130.
41. Ibid., pp. 131–133.
42. As noted earlier, the eastern part of Charlemagne's empire became the German Empire after Charlemagne's death, but the Thirty Years War, 1618–1648, resulted in the split of Germany into many small principalities and kingdoms. It was fascinating on my visit to Frankfurt an der Oder to have the weathervane symbol on top of the town hall pointed out and be told that this city was once part of one of the Hanseatic states—an upriver point for shipping of food-stuffs and raw materials from the hinterland.
43. Deutsches Institut fuer Wirtschaftsforschung, *DDR-Wirtschaft: eine Bestandsaufnahme*. Frankfurt am Main: Fischer Taschenbuch, 1974, p. 36.
44. Steele, op. cit. (note 26): 11.
45. Ibid., p. 134 and pp. 147–148.
46. Vicente Navarro, *Social Security and Medicine in the USSR: A Marxist Critique*. Lexington, MA: D. C. Heath, 1977.
47. M. Kaser, *Health Care in the Soviet Union and Eastern Europe*, op. cit. (note 23): 152.
48. Ibid., p. 154.
49. Kurt Winter, Health Services in the German Democratic Republic Compared to the Federal Republic of Germany, in R. Elling (ed.), *Comparative Health Systems*. Supplement to *Inquiry 12*:63–68, June, 1975 (quote from p. 66).
50. Tim Empkie, The Organization of Tuberculosis Prevention: A Comparison of the Federal Republic of Germany and the German Democratic Republic." Required research paper, UCONN Medical School, Spring 1975.
51. Craig R. Whitney, Rising Costs Strain West German Health Insurance, *New York Times* (July 20, 1975). Numerous other citations could be given on this point including works by Manfred Pflanz and Ulrich Feissler; Christa Altenstaetter, and Deborah Stone.
52. Kaser, *Health Care*, op. cit. (note 23): 164, shows total health and social welfare expenditures in the German Democratic Republic rising in millions of Marks as follows: 5,737 in 1968; 6,186 in 1970; 6,499 in 1972.
53. *Materialien Zum Bericht Zur Lage der Nation*. Bonn, 1974, p. 479, gives a figure for the Federal Republic of Germany of 807 marks spent on health

per inhabitant, amounting to nine percent of GNP (9 Prozent des Bruttosozialprodukts) for 1968. The figure given for German Democratic Republic in the official West Germany Publication, *Materialien Zum Bericht . . .*, p. 479, is 336 Marks per inhabitant. No figure is given in this source for percent of GNP spent on health in the G.D.R., but Kaser set this at 5.7 for 1968, considerably less than the nine percent for F.R.G.

54. Maxim Zetkin (ed.), *Health Care in the German Democratic Republic*. Berlin: Institut for Social Hygiene and Health Care Organization, May, 1981, p. 33 and "West Germany: The Crumbling of Social Democracy," pp. 110–112 in *World View 1982*. Boston: South End Press, 1982, Table 1.
55. Ibid., p. 33.
56. Winter, op. cit. (note 49): 66. However it should be noted that on this visit to G.D.R. I was told several times that their national average of about 8 percent sick absences among workers every day was very high compared with the figures for F.R.G. People in G.D.R. said the figure worried them and they were trying to reduce it. Some said the higher unemployment rate gave the F.R.G. a lower sickness absence rate. Yet in F.R.G., I was given estimates ranging from 6–8.5 percent for the national worker sickness absence rate. The function of OSH physicians in attempting to "police" sickness in an effort to cut absences seems problematic. I accept that these are official figures and that they may be subject to all the influences I detailed in Chapter 4. Nevertheless, I believe on many other grounds (including the figures cited above by Steele from an official F.R.G. source) that they reflect real differences in favor of the G.D.R.—whether of exactly this magnitude or not, I can not argue.
57. "Report on a working group on Evaluation of Occupational Health and Industrial Hygiene Services, Stockholm, 15–17 April, 1980." "Appendix: Description of Occupational Health Situation in Selected Countries." Copenhagen: WHO/EURO no date (unpublished report).
58. The section on OSH in Zetkin, op. cit. (note 54) is helpful as is Bodo Schwartz, "Health Protection and Labour Safety in the German Democratic Republic." Berlin: State Secretariat of Labour and Wages, 1977. Subsequent laws are of great importance, for example, the one requiring a national survey and follow-up of work hazards. These will be mentioned specifically below.
59. But exactly what group is responsible for what to which degree seems unclear, as the Labor Safety Ordinance of 1962 makes managers of undertakings legally responsible for assuring the health and safety of workers, (see below). Probably the managers are legally *accountable* and the unions are responsible for oversight of protection measures.
60. The most recent is from 21 April 1981. See Gesetzblatt der DDR, Berlin 6 March 1981, Teil I, Nr. 12, pp. 139–142.
61. "Arbeitshygienische Zentren und Arbeitshygienische Beratunsgstellen," *Gesetzblatt der DDR* Berlin, 11 February 1980, Teil I, Nr. 5, pp. 41–43.
62. "Verordnung ueber die Verhuetung, Meldung und Begutachtung von Berufskrankheiten" in two parts *Gesetzblatt der DDR*. Berlin, 6 March, 1981, Teil I, Nr. 12, pp. 137–139 and 25 September 1981, Teil I, Nr. 28, pp. 337–340.

63. I am unsure of the time period because this number of exams divided by the number of physicians involved part time and full time in OSH (3590) yields 1114 such exams per physician per year. Probably physicians are not involved in all these exams, some being carried out by auxiliary personnel. Also a large number are laboratory tests (see Appendix to this chapter).
64. Though another estimate I heard was 820 fully qualified works doctors and another 100 in preparation in 1981.

CHAPTER 10
West Germany

The reader who may be "dipping in" here with a special interest in the Federal Republic of Germany (F.R.G.) is advised that the historical background of pre-WW II Germany in terms of the land, the people, and succeeding governing structures was very briefly presented in the first section of Chapter 9 pp. 229–234. The political economic background, including a discussion of the workers' movement followed on pp. 234–239 and the pre-WW II health system was described briefly on pp. 239–241. This background is essential to understanding both present-day Germanies. It will prepare the reader to recognize deep common roots as well as current striking differences in the two political economies and workers' movements, and their effects on OSH conditions and services as well as the general health systems.

There is no satisfactory way of summarizing such a complex history here (thus the reader is urged to read or review the sections just referred to). But some of the key themes should be retained in consciousness as the development of F.R.G. is considered:

- It's not new for Germany to be divided. A unified modern Germany began rather late, only after 1862 when the Prussian power under Bismarck defeated Austria, then France, and quickly grew to the dominant power in Europe before WW II.
- Capitalism in Germany was no kinder than elsewhere. In fact, one could argue that the failure of the bourgeoisie to assert itself, leaving a tradition of hierarchism and the aristocracy in control, brought little social mobility to the but recently freed serfs, and engendered a rawness of abhorrent proportions which Virchow found at fault in his 1848 report in the typhus epidemic in Upper Silesia.
- The failure of the revolution of 1848 set the stage for a vicious rightist repression which sent many German communists, socialists and liberals into exile or emmigration to the U.S.A. and elsewhere (Karl Marx fled to impoverished circumstances in London).
- The first National Health Insurance (NHI) in the world was born of and in class struggle when Bismarck adopted the measure in 1883 along with certain

other social welfare legislation "to cut the legs off the socialist workers' movement."

- The labor unions got a late start and grew rapidly only after 1896, but then were divided into socialist, Christian, and company unions.
- The revolution of 1918-19 came close to establishing a system of workers' and soldiers' soviets in one of the most advanced capitalist societies, as Marx had predicted, but disunity on the left and opposition from the conquering capitalist Allies led instead to the Weimar Republic as a kind of liberal-right compromise which spawned violence and the rise of the Nazis in the face of raging inflation, followed by the unheard-of levels of unemployment as The Great Depression set in.
- The Nazi terror was resisted by communist and socialist workers as well as other liberal forces but was met by decimating, murderous repression. The vast destruction and utter defeat of Germany in WW II left an exhausted, disheartened people with communist and socialist labor supported by socialist-oriented Soviet forces in the East and the same kind of business unionism found in U.S.A. supported by capitalist occupation forces in the West. When there was any support for organized labor at all, it developed only through workers' own efforts, and some support through the Social Democratic Party under Willy Brandt, who came to power after the Christian Democrats and Konrad Adenauer. The development and "economic miracles" in socialist as well as capitalist Germany are indeed remarkable considering the depth of destruction and despair in 1945.

POLITICAL ECONOMIC AND CULTURAL CONTEXT OF F.R.G.

During the military occupation (concluded with the formal declaration of the end of a state of war in 1951) the Western powers established a relatively decentralized federated system in the bulk of Germany, an area which today has 61.3 million people (versus the 16.8 million of the G.D.R.). For most of the postwar years, the F.R.G. was governed by the right-leaning Christian Democratic Party until the Social Democrats under Willy Brandt gained a clear majority along with the Free Democratic Party. In 1976, this majority was narrowed, but the Social Democrats continued to govern until 1982 when the Christian Democrats again assumed control under Chancellor Kohl.

In the East, an area less than half the size of West Germany (40,646 square miles in the G.D.R. versus 95,815 in the F.D.R.), the Soviets, who ended the state of war in 1955, established a relatively centralized, socialist-oriented system under control of the German Communist Party. The economies contrast accordingly.

The F.R.G. has a monopoly capitalist, market economy with a few nationalized industries. Agriculture is entirely private. The G.D.R. has a predominantly state owned and planned economy with agriculture organized into large collectivized farms.

Both countries have experienced a so-called "economic miracle." The currency of the F.R.G. is now one of the strongest in the world. But, as in other highly stratified, class-based capitalist societies, the economic wealth is not generally shared:

> Adolphia Kestane is poor. Her poverty makes her one of hundreds of thousands of people one never hears or reads about in West Germany.
>
> In Western Europe's richest country, she lives in a dirty, dilapidated, barracks-like "temporary" housing settlement in Am Grauenstein, stuck away in the weeds in the Poll section of Cologne.
>
> Mrs. Kestane has been there for six years, paying the city less than $30 a month in rent. A neighbor's daughter has known no other home for all her seventeen years.
>
> There are no trees or flowers in the dingy courtyards and no toilets in the apartments. Mrs. Kestane shares one bedroom with her Turkish husband, Mustafa, and their two small children, Sherifa and Mustafa Jr.
>
> Mrs. Kestane is not, like her husband, one of the 100,000 so-called "guest workers" who collect Cologne's garbage and do its hard and dirty work—tolerated but shunned socially by the Germans. She is a blonde, a German, and she speaks a rich version of the Cologne dialect. But in a society where the middle class dominates, her poverty makes her an outcast.
>
> In West German society social appearances count for much. The housewives who wash their windows once a week and keep their white lace curtains spotless have nothing but contempt for people like Mrs. Kestane. They have a word for her condition—"asozial." In a country where the average weekly income of a worker before taxes is $175 they reason there must be something wrong with anyone who cannot get in on the "economic miracle."
>
> How many did miss it is hard to find out. The cost of living is so high in West Germany that even an average worker has to cut corners and live modestly to get by . . .
>
> "I tried a while ago to get a job as a cleaning woman," Mrs. Kestane said the other day. "Everything went fine until they asked where I lived." I said, "Am Grauenstein." They told me "Sorry the job has been filled."[1]

Some World Bank figures were given showing the distribution of income to be much more disparate in F.R.G. than in G.D.R. (see p. 244 and note 30 on pp. 278-279). This lesser disparity in the G.D.R. was even perceived by a representative sample of voters in F.R.G. when they were asked if "there is a large difference between rich and poor." Forty nine percent said this "applies very much" to the F.R.G., while only 24 percent felt it applied this strongly to the G.D.R.[2]

As has been suggested in Chapter 9, the class struggle in Germany reaches back to Bismarck's time, with the owners seeking to "pull the plug" on the workers' movement (das Wasser abzugrabben)[3] through various state-assured social welfare measures. But several major breaks have occurred–WW I; Nazi Germany; and WW II. Thus the history of struggle and legislation relating to OSH is quite recent. Perhaps it is this recency, but perhaps also a German history of patriarchy, hierarchism and cooptation which have kept the workers' movement relatively weak.

It is important throughout to have a conception of state power and state sponsorship of various measures. In Lenin's view, the state in capitalist society is a kind of executive committee acting on behalf of the ruling bourgeois class. Others argue that while this is fundamentally true, the complex of state programs and services at any moment must be seen as the outcome of conflict between the working class and the bourgeois class.[4] The case of Bismarck and social welfare legislation in Germany is an excellent example of state response to workers' struggle. But the German case may be especially interesting because various measures–the first NHI in 1883 as well as some of the quite elaborate OSH apparatus of today–may have been adopted more in anticipation of than in response to struggle. In short, things may have been given and labor coopted more than in other countries; whereas if labor had taken these things in solidary struggle, the movement might be stronger than it is today.[5]

The workers' movement in the F.R.G. may be slightly stronger than in U.S.A. for at least three reasons. First, even if a narrow business unionism[6] has been encouraged by the capitalist owners, under the sponsorship and hegemony of the occupying Western forces, particularly the U.S.A., which emerged dominant in the capitalist world-system, following WW II, there is a great left labor tradition in Germany with prolonged, sometimes heroic struggles in its collective consciousness. After all, The Communist Manifesto first appeared in Germany in 1848. Second, for about half the years since WW II, following a long period of control by the conservative Christian Democrats (now back in power since 1982) the F.R.G. was governed by a labor-oriented party, the Social Democrats, a mainstay party of the Socialist International. Third, as in Sweden, the F.R.G. labor law provides for a single union to organize a given plant or workplace. But estimates indicate that no more than 30 percent of nonagricultural workers belong to unions. There are seventeen unions with about 8 million members. I.G. Metall, the metalworkers' union which has also organized the engineering and auto industries has 2.5 million members and may be the largest union in the world. Also workers cannot be said to have had their own political party in power, even though the Social Democrats have been more favorable toward labor than the Christian Democrats. And the severe recession of the world-system brought high unemployment: 5.6 percent in February 1981; 8.1 percent in February 1982; 10.2 percent in February 1983. Although it began to fall in late 1983 in response to the world economic recovery, unemployment in F.R.G. still stood at 9.5 percent in December 1983 and lay-offs were on the increase.[7] These conditions have brought severe health insurance and social welfare cuts and in general have not made matters better. [8] In 1981 the Social Democratic regime (billing itself as labor's partner) asked labor to take cuts, make sacrifices, and was going to cut a job creation program. But the reaction from labor was sharp including the observation that capital wants more tax free money to export it to more profitable foreign investments, not to create jobs in F.R.G.[9]

The unions have begun to fight back, particularly the largest and best organized, I.G. Metall. It will help to understanding strengths and weaknesses of the workers' movement in F.R.G. to examine the case of the most important and most recent struggle in many years. The workers' demand is for a shorter, 35-hour work week at the same 40-hour pay to hold the level of living, but at the same time create more jobs. A long strike began on May 14, 1984 with 13,000 workers going out in fourteen plants in the southwestern state (Land) of Baden-Wuerttemberg. By the first week in June, 400,000 workers were idle and virtually the whole auto industry was shut down. There has been wide support from other unions, particularly the printers, and from the left wing of the SPD (Social Democratic Party now out of power),[10] though it was not easy to get a strike vote in this time of high unemployment and uneasiness among workers. The miners and construction workers settled for higher wages and earlier retirement, rather than a shorter workweek.[11] Hans Mayr, I.G. Metall's national chairman called this struggle "the greatest social dispute in (West German) postwar history."[12] The surprizing strength of the workers' movement in this struggle related not only to the three factors mentioned (a long tradition of vigorous struggle; political support from and through the SPD; and closed-shop, single union organizing) but to other influences, one of which has mixed effects. On the one hand, many employers fear the effects of a very high level of unemployment and they are worried by the model offered by the G.D.R. and the risings of the 1960s:

> The ghost of Hitler is still haunting us. We all learned in school and partially even from our own experience that mass unemployment cleared the way for Hitler to seize power. It will be difficult to accept the idea that unemployment is a more or less permanent situation. As layoffs increase so does the fear. We fear unemployment because we have experienced its dreadful political consequences.
>
> Then there is the division of the two Germanies. The unions exploit the fact that East Germany has full employment, presenting it as a better model for the worker than West Germany is. Finally, there was the ideological turmoil of the 1960s when angry student revolts and the so-called Extra-Parliamentary Opposition, the radical left wing, wanted nothing more than the overthrow of the existing structure of German society.[13]

But the history of violent fascist repression is still remembered by unionists:

> On the picket line outside Gate 6 of the Mercedes plant, Mr. Lutsch and three colleagues try to warm their hands on plastic cups filled with tepid coffee . . . "If the union is weak, it'll go kaput," says Mr. Lutsch. He lowers his voice for effect: "We've experienced that once before, in '33 and we are not going to let it happen again." Memories of the way Hitler crushed the divided unions on May 2, 1933–one day after the Nazis pledged whole-hearted support for them–are still strong.[14]

Another factor is the whole society's changing view of work and changing concepts of self in relation to work. On the one hand technology is seen as having robbed workers of some freedom to make decisions on the job. On the other there is an increasing desire to enjoy the fruits of one's labor and a declining tendency in surveys for Germans to identify their single best characteristic as "industriousness." Elisabeth Noelle-Neuman, a social scientist, says, "We are not living through a post-industrial but a post-industrious phase."[15] At the same time capital is fearful of the effects of this demand for a shorter work week on their ability to compete with other producers in the world export market. This fear tends to harden their position.

By mid-June, both labor and management had agreed to arbitration.[16] By the end of June, it appeared that a compromise would be accepted–a 38½-hour workweek with no cut in pay, but phased in over the next two to three years.[17] This was a victory for the workers, even if a small one. It can lead to renewed spirits and efforts and certainly offers a new goal for which workers in other countries are likely to struggle. Thus, on balance, one must admire the history of struggle and present strength of organized labor in F.R.G. Certainly it is stronger than in U.S.A. But there are many shaky aspects and uncertainties as well; thus it cannot be rated as strong as the workers' movements in Sweden or even Finland, and probably falls slightly under the strength of the workers' movement in the U.K.

Many of the important decisions about health, education and other matters of daily life are reached at provincial level (the Laender) thus the nation's authority structure is quite decentralized. But there are many more signs of fractionation in F.R.G., as compared to Sweden–be it the young people's malaise; a quite massive Peace Movement;[18] continuing class conflict (even though organized labor is not always leading the way); sometimes it is the Greens; regional and religious differences; etc. If Sweden is comparatively decentralized and concerted, and the G.D.R. is centralized and concerted, then F.R.G. tends to be decentralized and fractionated–similar to the federal system in U.S.A.

The Health System of F.R.G.

Again the reader is reminded to look at Chapter 9 where a number of points common to the health systems of F.R.G. and G.D.R. were given as well as several points of direct contrast (see pp. 239-241 and 253-257). Consonant with its federal governing structure, private economy, and "voluntary way" (somewhat as in the U.S.A., though voluntarism is less emphasized), the major characteristic of the F.R.G. health system, even though it has a national health insurance law (the first in the world) is one of fractionation and division. The various controlling entities or sponsorships involved include: industries; voluntary associations' carriers of social insurance; physicians' associations: states (Laender); and the federal government. There are eleven states, and the number of insurance carriers (sick banks or Krankenkassen) seems to vary according to: who is counting; how; for what period of time;–but there are a great many. If I choose the number of 1,851 sick funds in the mid-1970s, it is only because the indicated source goes on to offer the breakdown in Table 10 which is helpful for understanding the multiplicity of sponsorships and interests involved in the F.R.G. system. In short, although the population coverage is very complete (some 90 percent) and coverage for most kinds of care is very comprehensive, the system is highly complex and expensive. Physicians, for example–the main

Table 10. Sickness Funds in the Federal Republic of Germany

Type of Fund	*Number*	*Membership Total*	*Average size of funds in thousands*	*% of all members in each type*
1. Local sickness funds	401	15,715,000	39.18	52.6
2. Rural sickness funds	102	438,000	4.29	1.4
3. Guild sickness funds	179	1,385,000	7.73	4.6
4. Company sickness funds	1,147	3,995,000	3.48	13.3
5. Seamen's sickness fund	1	76,000	76.00	0.2
6. Miner's sickness funds	6	1,126,000	187.66	3.7
7. Substitute sickness funds*				
(a) for blue-collar workers	8	328,000	41.00	1.0
(b) for white-collar	7	6,783,000	969.00	22.7
TOTAL	1,851	29,845,000	16.12	100.0

*These are mutual insurance associations registered before 1st April 1909 and officially authorized as sickness funds.

Source: Allan Maynard, *Health Care in the European Community*. Pittsburgh: University of Pittsburgh Press, 1975, Table 1, p. 7.

determiners of types and amounts of use of drugs, hospital, and other care—were earning considerably more than the $63,000 physicians earned on the average in the U.S.A. in early 1978. In fact, the position of physicians has steadily improved in the Federal Republic of Germany, "especially since 1955 when they were granted a monopoly over ambulatory medical care. Specifically, neither sick funds nor hospitals (except university hospitals) can run their own outpatient departments."[19] The net average annual incomes of physicians in practice have risen from DM 35,290 in 1959 to DM 115,580 in 1971. In the same period, the ratio of physicians income to that of the average employed person rose from 5.46 to 6.52 (in the U.S.A. it remained fairly stable at 4.45 to 4.72 in the same period). It is not as extreme a situation as in Brazil, where the disparities are obscene, but we could paraphrase the president of Brazil who said, "Brazil is doing fine, but the people are not" and say of the F.R.G., "The physicians are doing fine, but some of the people are not."

It is helpful to see how the major elements of the system relate to each other. Figure 4 graphically indicates that there is no direct fiduciary relationship between physicians and patients.

As for the process of patient care, there are no careful comparative studies between different systems. One gap in the coordination of patient care occurs between the general practitioner and care in the hospital (as is true in the U.K.) for the GP is restricted to the community and specialists reign in the hospital.

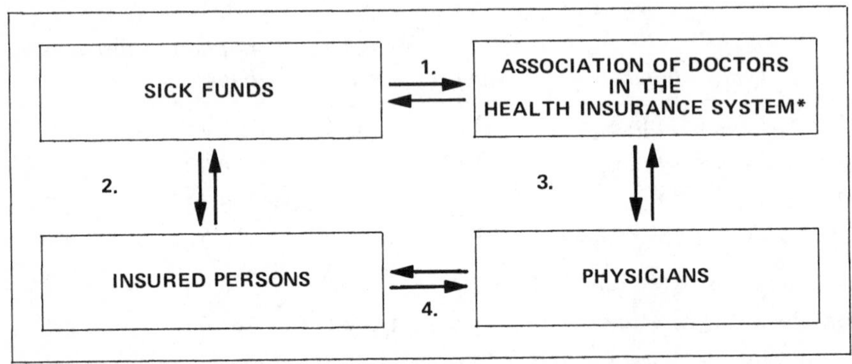

* Kassenaerztliche Vereinigung: 1. Relationship of collective negotiations; 2. Compulsory membership; 3. Compulsory membership; 4. Individual patient-doctor relationship.

Source: M. Pflanz, The Two Faces of the Patient-Doctor Relationship in a Changing Welfare State: The Federal Republic of Germany, in *The Doctor-Patient Relationship in the Changing Health Scene,* E. Gallagher (ed.), DHEW Pub. No. (NIH) 78-183, U. S. Government Printing Office, Washington, D.C., 1978.

Figure 4. Relationships between insured persons, physicians, and sick funds in F.R.G.

For cultural, financial, availability, and certifying (Bestaetigungen) reasons, the use of physicians as noted earlier is very high—some twelve visits per person annually.

A picture of rapidly rising costs for health and medical care was given for F.R.G. and compared with lesser rises in G.D.R. (see p. 255 and notes 51-53 on pp. 280-281).

> In 1980 the cost for health services amounted to 200,520 million DM. That total corresponds to 13.44 percent of the gross national product and about 41 percent of the total social budget (DM 403,477 million in 1978). Prevention and medical assistance account for 64.4 percent of the total cost; 58.4 percent for medical services in the area of ambulatory care and hospital care; 30.6 percent went for services occurring as a result of an illness (professional and social rehabilitation, loss of income, pensions owing to unemployment, or incapability to work); 1.5 percent went for training of health personnel and health research; and miscellaneous expenditures account for another 3.7 percent. The total cost for health services is financed as follows: 14.3 percent from taxes, 43.7 percent sickness insurance, 7.4 percent pension insurance, 3.1 percent accident insurance, 4.4 percent private sickness insurance, 18.8 percent employer, and 8.3 percent private households.[20]

The escalating costs of hospitalization and medical care in F.R.G. continued through the worldwide capitalist recession of recent years. At the same time there were pressures from U.S.A. and NATO to raise the amount spent on armaments. With outlays rising for unemployment insurance, social services, and the health sector—as well as "defense," and state income diminishing because of the economic recession and lower tax revenues—the reaction to cut social and health services began in 1982 even while the SPD was still in power.[21] The CDU's (Christian Democratic Union) accession to power in a conservative coalition in late 1982 only exacerbated the cuts, placing the burden of them disproportionately on the working class and those most in need. A new medical-public health cultural hegemony evolved to justify and rationalize these moves. The essence was to place the blame on the victims themselves, emphasizing personal health behavior (smoking, drinking, overeating, etc.) rather than mass unemployment, poor living conditions, and hopelessness. The election campaign had also brought charges of "misuse" and "overuse" of services from the conservatives. These charges appeared grotesque and cynical to many workers and others in the face of evidence that physicians' prescriptions of rest cures (Kuren) had been declining as early as 1970, that hospital stays were shortening, especially for older workers, that only a portion of those who could be said to need ambulatory care actually obtained it,[22] that in 1981 only two-thirds of registered unemployed persons received support, and that many of those who could have qualified for social services and support failed to receive such help.[23] Nevertheless, these charges served their function:

> This function was to put the blame on the very people hardest hit by the risks and dangers from capitalist economic and societal order, making their social situation an outcome of their personal behavior. The individual person is given the responsibility for the society's crisis. The debate focuses not on the causes of mass unemployment (economic crisis) but on the unemployed and the sick themselves.[24]

Some of the specific cutbacks in the realm of health services have included increased self pay:

- from one-half Mark to two Marks for every prescribed medicine;
- five Marks for every day of the first fourteen days of hospitalization;
- 10 Marks per day for rest cure stays in a resort (Kuren) (a traditional form of care much cherished as a part of the modern health systems in both the G.D.R. and F.R.G.);[25]
- so-called, "petty medicine" (Bagatellarzneimittel") such as those against travel sickness, coughs, colds, hoarseness, no longer covered by the sickness insurances (Krankenkassen);
- retired persons to be denied some of their projected cost-of-living increases and forced to contribute to their sickness insurance on an increasing scale: 1 percent in July 1983; 3 percent in 1984; and 5 percent in 1985.

In addition, the staff and rules for controlling physicians' "writing someone sick" to be excused from work were significantly reinforced. There was also much discussion among health facilities planners concerning capital investments in hospitals, pharmaceutical production, etc. but not much action (effective action in this realm would harm capital's interests in accumulation through investments in the health care sector—indeed, there is some built-in ambivalence and conflict within the ruling class regarding medical care cost controls for this reason).

While Deppe attributes these cutbacks to the recent recession and the particular political dynamics in F.R.G. reference, (note 8), the observation that the same sort of thing can be seen in Thatcher's U.K. and Reagan's U.S.A. makes one realize that something is going on in the wider world-system. Indeed, the fact that these moves began in West Germany under SPD (and were only sharpened under the CDU) suggest that the system is in crisis. These cutbacks are not simply moves of conservative political parties. Such parties may accentuate them. But as I have tried to make the case elsewhere,[26] drawing on O'Connor's work,[27] monopoly capitalism itself makes ever greater demands on state funds for its mammoth projects (military and other) and the pressure develops from the ruling class to cut human services to the working class and disinherited. This is the long term fundamental origin of the "fiscal crisis" of the modern capitalist state.[28]

With regard to other aspects of the health system, in 1975 the F.R.G. had slightly more physicians per population (1 per 522) than did the G.D.R. (1 per 548). The growth in the physician population ratio has been:

1960–1 physician per 703 inhabitants
1970–1 physician per 612 inhabitants
1980–1 physician per 442 inhabitants

Physicians in private practice made up a sizable proportion of these, rising from one private physician per 1858 inhabitants in 1960 to one per 1,028 in 1980.[29]

A similar comparative picture presented itself as regards hospital beds—a few more per population in F.R.G. than in G.D.R. in 1975 (one bed per 85 people in F.R.G. versus one per 91 inhabitants in G.D.R.). In 1979 the F.R.G. (including West Berlin) had 3,286 hospitals and 712,055 beds or one bed per 86 inhabitants. (General comparative indicators of health status were presented on pp. 256-257). Health indicators for F.R.G. compare unfavorably with those for G.D.R.

According to a set of ten interwoven ideals of regionalization of health systems which I have developed and discussed elsewhere as a way of measuring the potential of health systems to deliver effective care in an efficient way (with an emphasis on PHC action at the periphery),[30] regionalization is considerably more in evidence in G.D.R. than in F.R.G. There are distinct shortcomings in the G.D.R. too. For example, there is too little direct citizen involvement and control of the health system at every level—other than for the portion of the citizenry belonging to the Communist Party. But in general, on a regionalization scale from 0 (low) to 10 (high), I would place the G.D.R. at 7 or 8 and the F.R.G. at only 2 or 3.

SUMMARY AS REGARDS THE HEALTH SYSTEMS OF F.R.G. AND G.D.R.

Demands for adequate health care and costs are rising across the world and perhaps especially in the "developed" countries.[31] These demands place heavy pressures on governmental leaders and health service providers to devise more effective and efficient health service systems which provide coverage to all parts of their populations. Even though there are common widespread problems found in most national health systems, the extent and approaches to these problems differ markedly between "developed" countries, especially between the core capitalist nations of the world-system and the semi-independent, socialist-oriented nations. A comparison between the West German and East German health systems allows a more complete understanding of relations between health systems and their political-economic contexts. Evidence suggests that the more fully regionalized national health service system of the G.D.R. is associated with lower absolute and proportionate costs (percent of GNP spent on health). However higher health status measures of several sorts (infant mortality, maternal mortality, work accidents and illness) exist than is true for the poorly regionalized national health insurance system of F.R.G. People are or can be in charge. As Rudolf Virchow, the famous German physician, theorist, and practitioner of social medicine put it in 1849, "Politics is medicine on a larger scale."

THE OSH SYSTEM AND SERVICES IN F.R.G.

The History

A regulation from 1869 mentioned measures to protect workers from some of the more extreme forms of exploitation found in early industrial capitalism. Accidents at work were the main concern of this regulation. But the basic OSH law (para. 120a) was passed at the end of the 1800s along with a state authority, the works inspectorate (para. 139b–Gewerbeaufsichtsbehoerde) to carry it out. This law made the employer responsible for providing and maintaining workrooms, machines and other arrangements so as to protect the worker from dangers threatening his life and health. A temporizing phrase was included, however: "As the nature of work permits" (wie es die Natur des Betriebes gestattet). Perhaps for this, as well as many other reasons (I would say, most importantly, the comparative weakness of the workers' movement itself) "it must be admitted that this 'basic law' has not been effectively implemented until now."[32] As one especially knowledgeable informant put it:

> Occupational safety and primary care, your subjects, are both in West Germany underdeveloped. Occupational safety and medicine increased at the beginning of the seventies for we got a new law (Arbeitssicherheitsgesetz 1974). Primary care and preventive medicine are foreign words in our country. But both–occupational safety and primary care–are traditional demands of the labor movement and progressive scientists.[33]

What were some of the steps between the basic law of 1900 and the Work Protection Act of 1974?

As suggested earlier, several major breaks or gaps occurred after 1900 in German labor history (the revolution of 1918-1919 with the system of workers' soviets defeated by division on the left and opposition of liberal-conservative forces; Hitler's crushing of the unions on May 2, 1933; etc.); thus few noteworthy developments, especially as related to OSH, took place until after WW II.

The modern movement for improved OSH can probably best be dated from an I.G. Metall-sponsored Work Protection Congress held in Duesseldorf in the late 1950s. Criticism was leveled there against the accident insurers for the steel industry.[34] An awareness had grown before this on the parts of workers themselves, labor leaders and others that "the economic miracle" was bringing an awful toll in workers' lives lost. The 1960s saw 2.5 million work accidents per year or one for every 10 workers (even not considering unregistered accidents)—a higher rate than for any other advanced capitalist or socialist society.[35] Up until this time, the ideology of "accident prone" workers prevailed. But now workers saw through such "victim blaming" and began to move collectively against work conditions which helped (or made it inevitable) for individuals to have accidents. The catalogue of developments since that time is impressive—even if some of it has more shine than substance and even if much remains to be done:[36]

- Accident Insurance Right (Unfallversicherungsrecht) newly issued, 1963. This *made accident prevention the first priority or duty of the workplace insurers* (Berufsgenossenschaften) an especially interesting, unique aspect of the OSH system in F.R.G. about which I say more below.
- A law calling for the manufacture and purchase (also from foreign suppliers) of safe machines and implements (Maschinenschutzgesetz) passed in 1968 and was improved in 1977 (with machines and materials thereafter labeled GS—Gepruefte Sicherheit) and further improved in 1979.
- A similar law to protect against dangerous chemicals and other work materials (Arbeitsstoffverordnung) passed in 1971 and sharpened subsequently, the last time in 1980.
- A most important achievement was the assurance of the worker's right to know and give his agreement (co-determination) to OSH measures for his workplace (Betriebsverfassungsgesetz) 1972. More will be said about this, but *it is important to note that this law does not mention unions, only employees.* Where there are strong unions in a workshop—fine. Where there are not, the worker can be top easily intimidated—regardless of what this law says.[37]
- the establishment in Dortmund of the National Institute for Work Protection and Accident Research (BAU—Bundesanstalt fuer Arbeitsschutz und Unfallforschung) 1972.
- the Work Safety Law about Work Doctors, Safety Engineers and Other Specialists for Work Safety (Arbeitssicherheitsgesetz) 1973. Under this law employers were required for the first time to hire occupational physicians and safety specialists and provide the necessary material support to OSH physicians—office, staff and equipment—to allow them to aid in worker protection and accident prevention. I'll comment in more detail below on the ways in which this law has been implemented. It is not as strong as the Finnish law.
- Government's Research Program to Humanize Work Life (Forschung zur Humanisierung des Arbeitslebens) was set forth in 1974. Again, more will be said about this and its possibility of having any real effect where the worker is in fact not in charge of his work. Some see it as a new obfuscating ideology.[38]
- the Work Environment Law (Arbeitsstaettenverordnung) of 1972 together with the machine safety law of 1968 as amended and the work materials (chemicals etc.) law of 1971 as amended gives the basic ground for the work of OSH personnel.
- the setting in force in 1977 of regulations for one of the most important components of the OSH system in F.R.G.—the insurers (Berufsgenossenschaften).

Many other national statutes could be mentioned—some quite specific, like the "Flammable Liquids Ordinance" and others more general, dealing with "Dangerous Work Materials" or categories of workers, such as the "Protection of Young Workers Act" or the "Maternity Protection Act." Of equal importance are regulations issued by the insurers (Berufsgenossenschaften) but this leads us to the structure of the system at present.

THE CURRENT OSH SYSTEM[39]

As just mentioned, legal mandates come from two sources—the laws or ordinances of the national state and regulations issued by the three types of Mutual Accident Insurance Associations (the Berufsgenossenschaften). The latter began as self-help insurance-advisory agencies of the employers. Today these organizations are legally public mandated corporations to cover work accidents and sickness in industry groupings—steel, shipbuilding, mining, etc. They are grouped under three categories: industry (36), agriculture (19) and a number for public service. There is a 50-50 representation of employers and employees on the controlling boards of the Berufsgenossenschaften.[40] Thus they are a major channel through which the workers' movement can have an influence on OSH. Generally, these laws, regulations, etc. from these different sources are said to be non-overlapping. They proceed from more general to more specific measures as shown in Figure 5.

The emphasis on law and formal arrangements is in striking contrast to the system in Sweden which has been largely built up through negotiations between SAF, representing employers, and LO and PTK, representing the unions, and only subsequently confirmed (given the seal of public approval, so to speak) in law. The U.S.A. system has developed through both law and negotiations, but probably mostly through legal battles and court decisions.

The F.R.G. law makes the employer responsible for assuring OSH and requires him to appoint health and safety specialists to assist in this duty. Within this legal framework, regulations issued by the insurers (Berufsgenossenschaften) spell out what kind of OSH staff is required according to type of industry, size of the plant and kind of hazards known to be involved. For example, the corporate medical director of a large worldwide chemical firm illustrated the requirements this way: There are four levels of hazards. If a chemical firm has 4,000 workers evenly distributed in the four levels, then 1,000 is multiplied times the required staffing ratio (say for OSH physicians) for each level: 1.2 for the most hazardous group I; .6 for II; .25 for III; and .15 for IV. The results are added to give 1.3 OSH doctors for this chemical factory. "Probably the company would employ two doctors to be on the safe side," said this informant. But a woodworking mill which has a lower hazard rating than a chemical plant could end up with forty-five hours of an OSH physician's time per year to cover a similar number of workers. As this medical director put

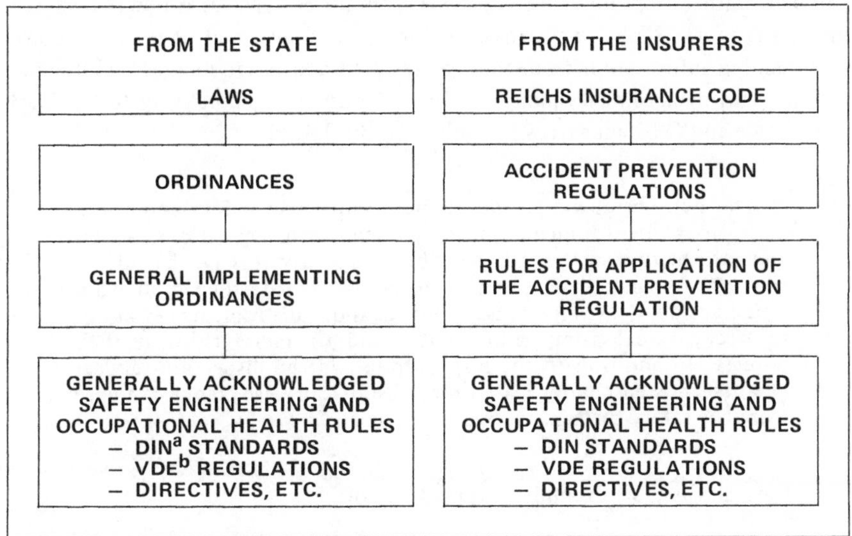

Note: [a]DIN = German Industrial Standard;
[b]VDE = Association of German Electrotechnicians.

Figure 5. Parallel structure for OSH laws and regulations in F.R.G.

it, "The unions aren't happy about this." I will come back to the numbers and types and arrangements for training, employing and paying the OSH physicians, since this expert competence is so key in the F.R.G.'s OSH system.

Also important, in the law or on paper, so to speak, is the in-plant structure for OSH other than medical and safety personnel. I am quite convinced that the effectiveness of these provisions depends heavily on whether the workplace is unionized and how strong the local union is. In any case, in all plants with more than twenty workers the employer must appoint safety stewards–one for each department or workshop in medium to larger firms. The employer must also set up an OSH Committee (Arbeitsschutz-Ausschuss) but this is only mandatory in plants which have appointed a physician and/or safety officer. This committee must meet at least quarterly. In smaller plants without a physician, an OSH Committee must be appointed when there are three or more safety stewards and must meet monthly with the Works Council (Betriebsrat) represented at the meeting.

The Works Council is an important body (especially where there is a strong local union). Whether there is a union or not, the Work Constitution of 1972 provides for the election of a Works Council which is supposed to act on behalf of workers' interests by: 1) acting as a watchdog over enforcement of laws and regulations benefitting the worker, including OSH; 2) making suggestions; 3) co-determining general production plans, including OSH measure to be taken (though neither this body nor the OSH Committee have veto power over the hiring of an OSH physician, for example, as a similar body has in Sweden), 4) obtaining information from the management and other sources on an equal footing with management; and 5) lending support in a general way to the OSH Committee and OSH endeavors generally. On this last point:

> The Betriebsrat must support the occupational safety and health authorities. It is obliged to assist them with suggestions, advice and information in the course of their work. Conversely, the authorities are obliged to let the Betriebsrat participate during all plant visits, accident investigations, and discussion of problems pertaining to occupational safety and health, and to advise them at their request. In addition to the inspectorate and the inspection services (Berufsgenossenschaften) the Betriebsrat can also request advice from the safety officers and plant physician.[41]

In general, the OSH Committee is made up of:

- the employer or his representative
- two members of the Works Council (Betriebsrat) appointed by the Council
- the plant physicians
- the safety officers (safety engineers)
- three safety stewards.

The Committee is only a consultative advisory body. To fulfill this function it receives inputs as diagrammed in Figure 6.

As appointees of the employer, and as a matter of carrying out "on the spot" OSH work, the safety stewards (not to be confused with the term "union stewards" familiar to unionists in U.S.A.) meet independently of the OSH Committee, especially in larger plants where there are more than three safety stewards. These meetings are not required by law.

One of the more complex and confusing aspects of the F.R.G. system is that it has two types of labor inspection systems (OSHA-type agencies in U.S.A. terms, at least as far as the inspection function goes):

- labor inspectorates of the ten federal states (Laender or provinces) plus Berlin. These function on a geographic basis in relation to all kinds of industries.
- the inspectorates of the Berufsgenossenschaften (the legally mandated insurers). These are of the three types mentioned earlier:
    1) industrial with thirty-six different ones, each oriented toward an industry, such as, metal working, mining, chemicals, textiles and leather, etc. Some 1,000 personnel are employed in these;
    2) agricultural with nineteen different ones and some 300 staff;
    3) public service with forty-one and about 120 staff.

This totals to 106 inspection agencies with 3,720 personnel for the country. There is a total of some eighty OSH physicians employed in these 106 inspection agencies. I am unsure of their level of qualification. The unions and some others

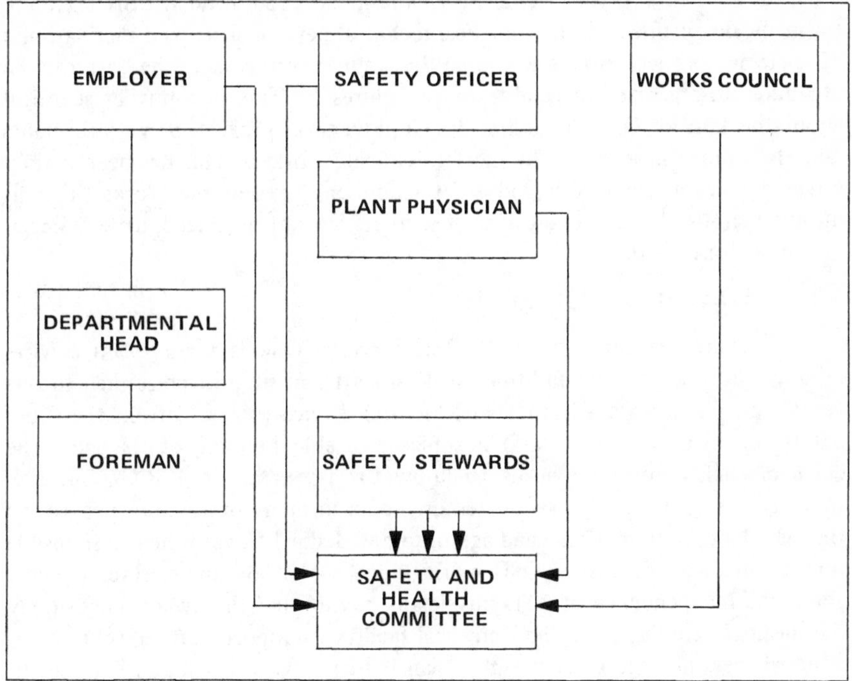

Figure 6. Inputs to Plant-Level OSH Committee in F.R.G.

are unhappy about this situation. Clearly, some of these are little more than paper organizations and how they avoid overlaps and gaps and cancelling each other out is anybody's guess. Theoretically, they are responsible for visiting plants and monitoring compliance with OSH laws and regulations. In this process they must seek to involve the Works Council as noted above (this involves notification ahead of time that an inspection will occur–I am unsure of the amount of warning time involved which could be important for some violations which might be cleaned up just for the visit). These various inspectorates take action by issuing inspection reports and notification of measures required to make corrections. When serious violations are found they have the power to shut down a particular operation or a whole plant (this is a far cry from the power of union safety people to take the same action in Sweden, even with long-term and not only immediate hazards). Clearly, the health and safety of workers in F.R.G. can not be left to the many splintered labor inspectorates. Two other primary elements then take on great importance: *the OSH education of workers themselves and the OSH services*.

The employer is responsible by law for educating "his" employees to avoid dangers. The prevailing ideology still is that employees themselves are responsible for most accidents. Until recently, even the most recommended OSH textbook for medical students and other medical-technical personnel claimed that "around 70 percent of accidents are caused by human error, and the rest can be attributed to organizational and technical failures."[42] Thus one must be skeptical about the kind of OSH education the employer will provide. Some informants said the unions have done the most to educate workers. This has been mainly directed toward the health and safety representatives of the Works Councils in union shops.[43] Since two-thirds of workers are not organized, little has been or can be done for them.

What of the OSH Services?

There are three types of OSH services. One is the in-plant service, usually with one or more full-time medical staff and support personnel. In one very large plant I visited, the OSH Medical service (separate from the Chief Safety Engineer and his staff) occupies a sizable building of its own. The chief physician not only needs to know the laws, etc. in F.R.G., but presumably has to keep his eye on requirements in many other countries, as the firm is a large multi-national and as Corporate Medical Director he is responsible for the medical aspects of OSH world-wide.[44] The law makes several points about the independence of OSH physicians, saying that the owner must supply the means (including salary and physical facilities, support staff, travel to international meetings, etc.) but the physician is free to give his views as he sees fit in accordance with scientific and professional standards, and he must not share information on his worker patients without their consent. I was especially curious to ask how this works out in practice. I was given many firm assurances that the medical and technical people are able to function strictly on the worker's behalf—albeit only in a preventive way (more on this key point in a moment). Then a curious thing happened. A radio station called and asked this Corporate Medical Director if he would talk on a live early morning talk show on shift work stress and OSH. He has a reputation for work on this problem and was somewhat flattered and excited by the prospect and said "Yes." Then after he had hung up, he remembered that his contract says he will check with the management's chief of public relations about any activities with PR implications (legally he is to report directly to the owner only). So he did this and the chief PR man said "No." So this Corporate Medical Director had to call the radio station back and withdraw—so much for OSH physician's independence in F.R.G. in my experience.

But this is a rather trivial example. According to other informants, the situation is actually worse. One knowledgeable person said companies always offer physicians more money than is provided for on a schedule required by law. If the physician does not perform satisfactorily, the extra salary (or other benefits) may be withdrawn. Then one of the major functions of the OSH service turns out to be the screening and excluding of prospective new employees for handicaps and health conditions which could conceivably lead to payment of worker's compensation at a later point; thereby functioning to save the employer money, rather than to assure the prospective worker a safe job.[45]

A medical student told me about one of the GPs qualified for industrial medicine by having taken a two-weeks course (see below). He lectured to them about his work serving several small plants as an OSH physician. When students pressed him to arrange a visit to one of the plants, he resisted, then finally agreed. When the students got there, the doctor acted like he didn't know them and on a hastily arranged tour with no depth at all kept referring to "our plant" with the clear implication he was firmly identified with management. This concern regarding de facto independence of OSH personnel of course applies to all three types of OSH services.

The second type is the OSH Center (Ueberbetriebliche) serving a number of workplaces. These have mainly been arranged by the Berufsgenossenschaften and may be the most promising type with fairly well trained OSH physicians and supporting staff. In saying these may be most promising, I mean only that the personnel may be better qualified and supervised. There is no vehicle for worker control of these any more than the others.

By far the most numerous type is that offered by the GP or other practitioner who "goes around the corner" ("geht um die Ecke" as one informant put it) to give his services part time. This is usually a solo practitioner who has received a basic OSH qualification sometime since the law requiring OSH services for all workplaces took effect on December 1, 1974.[46] The first ones received two weeks training in OSH. This has recently been upgraded to four weeks. Several points of vigorous critique were offered about this kind of service. The first point was raised by a Labor Department official at national level. He said, one of the prime motivations of these physicians is to increase the size of their practices by "attracting the patients out" ("die Patienten auslutschen") of whatever plants they visit on a part time basis. Another informant said that in the context of entrepreneurial medicine in F.R.G., time is money to the part-timer. So he will do very cursory periodic follow-up exams and work placement exams. These are only two of the major mandated functions. Others are pre-employment exams; first aid and training of 2 percent of employees as first aid helpers; diagnostic exams; health education; workplace monitoring, and advising of employers and works councils on physiology, psychology; hygiene, and ergonomics. Note especially that *there is no mandate for curative care*. The most general problem is the levels of training and of commitment to OSH concerns. At one OSH Conference with some of the leading OSH figures in the F.R.G. present, a national union official involved in OSH in his union for the last twelve years put it this way when addressing the representative from the physician's association:

> In my experience it is evident that particularly the part-time physician in small industries is not able to give the necessary care. . . . We initiate contacts with all the kinds of physicians—whether they are full-time in a plant or in a center outside the plant, or the part-time ones. We have had conferences with these physicians. All the part-time physicians confirmed that they are not sufficiently trained and not familiar enough with the work of the firms they are serving. The fourteen days basic course and additional learning through experience is not sufficient to be a competent advisor to the company and workers. . . . Colleagues who are working elsewhere as well as the doctors confirm this. I ask you Mr. Hess to take this discussion dealing with facts closer to the actual practice into the Aerztekammer.[47]

Thus overall coverage is poor. If one allows that industrial physicians spend on average just fifteen minutes a year per worker (a not ungenerous estimate according to studies of actual behavior) then with 26 million full time employees in F.R.G. working forty hours a week in 1980, some 15,000 OSH physicians would be needed, concentrating their main work in this field. There were some 15,500 physicians with some OSH preparation in 1983. But there was little information on the depth of this training or the amount of time devoted to OSH work. Only 969 OSH medical specialists had the longer ("large"—"Haupt") qualification. Most of the rest may have had training primarily in treatment and prevention of work accidents—little on disease. Possibly 10,000 have had the two-week BAU course, but, again, the amount of time they actively devote to OSH is unknown. In general, the estimate is that no more than 50 percent of F.R.G. workers are covered by OSH services, with the medium to small-sized plants much less well covered than the large ones.[48]

In terms of qualifications, the latest figures obtained from BAU in Dortmund (a kind of information research equivalent of NIOSH in the U.S.A. or the National Institute in Finland and G.D.R.) show that in December 1980 there were 14,420 physicians qualified to some degree in OSH. The full qualification (Facharzt fuer Arbeitsmedizin) was held by 932. These are specialists with two years in OSH work divided into three months of theory and twenty-one months of practice. The three months cannot be divided into more than three segments. Five university-connected institutes qualify to give this training, though almost all of the medical schools are said to have a chair in occupational medicine (but some schools, e.g. Hannover, do not have such a chair). These physicians may work for firms or they may be in universities or various institutes, etc. This is sometimes called the "big qualification," or "Doctor of Occupational Health." The "little qualification" (Betriebsarzt or Factory Doctor) was held by 2,488. This is obtained by one year in clinic practice (not necessarily an OSH setting) plus one year in one of the five academic OSH centers divided into three months of theory (again, in not more than 3 segments) and nine months of practice. The large bulk of some 11,000 OSH physicians had a certification (Fachkundebescheinigung) obtained from a two-week course or more recently four weeks. Many physicians have obtained this "in case" they'd like to do OSH work on the side. This is not bad, for it may spread OSH knowledge among PHC providers. But it is not known how many physicians are actually engaged in OSH practice.

Training for other OSH personnel is not as well defined, but is under discussion. Advanced training for nurses is being developed. Safety Engineers are recognized and defined. Industrial hygienists are not recognized as in U.S.A.

Many other aspects of the OSH system could be detailed. The BAU has been only briefly mentioned and would deserve more space as it is a major resource. Yet it is not exactly like a National Institute, being much more of a practically oriented information developing and transmitting center (it has an outstanding OSH library) but is neither a treatment nor research center. These are distributed much more in F.R.G. than in any other study country: Hamburg, Dortmund, Munich to name only three. In fact, one of the overall characteristics is the system's dispersed nature. The major national governmental responsibility for OSH is within the Ministry of Labor and Social Welfare (Sozialordnung) (no contact seemed necessary with the Ministry of Health). Yet this would appear to be mainly a legislation preparation and reporting arm, for much of the action is in the Berufsgenossenschaften, the BAU, the University Centers, and the state Labor inspectorates, and the OSH services of the different types mentioned above. Other detailed matters could be mentioned, for example, fifty-five substances have been officially listed as cancer causing in humans and there are some fifty-five officially recognized compensable work-related diseases. But the purpose here is to lay out enough of the OSH system so we can see how it relates to PHC.

OSH-PHC LINKS IN F.R.G.

One of the most important things to understand about the OSH-PHC relationship in F.R.G. is that because of fear of economic competition in this entrepreneurial (even if NHI) medical care system, the GPs and practicing doctors in general have succeeded in having it written into law that OSH services will be strictly preventive.

I had the fascinating experience of going with a GP to visit a young full-time OSH physician in a subsidiary of a U.S.A.-based multinational firm. The older GP almost disrupted the rapport of the interview by grilling the young fellow on whether he treats any patients privately. At first he fully denied any such thing. Then after telling us of his work—including 20 percent sick absence on average in some departments and people throughout the firm under terrible pressure and stress which he attributed to aggressive, frequently changing

American management—he did admit that he sees some of the management people privately (another instance of closeness to management).

Only emergencies and such are supposed to be treated by the OSH services. Thus one can be sure that some form of PHC offers the point of first contact to over 90 percent of workers. Thus the interest and information held by PHC personnel is crucial for proper treatment of individial cases and preventive follow-up as regards other workers who may be similarly exposed.

About five years ago, OSH was made a special required area of study (Fachgebiet) in medical schools and included in the qualifying exams. Hours vary greatly in the curriculum from one school to another, however. Two medical students from one school who have a special interest in the subject (so I believe they are well informed) told me they only have twelve hours. One said, "Three afternoons in six years of medical school; that's really something isn't it?" Still it's good to have something (thinking of my own state medical school in Connecticut where I teach an intensive elective course but there are no, that is zero, required hours of OSH education for medical students). Also, the subject is included in the exam everywhere, so, however many hours there are, students are generally advised to study the revised text by H. Valentin, *et al.* (see note 42). Sadly enough this modest amount of required OSH education in medical schools, puts the F.R.G. in the forefront of study countries included here, though nowhere near the thirty-six required hours of OSH study in medical schools of the G.D.R.

I did not learn about OSH hours in basic nursing education, though it is an important concern, since nurses make up a large proportion of OSH personnel.

For practicing physicians, very few of whom had any OSH education in medical school, little is done. One informant in the BAU said that in the Ruhr area where silicosis was once such a plague, the area itself and its problems and the workers taught the doctors. "So it isn't always necessary to give courses. Silicosis is almost gone now." The logic here seems dubious and in any case it seems like doing it the hard way—the hard way for workers, that is. Although some informants claimed that courses are offered to practicing GPs and other physicians, two GPs I spoke with said the local physician's Association puts on a program almost every week, but they couldn't remember one on OSH. Of course there are the 11,000 some doctors who have an OSH certificate, some of whom work part-time as OSH physicians, so this is an important achievement.

There is a reporting requirement so that PHC personnel who do see a worker with a work-related problem should report it, but again we are back to how aware and informed these personnel are.

At both the policy and organizational levels, I did not find the kind of awareness and concern for reorienting the overall health system toward the PHC level as was evident in Sweden and Finland. Because this policy thrust is lacking and because almost all the energy for improving OSH is in separate channels (that is, apart from the general health services) and is heavily concentrated on how to upgrade the special OSH services so that more workers will be adequately covered, there seemed to be little evidence of anyone trying to work on and better figure out the OSH-PHC links organizationally.

Yet it is evident that not only the workers, but also the GPs need help. I spent some time with a small town GP to learn what OSH problems come to him that he is aware of. There are many, some from the farm population (accidents, strains, and sprains, etc), some from small industries which have increasingly located in this area (he mentioned a patient who had hot flashes from breathing some gas which comes off in a metal dipping process in which TRI is used), and people under stress from work. Then there are many problems he suspects are there but doesn't know. He has many connections to OSH for which he is little prepared, though not uninterested. Quite often he gets a commission from one of the Berufsgenossenschaften (insurers) to follow-up a patient who has been injured or made sick at work. One current example is a worker who had his neck broken by a swinging block hitting him in the back of the neck. "He couldn't sit down or move his head–it was serious." The patient was sent to the hospital and on discharge this GP does the follow-up. The patient still gets frightening and painful cramps in his arms, so this doctor may be asked to rush to the patient's home at any time. The doctor does not send the sick slips to the Krankenkasse, his usual way of being paid (some doctors are called "Krankenkassenloewen–sick slip lions–because they have a busy practice and collect many slips which brings them more money, with decreasing reimbursment after a certain level). In this case, he sends the slips to the Berufsgenossenschaft. Possibly the biggest connection this doctor has to OSH is that he treats the workers in the area. OSH physicians could not treat patients, even if there were any (none of the local plants have any OSH coverage as yet).

So all workers in the area go for first contact to this or another nearby GP and have to go on the third day of illness if they want to be excused from work. Even if he recognized a work-related problem, he is relatively powerless. In the case of the patient suffering from hot flashes after working with TRI, he told the patient to "tell the boss you need some better ventilation." If nothing happens, and the patient comes to him again, he said he could call in "the public health inspectors" from the nearby city. (He did not know the OSH system and how to use it.) Also, few workers are protected by unions in this area, so "telling the boss" may not be so easy a matter. This GP's relative helplessness is perhaps best illustrated by another case:

He mentioned seeing an engineer from a U.S.A.-owned subsidiary. "The man is a wreck, but he didn't want me to send him on a rest cure or even to write him sick, "Because, he said, "my job may still be there when I get back, but it may be gone a little while later." Thus he could only handle the patient symptomatically and could not get into the work situation to have an influence. Nor could the young full-time OSH physician of this plant (mentioned earlier) treat this patient as his activities are limited to preventive exams, etc. Furthermore, there are no OSH standards regarding pressure and stress and little study and action around the problem in spite of the supposed big push for "humanizing work life." One of the major suggestions this concerned and committed, yet relatively poorly informed GP had was that OSH physicians should send more complete information about the work situation and the patient when they refer a patient. He has had such referrals and gets practically no information. This was one of the GPs I talked with who said their Physician's Association gives evening programs almost every week, but none had been on OSH. "I think I'll suggest it," he said.

The problems with financing and sponsorship and control have been mentioned. The employer pays for OSH services and there is little worker control. Furthermore, for financial reasons, the practicing physicians in the general health services have held the OSH services to a strictly preventive role. Thus there is an appalling set of disjunctures between PHC treatment and OSH follow-up and prevention in the F.R.G. system.

SELECTED OBSERVATIONS, QUESTIONS, AND RECOMMENDATIONS

I did more criss-crossing and traveling in F.R.G. than in any other study country. This was rather pleasant as the train system is so excellent. The OSH-PHC "non-system" should take lessons from the train system. The major overall impression I have is of an OSH system which lacks coherence within itself and certainly lacks effective links to PHC. This is especially serious in this system, since the care of workers is more determined by PHC in F.R.G. than in any other study country—except perhaps the U.S.A. The one strong point in F.R.G. is the inclusion of a significant amount of OSH material in the qualifying exam for medical students. The central position of the insuring organizations, the Berufsgenossenschaften, is also unique and interesting.

The major problem is that workers have themselves not widely organized and those who have are too dependent upon an expertise which does not firmly adopt the worker's point of view. Labor itself is too weak and divided.[49] A seeming great deal has been achieved in a short time, but the elaborate apparatus which is in place does not really get in there and dig out and set aside the problems. For example, the medical director of a large OSH service for a very large firm told me he and his staff devote the large bulk of their time to routine employment, screening, and periodic follow-up exams. He said he has a hard time doing epidemiologic studies. "If a forty-year-old dies," he said, "I get the diagnosis. If a retired person dies, I only get a death notice. I can and do call the deceased person's doctor, but if it's a doctor in Garmisch where the worker went to retire, he may use the 'data protection law' (Datenschutz) and won't tell me the diagnosis. If it's a Turk and he has gone back to Turkey, I'll never find out." Another informant said these are all excuses, that an OSH director with so many resources really could do good epidemiologic follow-up if he was seriously committed to it. The real point is that a strong union-controlled Works Committee with its own sophistication in such matters would see that this kind of work was done.

The system has so many resources, yet they are so dispersed and divided that I hardly know where to begin with any specific recommendations. One thing is very clear though about resources. Unions themselves need greater help and expertise. The head of OSH for one of the major unions said he has little staff, cannot easily counter management-oriented experts with their own, and has a hard time keeping the membership fully informed on new scientific findings and other developments which come across his desk each day. Also I can see a need for closer links between the Ministry of Health and the Ministry of Labor and Social Welfare regarding OSH and, especially, on the OSH-PHC connection. Perhaps a study commission should be set up on this subject with an important degree of union involvement as well as other interested parties.

A great deal more should be done to provide OSH education to GPs and other practicing doctors. The trend toward strengthening OSH education in basic medical education should be enhanced.

But, in general, the systemic level can not be corrected with a cafeteria-type selection of such specific suggestions. This simply constitutes managerial-type tinkering which I do not favor. In this regard, I think the concluding section of Professor Deppe's critical article on the OSH system in F.R.G. says most of what should be said:

> In conclusion, I believe that there are a few important demands which should be made in regard to occupational health and safety in West Germany:
>
> 1. Industrial physicians must be cut off from their present dependence on the private employer. This area of medicine is too important to leave open to considerations of profit if medicine is to live up to its humanitarian pretensions. To escape the cross-fire of competing interests, perhaps it would be best to have industrial physicians employed by professional insurance associations (Befusgenossenschaften) or similar agencies. The unions would have to assure that these agencies would be under the control

of the workers. Under such circumstances, we would welcome the repeal of the present prohibition of treatment by company physicians.

2. A greater stress on prevention is imperative, an emphasis which must go beyond mere early detection of disease. This is the only way to reduce the number of work-related casualties.

3. There must be a great expansion of the use of engineering techniques as the first line of defense against work-related diseases and injuries. Personal protective devices such as respirators should be used only temporarily, or as a last resort.

4. The present standards for the training and selection of occupational physicians and safety specialists should be raised and made more specific. It is not enough to take doctors from some other specialty, rush them through a two- or four-week course in occupational medicine, and expect them to perform adequately. The specialty of occupational medicine must develop in new directions, and this will be possible only if salaries are raised to competitive levels.

5. The range of officially recognized compensable diseases must be broadened to include psychosocial diseases such as stress.

6. All occupational injuries and diseases must be recorded in the official statistics, not just those which cause over three days absence from work.

In order to understand the relationship between work and health in a capitalist society, it is not enough to limit the area to a consideration of physiological problems and individual psychological characteristics. There must also be a careful examination of the concrete influences of the structural relations of property and surplus value appropriation in society. And it is hardly sufficient to limit medicine to a curative role. Medicine should be required to broaden its concern with prevention in order to ensure the health of working people. For this to occur, prevention must go beyond the superficial relief of particular symptoms to an attack on the technological and social roots of disease. Without such a fundamental strategy, occupational medicine in the Federal Republic of Germany will continue to find itself trapped on a sometimes fascinating and always dizzying merry-go-round of activities which has little to do with the humanization of the conditions of work and life.[50]

To move in the direction Deppe suggests, some new source of social energy is needed, especially as the economy turns down and organized labor feels even weaker than it already is. Thus I would recommend the formation of a number of COSH-type groups throughout this somewhat decentralized, fractionated land. Such groups of activists drawn from academia and from activist workers should have at least the informal support of organized labor, since such a COSH group, at the very least, can act as a gadfly which will aid the unions to achieve their OSH aims.

THE OSH AND OSH-PHC MODEL AND THE TWO GERMANIES

The dimensions of an ideal way of providing for OSH and OSH-PHC connections were briefly presented in Chapter 6.

Policy Level

At the policy level, in formal terms (i.e. on paper) OSH has a high priority in both countries. Very impressive sets of legislation were passed in both countries in the late 1960s and early 1970s. Many of these have been listed and discussed above. Possibly, most impressive, is the G.D.R.'s law requiring a survey of hazards in every workplace and follow-through in terms of 1) plans for removal of hazards and 2) systematic follow-up exams for all workers known to have been exposed in order to detect any illness early and limit pathology by proper treatment.

Both countries place the primary legal responsibility for OSH on plant managements. In F.R.G. this responsibility must be seen within the context of the first priority of managers and owners in all capitalist countries—to make as high a rate of profit as possible in order to grow and outdo the competition wherever that competition is within the capitalist economic world-system.[51] In G.D.R., the companies are not private. They are called, Volkseigene Betriebe, or people's own enterprises (rather than G.m.b.H., Inc. or Ltd. as in the West). There are still the pressures of production goals within a plant and pressures to economize (recall as pointed out earlier that enterprises in socialist-oriented nations must to some extent compete on the world market—the special Kombinaten quasi-isolated from the domestic G.D.R. economy for just this purpose of competing on the world market were cited earlier).[52]

But the ideology of production for social rather than private gain has an ameliorating or enhancing effect on the actual pursuit of OSH policy.[53] This political-economic structural difference, plus another key aspect of the political economy—zero unemployment and a labor shortage in G.D.R. versus 9 percent official unemployment in F.R.G.—contribute most evidently to an actual pursuit of the high priority OSH policy in G.D.R. in many ways which were suggested in Chapter 9 whereas the realities in F.R.G. suggest more accomplishment on paper than in fact.

Another aspect of these two contrasting political economies seems to play a key role. This is the differing forms and strengths of the workers' movements. The high priority OSH policy appears to be generally applied in G.D.R., since all workers are in a union and it is, formally speaking, a worker-controlled state, at least to the extent that the Communist Party fulfills this promise. But there is the question of whether there is or is not a need for a really independent trade union movement in G.D.R. In F.R.G., the high priority OSH policy appears to be more effectively pursued in large, unionized enterprises and these represent a minority of workplaces and workers, since only some 30 percent of workers are organized. I recognize the complexity of this dimension and the need to elaborate its real meaning. For example, it has occurred to me that the relatively strong formal position of OSH in F.R.G., in spite of a rather weak workers' movement, stems not only from a Bismarckian kind of strategy to anticipate and take the wind out of the sails of the workers' movement, it may also represent keeping up with the G.D.R. in a situation where both countries watch each other closely.

With respect to policy and the OSH-PHC connection, it is clear that the problem is recognized in G.D.R. and various efforts and experiments are being carried out to improve these relationships. The geographic assignment of PHC units in this highly regionalized system provides a considerable advantage over the largely market-oriented structure in F.R.G. The combined policlinic-OSH center I saw in Grossraeschen G.D.R., comes as close to a model as anything I saw in any study country. But this was not general and there remain many problems in improving the OSH-PHC connection in G.D.R.

In F.R.G., the problem of OSH-PHC connections is little recognized, even among OSH specialists, and medical educators in university centers. True, there is some awareness, as medical students are required to pass a section on OSH in their national board exams. But national policy recognition, concrete arrangements, etc. suggests little attention to this aspect of the model.

In *summary of the policy dimension*, of our model, the G.D.R. shows more actual pursuit of an OSH and OSH-PHC high priority than is true for F.R.G., the latter reflecting high concern for OSH on paper and little awareness of the OSH-PHC problem, with actual pursuit of improved OSH primarily in strong unionized shops.

Sponsorship and Control

The dimension of sponsorship and control shows striking contrasts between the two countries. These relate to the key political economic differences just pointed out, as well as to differences in the organization of health services in the two countries. Most symbolic of differences in this sphere is the formal recognition of unions as playing a key role in determining OSH conditions and services in G.D.R. (whatever one decides about the actual independence and effectiveness of these unions) while unions are not mentioned in the F.R.G. law. Also, OSH services in G.D.R. have an independence of both management and unions in that the Ministry of Health is the employing and supervising agent. Management is clearly in control in F.R.G., though the law and ethical statements attempt to give professional independence to OSH personnel.

One unique and especially interesting aspect of the F.R.G. system is the strong legally mandated consultative and inspectorial and OSH service educational and staffing role played by the workplace insurers (Berufsgenossenschaften). These of course reflect industry's profit-oriented interests but have an important degree of worker control involved in them. They are independent of management and represent a strong positive OSH force, given this kind of political economic system.

In neither country did I see clear evidence of worker involvement in and control over PHC services as these might be expected to relate to workers' health. It appears that workers and their representatives have much to learn in all countries about taking on the largely physician-controlled health services so as to make them more responsive to workers' health and safety interests on the job.

In summary of the sponsorship and control dimension, there is more worker orientation evident in the G.D.R. than in F.R.G. Whether this reflects more actual worker control is uncertain. It appears rather to be an expert-based technocracy ideologically oriented toward workers' interests. In F.R.G. the workers' movement is quite weak, buffetted by high unemployment, anti-union moves, and a conservative government.

Organization

The organizational dimension of the model also reveals striking differences when held up against realities in the two countries.

In the G.D.R., the Ministry of Health sponsors, pays for and supervises the OSH personnel; thus lending an important degree of independence from management and unions. The coverage with one type of OSH services or another (in plant or multi-plant policlinics) is quite complete (though not universal as in Finland). The general health services are fairly fully regionalized and geographically oriented at PHC level to give PHC providers familiarity over the long run with workplaces and their problems, in spite of a lack of (and need for) continuing OSH education for PHC providers. The policlinic in Grossraeschen integrated PHC and OSH services and, except for a need for greater direct worker control and involvement, serves as a veritable model of OSH-PHC linkages.

In F.R.G., the OSH services are highly varied, fractionated and limited. By law, managements must provide OSH services according to certain standards as regards staff-worker ratios, etc. (In U.S.A., OSHA does not even have an OSH service standard). But the actual service may be provided by a GP trained for two weeks in OSH and paying occasional visits to the plant and not more than 50 percent of workers are covered (more in large plants, less in medium and small plants). Because of worries over competition in this largely market-oriented medical care system, the community and hospital physicians have had it enacted into law that the OSH services, whatever their actual form, will be strictly limited to preventive efforts. The OSH-PHC connection is thus seriously undercut and the system does not generally have a geographic assignment of PHC services which might amerliorate the situation in that PHC providers would get to know the plants and problems in their areas.

In summary of the organizational dimension, the system in the G.D.R. is possibly the best organized of all the study countries, even though there remain many problems.

In the F.R.G., multiple sponsorship and control of inspectorates, poorly prepared OSH personnel, legal limitation of OSH services to preventive efforts only, the lack of a regionalized, geographically assigned general health service system lead to an OSH service system which is highly varied, fractionated, and limited with very poorly developed OSH-PHC connections.

Education

The education dimension suggests further points of convergence and divergence between the two Germanies.

Both countries, as is true for several of the other study countries, are moving toward a two-tiered system of OSH specialization in medicine. A longer, more exacting preparation will provide researchers-teachers, and perhaps top OSH policy and inspectorate personnel. The shorter preparation will provide personnel for in-plant or plant-related OSH services. Beyond this broad similarity, there are many differences.

While the preparation of the highly qualified specialist may be similarly demanding in both countries, the preparation of plant physicians may range from two weeks upward in F.R.G. while two years are required in G.D.R.

Education of PHC personnel in OSH is inadequate in both countries. In G.D.R. there is a laudable effort to acquaint all medical students with this sphere of concern by requiring thirty-six hours of OSH in the basic medical curriculum (variously arranged in different schools) and the passing of OSH questions on exams. In F.R.G., no hours are required, but students know they must pass a section of OSH questions on their national boards.

In-service and specialty education of PHC personnel in OSH problems leaves much to be desired in both countries. While some OSH specialists claimed that they give programs for PHC providers in the G.D.R., in one instance where this claim was made, a nearby GP who was personally quite interested in workers' health problems said she never received any invitation to, or announcement of such programs. In F.R.G., a GP in a semi-urban area said the physician's association puts on monthly programs, but he could not recall one devoted to OSH.

Probably the most important aspect of the educational dimension is worker education for OSH. Here the G.D.R. has an extensive program, but it struck me as formalistic and paternalistic and filled with concern for documenting responsibility, etc. (the Arbeitsschutz-kontrollbuch or work hazard education diary, for example). In the F.R.G., there is an extensive worker education program jointly organized by unions and management with cooperation of BAU (a kind of NIOSH agency) but non-organized workers are hardly reached. In both countries, an additional element of worker involvement in and control of the educational process is needed to envigorate and creatively spark this key aspect of the OSH system.

Financing

Differences between the two countries do not appear to be as great on the financing dimension as on other dimensions of the model.

In both countries, OSH services are paid for by the managements (though in G.D.R., salaries for OSH personnel come through the Ministry of Health) and the general health services are almost entirely covered either by taxes (G.D.R.) or employer-employee premiums (F.R.G.). The OSH inspectorates and advisory services are tax supported and run through the Ministry of Health (represented by ZAM, the Institute for Occupational Medicine) in G.D.R. In F.R.G., some are tax supported and under the Ministry of Labor, while the majority of inspectorate and advisory functions are management financed but independently organized through the workplace insurers (Berufsgenossenschaften).

In both countries, the payment for OSH services by management may sometimes involve subtle, informal, but powerful ways of orienting the OSH services toward management interests. This problem is more serious in F.R.G. Also, the payment of OSH personnel in F.R.G. to only provide preventive measures seems unfortunate in terms of failing to offer a complete integrated OSH-PHC service.

In summary of the finance dimension, there are similarities and differences between the two countries, with a growing realization on my part that financing *per se* (at least in terms of sources of funds) may not matter so much if organization, sponsorship and control orient or motivate personnel to serve workers' OSH interests.

Information

The information dimension reflects clear differences between the two countries which I will only briefly highlight. One similarity across all study countries is the responsibility of PHC providers to report work-related illness to the OSH Institute or inspectorate. This is fine in theory, but breaks down when (as is usually the case) PHC personnel are poorly trained and motivated to pick up OSH problems.

The regionalized health system of G.D.R. facilitates a very comprehensive, geographically-oriented, population-based (or epidemiologically defined) information system. It is generally well known what disease problems there are (except for a continuing serious problem that work diseases may not be recognized by PHC providers) in relation to well defined denominators. Especially in the OSH field, the new system of surveying every workplace for hazards and systematically following up exposed workers is ideal.

By contrast, the fractionated system in F.R.G. with management control of information and arguments concerning "trade secrets" leads to an OSH information system which is hard to identify as such.

In summary of the information system dimension, the G.D.R. has a remarkably complete and integrated system, with shortcomings related to the lack of awareness of OSH problems on the parts of PHC personnel (and therefore only partial reporting and follow-up), while the OSH information in F.R.G. must be identified as partial, fractionated, largely management controlled and inadequate.

In general there are too many fascinating and subtle differences in the OSH and OSH-PHC arrangements in the two Germanies to make any overall summary statement very meaningful. However, it is clear that the sharp split after WW II in political economic arrangements in what had previously been one country, led to remarkable differences in the health and OSH systems of the two countries. While there remain a number of things which should be improved in the G.D.R. (e.g. removal of hazards requiring major capital investments, rather than continued exposure of workers with protection offered through continuous follow-up exams; and improved OSH-PHC connections) it seems clear, on balance, that this worker-oriented, even if expert-technocrat-determined system is much more protective of worker health and safety than is true for the profit-oriented capitalist system of the F.R.G.

NOTES AND REFERENCES

1. Craig R. Whitney, German "Economic Miracle" Bypasses Poverty Areas, *New York Times*:3, August 26, 1975.
2. *Materialien Zum Bericht Zur Lage der Nation*, 1974 (Bonn: Bundesministerium fuer innerdeutsche Beziehungen, August 1974): pp. 479–486; also a section of this same annual report (since 1968) giving results of an opinion survey in F.R.G. asking a representative sample of voters to compare various aspects of the two countries, including the health systems (pp. 99–106). For the cited figures see Table 23, p. 100.
3. "Gewerkschaftliche Standpunkte zur Arbeitssicherheit," Frankfurt am Main: I. G. Metall, (April 1981): 19.
4. See for example, Fred Block, "Marxist theories of the state in world-system analysis," in B. H. Kaplan, ed. *Social Change in the Capitalist World Economy*. Beverly Hills, CA: Sage, pp. 27–37, 1978.
5. Some analysts of the OSH scene in U. S. A. have wondered how the incredibly weak labor movement succeeded in having OSHA passed (currently only 18.8 percent of workers are organized). Admitting that OSHA itself leaves much to be desired, it still appears to be a great achievement. I would suggest that some Bismarckian strategy was involved in the Nixon Administration—give them a little to keep them from getting organized and fighting for and winning a lot.
6. Essentially this term means a form of organized labor which accepts, even defends, capitalism and enters into collective bargaining and adversarial relations with owners and management on very narrow grounds—wages and hours and working conditions *at work*. Few demands are made regarding the broader social order and little effort is put into establishing rule through a labor party. More will be said about this in the chapter on U. S. A. See Stanley Aronowitz, *False Promises, The Shaping of American Working Class Consciousness*. New York: McGraw-Hill, 1973.
7. Edelgard Simon, "Prospect of Pink Slips Turns German Bosses Blue," *The Wall Street Journal*: 29, January 18, 1984.
8. With inflation rising faster than wages and salaries, the monthly net buying power per worker sank to 1844 DM in 1981 and was expected to sink further to 1835 DM in 1982 from a high of 1880 DM in 1979 (in 1981

DMs). "Abstieg vom Wohlstandsgipfel," *Hannoversche Allgemeine Zeitung* (11. November 1981): 5. Also see Hans-Ulrich Deppe, "Operation '83: Staatschaushalt und Sozialversicherung in der Wirtschaftskrise," pp. 12–13 in *Medizinische Soziologie*, Jahrbuch 3. Frankfurt am Main and New York: Campus Verlag, 1983 for an excellent, detailed examination of the cutbacks in social and health insurance provisions disproportionately affecting those most in need.

9. "I.G. Metall-Loderer mahnt Vertrauensleute: 'Lasst die Kirche im Dorf'," *Westfaelische Rundschau* (17 November, 1981): 5. "Die steigenden Arbeitslosenzahlen treiben die Gewerkschaften auf die Barrikaden. 'Der Zustand ist unhaltbar', heisst es im Duesseldorfer DGB-Hauptquartier." Werner Muhlbradt, "Gewerkschaften und Unternehmen begehen schwieriges Gelaende." *Hannoversche Allgemeine Zeitung* (11 November 1981): 2.
10. Associated Press "Thousands Rally in Germany for 35-Hour Week," *The Hartford Courant* (May 29, 1984):A8.
11. Peter Gumbel, "German Strikers Strive for Solidarity," *The Wall Street Journal* (May 30, 1984): 37.
12. Ibid.
13. The person quoted is Friedrich Lohss who was in the process of closing his small zinc oxide plant, Hamburg Zink Weiss, and laying off 20 workers. Edelgard Simon, op. cit. (note 7).
14. Peter Gumbel, op. cit. (note 11).
15. Associated Press, " 'Weary' Germans Dissent from Famed Work Ethic," *The Hartford Courant* (May 30, 1984): A10.
16. "Labor, Management Agree to Mediation in West German Strike," *The Wall Street Journal* (June 18, 1984): 2.
17. Associated Press, "Shorter Workweek May End German Metalworker Strike," *The Hartford Courant* (June 28, 1984): A18.
18. The German Social Democratic Party (SPD) lost some of its credibility with the German left and the quite massive Peace Movement by its support, when it was in power, for the NATO "defense" plan and the placement on German soil of U.S.A. missiles with very short warning time (for Moscow and the Eastern Bloc generally) and first-strike capability. Now out of power the SPD has been freer to seek to recapture its credibility and has been vigorously fashioning an alternative defense strategy which differs sharply from NATO's which Chancellor Helmut Kohl's Christian Democrat conservative government supports. "The party (SPD) wants to develop an alternative to continued military confrontation and competition in Central Europe." Dan Charles, "German SPD seeks defense rapport," *In These Times*, (August 7, 1984): 15.
19. Manfred Pflanz, "The Two Faces of the Patient-Doctor Relationship in a Changing Welfare State: The F.R.G.," p. 46. In Eugene Gallagher, ed. *The Doctor-Patient Relationship in the Changing Health Scene*. DHEW Pub. No. (NIH) 78-183 (Washington, D.C.: GPO, 1978).
20. Siegfried Eichhorn, "Health Services in the Federal Republic of Germany," pp. 286–334 in Marshall W. Raffel, ed. *Comparative Health Systems, Descriptive Analyses of Fourteen National Health Systems*. University Park, PA: Pennsylvania State University Press, 1984, quote from p. 307.

21. The facts and analysis of these recent developments are taken essentially from Hans-Ulrich Deppe, "Operation '83: Staatshaushalt und Sozialversicherung in der Wirtschaftskrise," op. cit. (note 8).
22. "Modell einer allgemeinen Vorsorgenuntersuchung," Arbeits- und Sozialministerium Baden-Wuerttemberg, 1970.
23. Deppe, op. cit. (note 8): 15–16.
24. Ibid, p. 16, (translated from the German). Here Deppe cites "Unternehmerverhalten in der Krise und gewerkschaftliche Gegenwehr." Frankfurt am Main: Industriegewerkschaft Metall (I.G. Metall) 1983, especially from p. 64 on: "Jagd auf Kranke–Aussonderung von Arbeitnehmern," (Persecution of the sick–the elimination of workers). This strategy of course is not new. See William Ryan, *Blaming the Victim*, revised edition. New York: Vintage Books, 1976.
25. Paul U. Unschuld, "The Issue of Structured Coexistence of Scientific and Alternative Medical Systems: A Comparison of East and West German Legislation," *Social Science and Medicine* 14 B (February 1980): 15–24.
26. Ray H. Elling, "The Fiscal Crisis of the State and State Financing of Health Care," *Social Science and Medicine*. 15C (1981): 207–217.
27. James O'Connor, *The Fiscal Crisis of the State*. New York: St Martin's Press, 1978.
28. Of course other conditions are entangled with what has been called "the capitalist work crisis." The revolt of peripheral and semiperipheral peoples and the small likelihood that the more than $700 billion owed to Western banks by these countries will be paid back are part of it. For a review of Frank's and other recent work on the World Crisis, see Gregory Shank, "The World Economic Crisis According to Andre Gunder Frank," *Contemporary Marxism*. No. 5 (Synthesis Publications, San Francisco) (1982): 147–153. Also Andre Gunder Frank, *Reflections on the World Economic Crisis*. New York: Monthly Review Press, 1981.
29. Eichhorn op. cit. (note 20) pp. 289–290.
30. Ray H. Elling *Cross-National Study of Health Systems, Political Economies and Health Care*. New Brunswick, NJ: Transaction Books, 1980, esp. pp. 100–101.
31. Robert Maxwell, *Health Care, The Growing Dilemma: Needs Versus Resources in Western Europe, the U.S., and the U.S.S.R.* New York: McKinsey Associates, 1974.
32. "Gewerkschaftliche Standpunkte. . ." (note 3 above).
33. It is not an indication of the source of this quote that the spirit of committed support to the working class has been shown by a number of German physicians from Virchow and others of his time on. See Kurt Kuhn, *Aerzte an der Seite der Arbeiterklasse, Beitraege zur Geschichte des Buendnisses der deutschen Arbeiterklasse mit der Medizinischen Intelligenz*. Berlin: VEB Verlag Volk und Gesundheit, 1973. However, it is sad to note that as far as I can learn there is no COSH-type group in F.R.G. as there is in Denmark, Finland, U.K. and U.S.A. There are local groups of faculty and students working with union people, including giving presentations, etc., but no formal organization has evolved. I doubt that this is for the same

reason there appears to be no such group now in Sweden (a group named SARB used to exist there) namely, because the unions have been so strong and successful in getting OSH measures adopted and in gaining control over OSH conditions that there is less need for COSH groups.

34. Hans-Ulrich Deppe, *Industriearbeit und Medizin, zur Soziologie medisinischer Institutionen am Beispiel des Werksaerztlichen Dienstes in der BRD.* Frankfurt am Main: Fischer Athenaeum Taschenbuecher, 1973, p. 17.
35. Ibid, p. 12. The latest official figures for 1980 were issued in October, 1981 by BAU (see below). These show reported work accidents down to 2.1 million in 1980 from 2.7 million in 1970. Deaths from work accidents *and diseases* (the problems of diagnosing and reporting and following up on work-related diseases are part of what this study is all about, especially when PHC facilities are involved, thus these figures must be taken with a grain of salt) were down from 4,275 in 1977 to 3,998 in 1980. The suggestion is that these improvements are due to the training and placement of large numbers of OSH personnel and other OSH laws and measures in the last decade. For a critical analysis and rejection of such figures and claims and a suggestion that the figures may reflect more a shifting work pattern from heavy dangerous work to lighter technically supported work and other factors, see the penetrating article on the F.R.G.'s OSH system: Hans-Ulrich Deppe, "Work, Disease, and Occupational Medicine in the F.R.G.," *International Journal of Health Services* 11: 191–205, 1981.
36. This list is taken largely from "Gewerkschaftliche Standpunkte. . ." op. cit. (note 3) pp. 21–23. But see also Rolff Wagner, "Die Arbeitsmedizin in der BRD–Stand 1979," *Arbeitsmedizin Aktuell.* Stuttgart: Gustav Fischer Verlag, Lieferung 3 (November 1979): I–29.
37. Hauss and Rosenbrock note that the co-determination law in West Germany tends to co-opt the worker in what is called "cooperative conflict" and leaves OSH almost exclusively in the hands of experts. "Consequently the rights of the individual worker or of a group of workers are less developed in F.R.G. than in comparable countries." Friedrich O. Hauss and Rolf D. Rosenbrock, Occupational Health and Safety in the Federal Republic of Germany: A Case Study of Co-Determination and Health Politics." *International Journal of Health Services* 14: 279–287, 1984. Even in strong union shops there is considerable potential for cooptation. Writing the praise of the *Mitbestimmung* law as regards labor peace and productivity one management-oriented observer writes:

> *Mitbestimmung* was given its greatest test during the economic downturn of the mid-seventies. At Volkswagen, management wanted to cut back the labor force and shift one-third of production to the United States. Despite evidence from its own economists that car sales would soon improve, the union reluctantly agreed to the eventual dismissal of 25,000 workers, which could have ultimately cost the company as much as $100 million in severance payments. When demand for cars soon picked up and the company was forced to rehire all the laid-off workers, vindicating the union's original

> advice, labor representatives on the board gained new stature. The decision to build Volkswagen Rabbits in the U.S. was handled with equal cooperation. The union wrested from management some concession on investments to maintain Volkswagen employment in Germany and then bowed to the realities of the market, which dictated shifting production abroad. It is hard to believe that such difficult adjustments would have been made peacefully without the full participation of labor representatives on the supervisory board.

Bruce Stokes, "Answered Prayers," *Master in Business Administration* (MBA) (December 1978–January 1979): 12–23, quote from p. 21. Here Stokes cites Alfred Thimm, "Decision Making at Volkswagen," *Columbia J. World Business* (Spring 1976) and Robert Ball, "The Hard Hats in Europes Boardrooms," *Fortune* (June 1976).

38. Hans-Ulrich Deppe, "Von 'Humanisierung der Arbeitswelt' kann noch keine Rede Sein," *Frankfurter Rundschau* Nr. 229 (3 Oktober 1981): 14–15.
39. There are many sources for this account, including my field notes, but one clear (even if uncritical) overview in German, English, and French deserves special mention: Alfred Mertens, "Occupational safety and health in Federal Republic of Germany" Dortmund: BAU, 1979 (reprinted from "Sicher ist sicher"). The 10 states or Laender plus West Berlin also pass laws and regulations in this field as they do in education and health and other areas, in keeping with the overall federated structure. They do this within the national framework. Thus, for example, there are a number of different patterns for organizing the regional industrial inspection by OSH physicians (Landesgewerbeaerzte). Wagner, op. cit. (note 36) diagram p. 4. This further complicates an already complex and divided system, though the element of local initiative may have some value.
40. "Berufsgenossenschaften" pp. 165–167 in *Woerterbuch zur Humanisierung der Arbeit*. Bremerhaven: Bundesanstalt fuer Arbeitsschutz, 1983.
41. Mertens, op. cit. (note 39): 9.
42. H. Valentin, et al. *Arbeitsmedizin*. Stuttgart: Georg Thieme, 1971. I do not know if a 1979 two volume revision of this work offers the same "blame the victim" ideology. In any case, such a figure is a complete fiction, since the detailed generally validated scientific work on indicators of responsibility for accidents has not been done.
43. The head of OSH for a large union spoke with some pride of the number of pamphlets, posters, courses, etc. they have produced. Indeed a lot has been accomplished. But the union does not have its own centers or courses. Rather, these are organized through the Berufsgenossenschaften or BAU and are jointly supported with owners. A certain naivete or grateful desperateness in the conversation made me uneasy. It was clear that the unions do not have power and resources to control the situation and offer their own independent worker education for OSH. Anyone involved in this

field will realize that the slant placed on problems (for example, use of personal protective equipment versus engineering controls which might require major capital outlays) will differ according to sponsorship and control of the education.

44. The job sounded important, even exotic and exciting, but I doubt it is very effective. Later an OSH epidemiologist doing research on dioxin (a component of Agent Orange which the U.S.A. Armed Forces used in Vietnam) told me this firm has done practically nothing with the volumes of data it has on workers exposed to different chemicals. This researcher, by the way, has had a recent shock to his scientific idealism and integrity. He has been an advisor to another major chemical firm which had a large cohort of workers exposed to dioxin. This researcher's work shows a higher than expected rate of cancer and other unfortunate effects. The company has censored his results, telling him he can only publish the (from their view) more favorable findings. He was dismayed and planning to resign as advisor to this company when I talked with him.
45. Whether this situation is related or not I can not say, but the proportion of handicapped persons among the unemployed has been growing rapidly in F.R.G. At the end of 1980 there were 82,700 seriously handicapped persons or more than 8 percent among the almost 1 million unemployed. At the end of October, 1981 there were 93,800 seriously handicapped among the unemployed. "Unter den Arbeitslosen immer mehr Behinderte," *Hannoversche Allgemeine Zeitung* 11 (November 1981).
46. As we will see, coverage is actually quite poor because the bulk of workplaces are served by general practitioners with little OSH training and interest. Thus this law cannot be seen as having the same meaning as the law for universal OSH coverage in Finland (Chapter 8) or the system of OSH services structured by the Ministry of Health in G.D.R. (Chapter 9).
47. *Arbeitsmedizin und betriebsaerztliche Taetigkeit.* Bericht ueber eine Arbeitstagung der I.G. Metall Gelsenkirchen, 31.3–1.4.1977. No. 20 in the I.G. Metall Series on Work Safety. Frankfurt am Main, 1977, quote from Alfred Fischer, pp. 80–81.
48. Private communication April 30, 1984, taken from an unpublished manuscript of Professor Hans-Ulrich Deppe.
49. Note that racism still exists leaving workers divided. For example, the work-injury rate is 2.5 times higher among foreign workers (the "Gastarbeiter" from Turkey and other low status southern countries) than among native German workers, a continuing form of racism in F.R.G. Deppe, op. cit. (note 35): 200. But the recent show of solidarity in winning the strike for a shorter work week with the same pay in order to create more jobs is encouraging.
50. Deppe, op. cit. (note 35): 203–204.
51. It is important to remind ourselves that it is the rate of profit and not just a profit which is in focus in capitalist enterprises. In the film "Controlling Interest," a large corporation with headquarters in New Jersey, threatens the workers and townspeople in Greenfield Massachusetts with the

closing of their single largest plant and source of employment. This plant has made a profit for 100 years. But the rate of profit for the same capital investment could be higher in South Carolina, or Taiwan or Venezuela.

52. Sometimes this world-system exposure takes on bizarre forms. Sitting with my guide in the bar of my rather swank hotel in East Berlin one evening, after a day of OSH clinic visits, etc., I was told that the alluringly dressed women at a nearby table were members of the world's oldest profession. I suggested that they might work for the secret service and be a good source of information. This possibility was not denied but my guide went on to claim that western business demand this kind of "service." He further stated that this service had been recently introduced at the world famous annual Leipzig industrial fair at the demand of western sales representatives. One can have a fieldday of intellectualizing and moralizing about the womens' movement and workers' rights under "socialism." I am only reporting what I saw and heard and interpreting it in terms of continuing exposure of the system to the morality and modes of competition in the capitalist world-system.

53. It has an effect in other realms as well. In the West, firms attempt to "externalize" as much of social welfare and other costs as possible, placing responsibility on tax payers and attempting to keep these outlays as low as possible: James O'Connor, *The Fiscal Crisis of the State*. New York: St. Martin's Press. 1978. In G.D.R., firms, especially large ones, are forced by law and general ideology and policy to assume much of the social welfare function. Thus, alcoholic or otherwise inefficient workers are difficult to fire and are put into maintenance programs long beyond what would be likely in the West.

CHAPTER 11
The United Kingdom

THE LAND AND THE PEOPLE[1]

The official name of this island kingdom, lying northward across the English Channel from the coast of continental Europe, is the "United Kingdom of Great Britain and Northern Ireland," here abbreviated U.K. On the east it is separated from Scandinavia by the North Sea. To the west lies Ireland, and its own part of the Irish island, that is Northern Ireland. To the southwest and west of Ireland lies the Atlantic Ocean which carried British explorers, and colonizers, and later dominant British sea power, to the farthest reaches of the Globe. Small islands near the main island are all British–the Hebrides, Shetland and Orkney groups off Scotland; and the Isle of Wight off England. The Channel Islands in the English Channel and the Isle of Man in the Irish Sea are British dependencies. This all makes up an area of 94,220 square miles, a bit over half the size of Sweden, with seven times the population in 1980 (56,010,000 in the U.K. versus 8,310,000 in Sweden). The main island is made up of historically and culturally distinguishable areas. Some have even spawned vigorous independence movements, though none successful thus far in breaking away from English domination as did the Irish after hundreds of years of struggle–still going on in its own unique set of class and religious conflicts in Northern Ireland. In the south and east is England (it is a mistake to refer to the U.K. as a whole as England, though this is sometimes done); in the west, Wales; and Scotland in the North.

In spite of its comparatively small size, a complex geological structure gives the British Isles a wonderfully varied landscape. In the Scottish Highlands to the north is found the highest peak, Ben Nevis, rising to 4,406 feet, about the height of some of the mountains in Vermont and New Hampshire in U.S.A. This highland chain of the main island arches southward through the Cotswold hills in western England, to the Cambrian Mountains in southwestern Wales.

Most of England is made up of low plains; those to the south are called the Downs and those to the east the Fens. The only lowland outside of England is in central Scotland running west to east between the Highlands in the north and the Southern Uplands. The most important rivers are the Thames exiting to the sea through London, the capital, lying to the east and south of England; the Severn which is joined by the Avon to empty into the Bristol Channel separating the long southwest arm of England from Wales to the north; and the Tweed and Tyne in Scotland.

The islands are warmed by the Gulf Stream. Thus the climate is temperate with winters mild and summers cool. The average temperature in winter is 40°F, in summer 60°F. The reputation is for cloudy, damp weather to prevail. The British are always shown carrying their umbrellas. Rainfall is moderate to heavy, with significant variation by area. It averages 120 inches per year on the west coast, but only 20 inches in the southeast. The average for the country as a whole is 40 inches.

The large cities, all heavily industrialized, are London, with some 12 million population in its metropolitan region; Birmingham in middle England; Liverpool, Manchester, Sheffield and Leeds in the north of England; and Glasgow and Edinburgh in the western and eastern ends of the Scottish plains respectively. But there are many other sizable cities and towns. In 1980, 91 percent of the population lived in urban areas.[2] On average, there were 230 people per square kilometer of surface area compared to 14 for Finland, 18 for Sweden, 24 for U.S.A., 155 for East Germany and 247 for West Germany.[3] The bulk of the population live in England, particularly the southeast, with slightly sparser settlement toward the west and north.

Early migrations across the channel peopled the British Isles with Celts. England was made a part of the Roman Empire in 43 A.D. Almost 400 years later, when the Roman legions withdrew, waves of Angles, Saxons and Jutes arrived. Later these peoples contended with Viking invaders. To describe one of his characters, a lower member of the still extant English aristocracy, a modern novelist says his face is

> . . . an example of that cross between the Saxon peasant and the Viking that has become over the centuries one of the well-bred English faces; Viking in the raids on convention, Saxon in the fundamental placidity and contentment. It was clear that, as in history, the Saxon had tamed the Viking.[4]

The last successful invasion was that of the Normans in 1066. They joined the country with their dominions in France. Although there is a present-day royalty and peerage, the United Kingdom prides itself on a strong democratic tradition. Resistance to royal authority forced King John to sign the Magna Carta in 1215. This guaranteed certain rights and the rule of law. The fable of

Robin Hood stems from this time and continues as an oft-repeated folk story—taking from the unjust rich to give to the deserving poor. Other parts of this tradition include the rule of common law and Parliament's struggle with the Stuart kings resulting in the Civil War of 1642-1649 and Cromwell's republic. Although the monarchy was restored in 1660, the peaceful but effective bourgeois revolt of 1688 confirmed Parliament's power and brought a Bill of Rights in 1689.

English is the universal language, though the Celtic languages of Welsh and Gaelic are spoken by some in the north and west. Anglican (Church of England) is the dominant religion, but other Protestant groups are important—particularly the Methodists, in terms of labor history—as are the Roman Catholics, particularly in religiously and class-divided Northern Ireland.

THE POLITICAL ECONOMY AND WORKERS' MOVEMENT

The U.K. is a constitutional monarchy, similar to Sweden in this regard. The King or Queen is the formal head of state (I do not present the long complex series of British monarchs). Actual executive power is wielded by a prime minister and cabinet responsible to the Parliament in which a majority party or coalition of parties supports their prime minister until a no confidence vote or an election which brings a change in the balance of parties' powers. Parliament has two houses—a popularly elected lower house, the House of Commons, and the (today) less powerful, hereditary and appointive House of Lords.

England is the oldest industrial capitalist nation in the world. While Capitalism began in Catholic Portugal and spread to the Low Lands,[5] it was in England where surplus labor from enclosed farm lands moved into urban areas and where capital and technology were combined with tool-less labor to develop the first factory forms of production. The Stock Exchange was founded in 1773. Of course the complex of factors involved was much more complicated, including the core's essentially extractive relations to a periphery found both in rural Britain and in a spreading overseas colonial network.[6] This network grew into the largest colonial empire in the world just before World War I.

Consonant with this political economic development was a vast and deep cultural development which can only be alluded to. London remains today among the two or three world cities. If one wanted to do research on the Egyptian pyramids, the best place to begin would be the British Museum. The Rosetta Stone, found in 1799 by one of Napoleon's soldiers during the campaign in Egypt, now rests in the British Museum. With its three versions of the same text—one in Greek, one in formal Egyptian script, and one in hieroglyphics—it provided the clues which finally allowed scholars to decipher ancient texts in hieroglyphics.[7] To mention the works of William Shakespeare (1564–1616), George Frederick Handel (1685–1759), Charles Robert Darwin (1809–1882), George Bernard Shaw (1856–1950) and Bertrand Arthur Russell (1872–1970) is to pick only gem-like highlights from an incredibly rich cultural production in literature, music, science, and philosophy. It would be presumptuous to attempt any kind of summary.

But the conditions which early industrial capitalism created for workers and their families were anything but idyllic:

> Such is the condition of the English manufacturing proletariat. In all directions, whithersoever we may turn, we find want or disease permanent or temporary, and demoralization arising from the condition of the workers; in all directions slow but sure undermining, and final destruction of the human being physically as well as mentally. Is it a state of things which can last? It cannot and will not last. The workers, the great majority of the nation, will not endure it.[8]

The workers would not endure such conditions. Thus the history of workers' struggles in the U.K. is also the longest among the six study countries.

Engels recognized class struggle as the principle dynamic force in socio-economic development. At first, the working class revolt was seen in the impoverished workers' individualistic actions to keep his body together, perhaps not body and soul, since his actions were defined by the establishment as "sinful" and "illegal." As Engels observed, the worker's "want conquered inherited respect for the sacredness of property, and he stole."[9] The bourgeoisie responded with police repression; it "defends its interest with all the power placed at its disposal by wealth and the might of the State."[10] But the impoverished masses grew and thefts grew. "It is thus no coincidence that the aim of early personal health services and general welfare legislation was the prevention of crime, or more appropriately, the prevention of the increasing disrespect for property."[11] These early measures included the Elizabethan Poor Laws and the important Poor Law Reform Bill of 1834. These measures represented more a supplement to the repressive responses of an uneasy and privileged elite to the potential outburst of an unorganized yet massive rabble, than they did a response to a well organized revolutionary working class.[12] This last has not really developed to this day in the U.K.

It is worth pausing here to observe at some length that there is disagreement as to the dynamics of these developments. Liberal as well as conservative analysts who seek to preserve the capitalist system, however much they seek to advance the working person's lot *within a capitalist structure*, see these advances and subsequent ones leading eventually to the first National Health Service in a capitalist state in 1948, as stemming from moral reform and an enlightened bourgeoisie. Even Shinwell falls into this view from time to time:

> Such social progress as there was came, not from the agitation of the sufferers, but from aristocrats inspired by sentiment and idealism—men like Lord Shaftesbury who championed the expansion of the Ragged Schools for the workers' children and the Ten Hours Bill for their parents.[13]

Hill[14] by contrast cites the following passage from Piven and Cloward for the U.S.A. experience:

> . . . relief arrangements are initiated or expanded during the occasional outbreaks of civil disorder produced by mass unemployment, and are then abolished or contracted when political stability is restored . . . expansive relief policies are designed to mute civil disorder, and restrictive ones to reinforce work norms.[15]

Further, for the U.K. experience, Hill says:

> Piven and Cloward are Americans and much of their book is con-concerned with the discussion of social policies in the United States, but similar views to theirs have been expounded in polemic form in Britain, by Jordan in his book *Paupers* and by Kincaid in *Poverty and Social Security in Britain.* But the most scholarly treatment of this theme with regard to Britain occurs in Gilbert's book *British Social Policy 1914-1939.* By drawing on Cabinet minutes and Secret Service reports that have now been made public, Gilbert shows how the extraordinarily erratic treatment of the provisions for unemployment insurance in the early years after the First World War were heavily influenced by reports of dangers of civil unrest. . . .
>
> The importance of Piven and Cloward's book is in providing an attack on the naive view that advances in social security provision depend simply upon a growth of humanitarianism, a growth of a "caring" Welfare State. Where Piven and Cloward are at their weakest is in treating as broadly similar forces towards change, social unrest perceived as a danger by ruling élites and social discontent reflected in electoral movements. The fact that a long line of Conservative and Liberal politicians—Disraeli, the Chamberlains, Balfour, Lloyd George, Churchill and MacMillan—were eager to "steal the clothes" of radical and socialist movements should not be treated as an élite response comparable to the nervous concessions of the founders of the Elizabethan Poor Law or the politicians of the Napoleonic period. It should be counted as one of the successes of the British Left that it has been so easily able to get conservative politicians to make concessions to the working class. Of course, "welfare capitalism" has come about partly because the capitalists have gradually and reluctantly made concessions to the working classes, for fear of worse alternatives. But it is equally the boast of democratic socialists that this is the road to social transformation which they have chosen. Peaceful change in our society has, moreover, also been facilitated by the extent to which more egalitarian

> ideas, and conceptions of the role of the State as an active protector of its citizens, have entered into the general philosophy of the age.[16]

This is a social democratic or conservative socialist view of social development accepting class struggle as "the motor of history," albeit with a somewhat softened view of the role of an enlightened bourgeoisie and the state. A firmer Marxist view would see the state as fundamentally protecting the class interests of the bourgeoisie, in part by integrating and controlling the working class through a variety of social welfare measures.[17] The phrase "general philosophy of the age" might be translated into the conception of "a cultural hegemony" which I have discussed in Chapter 5 as derived from Gramsci.[18]

This is not the place for a detailed recounting of the long struggle of the workers' movement in the U.K. Among other things it led to the adoption of the first National Health System in a capitalist country—this following World War II. But to achieve such developments the struggle had to evolve into more organized forms with the trade unions and Labour Party leading the way.

> Crime, however, was a *singular* action with little political potential. Unified class movements were much more potent means. In 1824, Parliament passed a bill enabling the working class the right of free association, making such unified movements a legitimate tool. Labor associations soon grew up in all industrial areas in England. The associations soon became trade unions, although they were in the beginning of only limited effectiveness against the powers of the bourgeoisie. Strikes, violence against capital and non-union (scabb) labor soon commenced, only to be outdone by the extreme force of capital in breaking the strikes. If nothing else, the struggle enlightened the workers, and a number of different factions, from the moderate to the radical, developed, all with the general aim of creating a republic with extensive democratic rights given to all classes. Their aim was not merely political equality, but also economic equality. Those advanced beyond chartism became socialists, and there thus existed three movements of the working class: (a) the trade unionists that were chiefly concerned with the material aspects of .work, (b) the chartists who worked for complete political franchisement, and (c) the socialists who wanted the abolition of class distinctions between bourgeoisie and the proletariat. To Engels, the future of the working class rested with the success of the socialists as they understood most of the bourgeois trappings—religion and education. Religion they abandoned and education they provided on their own terms through the development of "socialist institutes."[19]

But a socialist victory has not been realized to this day. The indicated divisions within the labor movement and other divisions within the working class, plus the responses of the ruling class, on the one hand several vicious and severe repressive moves, on the other hand, the development of ameliorative social legislation, plus the vast colonial network allowing Britain, as the then center of the capitalist world-system, to extract a surplus from even more exploited worker and peasant classes in other countries, kept the lid on, so to speak.

One of the long continuing working class divisions has been between the crafts workers who often did own their own tools and other workers who owned nothing outside of their own labor. In 1819 weavers in Carlisle, many of them women and children, were working 14–17 hours a day, six days a week for five to seven shillings. The craftsmen could not afford to let this plight of their sisters and brothers bother them:

> The conditions in the textile industry and in the mines—probably the two worst of all major industries—did not unduly worry the workers in the craft industries still unaffected by the machine. The skilled artisans—carpenters, metal workers, and some kinds of farm workers—were in whole-hearted agreement with the guilds of tailors, goldsmiths, shipwrights and other crafts that the labourers in the machine-powered factories were of a distinctly lower class—as indeed they were, when employers came to rely more and more on children, women, and men reduced to near-bestiality by poverty.[20]

Another dividing force was found in the nature of early capitalist development itself. Often the small enterprises of the time depended upon incentive pay systems which put one laborer in charge as a contractor for the work of others:

> Capitalism in its early stages expands, and to some extent operates, not so much by directly subordinating large bodies of workers to employers, but by subcontracting exploitation and management. The characteristic structure of an archaic industry such as that of Britain in the early nineteenth century is one in which all grades except the lowest labourers contain men or women who have some sort of "profit-incentive." Thus the engineering employer might subcontract the building of a locomotive to a "piece-master" who would employ and pay his own craftsmen out of the price; and these in turn would employ and pay their own labourers. The employer might also hire and pay foremen, who in turn would hire, and have a financial interest in paying such labour as did not work on subcontract. Such a labyrinth of interlocking subcontracts had certain advantages. It enabled small-scale enterprise to expand operations without raising unmanageably-great masses of circulating capital, it provided "incentives" to all groups of workers worth humouring, and it enabled industry to meet sharp fluctuations in demand without having to carry a permanent burden of overhead expenditure. (For this reason varieties of subcontracting are still widely used in industries with great fluctuations of demand, such as the clothing trade, and in primitive industries undertaking rapid expansion, such as the house-building boom of the 1930s.) On the other hand it has disadvantages, which have caused developed large-scale capitalism to abandon it for direct employment of all grades, and the provision of

> "incentives" by various forms of payment by result. Historically it may be regarded as a transitional stage in the development of capitalist management, just as the buying and selling of civil service posts and the hiring of armies by subcontract in the sixteenth and seventeenth centuries may be regarded as a transitional stage in the development of modern bureaucracies and military forces. I propose to call this phenomenon "co-exploitation," insofar as it made many members of the labour aristocracy into co-employers of their mates, and their unskilled workers.[21]

Some of the more violent events of the struggles included riots over the price of food from 1800 on and the anti-machine riots which began in the hosiery factories of Nottingham. Violent actions spread in the second decade of the nineteenth century in the form of the developing Luddite movement to the textile mills in Lancashire and Yorkshire, the shipyards of Sunderland, the iron foundries of Wolverhampton, and even to farm workers in East Anglia.[22] These and other acts were met with police and sometimes military force. The war against Napoleon's forces provided a convenient excuse to curb free speech and suppress undesirable political activities. After the peace of 1815, when protest marches on Parliament threatened, "A Secret Committee of the House of Lords convinced both Houses [the House of Commoners was the second, popularly elected through a system of limited, propertied suffrage] that the country was on the brink of revolution and various repressive measures, including the suspension of *Habeas corpus*, were put in force."[23] Widespread Methodist prayer meetings which gradually evolved into protest and political meetings were more difficult for the bourgeois authorities to combat. These evolved around 1835, along with remnants of the National Union (see below) and the Radical Association, into the Chartist movement whose somewhat vague goals were universal suffrage, annual parliaments, and vote by secret ballot. This movement became a haven for a variety of radicals bent on revolution without a clear plan for taking working class concerns into account. The strategy seems to have been to first change Parliament, then put the people's welfare right when visionaries were making the laws. During the nervous time when the revolutions of 1848 had culminated in Louis Philippe's abdicating the throne in France, the famous Chartist petition with a half million signatures (the organizers claimed 5 million) was to be carried to Parliament by a massive march of a hundred thousand.

> . . . the crowd never consisted of more than 25,000 even before the police and Wellington's troops broke it up.
>
> Soon afterwards, agents tipped off the police that the Chartist leaders were meeting at the Angel tavern, Blackfriars, allegedly for the purpose of planning a second fire of London. They were arrested and the Chartist movement was finished."[24]

The first union of any consequence, the National Union of Working Classes, grew out of the enlightened management-sponsored efforts of Robert Owen for worker participation in ownership, the so called, co-partnership theory. Owen had shown in his own cotton mill at Charlton that beyond an optimum working day neither the owner nor the worker benefited. His efforts led, with support of the cotton workers to passage through Parliament of a rather flawed version of his proposed factory legislation, the Factory Act of 1819. The National Union may have reached some 40,000 in membership. But it developed into a militant organization.

> When it plastered London with handbills calling for a national convention to influence Parliament the authorities had to act. The result was the so-called Battle of Cold Bath Fields in the neighbourhood of Gray's Inn Road when a lot of people got cracked heads and a policeman was killed. It was a futile clash with authority and sounded the death knell of what was the first workers' organisation of any national importance.[25]

There were more general repressive moves such as the Combination Act of 1825. For reasons of internal division and various forms of bourgeois repression the workers' movement was a long time getting effectively organized. Thus even though the workers' struggle was a long one in the U.K. the union movement as such did not begin to grow significantly until around 1850. One of the early powerful unions was the Amalgamated Society of Engineers formed in 1851. "It had full-time personnel, large subscriptions to maintain its highly efficient organization, and benefits in the way of sick pay and so forth."[26] The first important umbrella organization, the London Trades Council, came into being after unions in the building trades had shown their solidarity with each other in the strike of 1859. In 1868 similar organizations in Manchester and Salford demanded a national Trades Union Congress. The newborn TUC, which today is the most powerful labor organization in the U.K., could then speak for only 150,000 workers, "but they were in many trades and worked in every town in the country."[27]

The union movement formed with relatively narrow interests, without the close links with a socialist political party from the beginning as in Sweden (even if the whole development occurred later there) for it was only with the Acts of 1884-85 that universal adult male suffrage was achieved. Nevertheless, with an important degree of leadership from the socialist left,[28] and a latent political potential of immense proportions, realized by the ruling class, if not by unionists themselves, the impact of British labor would be felt far beyond the workplace itself. But this broad influence was really only possible after universal suffrage and, perhaps more important, the New Unionism, that is, organization by unskilled factory workers who saw the benefits being realized by the skilled trades unions affiliated in the TUC.

Table 11. Membership in Certain U.K. Unions 1892-1912 (in thousands); Average Annual Members Over Three- and Two-Year Periods

	1892–4	*1895–7*	*1898–1900*	*1901–3*	*1904–6*	*1907–9*	*1911–12*
All Unions	1,555	1,614	1,895	2,010	2,058	2,492	3,277
'General'[a]	76	69	88	81	67	74	186
'All-grade'[b]	19	22	29	34	45	60	109
'Local crafts'[c]	12	12	12	12	11	12	24

[a]London Dockers, London Gas-workers, Birmingham Gasworkers, NAUL, National Amalgamated Labourers' Union.

[b]London Carmen, Amalgamated Carters, Amalgamated Tram and Vehicle Workers Municipal Employees, Liverpool Dockers.

[c]London Stevedores, Thames Watermen, Cardiff Trimmers, Mersey Quay and Rail Carters, Winsford Saltmakers.

Source: E. J. Hobsbawm, *Labouring Men, Studies in the History of Labour*. New York: Basic Books, 1964, p. 180.

As seen in Table 11, the most successful of the "New Unions" were the "general societies" which recruited from different industries as well as "all grades"–the Dockers, Gasworkers, Tyneside Labour Union and a number of others, including the National Federation of Women Workers formed in 1906. "Most of these have since merged to form the two giant unions of Transport and General Workers and General and Municipal Workers which today include something like a quarter of the total British trade union membership."[29]

Table 11:

> . . . gives a brief comparative picture of the fortunes of various types of "new" unions (all composed of conventionally "unskilled" men) between 1892 and 1912. It is clear that we have here three patterns: the "craft" societies with their stable (and restricted) membership; the General Unions, fluctuating, but without any marked upward tendency, yet climbing almost vertically after 1911; and the "industrial" or compound unions, growing steadily from 1900, though rather faster after 1911. In, say, 1910 it might have seemed that the second group was destined, if not to replace, then increasingly to overshadow the third. But in fact, the opposite has happened. Both groups (a) and (b) have merged to form the two vast general unions of to-day. We can thus distinguish three phases in the development of the general unions: the expansion of 1889–92, the relative decline in their importance between 1892 and 1910, and their renewed, and as it turned out, permanent, expansion after 1911. Each of these phases developed its peculiar forms of organization and policy.[30]

The ups and downs of the economy seriously affected the growth and strength of the unions. A number collapsed badly in the depression of the early 1890s and did not recover until the renewed expansion of 1911–1914. Also, it was not until 1906 in the age of "Robber Barons," when capitalist power was without accountability, that Labour as a party succeeded in having its first members elected to Parliament–twenty-nine "cloth-capped, rather unimaginative men who were without real political accumen."[31]

Generally, it was a reformist party and labor movement.[32] During the Great Depression, and the rise of fascism in Germany and Italy, for example, even after issuing a warning when Hitler made early victims of the trade unions and the German Social Democratic Party in 1933, the TUC and the Labour Party refused to join the United Front of Communists and other left forces devoted to fighting fascism.

But following the loss of Empire and vast destruction and misery of WW II, when the Labour Party first took power in July 1945, there was a swelling tide of democratic demands. On this tide, the party pushed through a sweeping program of reform before relinquishing power to the Conservatives in 1951. This program greatly expanded insurance for unemployment and worker disability, created the National Health Service (NHS) which provided relatively cost-free medical care to every resident (even non-citizen visitors) and it nationalized the Bank of England, the coal industry, and the railroads, among others.

The Labour Party once again assumed power in 1964. During this period anticolonial liberation movements and the U.K.'s loss of dominance in the world-system were realistically confronted. Tight controls were placed on wage increases as well as prices, private spending abroad was curtailed, overseas defense expenditures were shortened, the steel industry was added to the list of nationalized industries, and the pace of granting formal independence to the remaining British colonies and dependencies in Asia, Africa and the Caribbean was quickened. The once proud lion of world empire was fast turning into an ordinary European cat. On several attempts seeking entry to the European Economic Community (the Common Market formed in 1957) Britain was refused. The currency was again devalued in 1967 and further austerity measures adopted leading to Labour's loss of the by-elections of 1968. The party held shakily on to power until 1970. The Torys then had four years in power, with Labour coming back until the Conservative victory in May 1979 brought Margaret Thatcher, "The Iron Lady," to power as Prime Minister.

It has been a long struggle, as noted, the longest workers' struggle among the six study countries. Certain strengths come from such longevity, for example, a level of consciousness and collective identity which can not be matched by younger movements as in U.S.A. But all in all the length of the struggle does not mean it has achieved complete success. Nor does a long struggle necessarily mean a strong one. In fact, the workers' movement must be judged intermediary in strength among study countries. Certainly it is stronger than those in the F.R.G.

and the U.S.A., but it is not so encompassing or solidary as that in Sweden or perhaps even that of Finland. Although there have been Labour Governments, following WW II, the Conservatives have been in power in recent years, and unions have succeeded in organizing no more than 55 percent of the work force. Those who are in unions are not as solidly joined with one another or with the political party as is the case in Sweden. There are more similar unions competing over similar work groups for membership and the TUC shows some of the same sharp divisions which plague the Labor Party (now being partially dismantled by the Liberals who fear the growing strength of the Left).[33] There is also a problem of democracy in the trade unions, though the current government overdraws this problem to further weaken the unions.[34] Indeed the attack on organized labor by the Thatcher regime has been severe and broad enough to recall the repressive history of the 1850s as a recent case showed:

> Prime Minister Margaret Thatcher has struck a new blow, perhaps her toughest yet, in her conflict with British organized labor. She has prohibited 5,000 Britons from belonging to unions at a state intelligence-gathering headquarters.
>
> The Conservative prime minister says she is motivated by national security interests in acting to end union membership at General Communications Headquarters, a top-secret listening post at Cheltenham, southwest England.
>
> The ban on membership in any of seven civil service and industrial unions there would take effect March 1, and has inspired a storm of protest from trade unions and the left wing in general. The Labor Party, the chief opposition to the governing Conservative Party, is largely funded by labor unions.
>
> The government cited three occasions when union-sanctioned disruptions at the center occurred at times of international crisis.
>
> But some union officials claim Thatcher's move may be part of a broader drive to outlaw strikes in other essential services.
>
> "The government's decision is a direct attack on the right to trade union membership and is but the thin wedge as far as other workers are concerned," read a statement by the Council of Civil Service Unions.
>
> The left has also suggested that Thatcher acted under pressure from the U.S. administration, something the government denies. Washington was already worried by the discovery last year of a Soviet mole, Geoffrey Prime, who had worked at Cheltenham. He is serving a thirty-five-year prison sentence.
>
> The top British union leadership, including Secretary-General Len Murray of the 10.5-million-member Trades Union Congress, went to see Thatcher Wednesday hoping to change her mind. They said they came away not optimistic, but faintly encouraged by Thatcher's agreeing to have more talks.
>
> "We have here a unilateral and arbitrary decision by the government to refuse to allow a group of working people to join a union," said Murray. "You would have to go back to the 1850s to find a precedent for that. A wide range of unions are now asking: are we next?"[35]

But there are signs that the conservative swing may have reached its limit. The widespread dockers and miners strikes have brought a worried, albeit repressive government response.[36] Possibly the widespread solidarity with the miners,[37] and the miners determined militancy[38] have done more than anything to begin turning the situation around (though the miner's strike has finally been crushed since the following was written):

> In the gospel according to Thatcher, mass unemployment was supposed to put an end to strikes. It hasn't. In mid-July, with unemployment close to 13 percent, Britain's miners were in the fifth month of their strike and going strong. The dockers were on strike, blocking most of Britain's trade. Seamen were on strike against the Tory plans to sell off Channel transport to private industry. Railway workers were refusing to move coal trains. And there were numerous smaller disputes.
>
> The significance of all this is three-fold:
>
> • The miners' dispute is of enormous significance in terms of the balance of class forces. If other important groups of workers are walking out as well, this can only tilt the balance in favor of the working class.
>
> • The previous time other groups of workers (teachers, nurses) appeared to be on the point of joining miners, the government bought them off to keep the miners isolated. This time, the nature of the disputes has limited the possibility for bribery. The disputes are political rather than economic and are more militant for being based on anti-state feeling.
>
> • A new dimension to the miners' strike is the spread of violence and rioting specifically against the police. For the first time in living memory, anti-state violence is beginning to be seen as a legitimate feature of British working class struggle.[39]

Thus, the workers' movement has considerable continuing strength; and so much has been achieved. Under pressure of the working class struggle, the U.K. has moved through the phases of capitalist development from raw competition, to colonial imperialism, to neo-imperialist monopoly capitalism and has become a, so-called, social welfare state within the core of Western capitalist nations, continuing to draw a surplus from former political colonies, now become economic colonies.[40] But this relationship to former colonies and position within the capitalist core has not continued to be strong enough since WW II to assure Britains the high level of living they were used to under the British Empire. In fact, one could argue that recovery has never been complete following the massive destruction and loss of life and loss of political control of former colonies during and following WW II. Competition from renewed industries in Germany and Japan and from the U.S.A., coupled with the general crisis of capitalism, and Conservative cut backs in publicly financed programs, has made for a high unemployment rate, running at 12.2 percent in 1981 (recall Sweden being upset by 2 percent) with inflation of 11.7 percent and close to 20 percent in recent prior years.[41] This high level of unemployment has persisted with 12.9 percent reported in July 1984, down from the peak of 13.5 percent in January, 1983! There were 3.1 million workers unemployed in July 1984, the highest July figure ever recorded according to the Department of Employment.[42] The level of living is at the low end among study countries, with per capita Gross Domestic Product at $1,830 in 1965, $4,120 in 1975 and $9,220 in 1980 as compared with the West German figures for the same years: $1,960; $6,800; and $13,390.[43] Recent returns from North Sea oil may have helped prevent further relative decline.

These comparatively tight resources would probably be quite adequate if equitably distributed. In spite of the U.K. being characterized as a social welfare state, such is not the case. Class and ethnic divisions are still sharp. Whereas in 1970 the top 5 percent of East Germans received just 9.2 percent of household income; the same figure for the U.K. in 1968 (nearest available year) was 13.7 percent; for the U.S.A. it was 16.6 percent. The bottom 20 percent in the G.D.R. got 10.4 percent, while the same group in the U.K. got just 6.6 percent; in U.S.A. 3.9 percent.[44] One finds this maldistribution even in relation to the enjoyment of the NHS which was supposed to equalize health care benefits if not risks.[45] Of the 55 plus million people in the U.K. in 1977, approximately one quarter, 14 million were in poverty and other disparities such as unemployment, poor educational opportunity, poor health care, etc., hit especially heavily on the sizable population of blacks and other minority ethnic groups from former colonies.[46] These facts were known long before the widespread urban riots in April of 1981. Since that time, in one area, Brixton, Mr. Ted Knight, leader of the Black Lawyers of England and Wales, says the unemployment has risen by 62 percent and he criticized the recently issued Scarman Report as offering "no hope to the people of Brixton,"[47] By late summer 1985 the police were being armed in a bid to put down widespread urban riots.

In terms of overall governance and authority structure, the divisions and disparity just mentioned, and regional differences and divisions between Northern Ireland and England, Scotland and England, and Wales and England, as well as the class-religious strife within Northern Ireland,[48] make me assess the U.K. as quite fractionated. How it stands on the dimension of centralization seems more difficult to say. There is the tradition of democracy mentioned earlier, but a great deal of what one needs to learn about any aspect of the society can be found out in London, whereas in F.R.G., informants in the capitol, Bonn, seemed able to provide only a very partial picture. But the U.K. is by no means sharply centralized like some military dictatorships. Nor is it as centralized as the G.D.R. While there are counties and other local units of governance, it is not as locally determined as Sweden or F.R.G. It seems to fall about mid way among study countries. Thus I see it as having a comparatively fractionated, somewhat centralized authority structure.

THE HEALTH SYSTEM

The NHS is too well known to call for more than a brief refresher description here. An NHI was adopted in 1911, but it had many limitations. The desperate situation of WW II which brought Britons of all ranks and means close together in combating the common enemy of fascism brought also an intensely optimistic mood at the close of the war. The Beveridge Report of 1942 on "Social Insurance and Other Services" had identified Want, Disease, Ignorance, Squalor, and Idleness as the five giants which had to be fought if such holocausts as WW II were to be avoided in the future. Many felt that the democratically elected post-war Labour government would take these giants on. The goal in health care had been set out in the White Paper of 1944 which accepted the idea of an NHS:

> . . . to ensure that in the future every man, woman, and child can rely on getting all the advice, treatment and care which they may need in matters of personal health; that what they shall get shall be the best medical and other facilities available.[49]

"On the appointed day" in 1948 the NHS started functioning to attempt to give comprehensive free care to all on an equal basis. The system started as a tripartite one—hospital service and GP service which were national but separate and the local authorities giving public health services.[50] The hospital physicians with the most bargaining power resisted the idea of coming under local authorities, thus the hospital system was nationalized and run through regionalized systems of management and hospital physicians were placed on salary at acceptable levels.[51] GPs were left pretty much as they were, continuing to serve as independent contractors, though there were some important changes, including geographic assignment and service to panels of patients for which they would be reimbursed according to number and location.[52] This conception of the PHC level should have led to personal, continuous, anticipatory care for defined populations with easy referral to hospital for specialized treatment and proper follow-up on discharge. There did develop some innovative practices which came close to such an ideal.[53] But the gulf between hospital specialists and GPs seemed as deep and at least as wide and stormy as the English Channel. Also, pressures for economy led to an emphasis in group practices and health centers on business efficiency rather than patient satisfaction.[54] There were other problems, for example, long waiting times for elective surgery as the hospital building program lagged in replacement of war damaged and outmoded facilities; an overemphasis on passive reception of cases for treatment and a relative deemphasis on prevention and public health (which was in any case separately administered under local authorities.)

To point to problems should not be taken as a rejection of the NHS. The system has been immensely popular and removed the basic anxiety which many in U.S.A. still face today: "How can I pay and where do I go if I get sick?" Still, twenty-five years experience with the problems led to what one expert on the British system called "a legislative orgasm," bringing the Reorganization which went into effect in 1974.[55] This was supposed to bring about a more integrated regionalized system with greater emphasis on community medicine. But the Reorganization plans were sharply criticized at the time,[56] and few would argue that it has been a great success.[57]

In the meantime, the crisis of capitalism, particularly the fiscal crisis of the state under monopoly capitalism[58] has led to a serious set of cut backs and moves to privatize and partially dismantle the NHS.[59] To save money in the face of rising costs, some hospitals are being sold to private groups, some are being closed, patients are being asked to pay for parts of care which were earlier free,[60] and private insurance care packages are being increasingly supported. On this last point, while interviewing a former Dean of a medical school now doing OSH-related environmental hazard research, I mentioned the surprising article on the NHS of that very morning,[61] and asked if the NHS is in such bad shape that a political party could run on a plan to dismantle it. His answer:

> I don't think the NHS is in such bad shape. But this government is very far right. They have already craftily undercut the system by giving tax breaks for private health insurance care arrangements. This has even got to the shop floor with some (I think he said 4 million!) union members having such insurance in their contracts now.

Other signs of attack on the NHS: While riding "the tube" in London I saw a young black worker with flyers in his hand titled "St. Mary's Hospital Workers Fight Closure." He said he was an orderly in that hospital and was going to a rally to develop support for a December 4th demonstration. While interviewing a young professor of public health administration we were interrupted at length by a call filling him in on a town council meeting for that night at which a U.S.A-based, multi-national hospital corporation would present a high gloss proposal to open a private hospital. This professor had been elected to the town council and was trying to mobilize support to oppose this idea. "It would simply skim the cream off—the profitable, quick procedure, high turnover stuff—and leave the NHS with the chronic, complicated, and long term costly stuff, thereby leading to even more deterioration of the system."

In January 1978 the Government issued a White Paper emphasizing prevention as the prime direction for health actions in the future in response to a report on the subject by a House of Commons Select Committee. But as subsequent statements show, this policy level push places much of the onus for good health on "individual health behavior."[62] This has some good aspects to it, in that it does emphasize prevention, but insofar as it leads to ignoring basic societal conditions leading to poor health such as poverty, high unemployment and poor housing, it amounts to no more than "blaming the victim."[63] The

General Council of the TUC "welcomed the government's commitment to preventive medicine generally, which was in line with the TUC evidence to the Royal Commission on the NHS reported to Congress last year [1977], but stressed that this section of the NHS was seriously under-financed."[64]

With all its faults, the NHS meets most medical care demands of the population and is fundamentally immensely popular in spite of chronic grumbling. This fact, coupled with the strong control of costs in this rather fully regionalized system and the comparatively healthy state of the U.K. population (that is, better than U.S.A. as reported by Maxwell according to some ten different indicators,[65] though not more healthy than Sweden—see below) has recently led Maggie Thatcher to back off from the Tory attack on the NHS and say "The NHS is safe with us."[66] As to costs, these have risen from £314 million in 1949 to £11.1 billion in 1981-82. Allowing for inflation, the real expenditure has more than doubled. But in terms of overall resources spent on health, this has risen from just under 4 percent of GNP in 1949 to about 6 percent now.[67] This is only about half of the health expenditures as proportions of GNP in Sweden, F.R.G., and U.S.A.!

As to health status, the work just mentioned (note 65) offers detailed comparisons with Sweden and gives other international comparisons. In general, the U.K. shares the relatively high health status of other industrialized countries. It is comparatively not as healthy a population as in Sweden: Life expectancy at birth in 1976 for males was 69.5 years, for females 75.8 years against Sweden's 72.1 for males and 77.6 for females. Infant mortality was 13.7 per 1000 live births in 1977 against Sweden's 8.0. Also, there are much grosser, class, ethnic, and regional differences in health levels in the U.K. than one finds in Sweden and some other countries.[68] As one London resident put it, "The worst health problem is the housing and the rents, but they give you a pill for it all the same."

This is not the place for an overall detailed analysis of the NHS. Two important works go beyond the usual detailed description[69] and offer such an analysis while considering the larger functions of the health service in relation to the integrative and social control needs of this capitalist political economy as well as the need to assist in the reproduction of the classes.[70]

THE OSH SYSTEM AND SERVICES

The only directly comparative study of OSH systems in the U.K. and U.S.A., offers as a major finding the much more consultative atmosphere in the U.K.—labor and management sitting down, sometimes with relatively neutral government representation from the Employment Medical Advisory Service (EMAS) or some other part of the Health and Safety Executive, "to work things out"—while in U.S.A. even minor disputes go immediately into the court system in a sharply adversarial atmosphere.[71] I believe this difference is real. It no doubt has a complex background to it including cultural factors as well as traditional negotiating patterns, and differences in the relative strengths of labor and owners. But with labor more on the defensive in the economic downturn, things may be changing. There may be both more of a reluctance to call for stronger OSH measures and enforcement—several informants on all sides said this was the case, for fear of loss of jobs (back to the vicious non-choice "your job or your life)—and at the same time there seems to be a sharpening of labor's criticism of the present system and recent proposals for change.[72]

> In contrast to these theoretical developments, however, not only has the expansion of occupational health services in practice been slow, but as the results of research quoted in *Occupational Health Services—The Way Ahead*, reveal, the vast majority of working people in the United Kingdom are still not covered by occupational health services which are either adequately staffed or in which the main emphasis is placed on prevention. Furthermore, over a third of the working population enjoy no occupational health service provision whatsoever.[73]

To understand this sharpening of the atmosphere, we need to go more into the background and current structure of the OSH system.

History of the OSH System in U.K.[74]

One of the earliest identifications of an occupational disease was Percival Pott's naming of soot as a cause of scrotal cancer in London chimney sweeps in 1775. As already suggested, general work conditions were so appalling in early nineteenth century capitalism that work hazards rose to a scale unimagined previously. The long hours and child labor were the focus of concern in the first factory legislation of 1802. As already noted, Owen's efforts and those of the National Union led to the Factory Act of 1819. It was not until 1833 that the Factory Inspectorate was set up. "Four inspectors policed 2000 mills to ensure that child labor was not overexploited."[75]

> Following this, regulations were passed in the 1840s concerning machine guarding and, in response to the Chartist's movement, the length of the working day. Although the Chartists campaigned for over ten years to reduce hours for *all* workers, the law ended up in 1847 being applied only to women and children. This was to be the beginning of "protective" legislation, whereby women are treated differently from men. It has been a source of disagreement within the women's movement ever since as to whether such protection should be extended to men or whether it should be abolished altogether because it is discriminatory. Certainly men felt that their jobs were protected by the new legislation.[76]

The Factory Act of 1844 required "Certifying Factory Surgeons" whose sole function was to verify that children were old enough to work. In 1855 other functions including investigation of industrial accidents were added. In spite of Percival Pott's work and popular knowledge among workers themselves, it was not until the 1860s that the medical establishment began to recognize that factory conditions were causing widespread disease. A Royal Commission found an excessive incidence of lung diseases among tin miners in Cornwall and lead miners in the north of England. Dr. Arnold Knight found that fork grinders in Sheffield often died before the age of thirty due to silica dust in their lungs.

> In response, liberal reformers in Parliament established the Factory Extension Act of 1864, which required that "every factory to which this Act applies shall be kept in a cleanly state, and be ventilated in such a manner as to render harmless so far as is practicable any Gases, Dust, or other Impurities generated in the Process of Manufacture that may be injurious to Health." (The Factory Act of 1961 was to sound remarkably similar!) These requirements applied to factories making earthenware, matches, cartridges, and staining paper, and contained rules to make sure that ventilation was not impeded by "the wilful Misconduct or wilful Negligence of the Workmen." The emphasis was clearly on *cleaning up*.[77]

During the rising economy of the late 1880s and early 1890s, worker agitation for safer workplaces grew.[78] This renewed worker resistance included the famous matchgirls' strike over wages and working conditions in London's East End. Gertrude Tuckwell made known the hazards of lead faced by women trade unionists in the pottery industry. And miners, railroad workers, and pottery workers forced the owners to appoint safety representatives. Even if these functionaries were mainly symbolic, it was an important advance.[79] These pressures brought several developments. The Factories and Workshops Act of 1891 empowered factory inspectors to issue notice to employers to improve ventilation. However, it also carried a still continuing management counter attack—an emphasis on individual worker responsibility to wear personal protective equipment, such as respirators, gloves, head coverings, and overalls (as distinct from engineering controls which would remove hazards, but would require larger capital investment).[80] Thomas Legge, the first Chief Medical Officer for factories, was appointed in 1896. His observations over the course of his career offer an important contribution to occupational medicine.[81] The first Workman's Compensation Act of 1897 led some larger employers to appoint physicians to protect themselves against claims (not to protect employees). This was "an unhappy introduction of medicine to industry."[82] Thus, as in other countries, the early OSH services were in large plants. As one informant put it, "Some larger factories have had their own docs for donkey's years."

While these early workers' struggles produced a mixed lot of quasi advances in provisions for OSH, they were not always led by the trade union leadership.

> It seems that British unionists felt that "self help" was the only effective means of attaining labor reform, whereas French and other European union leaders hoped for permanent improvement of their conditions through legislation and state cooperation. The trade unions had no overall policy regarding accident prevention, initiating very little legislation and focusing instead on merely improving existing legislation and getting collective bargaining agreements regarding safety.
>
> These confused attitudes reflected themselves in the piecemeal advances up to the start of WW I. There were many Royal Commissions, reports, inquiries, and regulations, especially in the pottery and mining industries, and later in the munitions industry. After the war, the important Whitley Report of 1919 accepted the idea of joint councils of workers and employers, which were then recognized by the Factory Inspectorate as including safety issues within their bounds.
>
> It was not until 1924 that there was a unique but significant TUC publication, *The Waste of Capitalism.* It stated categorically that workers should control their own working environment and challenged the value of joint negotiating committees where workers have no real power. "Joint Control is simply employers' control plus workers' advice—which may or may not be taken. The worker cannot be expected to take an interest in production so long as he is denied the elementary right to determine, in cooperation with his fellow workers, the conditions under which he labours."[83]

The NHI adopted in 1911 could be seen as part of the background, even if it was in the general health services, as it provided the care of GPs to workers and their families through a system nationalizing the local insurance club and physician contract kind of practice. But it had nothing to do with in-plant care or prevention. Thus it was the large private industries and state-owned industries which continued to have the only OSH services, and these were strictly management-oriented, an inheritance from which U.K. workers suffer to this day.

The failure of the General Strike in the late 1920s and massive unemployment of the Great Depression dealt serious blows to union strength. From 37.6 percent of the total workforce being organized in 1921, union membership fell to 23.9 percent of the workforce in 1931. Union leadership was further bureaucratized as rank and file participation fell off and the leadership became wary of OSH initiatives for fear of increasing unemployment. They emphasized hazard pay ("dirt money") and increased workers' compensation for disability due to accidents instead. "Compensation had the added bonus that it was a *visible* attack on dangerous working conditions, which was good for recruitment, whereas accident prevention had no such 'selling points.'"[84] Unfortunately, some of the same policy continues today.[85]

While still unable to force effective clean up of the hazards of the past, workers and unions were faced with a host of new hazards before and into

WW II—the asbestos, rubber, automotive, and petrochemical industries had all developed rapidly.[86]

> The "Three-Generation Law" was now well established in the mills, potteries, and mines. In one generation a hazard is introduced; in the next the hazard may be recognized; and in the third a law may be introduced to control it. It then may take a fourth generation before the hazard is properly controlled, but whatever the event the law is always late arriving. Asbestos was in its second generation between the wars. Writing in 1934, Sir Thomas Legge (who retired from the Factory Inspectorate over the Government's refusal to ratify the Geneva White Lead Convention) said of asbestosis: "Looking back in the light of present knowledge, it is impossible not to feel that opportunities for discovery and prevention were badly missed." Many cases of cadmium poisoning had been reported, but it was not to be a recognized occupational disease until 1953. Beta-naphthylamine was known by Imperial Chemical Industries to cause bladder cancer in rubber workers, but they continued to make it. They were eventually fined £20,000, but even today it is recognized that all the people who were exposed have not been contacted and checked. Polyvinyl chloride (PVC) plastic is another example of a substance which was manufactured without any attention to its potential damage to health.[87]

WW II brought with it worker shortages and the massive movement into the employed labor force of women from the unrecognized reserve army of the working (in the home) but unemployed. This trend held after the war so that women made up 40 percent of the employed workforce by the end of the 1960s. With worker shortages came enhanced safety campaigns as in other study countries involved in the war. There was also a sudden spurt in support for research. The Medical Research Council (MRC) established the Industrial Health Research Board while pointing out that "until June 1943 there was no complete organization for research in industrial medicine in this country."[88] This board established research units on pneumoconiosis in 1945, toxicology in 1947, and air and water pollution in 1955. Many universities also established OSH research and teaching units, "although there are now units only at Aston, London, Manchester, Newcastle, and Dundee, those at Glasgow and Durham having been closed."[89]

Probably the biggest achievement from WW II was The Beveridge Report[90] which led to the NHS as discussed (p. 335 ff.) and to improvements in workers' compensation. But, remarkably, in spite of TUC and Labour Party interest, the establishment of the NHS in 1948 did not bring with it a parallel, nor correlated, nor an integrated Occupational Health Service. This has been consistently blocked by the Government to this day.[91] Even the Health and Safety at Work Act of 1974 did not achieve what the Dale Committee had recommended a quarter century earlier: the establishment of "comprehensive provision for occupational health covering not only industrial establishments of all kinds, large and small, but also non-industrial occupations." In fact, amazing as it may seem, not even the NHS itself established an occupational health and safety (OHS) service for its own employees. Still to this day as the largest employer in all of Europe, it lacks such a service, though there may be some change in the near future. As the Health and Safety Officer of the Association of Scientific, Technical and Managerial Staffs Union puts it:

> The situation in the NHS is far from satisfactory. I have been on a Working Party that has recently produced guidance for setting up OHS services in the NHS and a copy of this was sent to you. However, as it does not have any *legal* force the response may be patchy.[92]

The hopes for an integrated attack on hazardous work conditions which was aroused by the Labour Party's landslide victory in July 1945 were not to be realized. Instead, the main source from which guidance for such a development should have come, the union leadership, stuck with its old pattern of depending upon the bargaining efforts of individual unions and only seeking government action to improve existing legislation, rather than proposing anything new. Thus there were some important gains, for example, in the mines and foundries, but no comprehensive unified approach. Only with the TUC conference report of 1959 was there a formal call for more factory inspectors, more inspections, increased powers for inspectors, more ressearch, and improved hazard standards.

> According to Grayson and Goddard, "Between 1950 and 1973, at least 15,533 workers met their death—two or three every working day. From 1962–69 there was a relentless increase in the Factory Inspectorate's statistics on accidents at work. However dubious these statistics may be, they had an important impact on TUC delegates."[93] They may have had an impact on TUC delegates, but whether they had an impact on the TUC itself is another question.[94]

Aside from medical services, other preventive OSH measures had developed piece meal—one hazard or one industry at a time, with a vast compendium of regulations, standards, etc.—very difficult for anyone but a centrally involved bureaucrat or other expert to figure their way through. Ordinary workers could hardly feel they were in command of the measures for their own protection. A partial list and description of some of the major pieces of legislation other than the 1974 Act includes:

The Agriculture (Poisonous Substances) Act, 1952. This Act makes provision through Regulations for the protection of workers against risk of poisoning by substances to which the Act applies in connection with their use in agriculture or on land on which the substances are being or have been used.

Included in the list of such substances are the dinitrophenols, organophosphorus compounds and many other agricultural chemicals similarly regulated.

The Agricultural Safety, Health and Welfare Act, 1956. This Act has provisions relating to protection from dangerous machinery and toxic hazards of pesticides and corresponds broadly with the content of the Factories Act. Following the 1974 Act, the Agricultural Safety Inspectorate is an integral part of the Health and Safety Executive (HSE) and the Ministry of Agriculture, Fisheries and Food (The Secretary of State for Scotland in Scotland) is answerable to Parliament.

The Mine and Quarries Act, 1954. This Act includes general welfare provisions and lays down the broad principles of good mining practice as currently understood. The Secretary of State for Energy is responsible to Parliament for matters arising under this Act but the Inspectorates are now part of the HSE.

The Factories Act, 1961. This is the principal Act dealing with safety, health and welfare conditions in factories. It is the last of a long series of Acts dating back to 1833 and is now in part amended or replaced by the 1974 Act. The Secretary of State for Employment is Answerable to Parliament for the enforcement carried out by the Health and Safety Executive (HSE) provided for in the 1974 Act (see Fig. 7).

The Offices, Shops and Railway Premises Act, 1965. This Act applies to these premises provisions similar to those applied by the Factories Act to factories. Now administered by the HSE, it too is ultimately a responsibility of the Secretary of State for Employment.

The Employment Protection Act, 1975. This Act, although mainly concerned with terms of employment and its termination, contains important requirements dealing with workers temporarily removed from work for which a statutory examination is required to be undertaken by the EMAS (see below). Such workers must be offered alternative employment without loss of basic pay or, if such work is not available, be paid their full wages up to a maximum period of six months so long as they are available for work. Women workers are entitled to full pay during the last six weeks of pregnancy, taken as maternity leave, and their job must be held for them for up to twenty-nine weeks after delivery, provided they have indicated their intention to return to the former employment.

The Social Security Act, 1975. This Act includes provision for a higher rate of sickness benefit for injuries occurring at work or for certain "prescribed" industrial diseases (currently around 50 in number) in which there is a "presumption" of them being brought about by occupation. Apparently, the workers' compensation insurance mechanisms fail to significantly influence employers to clean up the workplaces, since, as long as employers can get insurance to cover actual and possible outlays they seem unconcerned. Thus the insurers in U.K., as is true in U.S.A. and other study countries, do not play the

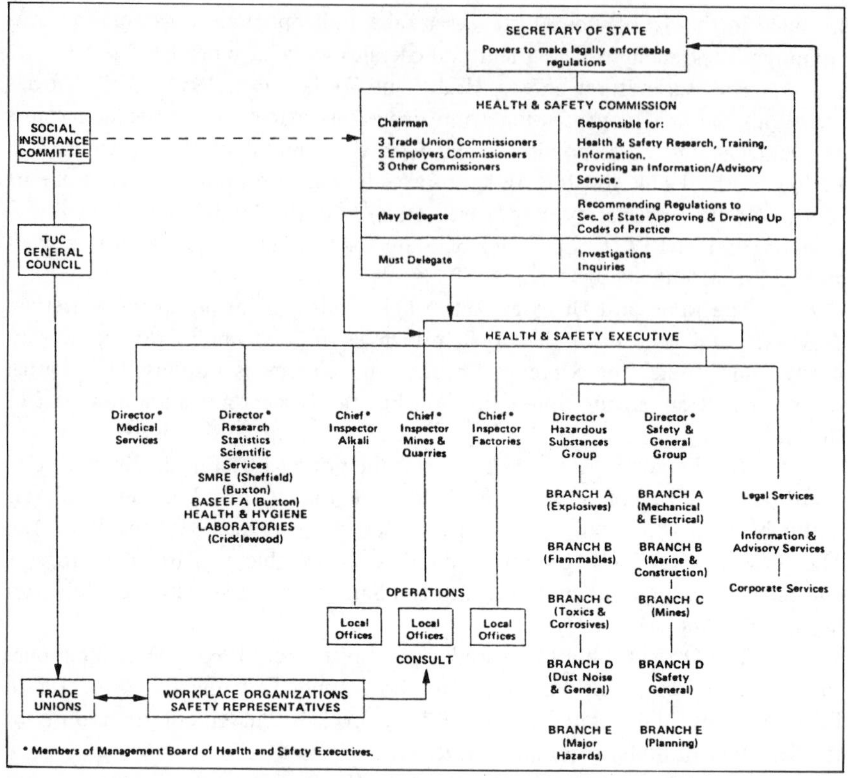

Figure 7. Health and Safety Organization Chart, U.K.

central role which the Berufsgenossenschaften do in influencing OSH conditions and provisions in F.R.G. (see Chapter 10). However, as in U.S.A. where recent estimates place the incidence of work-related disease of all types (skin rashes and others included—some of these being very severe) at 28.4 per 100 workers and Manville Company has declared bankruptcy in anticipation of millions in asbestos damage claims, the future may bring more influence from insurers.[95]

> There is little evidence that the new laws on health and safety resulted from pressure by elected union officials. Of course there were motions at TUC conferences, but there always had been . . .
>
> It seems more likely that the pressure for new legislation came directly from the shop floor. With the increasing power of shop stewards in plant-level bargaining, there was more and more pressure to remove health and safety hazards rather than to depend on worker's compensation to take care of the injured. The increase in unauthorized wildcat strikes may have put additional pressure on the State and Factory Inspectorate to redefine occupational health and safety as a "technical" issue above and beyond politics. The Factory Inspectorate realized that the maze of regulations and government

bodies was often self-contradictory, that many new industries were not properly covered, and that its own resources and powers were totally inadequate to maintain any semblance of doing the job it was supposed to do. The fact that occupational health and safety services were being reorganized in several other capitalist countries, particularly the United States, may have had a demonstration effect. Perhaps the interests of the multinationals could be better served by cosmetic changes than by no changes at all.[96]

The 1970s and Now

The Robens Committee, set up by the Labour Government in 1970, provided the immediate background of the 1974 Act.[97] This background was not particularly propitious.

> After receiving 600 pages of evidence from companies, industrial associations, many government departments, some unions, individuals, and a joint submission from the TUC and CBI (the Confederation of British Industry, the employers' association), he concluded that "the single most important reason for accidents at work is apathy." Apathy apparently resulted from too much complicated law. He also believed that there was "less conflict of interest over matters of health and safety than most other areas of industrial relations." After 150 years of maiming and killing, of struggles to improve conditions, to set standards and enforce them, and to organize research, all he could blame for dangerous working conditions was apathy! Robens believed that there was no need for an Occupational Health Service, merely that the law should be simplified and regulations replaced by voluntary codes of practice.[98]

The 1974 Health and Safety at Work Act at least offered a new concept, attempting to put OSH into an identifiable (not to say integrated) system. The key point is that regulations issue in 1978 gave a large role to union-appointed health and safety representatives. The overall organization of this system is suggested in Figure 7. The Act is guided by a Commission with equal representation from the Trade Unions (TUC) and the employers' organization (CBI) and representatives of local authorities meeting under a full-time Chairman appointed by the Secretary of State for Employment (the counterpart to the Ministry of Labor in other countries). The Commission's policies are put into effect by the full time Health and Safety Executive (HSE). There is a scientific research staff with administrative support totalling some 300 personnel as well as standard setting and inspection as well as other functions. As one official put it, referring to the U.S.A. system for OSH, "We're a combination of OSHA and NIOSH all in one box." The Act by the way does not apply to Northern Ireland which is only now considering the adoption of such a system.

There is almost no connection to the Department of Health and Social Security (Ministry of Health). As one especially knowledgeable union official commented,

> The NHS is the largest employer in the country; yet it was impossible to have it run the occupational health services because it hadn't even looked after the occupational health of its own employees. It is a good system but terribly under resourced.

An official of the NHS told me they are glad that OSH is under the Secretary for Employment.

> That way the Health and Safety Executive can lean on the NHS and say, "You're not doing your job." They can't lean too hard, but at some time they may have to go at it at ministerial level. They've done a review of the NHS which gives them the armor to give us a poke. But they're holding their hand saying one government unit can't knock another one over the head in public.

He admitted that the NHS's own OSH services are still woefully inadequate. He referred to some publications[99] and said, "Here we are in 1981 and nothing much has been done to say 'get on with it,' at least as a matter of achieving the basic minimum." The Tunbridge Joint Committee Report (just referenced) has been revised and is being "sent up for comment" with the suggestion that Regional Health Authorities at least be allowed to each hire a well qualified OSH physician and nurse to plan and carry out a preventive OSH program for NHS personnel and facilities in their regions. In this submission one point of special interest for this study is para. 5.2 which calls for the exchange of information between the OSH Services of the NHS and the GP with the consent of the patient.[100]

The Medical Services identified on the left side of Figure 7 is the EMAS, set up in 1972 under the Employment Medical Advisory Service Act, and made a part of the Health and Safety Executive (HSE) by the 1974 Act. EMAS organizes medical exams of persons employed in especially hazardous processes using its own staff and some 450 doctors employed full- or part-time (many of them) in industry who have been approved to do these statutorily mandated examinations. EMAS also has inspectorial and research powers and functions, but has more generally operated as an advisory service to both management and labor. Another of its important tasks from the perspective of this study is to advise GPs in connection with hazards their patients may face at work or health problems their patients' work may have caused. While this function is here viewed as extremely important, there has been little done about it. EMAS has only some 120 doctors and eighty nurses. Thus there is only about one physician or nurse in the EMAS for every 110,000 employed persons in the population.

Thus unions are concerned that EMAS *not* be regarded as a substitute for employers setting up adequate OSH services–something the Act failed to require them to do, but did create a certain amount of pressure encouraging the establishment of OSH services.

In fact, the whole resource and staffing picture of the HSE is disappointing to union officials concerned with OSH. By 1979, "with the new premises covered, the ratio of factory inspectors (700) to premises (400,000) is very similar to that in 1833!"[101] As one TUC official told me in late 1981:

> There are only about 900 inspectors for the whole country. As the Act came in, it was promised that the total personnel for the HSE would go to 4,400, but only 4,250 was reached and there have been cuts. So there's only 200 more than before the Act. Also, one of the ways they had of saving manpower was to integrate the inspectorates. But this is coming under attack. The mines and railways inspectorates never prosecute or issue a fine–even when deaths are involved. The factory inspectorate has been a bit better. We are not on fruitful ground at all these days; things are going backwards fast, and it's going to get worse. Still, our Congress this year adopted a very strong position on occupational health.

But no OSH system can work solely through inspectors, however many there might be, and there usually are too few, as in the U.K. The really important base for prevention in the system must be the workers themselves and their organized representatives. The unions were able to play a much stronger role in the mid-1970s following two national miners' strikes which brought down the Conservative Government. Under a Labour Government, the unions were able to assure that the health and safety representatives would not be held liable, that there would be adequate training, and that they would have at least eight hours a week free from their production jobs to search out hazards and perform other tasks. This is the significance of the 1978 regulations:

> The trade union movement has always been convinced that effective trade union organization at work is the key to reducing accidents and disease. That is why the TUC pressed so consistently for, and won, the new rights and functions that are contained in the Safety Representatives and Safety Committees Regulations. These new regulations have provided a framework within which the trade union movement can develop their organization on health and safety matters. This has meant that, up and down the country over 130,000 trade union safety representatives are now on patrol acting as safety watchdogs. They are acting as a major force for improvement at the point where accidents and diseases actually occur.[102]

While this arrangement leaves unorganized shops uncovered, since Safety Representatives can only be appointed by workers who are members of a recognized trade union (worker representatives and committees in an unorganized shop would not be effective anyway) this arrangement has the merit of strengthening the hand of organized labor. While these Safety Representatives haven't got the kind of power similar people have in Sweden (to stop production processes they perceive are dangerous, to approve or reject the OSH physician the company wants to hire, etc.) if they are well trained, they can effectively use the threat of calling in the HSE inspectors who can shut a process down and issue calls for corrective action with defined deadlines. Safety Representatives can have a great effect through their other general tasks: identifying hazards in a variety of ways; gathering information which would lead to an improvement of standards; educating fellow workers; receiving complaints from other workers; and organizing and negotiating for OSH. The sophistication of some of the unions in these matters is suggested by the following:

> Do you need a "Safety Committee?"
>
> Many places of work have a joint safety committee. You will have to decide if you should participate in such a committee, or if there is not one, whether you should ask for one to be set up. If two safety representatives ask management for a safety committee it has to be set up within three months.
>
> The dangers of a safety committee are:
>
> - It can be a talking shop. Safety Representatives and management may air safety issues but nothing gets done and time is wasted.
> - It may deflect the energy of safety representatives from their real job, which is representing their members on health and safety issues through the collective bargaining structure.
> - Management may use a safety committee to try and shift responsibility for health and safety matters from management alone to management and safety representatives jointly. They cannot do this legally and must not be allowed to try it in practice.
>
> A safety committee may however be useful for long term consultation to cover such matters as the study of accidents, diseases, statistical trends, safety audit reports, amendments to the Safety Policy, discussions on the contents of management health and safety training and communication on health and safety matters in the workplace.
>
> As safety representative you do not need a safety committee as your authority comes from your status as shop steward and the added legal powers given to safety representatives. Even if there is a safety committee, management must deal with you directly. A safety committee will not usually deal with matters quickly. Most health and safety issues will be dealt with sooner and more effectively if raised by an individual safety representative in the normal way.[103]

A number of unions have their own training centers which is probably a stronger pattern than the combined union-management centers in F.R.G. In any case, many thousands of safety representatives have been trained, though I did not get an exact figure.

However important the workers and their unions are, the 1974 Act continued to place the overall responsibility for OSH on the employer. But it did not require the employer to establish OSH services. In fact, in-plant OSH services are still mostly limited to large private and public industries. Other forms of extra-plant services–group practices and part-time GPs partially trained in OSH have begun to develop. But these are far from adequate and again serve mainly medium-sized plants, and not all of them. Since one half of all workers are in workplaces with less than 100 workers and only a minority are in places with 500 or more, most workers have no OSH service coverage. Some form of multi-plant service will have to be developed, but these have not grown as much as in a number of other countries.[104] In none of the forms does organized labor have anything like the co-determination rights it has in Sweden. Thus the old problem of OSH physicians upholding mainly the employer's interest is still very much in the U.K. picture. For example, one GP with no special OSH training who has served for several years as a part-time once-a-month OSH physician for a large battery plant told me he knows of research which shows that workers heavily exposed to lead are actually healthier than control groups not so exposed! He did not say if "the healthy worker effect" was taken into account or what the exact methodology was. Nor did he offer any specific reference. The point is the attitude he has, not the details of any purported study. He also told me that if the worker is feeling stress at work, "It's good to ask how the wife and kids are, since the problem usually starts there." When the plant started a new thin plate process for long life batteries, he prohibited any women from working on the process–not because he believed there was any real problem, but because "All it would take would be one hydrocephalic or spinabifida and the Company would be up to its ears in law suits." When I asked him what the relationship of trust is between himself and the union he said:

> Oh the union man may come to me and ask which way to play it with the company–say a compensation case or some new ventilation. They do this even though I'm paid by the company. But as the golden aura of the physician disappears, and unions get more and more into it, it's getting more and more difficult.

Perhaps this general management bias of OSH physicians and their poor level of training (though moves to change this have just been launched–see below) together with the union's comparative lack of control over OSH services, and the failure of the physician-dominated NHS to do anything at all about OSH have led (more than in any other study country) to strong union opposition to continued medical dominance of OSH services. It is worth quoting at some length from national health and safety officers of different major unions on this and related points. As one such informant put it:

> One of the main problems is the credibility of medicine, especially its biological monitoring in the workplace in general. When we do issue a policy paper—we've held off because of the economic situation; we can't be issuing Cloud 9 Proposals in this atmosphere—but when we do, we'll come out against the laying on of hands in the OSH services. This just puts them (the doctors) in the role the companies want them in: the worker comes to the doctor to get an exam and the doctor decides "no, nothing is wrong" and the company is whitewashed. If something is wrong, the workers are the last to find out. In one local, the workers refused to go along with the proposed safety-health system because they had no control over the information. Short of restructuring the whole occupational safety and health system, the main thing we have to do is raise consciousness about the inadequacy of medically dominated occupational health services—raise the level of scepticism. I'm pissed off to the back teeth! with going to meetings and hearing some ignorant doctor pronouncing on chemical effects or something else and being believed when he doesn't know the correct technical details!

Another national union spokesperson for OSH emphasized the teamwork aspect necessary for a good OSH program as a way of avoiding the hegemony created by medical manpower shortages in this field. This person went on to say,

> Part of the idea of not having occupational health as part of the NHS was to assure independence from that system which is completely medically dominated and curative. We want to see the NHS strengthened. It's a good system, but we can't depend on it for preventive action in the workplace. The traditional medical approach and the approach of personnel departments is to ask, "Is the worker fit for the job?" We want to ask, "Is the job fit for the worker?" We don't see medicine as the core, rather a team. Not just the doctor or microbiologist, etc. We're trying to get teams in relation to each work setting and have it assured in the statutes that all employees have adequate health and safety services. This has been adopted by the Dutch. We like something similar. We want the professionals to be advisors. This will distance them from management and for that matter from the local trade union. In the Dutch system, the factory doctor can't be dismissed except with the agreement of the local magistrate. I think it's similar in Scandinavia. The best use of scarce occupational medical people would be for epidemiologic studies, etc. not just routine exams.

With an awareness of the uneasiness of labor over the management bias of OSH physicians, an official of EMAS handed me a copy of the ethics statement of U.S.A. OSH physicians[105] as if this was some kind of answer to the problem—when everyone (well at least every union person) knows that the economic relationships, and not ethics statements, are determinative. Additional suspicion is raised when one reads the British Medical Association's statement and sees the space devoted to remuneration for each grade of OSH physician with very little space devoted to the expected content of the work![106] Aside from a long tradition of worker-OSH physician suspicion, there continue to be current episodes, even with the supposedly neutral EMAS. One Union OSH official told me that EMAS undertook a human epidemiologic study of Bis (chloromethyl) ether only after the union brought some animal studies to their attention. NIOSH (in the U.S.A.) undertook a study at the same time and finished it in two years. EMAS dropped their study saying the NIOSH study would do and they never went back to the factory to tell the workers what was what, even though the workers had supplied data and were worried.

The sharpening of the struggle for improved OSH in the U.K. includes then a certain amount of "parting of the ways" between medical and labor leaders in the field. While labor leaders speak vigorously of moving away from medical leadership of OSH, medical leaders of OSH speak of recent moves to upgrade the field in general and the training of OSH physicians in particular. In the past, some GPs have obtained a Diploma in Industrial Hygiene through the TUC Centenary Institute of Occupational Health at the London School of Hygiene and Tropical Medicine so that they might serve as factory doctors. This has been a three-month course. Usually there were twenty to thirty students taking it each year. This Institute has also given a one-year MSc in Industrial Medicine for top job people in the field. But now a new Faculty of Occupational Medicine has been formed, with the same kind of national structure as for other specialties–medicine, surgery, etc. There will be a one-year Associate degree which some say unions and others will eventually insist be the minimum for a factory doctor. Then the four-year program will be for the true specialist in the field, the Senior Registrar, as in other specialties. The first groups of Associates will be graduated in 1982 from several centers. Their preparation was described in very general terms as follows:

> A broad general training of about two or three years, as in general medicine or general practice, is desirable. Although not mandatory, experience in subjects such as respiratory medicine, dermatology and general practice is desirable also. Occupational medicine, like other branches of clinical medicine, is essentially practical and demands further training with a large content of supervised experience in approved posts under established practitioners. However, because much of the content of knowledge in this field is barely touched upon in the basic or general professional training, intending Members and Associates may require some formal training. The details have yet to be worked out, but it seems clear that the existing part-time courses will provide the models on which future developments will be based. In the longer term, it is hoped to develop teaching methods and materials which can help to make training available at local centres. A more immediate likely development is the structuring of some courses on the "block-release" rather than "day-release" pattern.[107]

As to other OSH personnel, there are some 9,000 OSH nurses in the U.K. and quite a lot of the work in the field is done by them. There are now two levels in this field too. The higher level—Occupational Health Nursing Certificate—requires an academic year of full-time study and relevant experience. The occupational hygiene field is not well developed or controlled, but an Examining and Registration Board issues certificates and diplomas for those working in the field who bother to qualify. The picture with safety specialists, OSH psychologists, etc. is similarly underdeveloped.

Another special characteristic of the U.K. OSH system (though similar to F.R.G. in this) is its lack of a research and treatment reference center in a national OSH institute as was found in Finland (serving also Sweden and other Scandinavian countries) and in G.D.R. There are strong centers, such as that in Birmingham, but no move has been made to establish one national center. The establishment of such an Institute, perhaps at Birmingham, seems to be under discussion. The head of a potentially competing institute said he wasn't in favor of one national center for the U.K.

> In Sweden or Finland, yes. But in France or Germany or here, countries of this size can afford more than one center. What would happen if you got a man to head it who declined in his early middle age. You'd have all your eggs in one basket and you'd be in very bad shape indeed.

Another important, though not especially noticeable component of the OSH system in the U.K. is the existence of COSH-like groups—groups of activist workers and medical/technical people. Such groups exist in Finland, Denmark, and U.S.A. and did once exist in Sweden. The chief health and safety official of one of the major unions was critical of one of these groups headed by "the old left." Established social democratic unionists feel this to be old fashioned in its demand for medical care for all workers—"everyone should be able to have their bunions treated and all that." This thinking only supports medical dominance and offers a very narrow view of OSH. Another group, the Work Hazards Group, part of the British Society for Social Responsibility in Science, has more recognition and respect. Its local groups are active in six major cities. Their *Hazards Bulletin* comes out five times a year and had been through twenty-seven issues with the appearance of the September 1981 number. This issue dealt with: "noise control—ear muffs or noise control;" the reaction of industry to a local council banning asbestos pipe; "training safety reps for when factories belong to us;" stopping radiation hazards; and letters from readers, including one from a secretaries' collective which has banned cloromethane-containing correction fluids, etc. Work Hazards Group achievements include, "An understanding of preventive measures, a lack of reliance on the factory inspectors, and confidence in a 'self-help' approach have been developed, all with enormous potential."[108]

> What of the future of self help OSH efforts?
>
> Probably the most significant development has been the establishment of area health and safety groups consisting of members from different trade unions. The Coventry Health and Safety Movement (CHASM) was the first, and it established an information center, newsletters, regular meetings, and a delegated structure that keeps it independent of other interests. Following CHASM has come WHAC (Work Hazards Advisory Committee in Southampton), HASSEL (Health and Safety in S.E. London), MASH (Middlesex Action on Safety and Health), LASH (Leeds), HASH (Hull), HASSARD (Doncaster), TUSC (Sheffield Trades Council), MASC (Manchester Advisory Safety Committee) Merseyside Health and Safety Group, and BRUSH (Birmingham Regional Union Safety and Health Campaign). If they are as promising as their names, we can expect much of these groups. They usually work with the Work Hazards Groups, and some have links with tenants and community groups fighting pollution. Although each differs in composition and aims, all are based on local trade-union activists, generally apart from the traditional union structures. Much will depend on funding and on continued educational facilities, which will in turn rely on shop stewards maintaining an interest and commitment beyond their own workplaces. These groups will obviously cause concern among full-time officers and national [union] officials, not to mention local employers, but there is no doubt they are going to be the core of any future health and safety movement in the United Kingdom. As more workers turn away from struggling around wage issues because increases are decided by the Government, these groups may provide the basis for the long-awaited organization to intervene on behalf of workers in the industrial war that has been raging for the past 150 years.[109]

The recent House of Lords Report called for occupational health and safety services as "necessary in British industry today to safeguard the health and welfare of workers and contribute to greater economic efficiency." But the action it recommended was "that the Health and Safety Commission should instruct the Health and Safety Executive to draw up a voluntary Code of Practice setting out the kind of service that should be provided in the various types of industry.[110] The reaction of organized labor is suggested by an especially well informed union Health and Safety Officer:

> I think you will have to take into account the House of Lords Report which has been badly received by most of the trade unions. It was a great disappointment and we are currently producing a response, but it is not yet final. Most people seem to regard it as a great opportunity missed and the only positive outcome may in fact be an improvement in research. There are no specific legal requirements in this country concerning occupational health services as such, although there are a number of legal requirements relating specifically to medical surveillance, e.g. lead workers.[111]

While other particulars of the OSH system in the U.K. might be mentioned, it is hoped that we have enough background to pass on to considering links to PHC.

OSH-PHC LINKS

Sadly enough, material for this topic is quite short. I could begin and end it by saying, "There are hardly any links" and I would not have significantly distorted the picture.

At policy level, I found little attention to the problem, though there was concern when it was explained that with less than half of workers covered by special OSH services, and other workers usually poorly covered, and all workers free to go to their GPs as first point of contact, the recognition and control of OSH disease must of necessity fall heavily to the PHC level of the NHS—whether it does the job right or not. One major union has issued a discussion paper on training of GPs for greater attention to OSH. And some knowledgable union OSH people have exchanged working outlines on what basic medical education should include in the way of OSH preparation. These are quite good.[112] Also the discussion paper now "going up" for ministerial consideration in the NHS mentions exchange of information between hoped for OSH services of the NHS and the GP involved in any case, if the patient agrees to such exchange (it should say "if the patient or his representative agree" to allow the unions to play their proper role with agreement of the worker). Other than these evidences of concern, there seems to be a serious policy vacuum on this matter. Where there are OSH services, a recent study in Nottingham showed "little contact between occupation health services and the National Health Service."[113]

However, the recent House of Lords Report, even if unsatisfactory in its dependence upon voluntary initiatives and private financing does for the first time give an explicit official voice to the OSH-PHC concern:

> In making their recommendations the Committee takes the view that occupational health should be regarded as an integral part of the primary care of patients, that is to say at the level where a person's health problems first come to the notice of the medical and nursing professions. An individual spends so much of his life at work that occupational medicine should play a more prominent part in the primary service which is available to him. More account should be taken of the effects of work on health. Both general practice and occupational medicine are concerned, after all, with the same patients.[114]

The OSH training of health personnel at all levels, especially physicians is as poor in the U.K. as in the U.S.A. and worse than in all other study countries.

As one national union health and safety chief put it, "The ordinary GP in this country wouldn't recognize an occupational disease if he or she fell over one."

Indeed there is very little OSH training at any level. At undergraduate level, one of the stronger schools was said to have five or six required hours. Another is known to have two hours. "Two hours in six years of school, isn't that something" said an NHS official concerned with the general low status of OSH in medicine. And, except for GPs taking on a part-time OSH job or going into the field, there appears to be no OSH included in the specialty training to become a GP. Several inquiries from people who should know brought the response, "None."

Organizationally the situation is no better. OSH is almost completely separated from the Department of Health and Social Security. And the NHS, the largest employer in the country, is one of the poorest when it comes to providing OSH services. At local level, there is the one strength of the system. This is the geographic assignment of GPs so that some argue that even in the absence of training they come to know the factories and workshops in their areas and the OSH problems of their patients. I am sceptical. I believe a careful national survey of GPs would reveal the correctness of the union official's view just quoted above. Still, this organizational element is of definite importance, for it offers potential for the time when GPs and other PHC providers do become concerned with and trained in OSH.

Financing of OSH services is a problem in the depressed U.K. economy. Without a legal requirement for employers to provide OSH services, extra-plant forms of service have not grown much and I heard of no experiments making local PHC centers responsible for OSH, such as I saw in G.D.R. and such as exists in Skövde in Sweden. Another unfortunate aspect of present financing is that it places the OSH physician too much under the control of management and in the part-time arrangements may orient them too much toward pre-employment exams to exclude presumed high risk workers as well as periodic exams, rather than preventive efforts. One can easily get the impression that the part-timer is in it to get a significant addition to his income simply by "passing by" the plant once a week and not much more. As one young professor put it, "Occupational health in the United Kingdom is a mess. It's a scandal. Only the big companies have any properly trained people. The rest are part time GPs who pass by the plant and say 'Ah, yes, O.K.'" No wonder one finds union OSH officials oriented strongly away from medically-centered OSH services for the future.

If links were to be made with PHC, better arrangements would have to be figured out than exist now for organized labor to have an authentic voice in the structure and running of the service. The reorganization of the NHS provided for local community and advisory groups–but only advisory and I don't believe organized labor has been significantly included other than in a few situations.

OBSERVATIONS, QUESTIONS, RECOMMENDATIONS

It is hard to say where to begin. An impressive apparatus has been set up by the 1974 Health and Safety at Work Act and the 1978 Regulations on Safety Representatives and Safety Committees. But there is not much sense of real accomplishment, particularly in the present political-economic atmosphere when things seem to be going backwards as far as workers are concerned. It might look like a cop out, another delay tactic, to recommend the establishment of a top-level commission to study the OSH system and its lack of links to PHC and bring in recommendations for action. Such a Commission should have strong TUC representation and hear fully from some of the workers and scientists in the Work Hazards Group. Practically every facet I ran across needs serious attention and improvement. I believe it would be gratuitous and out of context for me to develop a long laundry list of specific points. The reader can make up her or his own list from reading the foregoing sub-sections (pp. 345-355). The TUC has expressed concern over many things (the article by Bibbings was cited earlier–see note 73) including the need for more workplaces to be covered by team-based OSH services which have a primarily (though perhaps not exclusively) preventive orientation (recall the case of F.R.G. where the unions are unhappy that OSH services are exclusively preventive–though there the economics of the PHC system have forced such an orientation on the OSH services). Present law does not require employers to provide OSH services. Even the NHS, the largest employer, and a "health" agency has failed to provide for OSH services. The law only requires employers to protect the health and safety of workers in relatively unspecified ways–though many OSH-conscious employers have established such services to help them stay out of trouble with the HSE. In any case, a Commission which could build the facts and a head of steam for change when the time is ripe may be the best that one could do at this point.[115]

The links to PHC are especially important in the U.K. as one knows without a survey that a very large majority of workers are going to GPs as their first point of contact and the GPs are not trained to recognize or follow-up on work-related disease–neither in terms of the individual worker (rehabilitation, job-placement, etc.) nor in terms of the other workers who might be similarly exposed. Also OSH training in basic medical education is very inadequate.

To me the most curious thing about this case is that it strikes me as one of the poorest as regards OSH in spite of the workers' movement being the oldest and rather strong (not the strongest, but also not as weak as in F.R.G. and U.S.A.). I will puzzle over this for some time. As of now, the only hint of an explanation comes in what one union official said,

> There has long been a discussion whether occupational safety and health should be in the NHS or not. We fought long and hard for an

> NHS. When it came in, occupational health was not part of it, so we got the NHS, but we got no occupational health services.

Labor itself has been too narrow in its struggle for health services. It is waking up now and the struggle for OSH (hopefully properly linked to PHC) will sharpen in the future.

THE U.K. AND THE OSH AND OSH-PHC MODEL

In Chapter 6 the several dimensions of the model (only mentioned here) were developed in more detail.

Policy Level

At the policy level the 1974 Act expresses deep official concern with the need for improved OSH and the recent House of Lords Report gives for the first time explicit official recognition of the need for improved PHC-OSH connections. However, in spite of these good official words, the sense of policy enactment one gets after a careful look around is not especially inspiring. There have been many accomplishments and the 1978 Act assuring the training and conditions for effective union Safety Representatives is perhaps the major one. Still some 45 percent or more of workers are unorganized and most workers lack coverage by any OSH services, even the kind of part-time, management-controlled ones available in many medium to larger-sized shops. There are other shortcomings detailed above. In summary, there are official statements but little, if any, more concrete action than in F.R.G., or even U.S.A., and certainly not as much as in G.D.R., Sweden, or Finland.

Sponsorship and Control

Sponsorship and control of OSH services has not been achieved by workers, except for the laudable system of safety representatives available in union shops. It is a point of no small significance and strength that the unions control their own OSH training centers for safety representatives and other workers, rather than having joint union-management training as in F.R.G. There is no union or worker control over the hiring of OSH physicians and engineers as in Sweden.

As regards the general health services, there seems to have developed almost an antipathy between organized labor and physicians, in part because the physician-dominated NHS (which is greatly appreciated for its protection as regards ordinary illness) has done almost nothing about OSH—even for its own employees. Thus there is no worker control over the PHC-OSH connection, though some especially involved union OSH experts have begun developing discussion outlines on the need for OSH training for medical students and PHC practitioners.

Education

The education dimension of the model reflects both strong and weak points in the U.K. case. Union-controlled OSH education centers have prepared some 130,000 Safety Representatives. This is a powerful accomplishment. In addition, the Work Hazards Groups (COSH-like organizations) have trained other workers and continue to issue a Hazards Bulletin. The TUC has issued a number of important OSH publications.

The training of professionals is generally poor, but there are clear signs of gearing up for the preparation of many new factory-level OSH physicians, recently accepted as a specialty of the Royal College of Physicians. And there are a number of university research and teaching centers where more highly qualified OSH specialists prepare others.

The preparation of medical students, other health students, and PHC providers of all types in OSH leaves nearly everything to be desired.

Organization

Organizationally the U.K.'s OSH system seems especially poor. Maybe the outside observer is fooled and misperceives the usual British "muddling through" as poor organization, when it is in fact quite effective. But one inside, especially knowledgeable critic, for example, found in 1979 that the ratio of inspectors to workplaces was about what it was in 1833. And the staff of EMAS in no way can make up for the lack of OSH services in most workplaces. The NHS seems uninvolved with OSH, this last coming under the British equivalent of the Ministry of Labor. The one organizational strength (weakened by lack of OSH training for PHC providers) is the geographic focus of the generally regionalized NHS.

Financing

Financing is also problematic. The few OSH services there are, mostly in large to medium-large plants, are funded by the managements with a history of protecting against worker compensation claims. The NHS is a great boon and is publicly funded, but it is as yet uninvolved with OSH. In general, funding for public services, including EMAS, has deteriorated under the Thatcher regime. Funding for OSH research, however, has remained fairly strong, though not more so than in most of the other study countries.

Information

The system of well-trained union Safety Representatives helps assure the workers' right to know what hazards they are facing and what can be done about them. However, nearly half the workforce is not covered because they are not organized. There have been some widely effective BBC programs on asbestos and other hazards. Otherwise, much remains to be done as regards information flow to workers in the U.K.

The general OSH accident and illness reporting by the HSE is not especially impressive nor is it weaker than what exists in some other study countries—F.R.G. and U.S.A. particularly. The usual weakness is found: PHC personnel are not well trained to recognize and report work-related illnesses.

The information generated by special studies is quite impressive, especially since the Medical Research Council set up a special OSH unit which now has a number of special problem-focused components.

In conclusion, I believe Hobsbawm's summary of the workers' movement in U.K. helps shed light on the OSH and OSH-PHC situation as well.

> In Britain, where the working class has been for almost a century far too strong to be wished away by the ruling classes, its movement has been enmeshed in the web of conciliation and collaboration more deeply, and far longer, than anywhere else. In most European countries the decision to tolerate the labour movement and to operate with it rather than against it was taken not earlier than the end of the nineteenth century. In France it occurred in the 1880s, after the ending of the post-Commune hysteria, in Germany after Bismarck, with the abrogation of the anti-socialist laws, in Italy after the failure of the Crispi repression. In Britain, however, both the official acceptance of trade unionism and of the mass (and, in this instance, predominantly proletarian) electorate occurred in the middle 1860s. Recent historical work has thrown much light on the soul-searching among the ruling classes which preceded the deliberate decisions to do so. From that moment no systematic attempt to suppress the labour movement has been made in Britain, except by particular sections of business, never entirely backed by even the most conservative of governments. On the contrary, the fundamental posture of government and increasingly of the major industries has been that of the lion tamer rather than the big-game hunter. The most militant periods of labour-baiting or labour-smashing—notably in the later 1890s—were those in which labour policy escaped from the control of governments into the much more short-sighted one of private consortia of capitalists and ultra-conservative lawyers. The dream of a country free from unions and a Labour Party (let alone from socialists) still sweetens the lunchtime conversations of such businessmen as have no experience of the realities of industrial life—stockbrokers, bankers and their like—or of the helpless little businessmen, and finds an echo in the speeches of the more stupid politicians and the publications of the more feudal press-lords. But

even between 1921 and 1933, when the "fight-to-a-finishers" temporarily gained control and tried to force labour to its knees, they were, as we have seen, always held in some sort of check by the dominant and moderate wing of the Conservative Party tacitly assisted by the Liberals. At moments of fear and hysteria attempts to attack labour all along the line may still be made; but the rulers of a country 90 percent of whose citizens live by earning wages and two-thirds of which are manual workers, have been far too wise to indulge in them, even in the 1930s, when European fascism made the defeat of labour look tempting and possible.

There are, in such a situation, two factors which drive a labour movement to the right. On the one hand the mere technicalities of recognized trade union activity in a modern capitalist economy involve leaders, above the shop or works level, in a network of joint activities with employers and the state, and so does the mere existence of a working-class party which is a potential government or partner in government coalitions in parliamentary systems. This is a problem for communist as well as for social-democratic parties, as in France during the Popular Front and the immediate postwar periods. On the other, there are the systematic efforts of government and (normally big) business designed to strengthen labour moderation and weaken the revolutionaries. Communists have so far been largely exempt from these, because government and business have regarded them collectively as irreconcilable, but it is by no means impossible (especially since about 1960) that the same tactics will be applied to them.[116]

NOTES AND REFERENCES

1. Several sources have been used for this section and for the historical part of the next section on political economy and workers' movement. One of the more frequently consulted has been Dudley W. R. Bahlman, United Kingdom, pp. 2282–2292 in *The Volume Library*, Nashville, Tennessee: The Southwestern Company, 1974. Other sources will be referred to more specifically on special points.
2. Bernard Cassen, Great Britain: "Thatcherism" Eroded, pp. 112–114 in *World View 1982, An Economic and Geopolitical Yearbook*, Boston: South End Press, 1982.
3. Ruth Leger Sivard, *World Military and Social Expenditures, 1983*, Washington, D.C.: World Priorities, 1983, Table II.
4. John Fowles, *Daniel Martin*, New York: New American Library, Signet, 1978, p. 232.
5. Immanuel Wallerstein, *The Modern World-System I: Capitalist Agriculture and the Origins of the European World-Economy in the Sixteenth Century.* New York: Academic Press, 1974.
6. Immanuel Wallerstein, *The Modern World-System II: Mercantilism and the Consolidation of the European World-Economy, 1600–1750.* Wallerstein (p. 93) cites a reference which documents that Scottish coal miners and salters were reduced to *de facto* slavery by early industrial capitalism.

7. The Rosetta Stone, p. 962, *The Volume Library* op. cit. (note 1).
8. Friedrich Engels, *The Condition of the Working Class in England in 1844.* Stanford, California: Stanford Univ. Press, 1968. (original 1845), p. 248, as quoted in Ringen, see note 11 below. This was Engels' best work. Conservative writers and others attempted in every possible way to attack and discredit it. But as Hubsbawm says, "Rarely has a book been subjected to such systematic and painstaking hostile cross-examination. It can be said quite categorically that he comes out of it with flying colours; much better, in fact, than one might have expected. (I wish the same could be said of all subsequent attempts by Marxists to write accounts of working-class conditions under capitalism)." Eric John Hobsbawm, *Labouring Men; Studies in the History of Labour.* New York: Basic Books, 1964, p. 115.
9. Engels, op. cit. (note 8):250.
10. Ibid, p. 249.
11. Knut Ringen, "Liberalism and Its Negation: Sanitary Reform in England, pp. 140-201 in The Development of Health Policy: Norway, England, and Germany, Dr. P. H. Thesis, Baltimore: The Johns Hopkins School of Hygiene and Public Health, 1977, quote from p. 159. This work has served as a most valuable general source for this section and I want to express my appreciation and thanks to the author.
12. For the early poor law history and development of social welfare in Britain see Derek G. Gill, *The British National Health Service: A Sociologist's Perspective.* Washington, D.C.: U.S. DHEW, NIH Publication No. 80-2054, July, 1980.
13. Emanuel Shinnwell, *The Labour Story, Being a History of the Labour Party.* London: Macdonald, 1963, p. 24.
14. Michael J. Hill, *The State, Admnistration and the Individual.* Totowa, N.J.: Rowman and Littlefield, 1976.
15. Ibid, p. 36. Quote is from Frances Fox Piven and Richard A. Cloward, *Regulating the Poor: The Functions of Public Relief.* London: Tavistock, 1972, p. xiii (this book was also issued by Pantheon, New York, in 1971).
16. Hill, op. cit. (note 14):36-38. The full reference for Gilbert's work is B. B. Gilbert, *British Social Policy 1914-1939.* London: Batsford, 1970.
17. Ernest Mandel, The State in the Age of Capitalism, pp. 474-499 in *Late Capitalism.* London: Verso, 1975.
18. Antonio Gramsci, The Modern Prince in *Selections from the Prison Notebooks.* New York: International Publishers, 1971 (first published in Italian about 1923).
19. Ringen, op. cit. (note 11): 159-160. With regard to "socialist institutes and the labor movement" he cites Engels op. cit. (note 8): 249-277, especially 266-277.
20. Shinwell, op. cit. (note 13): 16.
21. Hobsbawm, op. cit. (note 8): 297-298.
22. Shinwell, op. cit. (note 13): 17.
23. Ibid, p. 18.
24. Ibid, p. 24.
25. Ibid, p. 20.

26. Ibid, p. 29.
27. Ibid, p. 30.
28. Shinwell, op. cit. (note 13) seems to deny this, but Hobsbawm, op. cit. (note 8) gives extensive documentation. See particularly p. 192 on Tom Mann's role and the "strategic recruiting" followed by many unions after 1906 in which attempts were serious to involve "all grades" and thereby create a unified working class as such with broad political as well as workshop goals.
29. Hobsbawm, op. cit. (note 8): 180.
30. Ibid, p. 181.
31. Shinwell, op. cit. (note 13): 61.
32. See Hobsbawm's penetrating overall evaluation, op. cit. (note 8): 336–337.
33. Cassen, op. cit. (note 2) observes (pp. 113–114):

> Quarrels within the Labour Party have for some time made the government's task easier . . .
>
> The left wing's success at Wembley in January 1981 gave Shirley Williams, David Owen, Roy Jenkins and William Rodgers their long-awaited opportunity to split from Labour and form a new party—the Social Democratic Party. Following its electoral alliance with the Liberal Party, the SDP had a series of successes in local council and parliamentary by elections, as well as persuading two dozen Labour MPs (and one Tory) to defect to them. Certainly its opponents do not underestimate the threat posed by the SDP. In the medium term, the introduction of proportional representation—which the SDP might impose as its price for joining a coalition government—may well transform the whole shape of British politics. Instead of the traditional pendulum, there might then be a stable centre bloc hinging on the SDP.

More recently, there are signs of a Labour Party comeback under its personable young new head, Neil Kinnock. Ben Lowe, A recovery or relapse for the Labour Party, *Guardian* (New York) (May 16, 1984): 14.

34. Keith Harper, Minister Warns of New Legislation on Unions. *The Guardian* (London) (December 1, 1981): 2. Also, Peter Osnos, Tories Offer Program to Undercut Unions, *The Hartford Courant* (June 23, 1983): A16.
35. Associated Press, Unions Worried by Thatcher Move to Insulate Spies, *The Hartford Courant* (February 2, 1984).
36. Gary Putka and George Anders, Strikes Prompt Thatcher to Call Cabinet Meeting; British Government Prepares to Be More Aggressive in Dock, Coal Walkouts, *The Wall Street Journal* (July 16, 1984): 23.
37. Associated Press, British Miners Seek Widened Strike Backing, *The Hartford Courant* (June 28, 1984): A17.

> Arthur Scargill, president of the National Union of Mineworkers, led the marchers from the Tower of London through the heart of the city to the Thames River, where a massive rally was held on the south bank.

"This union is not going to be defeated. Other governments give their coal industry subsidies and we want subsidies from (Prime Minister Margaret) Thatcher," Scargill, an avowed Marxist, told the cheering crowd.

Scargill said the union would meet Friday with steelworkers to appeal for support in the strike, now in its 16th week.

The Trades Union Congress, organizers of the march through London dubbed the "Day of Action," claimed about 30,000 people took part to show support for the miners. Police, however, estimated the marchers numbered 10,000 to 15,000.

Office workers leaned out of windows and lined the sidewalks as marchers passed by chanting "Maggie, Maggie, Maggie, Out, Out, Out," referring to Thatcher.

Many displayed yellow "Coal not Dole" badges and carried "MacGregor-Maggie's muppet" posters, referring to Ian MacGregor, head of the state-run National Coal Board.

About 130,000 miners went on strike to protest the coal board's proposed plan to shut down 20 pits and lay off 20,000 workers in an effort to make the coal industry profitable.

In support of the miners' cause, the National Union of Railwaymen called a one-day work stoppage, disrupting morning commuter service into London. Union leaders predicted the job action would cause worse problems late in the afternoon.

38. The Washington Post, "British Police under Fire as Strike Violence Grows, *The Hartford Courant* (June 1, 1984): D13; Associated Press, Striking British Miners, Police Brawl during Protest of Plan to Close Mines, *The Hartford Courant* (June 8, 1984): A10.
39. Ben Lowe and Alexandro Manning, Anti-Labor Tide About To Turn? *Guardian* (New York): 19. However, there may be some wishful thinking involved in this analysis, for very recent reports have the miners themselves split: "The dispute has split the union, with some 60,000 of the country's 174,000 miners having kept open 40 of 174 mines." Police, Miners Clash in Britain, *The Hartford Courant* (August 14, 1984): A2. In the meantime, such splits and further government pressure have brought about the collapse of the coal miners strike. Recent news suggests that the failure of the Labour Party to firmly back the miners' strike has further split the party: Miners' Strike Haunts Britain's Labour Party, *The Hartford Courant* (October 3, 1985): A 22.
40. Andre Gunder Frank, *Dependent Accumulation and Underdevelopment.* New York: Monthly Review Press, 1979.
41. "Britain to Increase Its Spending, Raise Some Charges to Help Pay," *Int'l. Herald Tribune* (December 3, 1981): 2. Incidentally one of the responses to unemployment and plant closings has been worker and community takeovers. See also the Gulvenkian Foundation Report "Whose Business Is Business?" referred to in Peter Draper, Unemployment Can Seriously Damage Your Health, *The Guardian* (London) (16 November, 1981): 9.
42. 3.1 Million Unemployed in Britain, *The Hartford Courant* (August 3, 1984): A2.

43. Cassen, op. cit. (note 2): 114 and the article on West Germany (F.R.G.) in the same work by Henri Menudier, p. 111. The comparative figures on this matter offer an example of the data problems I dealt with in Chapter 4. By some sources and figures the U.K.'s average level of living has been lower since 1977 than that of East Germany (G.D.R.) making the U.K. lowest among our study countries. However, Sivard, op. cit. (note 3): Table III gives per capita GNP (Gross National Product—whereas in the text I cited Cassen and Menudier's figures on Gross Domestic Product for the U.K. and West Germany) for 1980 as $9,213 for the U.K.; $13,399 for F.R.G.; and $7,226 for G.D.R.
44. Shail Jain, *Size Distribution of Income.* Washington, D.C.: The World Bank, 1975 (Table 27, p. 41 for G.D.R.; Table 76, p. 115 for U.K., and Table 77, p. 116 for U.S.A.).
45. Nicki Hart, Health and Inequality. Essex: Department of Sociology, University of Essex, March, 1978; J. R. Butler, *Family Doctors and Public Policy*. London: Routledge and Kegan, Paul, 1973; Julian Tudor Hart, The Inverse Care Law, *Lancet* Vol. 1 for 1971 (February 27): 405–412; Ellie Scrivens and Walter W. Holland, Inequalities in Health in Britain. A critique of a Research Working Party, *Effective Health Care, 1*:97–108, 1983.
46. John Downing, *Now You Do Know*. (An Independent report on racial oppression in Britain for submission to a World Council of Churches Consultation), London, March, 1980.
47. Scarman: We Have to Work for Peace, *Daily Express* (November 26, 1981): 2.
48.

> In Northern Ireland, hopes were raised in 1981 by the appointment of the conciliatory James Prior as Secretary of State, and by the agreement with Dublin to set up a joint Anglo-Irish council. These hopes were swiftly dashed. The Unionists were implacably hostile to the Dublin talks: Ian Paisley, accusing Mrs. Thatcher's ministers of having "blood on their hands," promised to make Northern Ireland "ungovernable." Meanwhile the Provisional IRA were clearly unimpressed by James Prior. They maintained their level of violence in the North, including the assassination of unionist MP Robert Bradford, and they also resumed their bombing campaign in London. Any sort of "settlement" seemed as far away as ever.

Cassen, op. cit. (note 2): 114.

49. As quoted in "cuts and the NHS," the politics of health group, pamphlet number two, London: typesetting Bread 'n Roses, printing Blackrose Press nd (about 1979).
50. If one is concerned with OSH services, as I am here, it should be called a quadripartite system, for OSH services were left outside the NHS entirely.
51. Harry Eckstein, *Pressure Group Politics: The Case of the British Medical Association*. London: George Allen and Unwin, 1960.
52. Some perks were offered for service in less desirable areas; these did not succeed in erasing inequalities. Butler, op. cit. (note 45).

53. Stanley R. Ingman, A Family Practice in England: Challenge to the Conventional Order in R. H. Elling, ed. *Comparative Health Systems*; supplement to *Inquiry, 12*:138–147, June 1975.
54. Julian Tudor Hart, Primary Care in the Industrial Areas of Britain, Evolution and Current Problems, *Int'l. J. of Health Services, 2*:349–365, 1972.
55. Roger Battistella and Theodore E. Chester, The 1974 Reorganization of the British National Health Service–Aims and Issues, *New England J. of Medicine, 289*:610–615, September 20, 1973. Curiously enough, this same year, the most important piece of OSH service legislation was passed; yet once again it was kept separate from the NHS.
56. Peter Draper, NHS: Candidate for Radical Surgery?, *New Humanist, 1*: 93–95, July 1972.
57. Organization Charts are presented for all levels and discussed in Gill's book, op. cit. (note 12): 158–176.
58. For the basic work on this see James O'Connor, *The Fiscal Crisis of the State*. New York: St. Martin's Press, 1978. For an application of this work to public financing of health services, see R. H. Elling, The Fiscal Crisis of the State and State Financing of Health Care–Toward a Conceptual and Action Statement for Improved Public Support of Medical and Health Care. *Social Science and Medicine 15C*:207–217, 1981.
59. "Cuts and the NHS" op. cit. (note 49).
60. In reaction to this program as proposed to Parliament by Sir Geoffrey Howe, Labour Party spokesman, Peter Shore, said, "You have reached a new low in your abysmal statement today," and he accused him of inflicting "sheer wanton damage on most of the nation and on all of the unemployed." "Britain to Increase . . ." op. cit. (note 41).
61. David Hencke, Tory Plan to Scrap the Way NHS Is Run; End of Financing by Tax-Action Would Boost Private Health Care. *The Guardian* (London) (December 1, 1981): 1.
62. Care in Action, A Handbook of Policies and Priorities for the Health and Personal Social Services in England, London: HMSO, 1981, esp. "Partnership with the Private Sector," pp. 41–43. While this most recent official health policy statement gives special attention to the elderly, mentally handicapped, disabled and impaired, and children, *there is no such treatment of workers' health.*
63. William Ryan, *Blaming the Victim*, (rev. ed.) New York: Vintage Books, 1976.
64. Rosemary C. R. Taylor, State Intervention in Postwar West European Health Care: The Case of Prevention in Italy and Britain, in Stephen Bornstein, *et al.*, (eds.), *The States in Capitalist Europe*. Vol. III of *The Casebook Series on European Politics and Society*. London: George Allen and Unwin, 1983.
65. Robert Maxwell, *Health and Wealth–An International Study of Health Spending*. Lexington, Massachusetts: D. C. Heath, 1981.
66. Barry Newman, Socialized Care; Frugal Medical Service Keeps Britons Healthy and Patiently Waiting, *The Wall Street Journal* (February 9, 1983): 1 and 27.

67. David Allen, England pp. 197–257 in Marshall W. Raffel, (ed.), *Comparative Health Systems, Descriptive Analyses of Fourteen National Health Systems*. University Park: The Pennsylvania State University Press, 1984, esp. p. 222.
68. Julian Tudor Hart, Primary Care . . . op. cit. (note 54): 363, gives figures showing a worsening of the relative mortality position of Class V men 15–64 and of coal miners and their wives between 1930–32 and 1959–63. Thus Class V men 15–64 were 11 percent above the mean mortality ratios for all classes in 1930–32, but 43 percent above in 1959–63!
69. The work by Allen, op. cit. (note 67) is particularly valuable in giving rich, updated information on costs, personnel, hospital beds, utilization, structure under the reorganization, health status and other matters.
70. Lesley Doyal with Imogen Pennell, *The Political Economy of Health*, London: Pluto Press, 1979; Vicente Navarro, *Class Struggle, the State and Medicine: An Historical and Contemporary Analysis of the Medical Sector in Great Britain*. London: Martin Robertson, 1978.
71. Elisha H. Atkins, Regulation of Occupational Carcinogens in the United States and Great Britain. The Politics of Science. Required research paper, UCONN Medical School, Program in Cross-National Study of Health Systems, Department of Community Medicine, 1977.
72. The key official document surveying the situation and asking for suggestions for future changes is "Occupational Health Services, The Way Ahead, A Discussion Document Issued by the Health and Safety Commission." London: HMSO, 1977.
73. R. E. Bibbings, Health and Safety at Work: The Trade Union View. *J. Soc. Occup. Med., 30*:90–97, 1980, (quote from p. 92). This is an abbreviated version of a longer official response to the "Way Ahead" document issued by the TUC: "Workplace Health and Safety Services, TUC proposals for an integrated approach," London: TUC, October, 1980. Among present OSH services shortcomings: it was said by a knowledgeable trade union official that "only 4 percent of the construction industry is covered, and half of that is the executives."
74. Some of the material in this section through March 1979 when the article was written is taken from R. Charles Clutterbuck, The State of Industrial Ill-Health in the United Kingdom, *Int'l. J. Health Services, 10*:149–160, 1980. I express here my appreciation and thanks to the author.
75. Ibid, p. 150. Clutterbuck quotes Marx on this series of Acts: "How could the essential character of the capitalist method of production be better shown than by the need for forcing upon it, by Acts of Parliament, the simplest appliances for maintaining cleanliness and health?" Karl Marx, *Capital*. London: Penguin, 1976.
76. Ibid, p. 150. On the point about men feeling their jobs were protected by restricting hours for women, Clutterbuck cites B. I. Hutchins and A. Harrison, *A History of Factory Legislation*. London: King and Sons, 1926.
77. Ibid, p. 151.
78. As the Webbs noted, "In the trade union world of today, there is no subject of which workmen of all shades of opinion and all variations of occupation are so unanimous and so ready to take combined action as the prevention of

accidents and the provision of healthy workplaces." Sidney Webb and Beatrice Webb, *Industrial Democracy*. Vol. I. London: Longmans, Green, 1911.

79. J. Grayson and C. Goddard, "Industrial Safety and the Trade Union Movement," *Studies for Trade Unionists, 1*:4, 1976.
80. One observer of the time recognized the trick: "This leaves alone the process of production. As long as dangerous processes remain, compulsory provision of a dispensary and free muzzle avails little." V. Nash, *Fortnightly Review* (February, 1893) as quoted by Clutterbuck, op. cit. (note 74): 151.
81. Thomas Legge, *Industrial Maladies*. Oxford: Oxford Medical Publishers, 1934.
82. R. S. C. Schilling, *Occupational Health Practice*. London: Buttersworths, 1973.
83. Clutterbuck, op. cit. (note 74): 152.
84. J. Williams, *Accidents and Ill-health at Work*. London: Staples Press, 1961, p. 342.
85. Associated Press, Britain Pays Workers at Nuclear Plant Who Suffer from Cancer, *The Hartford Courant* (January 5, 1984).
86. Legge, op. cit. (note 81): 191.
87. Clutterbuck, op. cit. (note 74): 153.
88. Industrial Health Research Board, *Health Research in Industry*. London: HMSO, 1945.
89. Clutterbuck, op. cit. (note 74): 153.
90. *Beveridge Committee on Social Insurance*. Cmnd 6404. London: HMSO, 1942.
91. Williams, op. cit. (note 84): 384–415.
92. Personal communication from Sheila McKechnie, 3rd May, 1984.
93. Grayson and Goddard, op. cit. (note 79): 2.
94. Clutterbuck, op. cit. (note 74): 154.
95. Richard Lewis, Compensation for Occupational Disease, *J. Social Welfare Law*, pp. 10–21, 1983, (esp. p. 16 and note 39).
96. Clutterbuck, op. cit. (note 74): 155.
97. *Safety and Health at Work*, Cmnd. 5034. London: HMSO, 1972, (The Robens Report).
98. Clutterbuck, op. cit. (note 74): 155.
99. "The Care of the Health of Hospital Staff." Report of the Joint Committee (Turnbridge Report) Scottish Home and Health Department, Ministry of Health. London: HMSO, 1971; Francis J. Cox, "Occupational Health Services for the Staff of the National Health Service: Current Policies and Problems," ISSN0141-2647, Working Paper No. 54, Health Services Management Unit Department of Social Administration, University of Manchester, August, 1981. Also, The Times Health Supplement of 13 November 1981 carried the results of a survey, "A Dangerous Place to be Ill."
100. But see the comment of a leading union Health and Safety Officer regarding the still unsatisfactory situation in 1984 (note 92 and associated text).
101. Clutterbuck, op. cit. (note 74): 156.
102. Bibbings, op. cit. (note 73): 90–91.

103. The unions have done a great deal through their national and regional safety officers. One of the largest has developed a very effective handbook: General and Municipal Workers' Union (GMWU) *Safety Representatives' Handbook*, London, 1978, reprinted January, 1980. The above quote is from pp. 49–50. Recently the GMWU has provided a "Model letter from Safety Reps to Manufacturers of Substances" to facilitate getting complete OSH information on machines, materials and chemicals affecting their union brothers and sisters.
104. J. C. McDonald, Four Pillars of Occupational Health, *Brit. Med. J.* V. 282, No. 6257:83–84, March 1, 1981.
105. Bernard J. Schuman, Physicians and Patients in the Occupational Setting: The Rules of the Game, *JAMA, 244*:2417–2418, November 28, 1979.
106. Twelve inches of text, plus a supplement giving actual suggested salary grades which could not be made available to me. The space given in the text to duties was eight and a quarter inches. The Occupational Physician, London, British Medical Association, 1980.
107. "Careers in Occupational Medicine," Royal College of Physicians, Faculty of Occupational Medicine, London, nd. (Probably January, 1980) (pamphlet).
108. Clutterbuck, op. cit. (note 74): 158.
109. Ibid, pp. 158–159.
110. "Select Committee on Science and Technology, Occupational Health and Hygiene Services," House of Lords, Session 1983–1984, 2nd Report. London, HMSO, December, 1983, p. 54.
111. Personal communication from Sheila McKechnie, Health and Safety Officer, ASTMS, 3rd May, 1984.
112. One of these is a nine page statement authored by Steve Watkins of the MPU. It is quite detailed and especially thoughtful. It would have been nice, if space permitted, to reproduce it here. It goes beyond medical students to consider OSH training for surgeons, GPs and psychiatrists, and considers what *The Lancet* and other publications might do.
113. James McEwen, James C. G. Pearson and Alison Langham, The Interface between Occupational Health Services and the National Health Service, *Public Health* (London) *96*:155–163, 1982.
114. Select Committee, etc. op. cit. (note 110): para. 8.7, p. 34.
115. Since this was first written as an informal preliminary report, the House of Lords, Select Committee, was formed and issued its Report (note 110). Unfortunately, the time was not ripe and the head of steam for real accomplishment was not there (see note 111 and associated text).
116. Hobsbawm, op. cit. (note 8): 336–337.

CHAPTER 12
The United States of America

This is the largest study country in terms of both population and geography. It is also the most diverse in terms of ethnic and cultural background, particularly among the working class. All these factors, and others to be discussed, have a bearing on the strength, or one should better say, right from the start, comparative weakness of the workers' movement and inadequacy of OSH conditions and provisions.

THE LAND AND PEOPLE[1]

The United States of America (U.S.A.) occupies the entire width of the North American continent from the Atlantic Ocean in the east to the Pacific Ocean in the west between the 30th parallel in the south to the 49th parallel in the north and also includes the large state of Alaska in the extreme northwest corner of the continent and the island state of Hawaii in the Pacific Ocean, some 2,000 miles southwest of San Francisco. It also controls a number of territories and dependencies. The sizeable island of Puerto Rico in the Caribbean has the formal status of "Commonwealth," while those involved in the active liberation struggle among Puerto Ricans, both in continental U.S.A. and on the island refer to it as a colony. Other territories and dependencies include the Virgin Islands, Guam, and Samoa.

The continental U.S.A. without Alaska is made up of forty-eight states and the District of Columbia in which the capital, Washington, D.C., is located. This area comprises 3,022,387 square miles. It reaches some 2,700 miles east to west and 1,300 miles north to south. With all fifty states included, it has 3,615,211 square miles, 38 times larger than the U.K. For comparison, the U.S.S.R. has an area 92 times larger than the U.K.

Neighboring countries are Canada to the north and Mexico to the south of the western half of U.S.A.; the Gulf of Mexico lies to the south of the eastern half.

An old mountain range, the Appalachians, stretches in a direction from southwest to northeast along the eastern part of the country, some 100 miles or so inland from the Atlantic Coast. Several peaks in this chain reach over 6,000 feet. In the west, the Rocky Mountains are a younger chain with many peaks reaching over 14,000 feet. The Rockies were a formidable barrier to early explorers and settlers seeking to expand European civilization to the Pacific.

The center of the country between these two mountain ranges is drained by a network of great rivers feeding into the Mississippi which runs north to south from Itasca in Minnesota to its mouth in the Gulf of Mexico, just south of New Orleans, Louisiana. Other major rivers in this network include the Missouri, the Arkansas, and the Ohio. The Ohio and Mississippi have served as major transport routes, playing an important role in development of the industrial heartland as well as shipment of agricultural products from the midwest.

A network of the vast fresh water Great Lakes formed by successive ice ages stretches along the border with Canada and drains into the Atlantic, via the St. Lawrence River.

The climate is generally temperate but ranges from the semi-tropics of southern Florida, southern California and Hawaii, to the frigid north of northern Maine, Minnesota and Alaska. The highest temperature ever recorded in U.S.A. was 134°F in Death Valley, California on July 10, 1913; the lowest was -76°F at Tanana, Alaska in January 1886.[2]

From a sparsely settled land peopled by numerous tribes of Native Americans before its "discovery" by Europeans,[3] the U.S.A. has grown through successive waves of immigration as well as natural increase to almost 230 million population as of 1980. Only China, India and the Soviet Union have larger populations. Each of the other study countries–U.K., Germany, Sweden and Finland,– contributed to the U.S.A.'s population. But there were other contributors. The present 230 plus million people are made up from wave after wave of settlers and immigrants from many lands, though not all contributed equally.[4] Ship after ship of African slaves and Chinese coolies were also brought over for the hardest dirtiest work on plantations and railroads. Eventually, all these peoples would take the whole land from the first Americans–the Indians whose ancestors may have come from Siberia some 40,000 years ago. The 1970 census classified 177,749,000 people or 87.5 percent as "White" and 24,463,000 or 12 percent as "Negro and Other." The total Black population was estimated at 24,426,736 in July 1975, so presumably "Other" are just over 1 million. The sizeable Hispanic population from Puerto Rico, Mexico and other Latin and South American countries were classified with Whites and may be some 5 percent of the total population. In 1972 there were 486,531 people reported as "Indians" living on or near reservations.[5] The total may be some half million both on and off reservations, down from at least 3 million at their height.[6]

In 1980 the U.S.A. had a density of twenty-four persons per square kilometer compared with 247 for F.R.G.; 230 for the U.K.; 155 for the G.D.R.; eighteen for Sweden; and fourteen for Finland. Just over three quarters of the people lived in urban areas (77 percent in 1980). The largest cities are New York with 16.5 million in its metropolitan area (7.5 million in the city proper); Los Angeles with 9.8 million in the metropolitan area (2.9 million in the city itself); Chicago with 7.7 million (3.0 million in the city); Philadelphia, 5.1 million (1.8 million); San Francisco, 4.6 million (1.6 million); Detroit, 4.4 million (1.3 million); Boston 3.7 million (.9 million); and Washington, D.C., 3.2 million (.6 million). Historically the center of population has moved southwestward from twenty-three miles east of Baltimore Maryland in 1790, to eight miles southwest of Cincinnati Ohio in 1880, to DeSoto Missouri in 1980.

The U.S.A. is a rich land in terms of natural resources with vast forests, coal and mineral deposits, and productive land and waters. While most would argue that today many of these resources are threatened from rapacious use and pollution, they have played an essential role in the U.S.A.'s development into a powerful part of the core of the capitalist world-system. It is the historical development of the U.S.A.'s political economy and workers' movement to which I now turn.

THE POLITICAL ECONOMY AND WORKERS' MOVEMENT

The U.S.A. is a constitutional bourgeois-dominated democracy. The American Revolution which, with France's help, established the independence of the country from King George III and Britain on October 19, 1781, with Cornwallis' surrender to Washington at Yorktown, was itself a bourgeois revolution. It led to the power of southern slave-holding landlords as well as northern businessmen, traders and later industrialists. The new nation was at first governed by the Articles of Confederation leaving most power in the hands of the thirteen original states. Only with the ratification of the Constitution on June 21, 1788 (when New Hampshire became the ninth state to ratify it, thereby putting it in force) was a system of popular elections established. The states' rights issue and fears of a centralized government kept the system a federated one with small states having an equal number of senators (two per state) and a proportional number in the House of Representatives. The laws passed by the two houses of Congress were to be executed by the President and his cabinet. A third independent arm of government, the Supreme Court and other levels of the judiciary was to judge the legality and constitutionality of laws and actions. Suffrage was limited to free white men of some property and majority age. For purposes of determining a state's number of Representatives, the population of slave-states was counted as including three-fifths of the slaves, though of course these three-fifths "non-persons" could not vote. Legally they were simply property.

I can not here go into a detailed history including the War of Independence from England (1775-1781), the Civil War (1861-1865) between the industrializing North and slave-holding rural South, and the rise of industrialism in the 19th century and waves of immigrants accompanying it.[7]

As already suggested, the U.S.A. is the most ethnically and culturally heterogeneous society included among our study countries and is surely one of the most heterogeneous in the world. It can be argued that this very heterogeneity has required that the society be held together with a very common, somewhat crass and simple culture–mainstreet U.S.A., if you will. Sinclair Lewis' *Mainstreet*, written out of his experiences growing up in a small Minnesota town is thought to have captured this common, but only skin deep culture, for an earlier age. Today one might find it best embodied on "Dallas" or some of the other TV serials which idolize new money and Cinderella-like success and big business and family connections in a flashy world of high technology and fast living. Such a view tends to ignore that some people still starve in the U.S.A. and many millions are unemployed and in poverty. Turner offered the idea of an ever-changing frontier as an almost unique American experience allowing a sense of open and new possibilities which might well cloud over the unpleasant realities of sweat shops and class exploitation.[8] Instead of affiliating with radical movements, that is, staying and fighting, the worker was admonished to "Go west, young man, go west." Turner's paper:

> . . . made individualism an interpretation of American history, by ignoring families and communities–that is, mutual aid–and tracing the secret of American uniqueness to the stoutest of all alleged individualists–the man of the frontier, as if there had been no women, or families, or communities, or books, or schools, or churches there.[9]

This history of escape from unpleasant realities through dreams of success in "new frontiers" while basic social problems go unsolved is laid out and well analyzed by Williams.[10] Following Gramsci's view, we can see that a ruling bourgeoisie encouraged such a cultural hegemony to help contend with potential opposition from the working classes. This culture usually emphasizes the common elements (apple pie and hamburgers; football and baseball; jazz and rock; individualism, free enterprise and patriotism) but sometimes plays groups off against one another to avoid a massive, unified working class movement.[11] Whatever view one takes of the formation and function of modern mainstreet culture in the U.S.A., the fact of underlying diversity, heterogeneity, and disparity is well known.

Although it is one of the wealthiest countries in the world, with a per capita income in 1980 of $11,347, the disparities are gross indeed. In 1977, the top 5 percent of households received 15.7 percent of income, while the bottom 20 percent received 5.2 percent of income. Among the bottom 20 percent of Whites in 1977 the top income was $8,690 per household, among the bottom 20 percent of Blacks it was $4,364. Stated in terms of extremes, rather than percentiles, while some are starving, and most people would be within a yard of the ground in a tower made up of child's blocks, each representing $1,000 of annual income, a very few individuals at the top would be higher than the Eiffel Tower![12] In 1977, 24.7 million people lived in poverty by official definition—11.6 percent of the population. Almost a third of Blacks (31.3%) lived in poverty.[13] During the last few years, with an especially conservative regime in power, the bottom 20 percent saw their after-tax income drop 8 percent, while the after-tax incomes of the richest fifth of households rose by 9 percent![14] Thus a poor family of four who had $10,000 income would have to get along on $9,200, while a family of four with $200,000 would have $18,000 more to "get by" on.

As in other capitalist nations, the unsteady rhythm of boom and bust has taken its toll in human misery. There have been many disasterous recessions and depressions. The biggest disaster of this sort came in The Great Depression of the 1930s.[15] Following WW II, while other capitalist nations were recovering from the destruction of the war, monopoly capitalism in the form of massive multi-national financial, manufacturing, mining, communications, and agri-businesses was developing in a long boom period. But in 1970 there was a recession, even as the Viet Nam War was still raging, again in 1975, and again the country was in a serious recession from 1979-1983. Only in recent months (late 1983 and 1984) are there clear signs of an interim recovery period developing.[16]

Unemployment reached 10.8 percent during the recent recession, when 12 million people were officially recognized as willing to work and looking for work but unable to find a job. Many more (perhaps half again as many) were discouraged, working only a few hours when they would want to work full time, or sick and disabled when employment might have made them well.[17] The rate of unemployment among young Black men was over 40 percent at its peak. Even well into the recent recovery, the overall unemployment rate was 7.5 percent.[18] Along with inflation running at about 13 percent for several years prior to 1981, unemployment has cut the effective buying power of the working class as in the U.K. and F.R.G. For some without a job, the decline is disasterous: A married woman with five children had worked as a secretary for thirteen months at a large airplane engine factory when she was laid off in May 1981. This factory is unionized, but it is an "open shop" with only some 60 percent of workers in the International Association of Machinists and Aero-Space Workers (IAM). This worker is still without a job, though she has been looking in all kinds of industries. As the same company recently announced another 935 layoffs, she and her husband were interviewed. He is a mason and was laid off in December, about the same time as her eligibility for unemployment benefits stopped.

> Elizabeth twists her hands as she describes her job hunt in words that come slowly. "I've been almost everywhere," she says, listing most of the major insurance companies and numerous industrial plants in the area. "A lot of them take the application and they say they'll let me know. . . . You go in and fill out the applications . . . some places I stopped in three, four, five or six times" . . . So she keeps looking, and they both keep worrying. "If push comes to shove I'll have to call my brother again (for a loan) but I don't like to do that" Winston said. "It really makes you frustrated. It makes me so mad I could go out and kill someone, but you go to jail for that. It just makes me sick inside. Everything you've got may go down the drain. . . . I feel like if I got sick now, I'd be better off dead—at least they'd get the insurance money."[19]

U.S.A. unions have long faced an uphill battle in protecting the interests of American workers. Aside from serious divisiveness between unions, and, as already noted, between workers of all kinds who could be played off against each other on the basis of racism and sexism, there have been involvements with organized crime in such fields as trucking, construction and shipping. And there have been other problems—not the least an anti-collective, pro-individualistic strain in the American cultural hegemony.

But possibly the most serious flaw developed with the worker's movement itself, albeit under pressures of waves of cheap immigrant labor, high unemployment during periods of recession and depression, and the cultural hegemony I have briefly discussed.[20] One wing represented by Eugene Debs (1855–1926) stood for a broad view of *labor in society and political organization*.[21] Debs polled 5.9 percent of the national vote as Socialist candidate for president in 1912.[22] By contrast Samuel Gompers (1850–1924) gave up the broad socialist view of his U.K. background and stood for a narrow view of wages, hours and working conditions—*labor in the shop*, often just the worker's own shop, without enough regard for workers in other situations. After Debs and fellow officers of the newly formed American Railway Union were jailed when the General Manager's Association of twenty-four privately owned railroads for the first time got a Federal court injunction against the Pullman strike in Chicago in 1894, momentum in the organization of American labor shifted to "Gomperism." With few exceptions it has remained that way.[23]

This brief report does not allow a detailing of the horrendous conditions and the slow and painful and even now only partially successful organizing experience of U.S.A. labor. As late as 1910, nearly 2 million children or 18.4 percent of those between ten and fifteen years of age were employed. This shortchanged their educations, ruined the health of many and took jobs from adults who would have demanded more money. Hundreds of children did back-breaking work in the coal mines.[24] It will help to fill in some of the objective conditions if we examine one very difficult, but eventually successful struggle.

Mostly local women workers were employed in the Lawrence Massachusetts textile mills when they opened. In later years they were joined by immigrants from many lands. The women were recruited by wagons sent into the countryside to find "rosy-cheeked maidens" for the mills. In time, these recruiting wagons became known as "slavers." The women worked from 6:00 a.m. to 10:00 p.m. six days a week and received $1.50 for a sixteen-hour day. Children too were sent into the mills, so the mill owners said, "to keep them from mischief." On January 10, 1860, the entire Pemberton mill collapsed, burying 670 workers, with hundreds killed and many more injured. By the turn of the century, the American Federation of Labor (AFL) had made progress in organizing workers with the slogan "eight hours for work; eight hours for rest; eight hours for what we will." To counter such organizing, the owners used many tricks, including enticing new immigrants with false promises and, for example, threatening to replace one entire department staffed by Polish workers with Italian workers.

When workers saw from their pay envelopes that they were suddenly getting less money (54 hours instead of 56) since the legislature had responded to public pressure to cut working hours, even though speed ups had been ordered to get the same production as with 56 hours, they left the mills *en masse* the morning of January 12, 1912. The companies hired goons to intimidate the workers with violence and used every other kind of vicious move. People began to go hungry and children had to be sent to working families in other cities. One morning women strikers, among them several pregnant women, volunteered to head the picket line to stop the brutal attacks of police. The police attacked anyway and two women had miscarriages. More than fifty militia companies were called in during the sixty-three-day strike. But as Joseph Etto of the Industrial Workers of the World (IWW) told the strikers "they cannot weave cloth with bayonets." Through these many struggles, tremendous national sympathy for the strikers was aroused and the strikers and their families provided for each other as best they could and stood together. It was called "The Bread and Roses Strike" for as Big Bill Haywood said, "The Women won the strike," and the women had sung "Give us bread, but give us roses too." On March 12 the mighty American Woolen Company, speaking for all Lawrence mills, surrendered. On March 14, some 25,000 men, women and children of many national backgrounds gathered on the Lawrence Common and solemnly voted on the settlement. This gave them wage increases on regular hours, time and one-quarter for overtime, premiums every two instead of four weeks, and "no discrimination to be shown to anyone." The workers of the world had united in Lawrence Massachusetts and carried through the first successful major industrial strike in the U.S.A.[25]

But without control over the means of production and the economy, working people were exposed to the vagaries of the capitalist system. In the intervening years, the world system of capitalism has had its effect and the textile mills in all of New England have closed and cheaper, unorganized labor was found in the south of U.S.A. and in underdeveloped countries. Thomas described the effects on the worker of management's efforts to rationalize and get more profit out of the capital input. With special reference to the much hated "stretch-out system" in the textile industry, he quotes a report of the Research Department of the United Textile Workers, written in 1932 at the depths of the Great Depression when some 20 percent of those normally gainfully employed were unemployed.

> In 1920 the maximum load anywhere in the industry was sixteen looms. . . . The predominant situation is that where the loom load formerly ranged up to sixteen, it is now around seventy-two looms, four and one half times as heavy as it was twelve years ago. . . .
>
> Even when unskilled workers are brought in to displace experienced weavers there is no decrease, but a tremendous increase in the weaver energy that goes into each yard of cloth. For example: a weaver running seventy-two looms must ceaselessly patrol a beat seventy-five yards long. In the course of an eight-hour day, he must walk between 15 and 18 miles. He does more patrolling than any two patrolmen on a city police force. He does this in addition to performing the variety of tasks connected with the weaving process. And he grinds out his 18-mile patrol in a shop with windows and skylights bolted shut, in order that no breath of fresh air may clear the intensely humid fog-like atmosphere, or reduce the 85 percent temperature at which the weave room is kept. These conditions mean more than complete exhaustion at the end of every day of labor. They mean permanent loss of weight, anemia, broken feet, varicose veins and finally, a complete physical breakdown.[26]

With its narrow shop-oriented, rather than broader political/economy-oriented view, and without a close and consistent tie to a political party of its own, as in Sweden and the U.K.,[27] the workers' movement in the U.S.A. has never been especially strong. In 1920 the AFL and independent unions in U.S.A. could claim only 18.6 percent of the non-agricultural workforce (unfortunately about what it is today for the AFL-CIO and other unions combined) while the TUC in the U.K. could claim 46.6 percent organized.[28]

Only in 1935 in the depths of the Great Depression when socialist and communist organizing threatened "the economic royalists," as Franklin Roosevelt called the most uncompromising members of the bourgeois class, did the relatively conservative American-style labor unions gain full and legal legitimacy with the passage of the Wagner Labor Relations Act. In 1937 the Supreme Court upheld the constitutionality of the Act in the National Labor Relations Board versus the Jones and Laughlin Steel Corporation. The act declared it illegal for employers to: 1) interfere in any way with the exercise of employees' rights to form unions; 2) promote company unions; 3) discriminate against union members in hiring or firing; 4) refuse to bargain collectively with legally formed unions. The NLRB, made up of three members chosen by the President, was to conduct union elections and make the final determination which union would serve as bargaining agent for the workers.

Other New Deal laws were also extremely important. The Social Security Act assured a minimal level of old-age pensions and included other provisions. It was this act which was amended in 1965 to become the first national compulsory health insurance, albeit limited to providing health care for those sixty-five and over under Medicare. The Fair Labor Standards Act of 1938 outlawed child labor in factories or shops producing for interstate commerce and introduced the minimum wage—set at 25 cents an hour then when some 15 percent were still unemployed; now at $3.30. The Works Project Administration provided government jobs for millions of workers—some even for parks, gardens, art, music, and writing. The New Deal sought to offer bread and roses. At the close of WW II, owners were feeling strong from their wartime profits and a renewed reserve army of unemployed (swollen not only by returning troops but also thousands of women who had entered work during the war, with the image of "Rosie the Riveter" to encourage them).[29] With overtime no longer necessary and paychecks shrinking, considerable labor violence and many strikes broke out. More work days were lost through strikes in the first half of 1946 than in any previous full year. John L. Lewis' defiance of the Federal government in the coal strike in late 1946 and the Republican victory in congressional elections brought the weakening of the NLRB Act through the Taft-Hartley Labor-Management Relations Act passed over presidential veto. President Truman called it "Shocking—bad for labor, bad for management, bad for the country." Union leaders called it "a slave labor law." Some of its major weakening provisions included the outlawing of the closed union shop and the outlawing of sympathy strikes; it also allowed states to adopt laws prohibiting the inclusion of union shops in labor agreements.

> Every southern state seized the opportunity to assure employers that it was prepared to cooperate with them to attract industry to its domain. The CIO Southern organizing drive, which had foundered in the late forties [in the face of the anti-Communist, anti-union atmosphere of the McCarthy period—see below], was to suffer almost complete disintegration during the following decade. After 1948 the trade unions found themselves on the defensive. Neither the swift expulsions of the CIO's Stalinist wing, nor the election of a Democratic President was sufficient to stem the antiunion tide.[30]

The anti-Communist histeria whipped up by the ruling class in the late 1940s and early 1950s through its agents, Senator Joseph McCarthy and the House of Representatives Un-American Activities Committee (HUAC), was disasterous for unions, especially for broadly-oriented and anti-business unionists who had provided important leadership in such key CIO unions as the United Electrical (UE) Workers. Shamefully, many union leaders collapsed and even capitalized upon this histeria.

> Clearly the CP and the left in the CIO could not have survived this coordinated onslaught under any circumstances, but it is important to note that the Communists had not prepared themselves to meet the purge. The party's reputation as a militant fighter for the rank and file and for blacks suffered during the war. Its radical change from militant anti-Fascism in 1938 to anti-war activism in 1939 and then to a 100 percent pro-war position in 1941 hurt the party's trade-union cadre. Communists found it difficult to present themselves as defenders of trade-union democracy when they had to follow radical changes in line dictated in Moscow. After the war, the party's preoccupation with foreign policy again hurt efforts to develop union support. Instead of concentrating on pressing issues of concern to union members, party activists were compelled to fight against the Marshall Plan, and to engage anti-Communists on Cold War issues. Under these circumstances, it was more difficult than ever for the party to build political support for socialist politics on a local level. . . .
>
> The Communists certainly made mistakes, but they had kept alive the practice of militant industrial unionism during the dark years of the twenties and early 1930s. And they performed much of the hard, dangerous work of building the key CIO unions. They were leading activists in the fight for black rights within the union movement. Though they compromised their position in some unions during the war, their absence from the House of Labor made it much harder for the civil-rights movement to find allies later on. Despite other compromises dictated from Moscow, the party tried to preserve the social unionism of the 1930s and extend it into the 1940s. And naturally, they tried to stem the tide of anti-Communism that helped many labor leaders reconcile themselves to conservative domestic policies and pro-business foreign policies. But the Cold War issue, which the Communists sought to defuse with an appeal to the old Popular Front, was the very issue that was used to destroy them. . . .
>
> As a result of the left purges and the counterattack launched by employers, CIO membership dropped from a wartime peak of 5.2 million to only 3.7 million in 1950. At this point, the old AFL claimed 8.5 million. Discussions had already resumed about a return of the purified CIO to the original House of Labor. Indeed, the rebel unions began to resemble those in the parent body more and more after they expelled the left and embraced the Cold War. Many had also retreated from aggressive, interracial unionism and from socially concerned unionism generally.[31]

Of course there would be many other battles to tell of—some successful, others not. But this is not a detailed history of labor in the U.S.A.[32] Today only 18.8 percent of the work force is organized, down 3 percent since 1980.[33] The strong points are in steel, autos, chemicals, mining, and other monopoly capitalist enterprises. But there are large unions of teachers, service workers, and government employees as well and unionization of health care workers is gaining. The ten largest unions are:[34]

– International Brotherhood of Teamsters: 1.7 million
– National Education Association: 1.7 million
– United Food and Commercial Workers: 1.3 million
– United Automobile Workers: 1.1 million
– American Federation of State, County and Municipal Employees: 1 million
– International Brotherhood of Electrical Workers: 1 million
– United Steelworkers of America: 800,000
– United Brotherhood of Carpenters and Joiners: 670,000
– Service Employees International Union: 750,000
– Communications Workers of America: 575,000

Other large unions which have been especially active in OSH, along with the auto and steelworkers unions, include the Oil Chemical and Atomic Workers and the International Association of Machinists and Aerospace Workers and the United Rubber Workers of America. Almost all these unions (the Teamsters excepted) belong to the AFL-CIO umbrella organization which represented some 15 million U.S.A. workers as of January 1, 1982. The AFL-CIO's organization is depicted in Figure 8.

Organized labor has been on the defensive in the U.S.A. for some time. But in recent years, especially under the conservative regime in power since 1980, the blows have been heavy indeed. Several weakening forces have permitted these blows. Not only did unemployment reach the official figure of 10.8 percent of the workforce during the recent recession (in Detroit and some other manufacturing centers it was a depression with unemployment reaching 20 percent or more in some areas) it has remained high in spite of the economic recovery of 1983-84. By March of 1984 it was still 7.8 percent with Blacks at 16.6 percent, Hispanics at 11.3 percent and Black teenagers at 46.7 percent![35] As noted earlier, the rate moved down in subsequent months but actually rose again to 7.5 percent in the summer of 1984 (note 18). A number of conservative analysts have begun to talk of high rates of permanent unemployment as if to shift the cultural hegemony from a pre-1970 expectation of three or four percent as "normal" to levels at which increased millions of working class members would learn to live with misery and hopelessness without taking to the streets and open rebellion.[36]

Important in this development has been the increased automation and computerization of U.S.A. industry in an effort to reduce labor costs and better compete with other capital centers in the world-system. The following comparison of organized labor's reactions to and fears of the introduction of labor-saving technologies in U.S.A. and Japan is revealing:

> In this country, as in Japan, robot enthusiasts dream of a bright new world in which the little fellows not only work in factories but

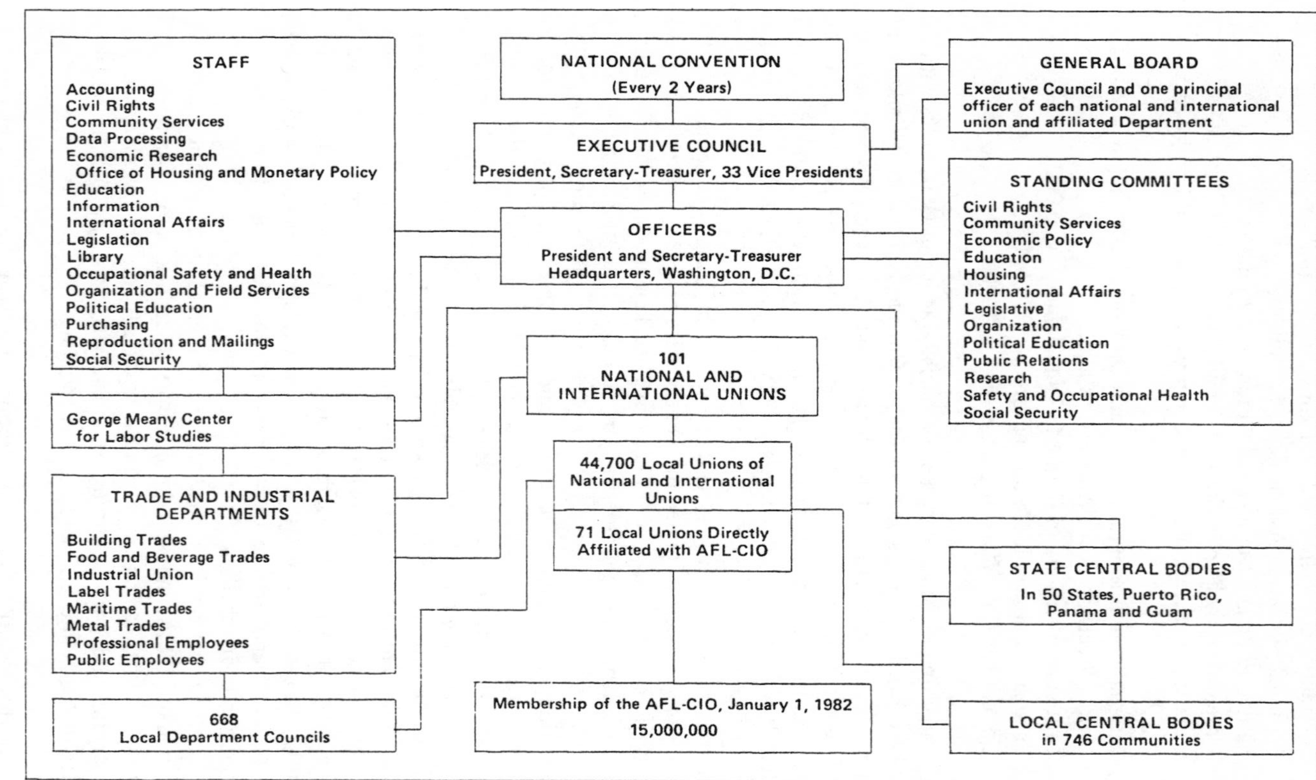

Figure 8. Structural organization of the American Federation of Labor and Congress of Industrial Organizations.

> also help housewives with the laundry, serve tea and fetch and carry for handicapped persons.
>
> As for the effect on workers, robot champions point out that, so far, robots have tended to be used for dirty, dangerous, or boring jobs that humans are well free of. In Japan, labor unions have actually welcomed the robots. Displaced workers have been retrained for other, usually better, jobs. And those who work with the robots give them pet names and treat them with affection.
>
> If Japanese workers do not see the robot as a threat, however, it is because they are guaranteed lifetime employment. No one is thrown onto the scrapheap because the robot can do his job better. Also, robots in Japan have been used in growing industries—autos, television manufacture and the like. And, perhaps most important, Japan has a homogeneous, slowly growing population. There is no large reservoir of hard-core unemployed, and no tide of immigrants that must be absorbed.
>
> Things are quite different in the United States, with its large-scale unemployment, shrinking auto industry, recession-plagued economy and a corporate ethic that accepts something less than full responsibility for the fate of workers displaced by advancing technology. A recent study by Carnegie-Mellon University concluded that state-of-the-art robots could replace 1 million workers by 1990 in the automotive and a handful of other industries. If robots grow more sophisticated, it said lost jobs might reach 3 million.
>
> Large companies, realizing that the robot revolution will progress much more rapidly with organized labor's acquiescence than without it, appear to be trying to avoid a confrontation on the issue. But there is no indication that they will give unions a say in robot-purchase decisions or offer contractual assurance that displaced workers will be assigned to other jobs.[37]

While the answer in Japan, for now at least, is a tradition of benevolent paternalism, labor cannot count on any such thing in the U.S.A. A most recent, especially ominous development is the use of computers and central video scanners to monitor worker productivity, literally from moment to moment.[38] The resultant stress and robotization of the person reminds us of Orwell's 1984.

Another weakening force has been the continuing displacement of capital and jobs to peripheral and semi-peripheral countries within the capitalist world-system where autocratic military regimes outlaw or repress unions and wages are extremely low. Lenin's view was that the surplus returned to core capitalist nations from such exploitation allows a better life for the working classes in those core nations and is therefore a weakening force which delays the revolution. But Szymanski shows that the surplus does not filter down, being captured by the ruling class for ever larger capital projects or various forms of self gratification, or being otherwise used up in military repression designed to keep "stability" in the neo-colonial empire. The real weakening forces in his view are lost jobs and lives lost in neo-colonial wars:

> The transfer of industrial capital and jobs out of the United States means that unemployment in the United States grows beyond what it would otherwise be. The larger the pool of unemployed (the bigger the "reserve army of labor"), the more workers actively seeking work. And thus the pressure on those with jobs not to press for higher pay, or even to accept decreases in their real wages, grows. The higher the proportion of the labor force that is unemployed because of jobs being exported overseas, the greater pressure is for generally lower wages in the United States. Thus, the greater the opportunities for the transnational corporations to invest overseas by taking advantage of cheap and compliant labor in countries with repressive regimes that restrict unions and strikes, the lower the general wage level is for all U.S. workers. Thus imperialism, through guaranteeing relatively labor-intensive investments in low-wage areas, results in lowering the material living standards of the working class of the advanced capitalist countries. The stagnation in the real take-home pay of U.S. industrial workers since 1965 (the purchasing power of their paychecks was approximately the same in 1965 as in 1980) must in good part be attributed to the pressure on wages in the most advanced imperialist countries due to both the export of jobs and the loss of export markets to their lower wage competitors. The stagnation in the wages of U.S. workers has occurred, it must be noted, at the same time there have been very rapid real wage increases for most other industrial working classes throughout the advanced capitalist countries.
>
> Last, it must be noted that the material effects of imperialism on the working class transcend economics. When wars are fought, such as that in Vietnam, it is mainly working-class people who are killed, injured, and permanently maimed. During the Indochina War 47,072 Americans died and another 153,000 required hospital care for their wounds (U.S. Department of Commerce 1979a, p. 375.) People from poor families were more likely to be drafted into the Army, were much more likely to volunteer (because of the lack of alternative job prospects), were more likely to be sent to Vietnam, to be assigned to combat, and to be wounded or killed. A study of those who were killed in Vietnam showed that GIs from poor families were about twice as likely to get killed as the average soldier.[39]

A recent example of prospective capital and job exports is Chrysler's overtures to Samsung in South Korea[40]—Chrysler which was bailed out of bankruptcy by massive government loans in major part on grounds of saving jobs for U.S.A. workers (as a part of the bailout, the UAW had to agree to cutbacks as compared with workers at General Motors, Ford and American Motors). As a metalworker union official said to me when I interviewed him in Helsinki, "Capital knows no home."

There have been other weakening pressures. Control of the government apparatus has afforded conservatives the chance to engage in all manner of deregulation which has not only hurt conservation and the environment but has hurt unions.[41]

At one point, a memo was written by a former Executive Secretary of the Young Republicans Federation, James L. Byrnes, now Deputy Associate Director of the Office of Personnel for U.S.A. government employees to the Director of Personnel recommending a certain "study" which

> . . . "would immediately divide the white-collar and blue-collar unions in the private sector as well as in government," Byrnes asserted.
>
> "We could create disorder within the Democratic House, pitting union against union and both against radical feminist groups," he said.[42]

As a result of such weakening pressures, unions have not only increasingly faced the union-busting tactic of more and more court battles (even when a law is clearly on the side of the union, but is untested in court, management brings suit to make the union expend its meager funds in this useless way)[43] but the conservatively stacked Supreme Court has been used to allow businesses to declare bankruptcy and reform under a new organization as a way of getting rid of unions.[44]

Threats of plant closings and moves of industries to low wage areas overseas and within the U.S.A.; subcontracts to small, nonunion shops (parallel production or outsourcing); automation and computerization; continuing high unemployment; other weakening pressures just suggested; as well as union-busting tactics make the union and working class struggle a tough one. Labor must unify and struggle. Fortunately there are some signs of this happening. Cooperation between unions is the word of the day. The United Auto Workers (UAW) with 1.2 million members (down 300,000 since 1979 because of especially high layoffs in the auto industry) recently rejoined the AFL-CIO. The Health Care Employees Union, especially District 1199 of New England, is urging merger with the Service Employees International Union, etc. Also, there are ways to fight against the givebacks and concessions of recent years.[45] Thus, in spite of the several weakening pressures mentioned above, some see a resurgence of organized labor in the offing.[46] And in spite of the low and dropping percentage of unionized workers, even the labor relations head for Ford Motor Company, the Jesuit-trained, as well as Georgetown Law–and Harvard Business School–trained, Peter Pestillo, sees organized labor as a standard setting force for wages and working conditions into the foreseeable future because they are strong in the core monopoly capitalist industries. "Our core industries are almost fully unionized, and labor will be a driving force in compensation in American industry, despite not having the majority."[47]

What is really required to achieve the kind of political transformation or "political exchange" which sets Sweden aside among our study countries as one in which labor is so strong that industrial peace generally prevails on its terms,[48] is the generation of and alliance with a left-guided political movement in which a labor-dominant political party gains control of the state apparatus so as to ameliorate both in-shop and community conditions. Unfortunately, the cultural hegemony against organized labor is so strong in U.S.A. that their recent almost unified and unprecedented pre-convention backing of the Democratic candidate (Walter Mondale) for President in the 1984 election almost turned into an embarrassment for the candidate, a sign that he is a "captive of special interests." Further, there is no sign of a broad left-oriented conception in this move:

> Leaders like former UAW President Douglas Fraser or even AFL-CIO President Lane Kirkland have warned that a class war is building—precipitated by management with the encouragement of the Reagan administration. But most unions have been reluctant to declare war in return. The AFL-CIO has not made a sharp move to the left in response to the Reagan attack. It has not called for strong and democratic planning, worker control on the job or an end to Cold War aggressiveness. Instead, it is trying to salvage and modernize an old social contract—the implicit agreement that has prevailed among government, business and labor since World War II, and that began to crumble in the early 1970s.
>
> The elements of this gentlemen's agreement are familiar: labor and management would fight over wages and benefits, but labor would not challenge management authority over products, investments and organization of the workplace. In exchange, management would not challenge the institutional status of unions. Management would accept a modest welfare state, and labor would not challenge capitalism or corporate power in itself. In foreign policy, both agreed to an aggressive U.S. battle under the cold-war banner against any anticapitalist or leftist movements or governments, and approved an expanding, permanent military.
>
> The choice of Mondale was a perfect expression of labor's political hopes for the return of a modest New Deal. Vastly superior though he may be to Reagan, Mondale offers not a bold alternative but a renewed gentlemen's agreement.[49]

This is not a unique view. Many observers, including foreign friends of labor have long observed that U.S.A. business unionism is too conservative and narrow.[50]

Aronowitz' prescription in his new book[51] calls for a realization that the "corporatist perspective" of the past in which organized labor provided social peace in exchange for its share, but a non-redistributive share, of an ever growing economic pie will not work. First, the U.S.A.'s position of dominance among core nations of the capitalist world-system is being challenged and the worldwide crisis of capitalism does not allow the assumption of an ever expanding economic pie. Second, the nature of production is changing to require more educated, technically prepared workers. Third, organized labor in itself is too weak and must form a broad political coalition with minorities, the poor, the women's movement, the environmental movement, and the elderly. It was just such a convergence of interests plus the ferment of the civil rights movement and the anti-Vietnam War movement, and not labor by itself, which carried OSHA into law in 1970.[52] Michael Harrington, co-chairperson of the Democratic Socialists of America, in his review of Aronowitz' in the *Guardian* book thinks the author is on the right track:

> All of this leads Aronowitz to his notion of a new "political bloc" based upon a labor movement that has regained—in precisely the context described in *Working Class Hero*—a class struggle perspective and that reaches out to the new social movements (minorities, women, environmentalists) that are "objectively aligned in opposition to the dominant neo-liberal and conservative fractions of capital." Is that pie in the sky? I think not. In September 1981, at the Solidarity Day demonstration in Washington, something like Aronowitz' "new political bloc" was assembled—but it did not go anywhere, and it was never mobilized again. Yet there was proof that a decisive call for such a mobilization would be answered.

While I am sympathetic and favor the general direction suggested, I am not sure the call is radically explicit enough in terms of the necessity of the working class and their allies to assume state power and gain control of the means of production for human social ends, rather than private gain.

Whatever one's view of the future of the workers' movement in U.S.A. even the brief background presented here should be enough to allow us to rate it the weakest among our six study countries.

THE GENERAL HEALTH "SYSTEM" IN U.S.A.

Working class struggles and poor people's movements have waxed and waned and grown again and faded and with each wave has come a state response of various social welfare and health programs many of which have then been removed or weakened when the ferment has subsided.[53] But the working class struggle has never been so prolonged or well organized and consistent as to have achieved the kind of comprehensive, fairly well regionalized system of health care represented by the NHS in the U.K. or the rather fully regionalized NHS systems in G.D.R. and Sweden. The U.S.A. has been said to have a "non-system." This is not sociologically correct, but it is descriptive of the fractionated congeries of programs and services which have been instituted over the years, with some growing and others left to languish.

It is a predominantly market-oriented system with care being provided primarily by private physicians and associated nursing and other personnel, under the physician's direction, in his private office. The 7,000 some hospitals are generally free-standing entities unto themselves in competition with each other.

Hospitals are sponsored by private, non-profit; private religious; government; and an increasing number of private, for profit organizations which "skim the cream" of cases which can be handled with high-tech treatments and are covered by insurance or well to do circumstances, leaving the chronic and poorly insured cases for other hospitals, especially those under public control.

The hospital staffs (except for physicians) are full-time wage and salaried employees, increasingly interested in unionization because of a tradition of low wages, arbitraryness of management and little worker control.[54] Curiously enough, the physicians whose real interest is in their own practices, centered in their private offices, actually determine the work of the hospital. Under this system, the hospital is regarded as "the doctor's workshop," rather than being considered the center of community health, within a regionalized network of hospitals. In this system as many as thirty "attending" physicians come in and out of a hospital ward to see their own one or two patients. They pass orders to the resident physicians and nursing staff, with considerable difficulties in communication and complexity of treatment styles. This complexity of treatment styles and of communication has been suggested as the main reason U.S.A. hospitals have almost double the number of hospital personnel per patient day as in Swedish hospitals (diagnosis held constant).[55]

The fractionated, market-oriented character of the U.S.A. health system is not only determined by the individualistic, free enterprise ethos of the American Medical Association which has struggled vigorously against any form of "socialized medicine,"[56] it is part of the federated system of governance and tradition of interpreting the Constitution so as to leave many powers with the states and above all, the private sector.

> Except in some narrowly defined areas, the delivery of health services has been the domain of state government and the private sector. The federal government's authority to operate a hospital or to provide medical care is constitutionally confined to personnel in the armed forces and their dependents, veterans, wards of the federal government (American Indians and Eskimos who live on native lands), and to a relatively small cadre of other federal beneficiaries such as those in the Coast Guard and the Public Health Service. The constitutional limitation is not explicit since the word "health" does not appear in the U.S. Constitution; the limitation is rather by interpretation and resulting tradition or precedent.[57]

In addition to Medicare and Medicaid coverage for hospital charges and for some parts of physicians' services for some people, a large proportion of the population has some form of private sickness care insurance. But direct self payments still made up 32.4 percent of the total for all ages (the rest coming from public funds, 40.2 percent; and philanthropy or industry 1.2 percent).[58] The reason direct self payments are still needed to cover such a high proportion of medical care costs is that the 2,000 or more different private insurances are usually very limited with partial payment of hospital charges, many exclusions, deductibles, etc. Union contracts often provide more complete health insurance packages. But one thing foreign visitors often remark on is that it is a fearful thing for ordinary people to become ill in U.S.A.–fearful as much for the cost as for the illness. As a cartoon depicts it, the sad-faced doctor standing outside a patient's hospital room tells the wife, "He's loosing his will to pay."

The proportion of federal, state and local governmental funds supporting medical care in the U.S.A. has grown from 27.2 percent in 1950 to 29.4 percent in 1966 (the year after Medicare came in, a compulsory medical and hospital care insurance for those sixty-five years and older, as well as Medicaid, a public voluntary program for the poor) to 42.2 percent in 1980. But it is still predominantly a system paid out of private funds.[59] Most of these are paid through many different private insurances with varied coverages and provisions, but, as noted, some costs are still self-paid. In his comparison of funding and costs of the health care systems in nineteen industrialized countries, Maxwell provided a classification of systems according to degree of government financing and control. The U.S.A. was the lowest on both dimensions, the U.S.S.R., other Eastern European countries, the U.K. and Sweden were close to the high ends on both dimensions.[60]

This form of "organization" leads to some startling contrasts with other systems. The U.S.A. has double the number of surgeons per population as compared with the U.K., so, of course, it also has double the number of operations per unit of population.[61] One estimate is that there are more neurosurgeons in Miami than in all of England. One study led to an estimate that some 4 million unnecessary operations are done each year in the U.S.A. with more than 11,000 deaths from the side effects of anesthesia alone, to say nothing of mistakes.[62]

Market forces not only influence the behavior of surgeons and others working in the system, the health arena becomes an attractive one in which to invest. In 1979, before the recent recession, the weekly TV show "Wall Street Week," giving well known investment advice, featured "an expert in health stocks." He said that whereas the market as a whole had risen 45 percent since 1965, health stocks had risen 500 percent! It is this motivating force, capital in search of profits, rather than some automatic "technological imperative," or a clear, simple desire to provide better medical care, which explains to some analysts such phenomena as the widespread adoption of unproven, expensive medical technologies[63] and phenomenal sale of drugs in the U.S.A.[64]

Since patterns of PHC are important if one is considering links between OSH and PHC, it is necessary to understand that in the U.S.A. much too large a proportion of people get episodic care from hospital emergency rooms in a quite anonymous way; few people can say who their family doctor is; and most of those who go to a doctor's office are first involved in self diagnosis, hoping to get to the right specialist. In addition, to add to the complexity of the picture of possible points of first contact for the sick worker or anyone else, there are walk-in surgery-centers, group practices, hospital outpatient clinics, public health clinics and neighborhood health centers (usually for the urban poor and most of these have been drastically cut in recent years).

The system is not at all regionalized. There are twelve CAT scanners in the Denver area alone, whereas the whole U.K. gets by with some fifteen. Nor is it a planned system. It has been called "a planning system" but not a planned system. There have been several, but sometimes competing pieces of health planning legislation—Hill Burton; then Comprehensive Health Planning and Regional Medical Programs together; then in 1974 the National Health Planning and Resources Development Act, setting up more than 200 Health Systems Agencies (HSAs) or planning areas and meant to merge earlier efforts.[65] Most recently a phase-out budget killing most of the HSAs was implemented by the Reagan Administration.[66] This sense of ferment with little real movement has been analyzed elsewhere.[67] Fundamentally, the explanation lies with the way control of the health system, participation in it, and enjoyment of its services articulate with the class structure.[68] It is a system seriously oriented toward the interests of the bourgeoisie but probably does not give them good overall and preventive care either—only good superspecialty, acute care.

At the same time as there are such overconcentrations of costly specialized elements in certain settings and places, there are great gaps in primary care and in preventive services for those without insurance. Of the 272,000 physicians in active office-based practice, both public and private in the AMA's 1980 census (but not some 104,000 active physicians, including interns and residents, in hospital-based practices) just 48,020 or 17.7 percent were in general or family practice.[69] As far as distribution goes, rural areas as well as urban slum areas such as the South Bronx are virtual medical no-person-lands. There is no such thing as a geographic focus at the periphery of the system, providing coverage to all parts of the population as in the regionalized systems in the U.K. and the G.D.R.

Physicians' incomes in 1982 averaged $93,000 in the U.S.A. These ranged from averages of $65,100 for pediatricians, to $72,200 for family or general practitioners, to $118,600 for both surgeons and anesthesiologists.

> Incomes above $250,000 per year are not uncommon, particularly among some specialties, and there is growing public concern as to the social acceptability of such high incomes. In view of the fact that the cost of their medical education was heavily subsidized, this social concern is particularly understandable.[70]

Overall costs for physicians' services, hospitals, drugs and other aspects of care have been rising steadily from 7.5 percent of GNP in 1970 to 10.5 percent in 1982 when $322 billion went for these services. It is estimated that in 1984 the cost per person will be $1,500 in U.S.A. compared with $500 in the U.K.

A researcher of public policy issues for Aetna Life and Casualty Company, one of the larger among hundreds of private insurers, writes as follows in order to generate public support for the Diagnostic Reimbursement Group (DRG) system, which has recently been adopted nationally for Medicare patients, as a prospective system of payment to hospitals (rather than the open-ended, retrospective system which has prevailed) and will spread by recent law in Connecticut to "all payers:"

> Clearly, the rate of growth in health-care expenditures is too high; health-care cost containment measures must be implemented. . . .
>
> Until recently, most private and government reimbursement to doctors and hospitals has been on a retrospective cost-incurred basis. Under such a system, hospitals and doctors providing more and more costly services are rewarded by higher revenues while those practicing in a more cost-conscious fashion are not.
>
> To encourage more cost-conscious care, hospitals should be paid under a prospectively determined pricing system whereby payments are determined in advance. Hospitals with operating costs less than the fixed reimbursement can retain the difference, while those that exceed it will incur a loss. . . .
>
> The healing arts cannot continue indefinitely to command a growing share of the GNP.[71]

For all the expenses, the U.S.A. population does not have a higher health status than other of our study countries. The infant mortality rate (deaths under one year per 1000 live births) for 1980 was 13 for the U.S.A. as a whole, compared with 13 for F.R.G.; 12 for the U.K.; 12 for G.D.R.; 8 for Finland; and 7 for Sweden. Life expectancy, males and females combined, was 74 years, compared with 73 in the U.K. and F.R.G.; 72 in G.D.R.; 73 in Finland; and 75 in Sweden.[72] These figures do not show any of the details of a generally more favorable picture for Swedish as compared with U.S.A. white males.[73] At the same time, disparities in health status are gross. The situation of Native Americans is particularly tragic.[74] The ratio of Black and other non-white infant death rates to white infant death rates has actually risen (though there have been overall declines in the rates themselves for all groups) from 1.6:1 (44.5 to 26.8) in 1950 to 1.9:1 (21.8 to 11.4) in 1979.[75] In 1979 comparative figures on cancer were issued. In the previous twenty-five years, the cancer incidence rate went up 8 percent for Blacks, while it dropped 3 percent for Whites. Cancer mortality increased 26 percent for Blacks, 5 percent for Whites. Studies also show that cancer in Whites tends to be diagnosed in earlier, death-preventing stages than is true for Blacks.[76]

With the U.K. spending only half as much on all health care as a percentage of GNP (and even less than half in absolute terms) and the U.K. rating higher on most of ten health indicators used by Maxwell—even the increasingly health-cost-conscious U.S.A. business interests are eyeing the U.K.'s NHS. A long article in

The Wall Street Journal cited these facts and quoted Prime Minister Thatcher as saying "The NHS is safe with us."[77] Indeed, as the D.R.G. system shows itself to be ill conceived and a failure, as it must, for each hospital cannot be an effective cost control center in an otherwise non-regionalized, market-oriented, competitive system, we may find some strange bedfellows—possibly big capital as well as big labor and a coalition of the elderly, the poor, and others fearful of becoming ill under the present system—pushing for something like the Dellums Bill to bring about a national health service in the U.S.A.—one which properly integrates OSH and PHC.[78]

While there would be many other aspects of the U.S.A.'s health "non-system" to present, including recent worries over a "glut" of physicians being turned out by the medical schools, and concerns for what this means in terms of the costs each one will generate, the main point here is to realize that the general health services cannot be very supportive of workers' health as presently structured. This point will be further detailed in the following sections.

THE OSH SYSTEM AND SERVICES[79]

The early history of OSH in the U.S.A. includes both safety and health disasters. Earlier, as part of the background to the "Bread and Roses" strike in Lawrence Massachusetts in 1912, I mentioned the collapse of the Pemberton mill in 1860. On March 25, 1911 fire broke out on the upper floors of the Triangle Shirtwaist Company in New York City. Nearly all the doors were blocked, or locked, or opened inward. There were no fire escapes.

> One hundred and forty-five of the five hundred employees, most young Jewish and Italian immigrant women, were burned to death or died jumping from the building. The New York State Factory Investigating Commission was formed in the aftermath of the fire.[80]

With more and more workers being killed or maimed on the job, unions agitating around these issues, and workers or their survivors, especially where unions could assist them, suing for compensation—occasionally for big awards—industrialists pushed for the establishment of the National Safety Council (NSC) in 1912 and a minimal workers' (then called workmen's) compensation system. Such a system had first been instituted in the steel industry in a context of union busting and corporate welfare paternalism meant to freeze unions out. It would take away the power of workers to sue their companies directly and would pension off workers at less than their regular wages. Even today, for example, in Connecticut, disabled workers receive only two-thirds of their wages up to a certain limit and the system generally in the U.S.A. is woefully inadequate especially for work-related diseases.[81]

As a way around this situation, workers exposed to asbestos and now suffering from asbestosis or lung cancer have used the well-documented asbestos coverup to bring suit against the suppliers of the materials. Many of these third-party cases are being won for sizable sums.[82] There are likely to be many such cases, since "more than 200,000 of the 9 million surviving workers who have been exposed to asbestos during the last forty years will die of asbestos-associated cancer by the end of the century," an internal Department of Labor report concludes.[83] In the meantime, Johns-Manville's response has been to declare bankruptcy, hoping to throw the burden of compensating damaged workers on the public.

> If there is any route that the Manville Corp. hasn't taken to avoid paying the full claims of the thousands of workers whose health has suffered from asbestos, then it probably hasn't been discovered yet.
>
> The corporation has already declared bankruptcy to stave off some 20,000 claims involving potential liabilities of $40 billion while it reorganizes; it has been pressing Congress to pass legislation that would require the federal government to pay half the costs of asbestos claims; and it has now sued the federal government for $1 million to recover money it already has paid in fifty cases.
>
> In short, Manville is thrashing about in every direction to avoid the total responsibility for workers exposed to its asbestos who are now suffering from asbestosis, mesothelioma, lung cancer and related ailments.
>
> Manville is being disingenuous when it claims it didn't know about the hazards of asbestos to shipyard workers during World War II, but there is some basis for assigning part of the blame to the federal government.
>
> Studies about the dangers of asbestos go back to the 1920s, and the Navy knew about them just as certainly as did the manufacturers. Yet the precautions that might have prevented the diseases that swelled up many years later, were not taken.[84]

Other asbestos supply companies have sought to form a combination with insurers to begin settling claims, if possible out of court at lower award levels.[85] Business interests are also attempting to push a "White Lung Bill" through Congress (named after the Black Lung Bill for coal miners, but not exactly modeled after it) to eliminte third-party suits (that is, against suppliers).

Another potentially even more interesting way around the restrictions of the workers' compensation laws has recently been found by workers harmed by dioxin and their lawyers. In these 170 cases the workers are charging that Monsanto Chemical Company knew the hazards of dioxin but went ahead deliberately and exposed workers and they have suffered higher than expected incidences of heart, liver, lung, nerve, and skin maladies. The worker compensation laws can not be used to hide behind deliberate exposures. These workers are seeking more than $2 billion in damages.[86]

Because compensation and awards for disease acquired in the workplace (as distinct from death or disability due to accidents) have only recently begun to develop, and because awards for accidents have been modest, capital has not paid much heed to the system, nor have insurers had the kind of organizing influence on OSH services which we found for the workers' insurers in F.R.G. (the Berufsgenossenschaften). But recently, with an increasing awareness of disease due to work, probably more through the Asbestos trials than anything else (though New Directions Programs in worker education, the unions, and COSH groups have undoubtedly had an impact—I will come back to these efforts) business interests have launched a campaign to cut back worker claims for disease compensation.[87] But I am getting ahead of the story into some of the most recent developments.

During the Great Depression a mass slaughter of workers occurred because of a disease (silicosis) contracted from the blasting, drilling and digging of a hydroelectric tunnel for a Union-Carbide subsidiary located near Gauley Bridge, West Virginia. The workforce was made up mostly of Blacks, desperate to have a job (first at 50¢ an hour, later 30¢)

> . . . the water tunnel was being cut through almost pure silica, and the dust was so thick that workers sometimes could see barely ten feet in the train headlights. Instead of waiting thirty minutes after blasting, as required by state law, workers were herded back into the tunnel immediately, often beaten by foremen with pick handles. According to the report:
>
> Increasing numbers of workers became progressively shorter of breath and then dropped dead. Rhinehart-Dennis contracted with a local undertaker to bury the blacks in a field at fifty-five dollars per corpse. Three hours was the standard elapsed time between death in the tunnel and burial. In this way, the company avoided the formalities of an autopsy and death certificate. It was estimated that 169 blacks ended up in the field, two or three to a hole . . .
>
> Toward the end of the project, some workers bought their own respirators for $2.50. The purchasing agent for Rhinehard-Dennis was overheard to say to a respirator salesman, "I wouldn't give $2.50 for all the niggers on the job." The paymaster was also heard to say, "I knew they was going to kill those niggers within five years, but I didn't know they was going to kill them so quick."
>
> Altogether, over 470 men died and 1,500 were disabled. Years later, when survivors sued Rhinehart-Dennis for negligence under common law (silicosis wasn't compensable), rumors of jury tampering were widespread, and a group of 167 suits were settled out of court for $130,000, with half going to workers' attorneys. Blacks received from $80 to $250 and whites from $250 to $1000 each. Later it was discovered that one of the workers' law firms had accepted a $20,000 side payment from the companies.[88]

While even this monstrous "incident," and the congressional hearings it provoked which ended in 1936, failed to galvanize public pressure and action for improved health as well as safety conditions on the job, industrial interests formed the Air Hygiene Foundation, later named the Industrial Health Foundation (IHF).[89] Its declared purpose was to clean up the air in industry. In fact, as a confidential report from the founding conference recognized, its main concern was to forestall worker claims stimulated by the Gauley Bridge disaster and subsequent congressional hearings.[90]

But one important piece of legislation did result from this period, even if not directly from the Gauley Bridge disaster. The Walsh-Healey Act of 1936 was the first major law requiring inspections and enforcement of standards on work under public contracts. In 1965, this was extended by the McNamara-O'Hara Act to workers supplying services or products to the government.[91]

Until the landmark legislation of 1969 and 1970, most of the action in OSH came from the IHF. WW II brought higher accident rates with the mass influx of new workers and a management consciousness to conserve manpower during a time of shortage. For the first time federal grants were made available to support state OSH programs. Berman mentions a number of significant strikes which occurred in the 1950s around OSH incidents and issues. But these had mostly a local influence.

Aside from the IHF standard setting, company efforts, which had earlier included significant safety measures around the time of the formation of the NSC when union action around safety issues was strong,[92] now revolved largely around educating workers to accept the "careless worker" ideology. Thus if a worker walked where a chain from a crane's boom broke, and the worker was crushed by a falling steel beam, the instructor analyzed it as the worker's fault for walking where he did, rather than an equipment failure preventable through proper safety engineering surveillance.

As in other countries, large companies had begun to employ physicians around the turn of the century. But a survey conducted in 1919 described industrial medicine, along with personnel, safety, and compensation as "specialties in the science of management."[93] The same survey found "a real medical director, with freedom, responsibility, and authority to which he was entitled" in only 4 percent of plants surveyed. Examinations for employment were exceedingly superficial and there seldom was any systematic follow-up of exposed workers. Mostly acute, emergency care was offered.

Unfortunately, the situation has not changed much. In May of 1984, the physician for a large manufacturing firm told my class of medical students taking an elective course in OSH that an important part of the job of industrial physicians is to arrange to have disabled workers spied upon outside of work. "They don't want to work and would rather get something for nothing," he said. "A worker may be getting compensation for a bad back and you'll find he's the star of a local basketball team." A much respected colleague with whom I once worked for several years was Medical Director and Vice President of Metropolitan Life. He told it like it is in a book of his published in 1961 to help the physician "orient himself in a new environment:"

> The physician's place in the industrial system is quite different from that to which he has become accustomed in private practice . . . he is not top man as he is in the hospital or his private office. His services are strictly ancillary to the main purpose of business: production at a profit. His value depends upon his willingness and ability to work with others to achieve that main purpose.[94]

In 1973, the Industrial Medical Association estimated that some 4,100 physicians practiced occupational medicine to some degree, sometimes in a private office or clinic, but more often in teaching, research or administrative settings. In 1969, it was estimated that industry employed 4,200 physicians, but it was not clear what level of OSH training these people had or what their functions were—that is, whether purely for acute emergency care or for hazard surveillance and other preventive measures. I do not have more recent figures. No doubt the training centers set up by NIOSH after 1975 have had an important and good effect, especially as companies have sought to protect themselves from OSHA inspectors. But in May 1980 the secretary for the Occupational Medical Association in Connecticut (a heavily industrialized state) himself a retired urologist with no prior OSH training or experience, now serving a major manufacturer as a full-time OSH physician, said to an elective medical school class of mine that only a handful of members of the association are full time, fully qualified specialists and most are part-timers serving one or two factories on the side without any OSH training. It is especially important to note that there is no general OSH services standard under OSHA. This contrasts sharply with what exists in Finland and the G.D.R. and even F.R.G., and what is planned in Sweden, according to the recent Wessman Commission Report (see Chapter 7).

Ashford found only one out of seven major unions employing a physician (note 79). This has improved since OSHA came into effect.

Berman and Ashford cover the requirements and manpower situation in other OSH fields—nursing, safety engineering, etc. The special purpose of this study does not allow such attention here. But it is important to make note that Industrial Hygiene is probably much better recognized and developed in U.S.A. than it is in other countries, though there are the same problems of management sponsorship and control. However, many industrial hygienists have served as OSHA inspectors and some with COSH groups (see below) and have made important contributions.

A totally false impression would be given if I did not recognize the contribution of courageous, independent-minded OSH physicians in the U.S.A. Perhaps the most inspiring figure of all was Dr. Alice Hamilton, the founder of occupational medicine in the U.S.A. Her many studies and clean-up efforts included a survey of mercury poisoning among the hatters in Danbury, Connecticut. Her well known "Illinois Survey" of conditions in the lead industry led to clean-ups in some situations and drew attention to the problem. But hers were the efforts of a courageous knowledgeable individual. She had no authority to enter the factories she did—other than the considerable informal authority a humanitarian physician can command. But if an owner or management ignored her findings and pleas, there was no recourse as the state laws for factory inspection adopted between 1900 and 1910 proved ineffectual.[95]

As mentioned earlier (see note 52 and associated text) it was not only the workers' movement which achieved the developments of the late 1960s and early 1970s. The middle-class-based environmental movement along with the anti-establishment ferment of the time which was associated with resistance to the Vietnam War, resistance to racism, to sexism, and to human exploitation generally, formed into a broad front of unionists and activists supporting the United Mine Workers to improve OSH conditions in the mines and receive compensation for black lung. After the disaster which took seventy-eight workers' lives in the coal mine explosion in Farmington, West Virginia on November 20, 1968, Congress passed the Coal Mine Health and Safety Act in 1969 and the Black Lung Benefits Act of 1972. "The nucleus of people who worked on health and safety issues eventually combined with the leadership of Miners for Democracy, a rank-and-file group which took control of the leadership of the United Mine Workers from the corrupt and murderous Tony Boyle clique."[96] Of course, such battles are never won forever. The struggle must always be renewed. In 1981, the death toll of U.S.A. miners from cave-ins, explosions, fires and other accidents was 153, the highest since 1975—this, while the Reagan Administration was cutting the budget and number of inspectors of the Mine Safety and Health Administration.[97]

Another important development of this period was the NLRB's decision in 1966 that OSH issues were subjects for mandatory collective bargaining. Prior to this, U.S.A. employers could refuse to bargain on these matters.

The same general forces, coupled with the curious need of Nixon to capture the blue-collar vote in the 1968 election (phrased in the rhetoric of "the silent majority" and "forgotten Americans")[98] led to a general OSH law, the Occupational Safety and Health Act (OSHA) signed into law December 29, 1970.[99] The most vigorous and clear support for the act came from the Oil Chemical and Atomic Workers (OCAW) under the OSH leadership of Tony Mazzocchi, Ralph Nader and his supporters, the US Steelworkers Union, the UAW, the AFL-CIO, and a few senators and representatives. It was quite a comprehensive and forward looking piece of legislation. For this reason, it seems worth quoting its intent and provisions at length from Section 2, paragraph b:

> The Congress declares it to be its purpose and policy, through the exercise of its powers to regulate commerce among the several states

and with foreign nations and to provide for the general welfare, to assure so far as possible every working man and woman in the Nation safe and healthful working conditions and to preserve our human resources—

(1) by encouraging employers and employees in their efforts to reduce the number of occupational safety and health hazards at their places of employment, and to stimulate employers and employees to institute new and to perfect existing programs for providing safe and healthful working conditions;

(2) by providing that employers and employees have separate but dependent responsibilities and rights with respect to achieving safe and healthful working conditions;

(3) by authorizing the Secretary of Labor to set mandatory occupational safety and health standards applicable to businesses affecting interstate commerce, and by creating an Occupational Safety and Health Review Commission for carrying out adjudicatory functions under the Act;

(4) by building upon advances already made through employer and employee initiative for providing safe and healthful working conditions;

(5) by providing for research in the field of occupational safety and health, including the psychological factors involved, and by developing innovative methods, techniques, and approaches for dealing with occupational safety and health problems;

(6) by exploring ways to discover latent diseases, establishing causal connections between diseases and work in environmental conditions, and conducting other research relating to health problems, in recognition of the fact that occupational health standards present problems often different from those involved in occupational safety;

(7) by providing medical criteria which will assure insofar as practicable that no employee will suffer diminished health, functional capacity, or life expectancy as a result of his work experience;

(8) by providing for training programs to increase the number and competence of personnel engaged in the field of occupational safety and health;

(9) by providing for the development and promulgation of occupational safety and health standards;

(10) by providing an effective enforcement program which shall include a prohibition against giving advance notice of any inspection and sanctions for any individual violating this prohibition;

(11) by encouraging the states to assume the fullest responsibility for the administration and enforcement of their occupational safety and health laws by providing grants to the states to assist in identifying their needs and responsibilities in the area of occupational safety and health; to develop plans in accordance with the provisions of this Act, to improve the administration and enforcement of state occupational safety and health laws, and to conduct experimental and demonstration projects in connection therewith;

(12) by providing for appropriate reporting procedures with respect to occupational safety and health which procedures will help

achieve the objectives of this Act and accurately describe the nature of the occupational safety and health problem;

(13) by encouraging joint labor-management efforts to reduce injuries and disease arising out of employment.

To administer the act, two major agencies were set up:

- the Occupational Safety and Health Administration (OSHA) intended to adopt standards and enforce them through a system of inspections and fines, and to provide for broad advisory and educational functions for both management and labor (as noted earlier, the large bulk of standards were those taken over from the industry-sponsored IHF). *No OSH services standard was adopted. None has yet been adopted.* Thus the large bulk of U.S.A. workers are uncovered by any adequate form of in-plant or extra-plant OSH services.
- the National Institute for Occupational Safety and Health (NIOSH) was set up in the Department of Health Education and Welfare (DHEW—now the Department of Health and Human Services since Education has been split off) to conduct research for standard setting by OSHA, carry out health hazard evaluations where some problem is suspected which is not covered by OSHA standards, and support professional and scientific education in OSH.

The U.S.A. law and the still vibrant movement of the late 1960s and early 1970s gave tremendous impetus to OSH improvements in the U.S.A. Although the law had serious weaknesses in its formulation and implementation—for example, there was no heat standard; there were only 350 chemical standards, most for acute effects (thus ignoring long-term effects such as cancer); other standards were weak or absent (noise and OSH services as noted); the inspection staffs were too small and encumbered by many "nit-picking" regulations, such as the exact height of hand railings on stairways (later eliminated by Dr. Eula Bingham, the courageous and effective head of OSHA under President Carter, so the important standards could be enforced more widely); the fines for violations were ludicrous;[100] and there were other weaknesses—the labor movement at least got a sense of backing and support for its own awakening efforts in the field. Whereas the Nader study group found in 1970 that only seven of thirty unions surveyed had any "safety staff" and only four had any "health staff" (and two had one person each for both) their survey in 1976 of the fifteen unions covering 7 million workers most exposed to chemical hazards showed great progress.[101]

This energy found an outlet in many ways, but especially through the New Directions Programs of OSHA. It was correctly reasoned among the leadership in OSHA that its small inspection force could never be the main instrument for enhancing OSH. Between 1970 and the first eleven months of 1975, workplaces with some 11.5 million workers had had some form of inspection from cursory to thorough. But there were 80 million workers and as many as 5 million workplaces. It was estimated that it would take seventy-five years just to get around to all workplaces and workers. Thus the strategy was developed to put as much understanding and information as possible in the heads of workers themselves. To be even-handed about it, several types of New Directions programs were supported—those run by management, unions, COSH groups, and universities. I have elsewhere described our own university-based, but worker-guided program in Connecticut.[102] Thousands of workers have been reached in a non-"blame-the-victim" manner. The focus has been on engineering or substance substitution to control hazards, with administrative solutions (for example, shorter shifts with less exposure) as a second choice and personal protective equipment as a last resort.[103] Further, many New Directions programs have included information on contract language and collective bargaining around OSH issues—for example, as in the UAW-General Motors contract which provides for a certain number of full time Safety Representatives paid by the Company, but chosen by the union. Of course the New Directions Programs could only have real success with unionized shops. Minority and women workers were hard to reach as most are in nonunion shops, often where hazards are horrendous. My own son worked for a time in a nonunion shop repairing gasoline pumps where seventy-five workers, most of them women, were exposed to a dangerous solvent—Magnum 855. No one could tell it was dangerous (although it smelled very bad) until I wrote to NIOSH and then, upon receipt of an answer in about ten days, set a medical student to looking up what the components do—one creates opacity of the eye lens, another kidney disease, and a third, liver disease.

But the move to the right of U.S.A. politics and government has brought OSHA and NIOSH under heavy pressure. Dr. Anthony Robbins, the courageous and much admired head of NIOSH in the Carter Administration, was given one hour to leave his office by the Reagan people. Dr. Robbins has since been honored by his election as President of the American Public Health Association. For most of Reagan's first four years OSHA was headed by a man whose only training in OSH was to have his construction firm cited by OSHA several times. He recently resigned. But a man equally objectionable to the unions was appointed in his place.[104] Currently the number of health and safety inspectors nationally, for some 5 million workplaces, is down to between 900–1000. A recent statement on OSHA prepared for a meeting of ConnectiCOSH (see below) recorded the following cut backs:

> *Update on OSHA*
>
> – Formal (written, signed) complaints for non-imminant danger will only stimulate OSHA to send letters to the employers in hopes that the employers will voluntarily correct violations. There are no provisions for routine inspections. Inspections will only occur if the person who complained calls back OSHA to say the hazard was not corrected. In response to non-formal complaints, when employers plan corrective action, no follow-up action will be taken by OSHA.
>
> – Previously, when a complaint was filed, the entire workplace could be assessed. Inspections now are generally limited to the specific area of complaint.

– Since employers now receive advance notice of inspection (this developed from the first policy mentioned above as well as the Supreme Court decision requiring OSHA to get a warrant when an owner, Barlow, refused entry to his private property–a factory with several thousand workers in Idaho), and since the inspections are restricted to areas of complaint, last minute clean-up scrambles are common.

– General safety inspections, scheduled with priority to high hazard industries may be cancelled if a review of the OSHA log for the last two to three years indicates a lost workday rate below national average. "Below national average" includes industries such as chemicals and aerospace. These logs are notorious for under-recording actual accidents.

– Penalty assessment. Penalties for violations, serious or otherwise, will be considered paid if the hazards are cleaned up in the amount of time OSHA gives. Follow-up inspections may or may not be conducted. This rule does not apply to penalties for willful, repeat, or failure-to-abate violations or violations involving catastrophes or fatalities.

– General Duty Clause. Previously, OSHA was able to cite the employer when an unusually high number of employees exhibited adverse symptoms in the workplace, even when no standard was available. This is now strongly discouraged.

– Budget cuts have also significantly decreased the number of inspections as well as their efficiency through inspectors' lay-offs, withdrawal of health standards, and numerous administrative changes. Before Reagan only 2 to 3 percent of all work places could be inspected, now even that percentage has been reduced.[105]

Some statistics concerning inspections are given in Table 12.

Table 12. Decline in OSHA Inspections under Reagan

	National	*New England*
Total inspections	-19%	-4%
Number of follow-up inspections	-69%	-35%
Number of serious citations	-30%	-26%
Total penalties	-44%	-38%
Backlogged complaints	+200%	+200%
Case hours per health inspector	-31%	-31%

Other significant changes:

– personal protective equipment (e.g., face masks, ear plugs) is defined as adequate as a first line of defense rather than engineering controls.

– walkaround pay for employees is no longer required.

– labeling standard to identify products used in the workplace has been withdrawn, and reintroduced in a weaker form with a larger trade secrets loophole.

– medical records and exposure access standard is being changed so that any "trade secret" information obtained under the standard be kept secret or a stiff penalty may be levied against the union by the company.

– changes in the Cancer Policy advocates cost benefit analysis and not publishing lists of suspect cancer agents.

Conclusion: These changes eliminate the *preventive* aspects of OSHA where before employers would clean up out of fear of an inspection, since now they will be notified before an inspection, will be less likely to be inspected, and can get out of fines by cleaning up only the things that OSHA says they have to *after* an inspection.

An analysis of cut backs of New Directions grants shows that "as a group unions were reduced by 54 percent from last year's funding level, universities cut back 75 percent, 'other groups' by 76 percent, but employer groups were only cut by 22 percent and most of that cut is attributed to the cancellation of one unsatisfactory grant program."[106] The overall fiscal year budget for 1982 for New Directions is \$6.8 million compared to \$13.9 million for 1981 and a requested amount of \$15.9 million for 1982.

Since the above rather dismal developments, OSHA has issued a "Right to Know" standard which is much weaker than state laws which it would preempt and it emphasizes protection of trade secrets.[107] Some call it a "Right to Hide" standard.

Of course, not all is lost. Much of the impetus at local level remains. Perhaps the most important kind of action is what unions themselves achieve in collective bargaining. A 1956 survey of 560 contracts registered by the New York Department of Labor found that 53 percent did not even mention safety. Of the 253 that did, 210 were merely pledges on the part of the employer to maintain safe conditions or to comply with regulations. Only thirty-two agreements provided for refusing to work on unusually hazardous jobs.[108] This situation is slowly changing. A survey by the Bureau of Labor Statistics (1976) found that of 1,724 major collective bargaining agreements covering 1,000 or more workers (7.9 million workers in all) some provisions on safety and health existed in 93 percent of the agreements, although many of these were just general policy statements or pledges of union-management cooperation. Joint safety committees were provided for in over one-fourth of the agreements, with 16 percent promising some kind of provision of safety information to employees.[109]

Another especially important sphere of continuing OSH action in the U.S.A.–a sphere of action unique among study countries in the extent of its development–is the COSH groups (see note 129). There are some forty to fifty such groups spread across the country. The oldest, best known and most effective have been in the large urban areas–the Chicago Area Council on Occupational Safety and Health (CACOSH) and Philadelphia Area Project on Occupational Safety and Health (PHILAPOSH). These groups combine some official union representation, rank and file workers, and activitists among OSH scientists and professionals in a variety of activities. Usually a newsletter is widely distributed. Courses on hazard surveillance in general or particular hazards are given and on organizing and bargaining for OSH. One of their main strengths is to engage in direct political action. For example, in pushing for a workers' "right to know" ordinance, PHILAPOSH representatives attended a Philadelphia City Council meeting on the subject with a canister of compressed air. When they turned it on and the hissing noise led several upset Councilmen to ask "What is that stuff?" the workers and others present answered, "That's what we'd like to know." A right to know ordinance was passed; one of the first in the country.

On October 24, 1981, a new COSH was formed in Connecticut (Connecti-COSH). Some 250 people attended the founding convention and the State AFL-CIO and Region 9A of the UAW, as well as many union locals were sponsors. Since that time, this group has succeeded in coalition with environmental groups, concerned citizen groups, and organized labor, to have signed into law a fairly adequate Right to Know bill (now threatened by the recent "Right to Hide" OSHA standard) and a bill establishing a new Worker Education Division in the Worker Compensation Commission. This last will replace with continuous local funding the New Directions Program which was cut back in the last few years and finally terminated in 1983.

Looking toward the future, PHILAPOSH has issued a very important "Job Safety and Health Bill of Rights" and a detailed position paper, "Revitalizing OSHA After Reagan."

OSH-PHC LINKS

From the above descriptions of the U.S.A. health and OSH systems (in preceding sub-sections) one would not expect to find much that is promising as regards connections between OSH and the general health services. Indeed the picture is worse than in any other study country.

At the policy level, there is not even an OSH services standard as a part of the OSHA Act of 1970; thus, there appears to be little awareness or concern, even among leaders in OSH, of the importance of the OSH-PHC concern. In short, policy is lacking, if not negative. Some signs of a change are beginning to appear. A widely distributed medical newspaper recently had a front page article calling attention to the need for PHC providers to be aware that their patients may be suffering from work-related diseases which were recognized as of high and widespread incidence.[110] The article also cited the important piece by Rutstein and colleagues identifying the key symptoms, case history characteristics, and tests for diagnosing the fifty most common occupational diseases.[111] Also, the current Director of NIOSH, Dr. Donald Millar, recently recognized that the reporting of occupational diseases is "at least seventy years behind" the reporting of other preventable diseases. He told a House of Representatives subcommittee that physicians should be required to report cases of work-related injury or illness, as they now do communicable diseases (though he apparently neglected to say that reporting of the latter does not work too well). He went on to recognize the tragic lack of public concern.

> "You have 4,500 cases of AIDS, and the country's going crazy about AIDS," Dr. Millar said in an interview. "We have 10,000 workers killed every year in occupational accidents and 100,000 who die of occupational diseases, and it's not news."[112]

As regards education, in my own medical school where we have had an exciting and active New Directions Program for several years, and a popular basic science elective in "Epidemiology, Carcinogenesis, and Air Pollution at Work," there is still not one required hour of OSH in the curriculum for medical students. This merely places us with the majority. A nation-wide survey of the 112 U.S.A. medical schools for the 1977-78 teaching year showed that in only forty-six (50 percent) of the ninety-six responding institutions was OSH taught at all. Only twenty-eight (30 percent) required it, usually in preclinical years. The median was four required hours among these twenty-eight schools. A variety of electives were given in thirty-five schools (38 percent).[113] This picture may improve under the influence of a very active "Task Force on Occupational and Environmental Health" of the American Medical Student Association. But the 1984 Preliminary Program for the Association of American Medical Colleges does not include one word on OSH.[114] A follow-up survey to the one just summarized shows some improvement in a five-year period.[115] With 111 (87 percent of medical schools responding regarding OSH teaching in the 1982-83 academic year, seventy-three (66 percent) replied that something was offered; sixty (54 percent) said it was a required part of the curriculum, the median time being four hours, "usually as lectures and/or seminars as part of a preventive medicine or community health course in the preclinical years." Fifty-four (49 percent) offered an elective course in OSH, or an elective OSH segment of another course not devoted to OSH. Thirty-eight schools (34 percent) reported no OSH content.

The class backgrounds and personal work experiences of the large majority of medical students would not give them any orientation toward or first-hand knowledge of workers' problems. In my elective class on OSH, one of the union board members of our New Directions Program usually asks how many of some twenty students (ones who have some interest since they have signed up for the class) have fathers who ever worked in a factory. Usually only one or two raise their hands. When he asks how many have themselves worked in a factory or been a member of a union, the number is usually two or three.

It may be that more nurses come from a working-class background, but recent changes in nursing education requiring a four-year college degree as well as an R.N. may be shifting this field toward bourgeois class orientations.

However, the current strong trend toward unionization among nurses (as distinct from "professionalization") may be an ameliorating factor (note 54).

As to practicing physicians, there is no hard data, but the situation is generally believed to be deplorable. Dr. David Wegman, then from Harvard University, told a class of Trainers in the Connecticut New Directions Program how they would have to educate their local doctors if they wanted them to pay attention to possible work-related diseases. The difficulties of distinguishing work-related disease from other disease are too great[116] for this problem to be so widely ignored in basic medical education, in post-graduate specialties, and in continuing medical education.

Organizationally, the U.S.A. case is poor indeed. In the U.K., Sweden and G.D.R. (also in Finland if one considers the public PHC centers) there are regionalized systems of care with a geographic assignment of PHC services. Thus even though there is less than adequate OSH instruction for PHC personnel in all these countries, there is at least the chance that they will come to know the plants, the workers, and their OSH problems. In U.S.A., where the PHC services are often provided through hospital emergency rooms or specialists in private offices oriented to a market and not toward a geographic area, and there are several other kinds of in-put points as already discussed (p. 385 ff.) this advantage of the other systems is absent. Unfortunately, the U.S.A. context is not right for such a system. In terms of overall governance, with its regional differences (North-South and East-West) its strong states' rights philosophy in a federated system of government, and considerable local determination of educational policies and programs, as well as public health and many other matters of daily life, the U.S.A. must be rated as comparatively decentralized.[117] Its many ethnic and class divisions make the U.S.A. quite divided.[118] Thus in an overall sense I would rate it as comparatively decentralized and fractionated—the most difficult kind of context within which to realize a coordinated, regionalized, PHC-oriented health system.[119]

Another organizational difficulty, as Dr. Millar has clearly recognized, is the absence of any seriously implemented reporting and follow-up responsibility for OSH disease (note 112).

Since there is no OSH services standard under OSHA, it is primarily in larger plants of multinational firms where full-time OSH physicians are occupied. Their activities are generally far from adequate. Most recently, with management concerned about the amount of money involved in the medical care part of contracts with unions, these physicians have got more and more into economic and administrative aspects of medical care and hospitalization plans in order to advise management (something for which they almost universally lack training). This generally draws them away from preventive OSH concerns. Smaller factories may or may not have a part-time physician, usually with no OSH training. In any case, the workers don't trust company-paid and company-controlled doctors. In courses we have given to workers through the New Directions Program, we have continually heard these physicians referred to as "company quacks" or "company stooges." The case history of the Tyler Texas Asbestos plant presented in Chapter 4 underlines what tragic dimensions this situation may assume. In that case, Dr. Lee B. Grant failed to protect the workers from well known serious hazards and failed as well to medically follow-up those exposed. He referred to the company as his patient.

In this generally bleak picture, it is nice to record that some innovations are occurring. I have elsewhere described how the Yale Occupational Health Clinic worked together with the New Directions Program, Connecticut, to provide a diagnostic and education-, organizational-, follow-up center for workers worried about work-related disease (see note 102). Most recently, a local community hospital has opened an OSH clinic serving shipyard and other workers in the Groton/New London area of Connecticut. This clinic has been opened with the understanding that patients must be referred through local physicians (no direct referrals will be accepted from the union or otherwise). It is rather transparent that this measure serves mainly to protect the money interests of local private practitioners. Still, the union is reasonably powerful in the area and the new clinic has good union representation on an advisory board, so the situation may develop in a good way. In any case, both the Yale Clinic and the Lawrence Memorial Clinic offer good possibilities for linkage both with PHC and the workplace. A little attention has been given in both to the PHC link, with one of the functions being to educate other health personnel in the community about OSH problems. As I understand it, such clinics have developed in twenty or so other places in the U.S.A. There is much variation among them, but they bear watching as a promising development. However, most lack significant labor input and I don't believe any is under union control.

This brings us to our last topic regarding the OSH-PHC relationships: sponsorship and control. The picture is simple and clear but disappointing. Labor has little voice in actual services in plant or out of plant. It also lacks a voice in such matters as medical education policy (as we saw in Chapter 7, it is possible for labor to at least comment upon and perhaps importantly influence medical education in Sweden).

OBSERVATIONS, QUESTIONS, RECOMMENDATIONS

If I had trouble making a specific list for certain other study countries, the U.K., for instance, because the list would be too long, I have even more trouble for the U.S.A. Although a list might be desirable, the needed improvements (for example, many particular OSH standards) are too many to make it feasible to include here. Thus only some more general observations and recommendations are offered.

The major effort must be for more workers to become better organized to recapture and perhaps surpass the positive mood which released so much creative human energy along with and following passage of the OSHA law in 1970. American unionism has been far too narrow and conservative to have made the kind of gains in OSH and in other areas one can see in a number of European countries. A "political exchange" or transformation is needed to orient the laws and governing structure primarily to the concerns of the working class,[120] albeit an expanded working class as Aronowitz suggests (note 51).

With such a transformation, the most important demand would be: "A decent-paying job or a living wage for everyone." This would tend to eliminate the primary OSH problem—choosing between keeping one's job or one's life. The G.D.R. guarantees everyone a job in its Constitution.

A second key demand would be for a National Health Service, comprehensive and fully regionalized, including geographic orientation of local PHC services, with OSH services provided independent of management and fully coordinated with PHC.

In the meantime, before such a transformation takes place, unions themselves should have much greater OSH competence with more OSH staff. Organized labor should join more closely to support the energy in the many COSH groups across the country and see to the founding of others.

U.S.A. workers who have had proper OSH training (as in New Directions courses or union courses) should have the right by law to halt the production process on their word alone when they perceive it to be dangerous, either immediately or in the long run, as is the case in Sweden.

All U.S.A. workers—whether unionized or not (but hopefully all will become union members and insist that the unions be run for and by the rank and file)—should have the right to refuse to work under hazardous conditions and to refuse without danger of reprisal. They must also have a total "right to know" what materials they are working with and what these materials may do to them if they are not protected.[121]

A great deal more attention should be paid to work-place democracy and the relief of stress by all parties concerned, but especially the workers themselves and their unions.[122] U.S.A. workers should all be covered by preventive, diagnostic and follow-up OSH services offered by truly competent and independent medical, safety, hygiene, and other personnel who, though independent, are responsible to and understanding of workers' needs and wishes. Hopefully, when such services are developed, they will be properly linked to a fully regionalized, PHC-oriented general health services system.

As regards education of PHC and other health personnel (wider efforts are required in U.S.A., given the fact that specialist are often the PHC providers)[123] the American Association of Medical Schools should take the lead in setting together a commission with strong national union representation, as well as representation from the Industrial Medical Association, the Section on Occupational Health of the American Public Health Association, the OSH Committee of the American Medical Student Association, and other relevant parties to establish recommended OSH curricula and teaching approaches at undergraduate, graduate, and continuing medical education levels.

A similar commission should be formed in nursing.

An adequate system must be developed to pretest chemicals and other hazards before they enter the workplace. The present one does not work, even for new chemicals,[124] to say nothing about the one-quarter million chemicals in workplaces before the system was adopted, and to say nothing about hazards of new machines and processes.[125] The time for workers to serve as the warning canaries for each other (as in the movie "Song of the Canary") is over!

A national survey should be mounted, as is being done in the G.D.R., to identify and correct the hazards in every workplace in the country and follow-up workers who have been exposed to assure their proper care and protection. An adequate ongoing epidemiologic OSH surveillance and information system must be established. This will be greatly facilitated by a regionalized NHS in which OSH and PHC are coordinated so that illness and accident reports can be related to a defined population base.

To aid in the planning of graduate and continuing education strategies as well as to assess the seriousness of the OSH-PHC problem, two kinds of national surveys should be conducted. One would tell what source of care workers first seek out when they first feel ill for any reason and would take a number of factors into account as suggested for Sweden and Finland. The other would tell us what the practitioners who see workers know about OSH and what they do and fail to do.

There is a long way to go in the U.S.A. Work in America is now too sad, discouraging, and detrimental to health and well-being.[126] But it is a country with tremendous resources and enough of a tradition of freedom and open exchange to help protect against renewed repressive measures and eventually allow working people to take charge of their own health, living and working conditions. If we remember the "Bread and Roses" struggle of 1912, it is clear that it can and has been done in part. It will be done in full.[127]

THE U.S.A. AND THE OSH-PHC MODEL

The reader will find a discussion of the idea of an ideal-typical model and the elements of this one as well as a rationale for those elements in Chapter 6.

Policy

On the policy dimension, the OSHA law of 1970 states that every worker in U.S.A. has a right to a safe and healthy workplace. But stating something on paper, even having the Congress pass upon it and the President sign it into law, does not make it a high priority direction for action. In fact, as the head of NIOSH recently stated (to paraphrase) there are 115,000 men and women dying every year from work-related causes and it is not news! There is considerable ferment today in the unions and COSH groups and other circles concerning OSH conditions and provisions, but, if anything, at the official level, especially under the Reagan regime, there has been a turning away.

Sponsorship and Control

Sponsorship and control of OSH services and conditions is essentially in the hands of the capitalist class. There is very little worker control. This is not to say that there is none. Some unions have succeeded through bargaining to have such important provisions as a certain number of company-paid, union-appointed full-time health and safety reps for every so and so many workers. These reps are free to walk around and look for hazardous conditions and initiate action to remove them. This is rare, but it exists (most notably in the UAW-GM contract). There would be other examples of worker control to mention, but, to choose a negative example, there is no say over the hiring and firing of in-plant OSH physicians and engineers as in Sweden. Workers have not even established the right to refuse hazardous work without reprisal. Nor is the right to know about chemicals and other hazards very broad, general (that is all states) or well established. There is a long way to go before OSH-trained U.S.A. workers will be legally able to stop the production process when they see a hazardous situation. As regards PHC, workers seem to have no voice at all, unless it is a union-sponsored medical care program which then usually has no right to concern itself with the in-plant conditions which may be making workers sick.

In general, because of a rather narrowly oriented and weak workers' movement, with no socialist or even social democratic party of their own, U.S.A. workers have less influence and control over all spheres of OSH and PHC than in any of the other five study countries.

Unless the workers' movement gains greater control over this sphere of concern, men and women who work or want to work could find themselves faced with a kind of bio-technical discrimination which could be very damaging of the human spirit and hard to fight.[128]

Education

The educational dimension shows more strength in U.S.A. than one might imagine considering the weaknesses in other areas. This is particularly true for OSH education with workers. The OSHA-sponsored and funded New Directions Programs have been particularly important, especially those sponsored by unions. Unions themselves and the many COSH groups across the country have also played key roles in this regard.[129] Thus thousands of workers have had a chance to ask if there aren't more things they can do to have "a job *and* your life" instead of a "job *or* your life." This is not to say that there is the same saturation as in Sweden where 110,000 health and safety reps have been trained with forty hours or more for a population of 8 million. It seems hard to contemplate that the U.S.A. with 230 million people would have to train 3,162,500 workers at an equal level to catch up.

Medical education and education of practicing PHC personnel leave a great deal to be desired. At latest report, only 54 percent of medical schools had any required OSH hours in their curricula and the mean of these was four hours. Training for practitioners (as in the Mora project in Sweden) seems to be totally absent, though some of the promising university-based OSH clinics, as at Yale, may begin experimenting in this direction.

The preparation of OSH specialists leaves much to be desired. There are several university centers for advanced OSH preparation in medicine, engineering and other fields. But the fields, especially industrial medicine, do not have high prestige and thoughtful people worry about management domination of their work. Thus there are far far too few fully prepared, board-certified OSH experts in medicine in U.S.A. The bulk of the field is made up of part-timers with little or no training in OSH. The U.S.A. has not adopted a clear strategy and sets of implementing steps to have two kinds of specialists as in several of the other study countries–the full specialization for university teaching and research and the lesser specialization for plant physicians. The situation is scandalous.

Organization

Organizationally the situation could hardly be worse. OSH services in plants are not at all respected by the workers as they are usually total creatures of management and often incompetent. There is no OSH services standard in the OSHA law. Every new standard is first contested by corporate experts in hearings and then brought to court as soon as it is promulgated, thereby contributing to a general union-busting strategy which seeks to impoverish the unions through court battles (maybe it would help if the U.S.A. had only 500 lawyers per million population as in Sweden, instead of the 2000 it has).

The general health services are highly fractionated, with no regionalization and no geographic orientation to the PHC services. The reporting of OSH diseases by physicians "is seventy years behind" the reporting of communicable diseases; itself not too perfect. There are no good examples, even on a project basis (as in Grossraeschen in G.D.R.) of coordinated PHC-OSH links.

Information

The above would suggest that the information dimension must be equally poor. In general this is true. There are no regional, population-related data systems (some states have cancer registries, a number of which are only now gearing up to include occupational data in an effective way–not an easy task). As already stated, the reporting of OSH conditions in general is poor.

The right to know is still being fought for by workers and their representatives. The variations between states in this regard is startling, but recent moves by OSHA to preempt state laws are no help because the proposed standard is weaker than the right-to-know standards in many states.

One strong point—possibly stronger than in any of the other five study countries—are special scientific studies. A lot of the most important OSH work in epidemiology and toxicology has been done in U.S.A. The influence of fairly generous funds available through the National Institutes of Health, the Center for Disease Control (CDC) and NIOSH, as an arm of CDC, have no doubt been key to this achievement. It is difficult to measure this aspect of this dimension. Clearly, the U.S.A. has not done or encouraged much work in the social sciences related to OSH, particularly the organizational and collective aspects of stress at work as has been done in Sweden (this will probably be the biggest sphere of concern for the future in all countries). And the works of the well-funded National Institutes for OSH in Helsinki and in East Berlin are very plentiful and impressive. Still, the U.S.A. is not far behind and may be ahead in this aspect.

What is done with the information from special studies is another matter. Workers and their representatives have little say in this in U.S.A. whereas they have an authentic voice in the design and use of scientific studies in some other of our study countries.

Financing

Financing is not structured so as to make either the OSH services there are (mainly in large plants), or the PHC services preventive in orientation as regards OSH hazards. To really solve this problem, the U.S.A. would need a National Health Service with a management-independent OSH service which is closely coordinated with the PHC level of the general health services.

NOTES AND REFERENCES

1. Several general sources have been used: United States, Vol. 24, pp. 1961–2082, especially Wilbur Zelinsky, Geography, pp. 1961–1969 and History, pp. 2021–2082 in *The Volume Library*. Nashville, Tennessee: The Southwestern Co., 1973; United States, pp. 2039–2042 in William Bridgewater and Elizabeth J. Sherwood (eds.), *The Columbia Encyclopedia*, 2nd edition, New York: Columbia University Press, 1950; Charles A. Beard and Mary R. Beard, *A Basic History of the United States*. New York: Doubleday, Doran and Co., 1944; Monique Fouet, USA: No Ebb, No Resurgence, pp. 152–158 in *World View 1982, An Economic and Geopolitical Yearbook*. Boston: South End Press, 1982; Ruth Leger Sivard, *World Military and Social Expenditures, 1983*. Washington, D.C.: World Priorities, 1983. Other sources will be cited on particular points.

2. Geographical Facts about the United States, p. 127 in *Universal World Atlas*, Census edition, Chicago: Rand McNally, n.d. (probably 1982).
3. There are disputes as to whether Columbus first made the "discovery" in 1492 or whether Vikings may have landed on the North American continent at an earlier date.
4. From 1820–1977, 75.3 percent of immigrants came from Europe, 18.2 percent from the Americas, 5.4 percent from Asia, .3 percent from Africa (hundreds of thousands of slaves brought in before this period are not counted); and .8 percent from elsewhere. *The World Almanac and Book of Facts*. New York, 1980, p. 200.
5. World Almanac, op. cit. (note 4) figures for Indians from p. 465; other figures from pp. 198 and 203. These figures can only be seen as close estimates. In the tables themselves there is a note that a calculation error of 23,372 too few was found after publication of the 1970 census so that the official total was 203,235,298 but calculations are on the lower figure of 203,211,926. Also it is well known that some 3 million Blacks and other inner city poverty area residents were missed in the 1970 count.
6. For an excellent historical account of broken treaties and the near genocide of Original Americans, see Ralph K. Andrist, *The Long Death; The Last Days of the Plains Indians*. New York: Macmillan, 1964.
7. A good general source is Charles A. Beard and Mary R. Beard, *A Basic History of the United States*. New York: Doubleday, Doran, 1944. See also Donald L. Kemmerer, and Clyde C. Jones, *American Economic History*. New York: McGraw-Hill, 1959, for a quite traditional, that is, non-Marxist view of economic development as if it occurred apart from the exercise of state power; nevertheless, very useful for dates and information.
8. Frederick Jackson Turner, *The Significance of the Frontier in American History*, a paper published by this University of Wisconsin history professor in 1894.
9. The Beards, op. cit. (note 7): 362.
10. William Appleman Williams, *The Great Evasion*. Chicago, Quadrangle Books, 1964.
11. The concept of cultural hegemony is developed by Antonio Gramsci in his essay The Modern Prince in *Selections from the Prison Notebooks*. New York: International Publishers, 1971. The other ways of keeping an essentially conflict-full capitalist society together is through state power and the use of law and raw force. The history of the struggle of labor in the U.S.A. is rife with examples of the use of everything from racism to raw force. Stanley Aronowitz, *False Promises: The Shaping of the American Working Class*. New York: McGraw Hill, 1973. One of the most outrageous repressions and continuing obfuscations involves the fact that Labor Day for the rest of the world originated in Chicago on May 1, 1886. Workers across the country, led by the IWW, had called for a general strike that day to agitate for the eight-hour day. The movement centered in Chicago where some 40,000 workers went out. There was no violence as police had predicted. But two days later police beat and shot striking workers at the McCormick Works. A meeting was convened in Haymarket to protest the police bru-

tality. The mayor attended to keep peace. He left when it started to rain and the crowd of 5,000 began to disperse. As soon as the mayor left, the police captain marched his men into the crowd clubs swinging. Someone threw a bomb. Seven policemen were killed and many wounded as were many protesters. To this day, it is not known who threw the bomb. But eight men were convicted by a prejudiced judge, a rigged jury and a prosecutor who said "anarchy is on trial." Seven of the eight were condemned to death, the other given fifteen years in jail. Six years after the executions, Governor Altgeld determined that the defendants were not guilty. A monument to the "Haymarket Martyrs" in Forest Home Cemetery in Forest Park Illinois carries the words of one of the martyrs, August Spies, "The day will come when our silence will be more powerful than the voices you are throttling today." Labor Day is celebrated in early September in U.S.A. and most U.S.A. workers do not know that they have been thus cut off from solidarity with their brothers and sisters in other countries. See Paul Avrich, *The Haymarket Tragedy.* Princeton, NJ: Princeton University Press, 1984.

12. A graphic depiction quoted from the economist, Paul Samuelson, in Vicente Navarro, "Social Policy Issues: An Explanation of the Composition, Nature and Functions of the Present Health Sector of the United States," *Bull. of the New York Academy of Med. 51*:199–234, 1975 (reprinted in his book *Medicine under Capitalism.* New York: Prodist, 1976).
13. *World Almanac* op. cit. (note 4): 201. Some categories were even worse off. In 1974, 52.2 percent of Black female heads of household were in poverty.
14. Associated Press, Study Says Income Gap Wider Under Reagan, *The Hartford Courant* (August 16, 1984): A. 6. See also, Associated Press, Census Finds Poverty Rise; 35 Million Officially Poor; Political Reaction is Swift, *The Hartford Courant* (August 3, 1984): A 1 and A 8. Reflecting on the recent government bail-out of Continental Illinois National Bank of Chicago one observer says, "As Michael Harrington well put it, our version of capitalism translates to socialism for the rich and free enterprise for the poor." Robert Lekachman, How the Mighty Are Saved, *The Hartford Courant* (August 10, 1984): D 9.
15. Stories of courage as well as despair are found in Studs Terkel, *Hard Times, An Oral History of the Great Depression.* New York: Pocket Books, 1970.
16. Critics of the Reagan regime argue that the recent recovery, which has developed without a return of high inflation is in fact false, being fueled by the incredibly large Federal deficit of some $195 billion in 1983 and an almost equally high deficit for 1984 which will escalate astronomically in terms of cumulated debt service unless taxes are raised and armaments and other expenditures cut. The cuts which have been made are in human services—food stamps, school lunches, health care, etc.
17. The work of Harvey Brenner on unemployment and illness is cited in Associated Press, Studies Link Unemployment Stress with Crime, Illnesses, Early Deaths, *The Hartford Courant* (June 28, 1984): A 10. This does *not* imply that work settings are healthy; only that being out of work can create terrible anxieties and ill health.

18. David Lightman, Jobless Rate Climbs to 7.5%, *The Hartford Courant* (August 4, 1984): A 1 and A 7.
19. Sarah Polloch, Pratt and Whitney Layoffs Dim American Dream, *The Hartford Courant* (January 26, 1982): A 1 and A 4.
20. These and other objective conditions which made labor's struggle more than a matter of individual leadership are excellently presented and analyzed in Aronowitz, op. cit. (note 11).
21. Nick Salvatore, *Eugene V. Debs: Citizen and Socialist.* Urbana, Il: University of Illinois Press, 1982.

    Following the crushing of his union by federal troops in 1984, Debs was forced to accept that struggle, not harmony, dominated worker/capitalist relationships. The rest of Debs's political career was an attempt to develop a socialist movement based on the class struggle of working people, but still guided by a political vision of small-town harmony.

    The new community had to be more egalitarian than the old: Debs did not accept the racism, the prejudice against immigrants or the sexism of U.S. society. Debs modestly said, "The Socialist Party has enabled me to be class conscious, and to realize that, regardless of nationality, race, creed, color or sex, every man, every woman who toils . . . every member of the working class without exception, is my comrade, my brother and sister." However, Debs was the only major Socialist Party orator who insisted on speaking to Blacks and whites together in the South, and he was a strong opponent of immigration restriction in a party that at best waffled on the issue.

    From Evan Kraft, Socialism with a North American Face, *Guardian* (New York) (February 22, 1984): 20 and 17, quote from p. 20.

22. *Historical Statistics of the United States, Colonial Times to 1957.* Washington, D.C.: USGPO, 1960, p. 682.
23. Again, I would emphasize that this struggle was and is not simply a matter of individual leaders. It is convenient to symbolize the different wings in terms of the leadership, but the objective political economic conditions of the times must be examined in detail to understand why one wing prevailed to establish what has come to be known as a narrow sort of business unionism in U.S.A. with no political party or clear ideology of its own. See note 21. Also, Michael Shalev and Walter Korpi, Working Class Modernization and American Exceptionalism, *Economic and Industrial Democracy*. Beverly Hills, CA: Sage. Vol. 1:31–61, 1980.
24. Harry A. Willis and Royal E. Montgomery, *Labor's Progress and Problems.* New York: McGraw Hill, p. 423, 1938.
25. This brief account is taken from a heart rending yet inspiring pictorial history and text: William Cohn, *Lawrence 1912, The Bread and Roses Strike.* New York: Pilgrim Press, 1980.
26. Norman Thomas, *Human Exploitation in the United States.* New York: Frederick A. Stokes, 1934 (quote from p. 168).

27. In the 1980 presidential election, the officials of the very large Teamsters Union even supported Ronald Reagan and the Republican Party, though it is important to note that there is a widespread rank and file move to clean up and democratize the Teamsters Union. The official Teamsters again supported Reagan in the 1984 election.
28. James R. Green, *The World of the Worker, Labor in Twentieth-Century America*. New York: Hill and Wang, 1980.
29. But the post-war situation of women and the labor market offers a perfect example of attempts to take the potential heat out of the system through media management of consciousness and self concept—the cultural hegemony in operation: "Near the end of the war, the Women's Bureau of the Labor Department found that 75 percent of the women working in industry expected to remain in the labor force after the conflict ended, including 57 percent of the married women. And 86 percent of these workers expected to stay in their wartime occupations. They would soon be disappointed. . . . Women's participation in the labor force dropped from 37 percent in 1945 to 31 percent in 1946. Married women were actively discouraged from working by a campaign to return "Rosie the Riveter" to the kitchen. . . . The "feminine mystique" wrote Betty Friedan strongly encouraged this trend by celebrating domesticity, romanticising motherhood, and revitalizing the traditional notion that women belonged in the home. Green, op. cit. (note 28): 191.
30. Aronowitz, op. cit. (note 11): 364.
31. Green, op. cit. (note 28): 301–303. One of the UE representatives who was at the CIO convention when his union was excluded says, "These figures are too conservative. The conservatives sought to minimize the damage. At least 1 million and a half workers were excluded that infamous day." See also Ronald Schatz, *The Electrical Workers: A History of Labor at General Electric and Westinghouse, 1923–1960*. Urbana, IL: University of Illinois Press, 1984.
32. The works of Aronowitz (note 11), Green (note 28) and Shalev and Korpi (note 23) are especially valuable for the details of the labor struggle in U.S.A. as well as important interpretive understandings which I will have occasion to come back to, particularly in the concluding chapter.
33. As reported by the Bureau of National Affairs, a private research firm: Associated Press, Union Rolls Shrink; Reagan Blamed, *The Hartford Courant* (September 25, 1984): C1. Also, James Warren, U.S. Unions—Weaker than Ever, *Chicago Sun Times* (September 4, 1983): "Views Section" 1 and 4–5.

    > At the end of July, the U.S. labor force (anyone older than 16, with or without a job) was 113.9 million. Of these, 103.3 million were employed, leaving an unemployed rate of 9.4 percent.
    >
    > Since 1970, the work force has grown by 34.2 percent and is bunched in six categories since the Labor Department just junked the traditional designations "white-collar" and "blue-collar."
    >
    > So one gets 22.4 percent in managerial and professional

specialty (executives, managers); 30.8 percent in sales and administrative support (technicians and clericals); 13.7 percent in service (waitresses, barbers, private security, etc.); 12.4 percent in precision production, craft and repair; 16.1 percent who are operators, fabricators and laborers, and 4.6 percent in farming, forestry and fishing.

In 1970, there were 51.2 million males in the civilian labor force, of whom 49 million (95.7 percent) were employed. This July there were 49 million, with 44 million (89.8 percent) employed.

Since World War II, there has been an ongoing rise in female "participation in the labor force"—the percentage of women who are in or are looking for jobs. In 1949, about a third of all women sixteen and older were in that category. By 1970, it was 43 percent. This July, it was 53.3 percent.

Since World War II, there has been a generally opposite trend for male workers. In 1949, the percentage of males sixteen and older who were employed or looking for work was 86.4 percent. In 1970, it was 79.7. This July, it was 78.8 percent. . . .

Union members are 20 percent of the work force.

34. Warren, op. cit. (note 33); but the BNA report cited in Note 33 observes that these figures may be affected by Canadian members; without them, the NEA is the largest union with 1.6 million and the Teamsters next with 1.58 million members.
35. Associated Press, March Jobless Rate Remains Flat at 7.8%, *The Hartford Courant* (April 7, 1984): C 7 and 9.
36. David Broder, Two Troubling Thoughts for Labor Day, *Chicago Sun Times* (September 3, 1983): Views Section, p. 13.
37. Ernest Conine, Dawn of the Robot Age, *The Hartford Courant* (January 26, 1982): A 13. This article cites Harley Shaiken, professor at the Massachusetts Institute of Technology, who has done the most to clarify the social effects of high tech sees the weakening effect of computers and robots on organized labor not only in terms of lost jobs but in terms of "telescabbing," the ability of companies to continue operating through use of elaborate circuitry in spite of a strike, as in the case of the strike of 675,000 phone company workers against AT&T in which managers kept the phone service going and the company extracted a very modest settlement from the union.

> "And we're not just talking about the phone company" said Shaiken. "The ability to operate during a strike situation has affected, and could affect, other industries. In any case, even in those industries where that's not possible, bargaining over technology is an increasing factor."

Also, it should be recognized that robots bring new hazards to the job: William M. Bulkeley, "Manufacturers Seek to Create More Safety-Conscious Robots," *The Wall Street Journal* (October 4, 1985): 23. Further, the Conine article gives a somewhat idealized view of work in Japan. For a different picture, see Anna DeCormis, "Made—And Enslaved—in Japan," *Guardian* (New York (October 9, 1985): 2.

38. Carol Bowers, Computers Monitor Employees in Race Against "Average Work Time," *The Hartford Courant* (April 24, 1984): C 1. Recently trucking companies have begun to use satellites to exactly trace their trucks through homing devices built into the trucks. Will the Teamsters finally rebel?
39. Albert Syzmanski, *The Logic of Imperialism*. New York: Praeger, 1981, pp. 489–490. The figures on losses of working class soldiers in the Vietnam War he takes from Maurice Zeitlin, Kenneth Lutterman and James Russell, Death in Vietnam: Class, Poverty, and the Risks of War, in *American Society*, Zeitlin, (ed.), 2nd edition, Chicago: Rand McNally, 1977.
40. Urban C. Lehner and James R. Schiffman, Chrysler, Samsung Weighing Details of Venture, but Korea Is Big Obstacle, *The Wall Street Journal* (August 22, 1984): 10.
41. John M. Barry, Reagan Administration Blasted on Unfair Treatment of Workers. Council Cites Assault on Protections, *AFL-CIO News* Vol. 29, No. 9 (March 3, 1984): 1. Ironically, even some businesses who invested in meeting previous regulatory requirements are developing resistance to further deregulations as they don't want their competition to get by more cheaply: Joann S. Lublin and Christopher Conte, The Rule Slashers, Federal Deregulation Runs Into a Backlash, Even From Businesses, *The Wall Street Journal* (December 14, 1983): 1 and 22. This article, incidentally gives the cuts and a few increases in budgets and staffs of the ten leading regulatory agencies. OSHA had a 6 percent increase in budget (not quite enough to keep up with inflation) from 1981 to 1983 and an 8 percent cut in staff. The article notes further: "OSHA, once perceived as a four-letter word by businesses, has changed its image from nit-picking cop to employers' ally. It has eased rules or proposals covering noisy workplaces and other safety issues to put more emphasis on worker protective gear than costly engineering controls." (p. 22).
42. David L. Perlman, Bid to Split Unions Follows Pattern of "Dirty Trick" memo, *AFL-CIO News* Vol. 29, No. 22 (June 2, 1984): 1 and 3.
43. Robert Georgine, From Brass Knuckles to Briefcases: The Modern Art of Union-Busting, in *The Big Business Reader, Essays on Corporate America* Mark Green and Robert Massie, Jr. (eds.), New York: Pilgrim Press, 1980, pp. 89–104. One of the main strategies is to tie everything up in litigation and make unions expend their small funds in defense of what is clearly theirs by law, but law untested in court. Many OSH cases are of this nature. See also Union-Busters Thrive under Reagan, *AFL-CIO News*, Vol. 28 (January 8, 1983): fifth article in a series.
44. Bryan Burrough, Continental Air Wins Defense of Chapter 11, *The Wall Street Journal* (January 18, 1984): 2.
45. Two Guide Books – Jane Slaughter, *Concessions and How to Beat Them*. Detroit: Labor Education and Research Project (P.O. Box 20001, Detroit, MI, 48220) 1983; and Workers' Policy Project, *It's Time for Management Concessions*. New York: The Project (853 Broadway, NYC 10003): 1983– are reviewed favorably in David Moberg, Guides Provide the Hard Ammunition for Fighting Concessions, *In These Times* (January 25–31, 1984): 19.

This attack against concessions should include exposure of incompetent management which has milked plants for short-term high returns without adequate reinvestment. Even many business people are beginning to realize this. Peter Behr, Insiders Take Hard Look at US Management Techniques, *The Hartford Courant* (January 25, 1982): C1. Also George F. Will, Birth and Death of America's Industry, *The Hartford Courant* (January 5, 1982): A17.

46. Lisa Gallant, "Rebirth" Predicted for Organized Labor, *The Hartford Courant* (January 6, 1982): D4. Unfortunately this article appeared just below one titled, "Torrington Co. lays off 60 employees." Also, see Schatz, op. cit. (note 31) who sees little hope for the future of unions in U.S.A. and sees it as a "society of individual consumers who identify with a corporation," a society likely to be corporate-controlled for a long time.
47. Warren, op. cit. (note 33): 5.
48. A very important comparative piece offering a theory of "political exchange" as an explanation of the comparatively favorable situation of Sweden's working class is found in Walter Korpi and Michael Shalev, Strikes, Industrial Relations and Class Conflict in Capitalist Societies, *British J. of Sociology, 30*:164–187, June 1979.
49. David Moberg, "Labor: Learning the Hard Way." *Mother Jones*, pp. 7–8, May 1984.
50. Carola Kaps, Amerikas Arbeiter Denken Konservativ. Zur Rolle der Gewerkschaften. *Frankfurter Allgemeine Zeitung* (5. Oktober 1981): 13.
51. Stanley Aronowitz, *Working Class Hero: A New Strategy for Labor.* New York: The Pilgrim Press, 1984.
52. Molly Joel Coye, Mark Douglas Smith, and Anthony Mazzocchi, Occupational Health and Safety: Two Steps Forward, One Step Back, pp. 79–106 in *Reforming Medicine, Lessons of the Last Quarter Century*, Victor W. Sidel and Ruth Sidel (eds.), New York: Pantheon, 1984. There were other conditions: the miner's Black Lung movement and the disaster at Consolidated Coal's No. 9 mine in Farmington, West Virginia in which seventy-eight miners lost their lives. These helped lead to the much improved Coal Mine Safety and Health Act in 1968 as well as Nixon's "blue-collar strategy" which explains why it was a Republican president who signed the OSH Act into law.
53. Frances Fox Piven and Richard Cloward, *Regulating the Poor: The Functions of Public Relief.* New York: Pantheon, 1971.
54. In fact hospitals and nursing homes are among the few workforce sectors rapidly growing in terms of unionization. The recent nurses' strike in Minneapolis showed great solidarity against all the hospitals who formed into an Association to fight the strike and then bargain. The nurses won most of their demands. The 20,000 hospital workers who jammed into Madison Square Garden to vote an end to their forty-six-day strike against thirty-three New York City hospitals and nursing homes showed great pleasure when Doris Turner, President of District 1199 of the Retail Wholesale and Department Store Workers told them, "I think we taught the hospital bosses a lesson. . . . The lesson is, 'Don't mess with 1199'." Associated

Press, Workers Vote On Pact to End Hospital Strike, *The Hartford Courant* (August 28, 1984): A 3.

55. Egon Jonsson and Duncan Neuhauser, Hospital Staffing Ratios in the United States and Sweden, pp. 128–137 in *Comparative Health Systems* special supplement to *Inquiry 12* Ray Elling (ed.), (June 1975). This overall picture remains correct in spite of the growth in recent years of health maintenance organizations (HMOs or prepaid group practices). In fact, unless the HMO controls a hospital and runs it with a closed staff, the picture is changed very little.
56. The blow by blow account of the AMA's payment of congressmen and struggle generally against the only compulsory national health insurance ever adopted in U.S.A., Medicare, which is limited to those sixty-five and older and was not adopted until 1965, is told in Richard Harris, *A Sacred Trust.* New York: New American Library, 1966. Medicare was an important achievement. Unfortunately, with rising prices, cutbacks and increasing deductibles, elderly people now spend as high a proportion of their incomes on medical care as they did before the Medicare system came in. In 1982, the initial hospital deductible rose from $204 to $260 and to $328 by 1984. The $60 deductible which Medicare patients have paid for out-of-hospital doctor charges will rise to $75. "Social Security Update" Coalition for a just society, (Middletown, CT) Newsletter #32, January 1982, p. 5. Proposals by President Reagan in his "State of the Union" address January 26, 1982 would move Medicaid to the Federal level while putting some forty social welfare and educational programs with the states. This is called "the new federalism." One Democratic congressman sees it as a distraction "While 10 million people are walking the streets looking for work, he's talking about who should administer the programs he's trying to starve out."

    More of the historical background and cultural and power hegemony governing the U. S.A. health system are given in Ray Elling (ed.), *National Health Care, Issues and Problems in Socialized Medicine.* New York/Chicago: Aldine-Atherton, 1971.
57. Marshall W. Raffel, Health Services in the United States of America in *Comparative Health Systems, Descriptive Analyses of Fourteen National Health Systems,* Raffel (ed.), pp. 520–586, University Park: The Pennsylvania State Univ. Press, 1984, quote from p. 522.
58. *The World Almanac and Book of Facts.* New York, 1980, p. 959.
59. Raffel, op. cit. (note 57): Table 17, p. 581.
60. Robert Maxwell, *Health Care, The Growing Dilemma: Needs Versus Resources in Western Europe, the U.S. and the U.S.S.R.* New York: McKinsey and Co., 1974. Maxwell also shows that physicians are overconcentrated in wealthy counties and scarce in poor counties, p. 21.
61. J. P. Bunker, Surgical Manpower, A Comparison of Operations and Surgeons in the United States and in England and Wales, *New England J. Medicine, 282*:135–144, 1970.
62. Unfit Doctors Create Worry in Profession, *New York Times* (January 26, 1976), first of a series on the medical care system continuing on January

27th and 30th 1976 which includes the reference to the study as first reported in the *New England Journal of Medicine*.

63. Waitzkin has examined the function of the medical arena as an economic sector for capital investment and accumulation. See Howard Waitzkin, How Capitalism Cares for Our Coronaries, pp. 317–332 in *The Doctor-Patient Relationship in the Changing Health Scene: An International Perspective*, Eugene B. Gallagher (ed.), Bethesda, MD: John E. Fogarty International Center, DHEW Pub. No. (NIH) 78-183, 1978, reprinted in Waitzkin, *The Second Sickness*. New York: Basic Books, 1983. Also see Howard Waitzkin, "A Marxian Interpretation of the Growth and Development of Coronary Care Technology," *American J. Public Health, 69*:1260–1268, December 1979. This view is critiqued in a companion article by Bernard S. Bloom printed in the same issue, pp. 1269–1271.
64. Richard Hughes and Robert Brewin, *The Tranquilizing of America*, N.Y.: Harcourt Brace Jovanovich, 1978; Milton Silverman and Philip R. Lee, *Pills, Profits and Politics*. Berkeley: University of California Press, 1974; D. Whitter, Drugging and Schooling, *Transaction-Society* (July-August, 1971):31–34; B. Blackwell, Minor Tranquilizers, Misuse or Overuse?" *Psychosomatics, 16*:28–31, Jan.–Feb., 1975.
65. The history of failed regionalization in U.S.A. is laid out in David Pearson's chapter in Ernest W. Saward (ed.), *The Regionalization of Personal Health Services*, (Rev. edition), New York: Prodist, 1977.
66. Numerous publicly supported health programs are under pressure. Nearly every day one sees articles with headlines such as this: David B. Rhinelander, Budget Ax Imperils State Health Clinics, *The Hartford Courant* (January 16, 1982): B 1 and B 2. Most health system students see this trend as a temporary aspect of a particular regime. But one finds the same cutback mentality in the U.K. and to some extent in other capitalist nation states. I have elsewhere examined this phenomenon as a more deeply-rooted part of the fiscal crisis of the state in monopoly capitalism, though I agree that the Reagan Administration exacerbates the trend. R. H. Elling, The Fiscal Crisis of the State and State Financing of Health Care, *Soc. Sci. and Med. 15C*:207–217, 1981.
67. Robert R. Alford, The Political Economy of Health Care: Dynamics Without Change, *Politics and Society, 2*:127–164, Winter 1972.
68. Vicente Navarro, op. cit. (note 12).
69. Raffel, op. cit. (note 57): 524–525.
70. Ibid, p. 525, other data from Table 12, p. 577.
71. Mary Ann Godbout, Diagnosing Health-Care Ills, *The Hartford Courant* (April 27, 1984): B11.
72. Sivard, op. cit. (note 1): Table 111. Note that other sources may give slightly different figures. Raffel, for example, gives the "estimated" infant mortality rates for 1980 and 1981 as respectively 12.5 and 11.7. op. cit. (note 57): Table 4, p. 570.
73. Richard F. Tomasson, The Mortality of Swedish and U.S. White Males: A Comparison of Experience, 1969–1971, *Amer. J. Public Health, 66*: 968–974, 1976.
74. S. J. Kunitz and J. E. Levy, "Changing Ideas of Alcohol Use Among Navaho Indians," *Quart. J. Alcohol Studies, 35*:243–259, 1974.

75. Raffel, op. cit. (note 57): Table 3, p. 570.
76. Edwin Silverberg and Cyril E. Poindexter, "Cancer Facts and Figures for Black Americans, 1979." New York: American Cancer Society.
77. Barry Newman, Socialized Care; Frugal Medical Service Keeps Britons Healthy and Patiently Waiting, *The Wall Street Journal* (February 9, 1983): 1 and 27.
78. "A Bill to establish a United States Health Service." H.R. 3884, 98th Congress, 1st Session. (The Dellums Bill). Although there will be ambivalence and conflict within the capitalist class on this matter, because some are anxious to continue an open system in which profits on drugs and equipment sales will be high, the impatience of large manufacturers carrying high cost health care benefit packages negotiated with unions makes such a push more likely than one might imagine. See Betty Leyerle, *Moving and Shaking American Medicine: The Structure of Social Transformation.* Westport, CT: Greenwood, 1984.
79. I would like to acknowledge the assistance I have had, especially in this section, but to some extent elsewhere as well, from the excellent course paper of a student and co-worker on the New Directions Program, Connecticut: Tim Morse, "Occupational Health Cross-Nationally: The USSR, Sweden, UK and USA." Cross-National Study of Health Systems, Sociology 305, Spring 1976. For the historical part (all too brief here) I have liberally used information from Daniel M. Berman, *Death on the Job, Occupational Health and Safety, Struggles in the United States.* New York: Monthly Review Press, 1978; especially the 1st chapter "Why Work Kills." Other key sources are Joseph A. Page, and Mary-Win O'Brien, *Bitter Wages, Ralph Nader's Study Group Report on Disease and Injury on the Job.* New York: Grossman, 1973; and Nicholas A. Ashford, *Crisis in the Workplace: Occupational Disease and Injury. A Report to the Ford Foundation.* Cambridge: MIT Press, 1976. Another useful general resource work on the history of OSH and current arrangements is *Protecting People at Work.* Washington, D.C.: USGPO, 1980. Also, Carl Gersuny, *Work Hazards and Industrial Conflict.* Hanover, NH: University Press of New England, 1981.
80. Berman, op. cit. (note 79): 9.
81. Peter S. Barth with H. Allan Hunt, *Worker's Compensation and Work-Related Illnesses and Diseases.* Cambridge, MA: MIT Press, 1980. Also Peter S. Barth, The Effort to Rehabilitate Workers' Compensation, *Amer. J. Public Health, 66*:553–557, June 1976.

    Berman notes that compensation replaces less than 10 percent of the expected income foregone by the victims. Further, there is great and discriminatory variation: "A Puerto Rican life is worth from $11.34 to $28.86 a week to the compensation authorities. A worker killed on the job in Mississippi will cost the boss's insurance company no more than a flat $15,000. A longshoreman's widow will receive from $79.60 to $318.38 per week as long as she lives. In Illinois, the family of a worker killed on the job may collect up to $205.00 a week. In North Dakota they add $7 a week to raise a child. Business complaints about the high cost of workers' compensation for death and disability conceal the fact that it replaces less than

10 percent of the income foregone by the victims. (See Table 5, Appendix 1)." Berman, op. cit. (note 79): 54.

82. In a court case now in Hartford against Johns-Manville, Uniroyal, Pittsburgh Corning, and Westinghouse, among other suppliers, involving workers exposed in the Electric Boat ship building yards in Groton, CT, the one-time Medical Director for Johns-Manville testified that he urged the company thirty years ago to put warning labels on its asbestos products, but it wasn't done. Expert Testimony Cites Health Hazards of Asbestos, *The Hartford Courant* (January 28, 1982): C4. On the cover-up in general, see Samuel S. Epstein, "The Asbestos 'Pentagon Papers,'" pp. 154–165 in *The Big Business Reader, Essays on Corporate America.* Mark Green and Robert Massie Jr. (eds.), New York: Pilgrim Press, 1980.
83. Caroline E. Mayer, Labor Report Sees High Rate of Asbestos-Related Deaths, *The Hartford Courant* (December 31, 1981): A3.
84. Editorial: Asbestos and Equity, *The Hartford Courant* (July 25, 1983): B6. For a view of the pot (the U.S.A. government) calling the kettle (Manville) black, see Harry F. Rosenthal, U.S. Charges Asbestosis Cover-Up, *The Hartford Courant* (September 17, 1983): A4.
85. "Negotiators for sixteen asbestos products companies and their insurers Friday proposed creating a non-profit office to settle thousands of asbestos-health claims that could reach into the billions of dollars." Associated Press, Asbestos Accord Proposed, Billions at Stake in Bid to Settle 25,000 Lawsuits, *The Hartford Courant* (May 19, 1984): A1 and A8. Also, Harry H. Wellington (Dean of the Yale Law School) Untangling Asbestos Litigation, *The Hartford Courant* (September 25, 1984): B11.
86. Tod Ensign, Steelworkers Sue Monsanto over Agent Orange Exposure, *Guardian* (New York) (August 22, 1984): 9. Associated Press, Monsanto Co. Brought to Trial As First Dioxin Lawsuits Begin, *The Hartford Courant* (June 19, 1984): B7. See also, Richard T. Pienciak, Papers Show Agent Orange Toxicity Known Before Ban, *The Hartford Courant* (July 6, 1983): A4.
87. Corey W. English, Cutting Costs, Abuses in Disability Insurance, *US News and World Report* (May 28, 1984): 80–81. Andrew Kreig, Job-Related Illnesses Mired in Workers' Compensation System, *The Hartford Courant* (May 7, 1984): A1 and A10. This was supposed to be a five-part series. The reporter and associates and a photographer worked for weeks on it. The owner of the paper said "There's no story here" and directed the editor to cut it. It appeared in one article with many pictures of individual experts (no hazardous work situations or damaged workers) and very little text. The reporter has since quit and is working on a book on this subject. In the meantime, another short but more revealing article has appeared: Andrew Kreig, UTC Goes Shopping for a Clinic, *Hartford Advocate* (March 20, 1985): 1, 6–8, 22. See also, United Press International, Compensation Laws Found Widely Abused, *The Hartford Courant* (March 4, 1984): A18.
88. Berman, op. cit. (note 79): 27–28.
89. When the Occupational Safety and Health Act (OSHA) was passed in 1970,

it incorporated only one new standard—that for asbestos. To understand the weakness of most standards under OSHA today, it is important to realize that some 500 were simply taken over from the IHF as produced over the years by this voluntary industry-sponsored group of engineering, hygiene and medical people with no control exercised by organized labor. Thus, for example, it has been estimated that 18 percent of workers will become deaf from spending a working life under the current eight-hour standards for noise of 90 db. Sweden's standard is 85 db. Such comparisons do not support Kelman's claim that OSH standards are about as protective in U.S.A. as in Sweden. Steven Kelman, *Regulating America, Regulating Sweden: A Comparative Study of Occupational Safety and Health Policy.* Cambridge, MA: MIT Press, 1981. Navarro offers many more examples of the more protective and effective Swedish OSH system in his telling critique of Kelman's work: Vicente Navarro The Determinants of Social Policy. A Case Study: Regulating Health and Safety at the Workplace in Sweden, *International J. Health Services, 13*:517–561, 1983.

90. Berman, op. cit. (note 79): 28–29.
91. Ashford, op. cit. (note 79): 52.
92. Berman suggests that the Red scares after WW I, the Depression, the unity forces of WW I, and the McCarthy witch hunt for communists took the starch out of union action over many issues, including OSH issues, producing a forty-year gap. The Walsh-Healey Act of 1936 suggests he is not entirely correct, but this impression of a well-informed expert is important.
93. Clarence D. Selby, "Studies of the Medical and Surgical Care of Industrial Workers." US Public Health Service Bulletin No. 99, 1919, from Berman, op. cit. p. 97.
94. William P. Shepard, *The Physician in Industry.* Saylem, New York: Ayer Co., 1961. In later work, we examined the backgrounds and relatively low prestige of people in the field: W. P. Shepard, R. H. Elling, and W. F. Grimes, Study of Public Health Careers: Some Characteristics of Industrial Physicians, *J. Occupational Medicine, 8*(3):108–119, March 1966. For a more recent analysis of the low prestige of the field within medicine, the inadequate training (but including discussion of a program of NIOSH fellowships placing physicians with major unions—a program cancelled by the Reagan Administration) see Susan Mazzochi, Training Occupational Physicians, Suppose They Gave a Profession and Nobody Came, *Health PAC Bulletin*, No. 75:7–10, 19–24, March/April 1977.
95. Alice Hamilton, *Exploring the Dangerous Trades.* Boston: Little Brown and Company, 1943. Berman points out a sad irony. One of her greatest successes, the National Lead Corporation, "recently attracted attention for poisoning a large number of workers in Indianapolis, then compounding the problem by 'curing' them of the lead intoxication with a medicine that ruined their kidneys." Berman, op. cit. (note 79): 200. The source he cites is *Spotlight on Health and Safety.* Washington, D.C.: Industrial Union Department, AFL-CIO, 1976. See also, Barbara Sicherman, *Alice Hamilton: A Life in Letters.* Cambridge: Harvard University Press, 1984.
96. Berman, op. cit. (note 79): 32.

97. Paul J. Nyden, "Miners, Mr. President, Are Not Slag," *The New York Times* (January 24, 1982): editorial page. This article also cites Dr. Lorin Kerr's estimate that each year at least 4,000 workers who at one time worked as miners die from the effects of coal dust. Also important to note: "76 percent of all [accident] fatalities were suffered by unorganized miners and foremen."
98. Again to emphasize the point of how the struggle for OSH must always be renewed, in the 1972 election, after OSHA was in force, Nixon's reelection committee circulated the now famous "Gunther letter" of the then Secretary of Labor promising to weaken the enforcement of OSHA in companies from which significant campaign contributions had been received.
99. Public Law 91-596, 91st Congress, to take effect 120 days after the date of signing. There are many provisions which I can not detail here. For example, every employer with more than ten employees must keep records of all work-related injuries and illnesses, and the worker, or her or his representative, has the right to review those records. Another important provision is the non-discrimination clause whereby a worker can not be punished or discriminated against for complaining to OSHA. While these mean little in a non-union shop, having them "on the books" may add to all workers' sense of OSH rights. For other provisions of direct interest to the worker, see "OSHA: Your Workplace Rights in Action; Job Safety and Health; Answers to some common questions." OSHA booklet 3034, Washington, D.C.: U.S. Dept. of Labor, Occupational Safety and Health Administration, 1979. See also "All About OSHA." Washington, D.C.: US Dept. of Labor, OSHA, revised April 1976.
100. An average of $25 for each of 724,582 violations listed in 140,467 citations resulting from 206,163 inspections between January 1975 and April 1977. Berman, op. cit. (note 69): 34. Instead of cleaning up, many managements came to see OSHA as part of "the cost of doing business." Thus it bothered big firms less than little ones as the per worker costs were less. Some argue that monopoly capital was given a further competitive edge by OSHA.
101. Public Citizen, Health Research Group, Survey of Occupational Health Efforts of Fifteen Major Labor Unions. Washington, D.C., September 6, 1976. However there remained many weak areas and staff needs. Few unions had in fact made OSH a key aspect of collective bargaining. As regards OSH physicians to work for unions, the only full time union physician recorded by this survey, Dr. Lorin Kerr of the United Mine Workers, said, "There exists a crying need for doctors in the unions."

    At this point I would like to make note of a curious focus of both union leaders and public health and medical care leaders on bargaining for curative medical care insurance and other treatment programs to the virtual exclusion of preventive OSH service programs. Certainly the necessary links between the two were ignored. I am at a loss to explain this unfortunate gap in the history of public health, medical care and OSH. I have been in correspondence with the best of medical care-public health leaders—Isadore Falk, Professor Emeritus of Medical Care, Yale University and the Director of the original Costs of Medical Care Study (1928-1933); Milton Roemer, Professor of Medical Care and Public Health, UCLA; and George Silver, Professor of Public Health and International Health, Yale University—but have no satisfactory answer. One volume suggested by Professor Falk is quite useful as regards the issues and considerations in establishing specialty OSH services in the workplace (albeit all from a management perspective) but there is no consideration of links to the general health services. The one union spokesperson quoted in the volume, Jerome Pollack, Consultant to the UAW, identifies medical care associated with workers' compensation as having a preventive focus (so much for health services in the workplace!) and gives the bulk of his attention to bargaining for medical care or insurance programs. Albert Q. Maisel (ed.), *The Health of People Who Work.* New York: The National Health Council, 1960.

Other relevant works are focused entirely on medical care and hardly, if at all, make the link to conditions in the workplace which generate the diseases for which workers need care. Joseph W. Garbarino, *Health Plans and Collective Bargaining.* Berkeley: University of California Press, 1960; Raymond Munts, *Bargaining for Health, Labor Unions, Health Insurance, and Medical Care.* Madison: The University of Wisconsin Press, 1967.

Perhaps both union leaders and medical care—public health leaders saw OSH as so completely under management's control that they thought only of bargaining for medical care which was politically popular, and a definite need of workers and their families. Still, the lack of a comprehensive set of demands linking OSH and PHC is curious to say the least. I am very much open to learning more on this matter, but can not devote more space to it here.

102. R. H. Elling, "Worker-Based Education, Organization and Clinical-Epidemiologic Support for Occupational Safety and Health (OSH): The New Directions Program in Connecticut." Paper summarized at the ILO/WHO Conference on OSH, Sandefjord, Norway 16-19 August 1982. Published in the conference proceedings. Geneva: ILO, 1982.

103. This policy has since been changed by OSHA under Reagan's regime. Now personal protective equipment is seen as perfectly acceptable as a first choice. Philip J. Simon, *Reagan in the Workplace: Unraveling the Health and Safety Net.* Washington, D.C.: Center for Study of Responsive Law, 1983. Other points from this very damning report are cited below under "OSHA Update."

104. "AFL-CIO Associate Director for Occupational Safety and Health, Margaret Seminario, said Rowland is 'unfit' to hold the position . . ." "Rowland to be nominated for OSHA Administrator Post," *Occupational Health and Safety, 53*(7):5, July/August 1984.

105. A January 15, 1982 mailing from the Department of OSH of the AFL-CIO in Washington, D.C., which has played an important watch-dog role, indicates that these figures for a year's time may hide a sharper monthly decline. Thus for October 1981 compared with October 1980: Total inspections down 45 percent; Complaint inspections down 49 percent;

Follow-up inspections down 81 percent; Number of serious citations issued down 40 percent; Number of willful citations issued down 94 percent; Number of repeat citations issued down 60 percent; Number of complaints filed down 40 percent; Complaint backlog up 208 percent. See also Simon, op. cit. (note 103).

106. Department of OSH, AFL-CIO, mailing of December 16, 1981.
107. "State Laws Under Fire; OSHA Issues Right to Know," RICOSH (Providence, Rhode Island) *Job Safety and Health, A Worker's Right.* (Spring, 1984): 1 and 4.
108. Page and O'Brien, op. cit. (note 79).
109. An exemplary set of union actions is described in Michael Alaimo, Union on the Move: The OCAW's Efforts in Health and Safety, *Job Safety and Health Magazine* (No. 1, 1978): 24–30.
110. Mary Ann Kahn, Community MDs Must Recognize Occupational Ills; "Nothing Else Will Work," *Medical News and International Report 8*(4): 1 and 8, April 16, 1984.
111. David Rutstein et al. Sentinel Health Events (Occupational), *Amer. J. Public Health 73*:1054–1059, 1983.
112. Bill Keller, U.S. Health Aide Urges Logging of Job-Injury Data, *The New York Times* (June 21, 1984). See also Kreig, op. cit. (note 87).
113. Barry S. Levy, The Teaching of Occupational Health in American Medical Schools, *J. of Medical Education 55*:18–22, January 1980. The school where Dr. Levy teaches, University of Massachusetts Medical Center in Worcester has some twelve required OSH hours.
114. AAMC "Challenges to Medical Education" 95th Annual Meeting, October 27–November 1, 1984, Chicago.
115. Barry Levy, "The Teaching of Occupational Health in American Medical Schools: Five-year Follow-Up of an Initial Survey," March 26, 1984, submitted to *JAMA* (reproduced) cited with permission of the author.
116. Barry Levy and David Wegman, "Recognizing Occupational Disease–How and When to Take a Complete Occupational History," chapter 3 in *Occupational Health: Recognition and Prevention of Work-Related Disease.* Levy and Wegman (eds.), Boston: Little, Brown and Co., 1983.
117. However, in broad economic policy, foreign affairs, defense, and control of the media one sees a comparatively centralized system: C. Wright Mills, *The Power Elite.* New York: Galaxy, 1959; William G. Domhoff, *The Higher Circles.* New York: Vintage, 1970. A new study, using official statistics, shows that .5 percent of the population who make more than $135,000 a year control 20 percent of the country's net worth. Further, almost no large fortunes are amassed in a single lifetime. Nearly all are inherited. Thus "the American Dream" of "rags to riches" has indeed become a myth. One half of everything produced in U.S.A. is produced by 150 corporations which are controlled by 250 families. From John Dalphin, *The Persistence of Social Inequality in America*, as reported in Andy Dabilis, No More Horatio Algers, Sociologist Says, *The Hartford Courant* (February 1, 1982): B8.
118. During my several months in Europe conducting the field work for this

study, people kept asking if there is no resistance to current rightist trends as there was against the Vietnam War. A massive march against U.S.A. support for a murderous regime in El Salvador and against social program cutbacks took place in Washington, D.C. on May 3, 1981. On September 19, 1981 350,000 people, by some estimates, marched in Washington, D.C. against Reaganomics. Jesse Jackson's "Rainbow Coalition" showed surprising strength in the recent elections.

> Jesse Jackson brought the ties between the rank-and-file labor movement and the struggle of Black people for justice another important step forward as he walked a militant picket line in Rahway, N.J. together with 3,000 workers, their families and supporters to protest the lock-out of the Oil Chemical and Atomic Workers union by the multinational Merck Pharmaceutical Company.
>
> Mounting the platform in front of the plant gates, Jackson was greeted with chants of "Labor for Jackson, Win Jesse Win!" by the mainly white crowd which was about 30 percent Black and Latin. "I want to express my solidarity with your struggle for justice, decent wages and working conditions and health and safety against a company that makes millions off your labor and then demands that you give up and give back benefits that took years to win," declared Jackson. Merck, one of the most profitable companies of the Fortune 500, confronted OCAW local 8-575 with a union-busting proposal, demanding over 100 concessions. When the leadership refused, the company locked the workers out. *Peoples Anti-War Mobilization and the All-Peoples Congress (PAM-PAC) Newsletter II*(6):6, June 1984.

119. B. Kleckzkowski, R. H. Elling, and D. Smith, *Health System Support to Primary Health Care*. Public Health Papers Series, No. 80. Geneva: WHO, 1984.
120. Korpi and Shalev, op. cit. (note 48).
121. These and other rights will not come out of the blue or some moral change in the hearts of capitalists. Citing Marx' realization that rights are defined in certain political economic contexts, Gibson recognizes that rights are achieved in struggles designed to change the context. Mary Gibson, *Workers' Rights*. Totowa, N.J.: Rowman and Allanheld, 1983, esp. 138–140. This work incidentally is primarily devoted to case studies of the violation of workers' rights by American Cyanamid (where five women were told they would have to get sterilized to avoid lead exposure of any potential fetus or take a non-exposure, but lower paying job) Occidental Chemical, Burnside Foundry, and J. P. Stevens.
122. A conference for health care professionals and workplace managers focused too much on internal, psychological stress. "New Developments in Occupational Stress." Cincinnati, Ohio: DHHS (NIOSH) Publication No. 81-102, December 1980. The Swedish experience and view seems much more promising: Bertil Gardell, Production Techniques and Working Conditions, *Current Sweden* No. 256, Stockholm: The Swedish Institute, August 1980.

123. James E. C. Walker and Norman L. Armondino, *The Primary Care Physician: Issues in Distribution.* Connecticut Health Services Research Series No. 7, North Haven: Blue Cross of Connecticut, 1977.
124. National Research Council: Only a Few Chemicals Tested for Health Effects, *The Nation's Health.* American Public Health Association: 1 and 3, April, 1984.
125. Associated Press, Dangers Lurking in High Tech Plants Worry Employees, Experts, *The Hartford Courant* (August 27, 1984): A3. This article identifies an arsenic containing substance, gallium arsenide, as the new substitute for silicon chips. The production process uses arsine gas which can be fatal in large doses and in smaller doses is a suspected cause of skin and lung cancer as well as mood swings and paranoia. At a recent joint meeting of MASSCOSH, RICOSH and ConnectiCOSH, one person on top of the situation, said the silicon chips are just as cheap and effective under normal conditions and the only reason for substitution is that silicon chips would be knocked out by radiation in a nuclear war, while the gallium arsenide is not so affected.
126. *Work in America, Report of a Special Task Force to Secretary of Health, Education, and Welfare.* Cambridge: MIT Press, 1973.
127. Tom R. Burns et al. (eds.), *Work and Power, The Liberation of Work and the Control of Political Power.* Beverly Hills: Sage, 1979.
128. Richard Severo, 59 Top U.S. Companies Plan Genetic Screening, *The New York Times* (June 23, 1982); also, *The Role of Genetic Testing in the Prevention of Occupational Disease.* Office of Technology Assessment, Congress of the U.S.A. Washington, D.C. G.P.O., 1983.
129. Charles Levenstein, Leslie I. Boden, and David H. Wegman, COSH: A Grass-Roots Public Health Movement, *Amer. J. Public Health 74*:964–965, 1984.

CHAPTER 13

A Summary Assessment According to the Model of OSH and OSH-PHC

In Chapter 6 the strategy of using ideal-typical analysis was discussed as developed by Max Weber who used this method to contrast the realities of bureaucracy with the ideals as he envisioned them. Analogously, some dimensions of an ideal OSH and OSH-PHC model were presented in that chapter. These were then used in subsequent country chapters (7-12) as a summary assessment measure, as a set of special glasses, so to speak, for viewing the workplace safety and health conditions, provisions for protection of workers at work, as well as for viewing the linkages (or lack of linkages) of the general health services, particularly at PHC level, to protecting workers against work-related injuries and diseases. In this chapter the interwoven dimensions of this model are used to provide a kind of comparative or contrasting summary assessment of the OSH-PHC situations in the six study countries.

Of necessity, this is a somewhat static and abstracted overview. The reader is urged to keep in mind the dynamic development of provisions for workers' health and safety in each country through the course of class struggle, union organizing around these issues, and other dynamic forces. The development and current strength of the workers' movement was presented, at least briefly, as background for understanding the OSH situation in each study country. In this chapter I will make only brief allusions to these explanatory forces. In the next, and final chapter I will probe more into these explanatory forces and their limits and potential for shaping more healthy working environments everywhere.

As to the organization of this chapter, I will take up the elements of the model one at a time–policy, sponsorship and control, education, organization, information and financing–by comparing the situation of the six study countries on each aspect of each dimension (for example, education relates both to worker education and preparation of health care providers, including experts in OSH and PHC providers) while offering at the same time some reminders of workers' movements and these dimensions of concern. In a concluding section, an overall comparative assessment is offered with reference back to our original working hypothesis that the historical and current strength of the workers' movements in these countries would explain their positions on both OSH conditions and provisions and OSH-PHC relations.

POLICY

It is quite fascinating to observe that the late 1960s and 1970s brought a heightened expression of formal concern for OSH in all six study countries—and as far as I have observed in many other industrialized nations. These formal expressions took the form of a rash of very important laws in each country—in U.S.A., the Occupational Safety and Health Act of 1970 (OSHA); in Sweden, the 1976 Working Environment Agreement between umbrella organizations of organized Labor (LO and PTK) and the employers' umbrella organization (SAF) which was then confirmed in law by the Working Environment Act of 1978, etc. (others and preceding developments are detailed in country chapters 7-12). No serious attempt has been made here to explain these common developments other than to suggest that they occurred during a high period for the world economy when workers' jobs were not threatened and there was considerable ferment in all study countries and others associated with the Anti-Vietnam War struggle and associated struggles of the times around women's issues, minority rights, and the awakening environmental movement.

These expressions of formal concern were impressive. For example, OSHA in U.S.A. was passed by a bipartisan Congress and signed into law by a Republican president (Nixon) "to assure so far as possible every working man and woman in the Nation safe and healthful working conditions and to preserve our human resources."[1]

But actions in pursuit of policies were not so impressive in every study country. Again, in U.S.A., while the law brought many advances and energy to a movement which is still very much alive today in unions, COSH groups, professional-technical associations (for example, recent policy statements and moves in the American Public Health Association),[2] and in scholarly and scientific work,[3] official enforcement of and support for OSH has generally gone downhill since the high point of the early and mid 1970s.[4] Thus our assessment of policy in each country must reflect both deeds and words. In terms of our model, these deeds and words should relate both to OSH and OSH-PHC.

The countries which stand highest on this dimension are Sweden, G.D.R., and Finland. In each, formal statements in law and other ways have accompanied or been followed by impressive concrete actions. From the perspective adopted here, placing special importance on the involvement of workers themselves in protecting their own health and safety at work, Sweden stands the highest. In that country, the strength of the workers' movement, in workshops, and in the political arena more widely, has led to what Korpi and Shalev have called a "political exchange" with capital.[5] Thus the union-dominated health and safety committees in every workplace of any size have been given responsibilities and powers by law which are truly impressive as compared with the situations in most other study countries. For example, they have a controlling voice, in the hiring and firing of any industrial physician or health service (a medium-sized or small-sized shop may obtain services from an OSH policlinic serving a number of shops). Union-identified health and safety representatives are empowered by law to stop the production process when they see a hazard—even a long acting hazard such as a carcinogen—until the perceived problem is clarified as no problem or cleared up if it is a problem. Sweden has weaknesses, particularly as regards OSH services coverage for small shops. And, until very recently, the problem of OSH-PHC linkages had received little recognition or attention. Now a very important report has been issued considering nearly all facets of this problem, but highlighting the need for cooperation between OSH and PHC services in the rehabilitation of sick or injured workers.[6] It remains to be seen if this OSH-PHC report will be followed by official pronouncements and subsequent actions.

Except for the important element of direct worker involvement, the G.D.R. must be rated as strong or even stronger than Sweden on the policy dimension. The shortage of workers and constitutionally assured full employment, together with a socialist-oriented (some would argue it is a state-capitalist) production system, and governing structure guided by the German Communist Party have led to a very complete system of worker protection and follow-up in case of exposure to hazards. Many examples could be given. Perhaps the most impressive is the 1981 law and actions to survey, measure and classify the hazards in every workplace in the country and adopt plans to remove hazards and follow-up exposed workers according to a set of priorities as to seriousness and urgency of risks. In the sphere of OSH-PHC connections, while there remain many problems, the G.D.R. is ahead of all other study countries. First, there is evidence of official awareness and concern for the problem (this only recently developed in Sweden as just indicated). Second, there are some thirty-six required hours of OSH education included in medical education across the country, with variations locally according to professorial preference. Third, there are some combined PHC-OSH policlinics which offer a near model of how the relations might best be worked out in practice.

Finland has the special strength of having adopted a law requiring OSH services coverage for every workplace in the country. Although difficult to enforce when there are not enough OSH personnel, the very impressive National Institute for OSH and other bodies are taking vigorous steps to develop enough physicians and other personnel for the OSH services. The workers' movement is strong, but not as strong as in Sweden or the G.D.R. Thus the role of private services, even in the provision of OSH as well as the general health services, does not involve workers themselves to the same extent as in Sweden.

At a lower level of achievement as regards OSH policy and its pursuit, the F.R.G. is especially notable for having an impressive paper facade but not as vigorous a set of actions as in G.D.R. or Sweden or Finland. It would appear that the facade itself represents a kind of "keeping up with the Joneses," that is, looking over the infamous wall of the G.D.R. to give workers in the West the impression that they are being as well looked after as are their former country men and women in the East. It may also represent a kind of Bismarckian anticipation as was the adoption of the first national health insurance in the world in 1883, designed to take the wind out of the sails of the workers' movement. Not everything in the F.R.G. system is sham. For example, OSH has become a definite section on the equivalent of national board exams for medical students, even though there are no set number of required hours in the curriculum as in G.D.R. More important, the real centers of the OSH system, the workplace insurers (Berufsgenossenschaften) have an impressive record of training OSH physicians and organizing OSH services as well as conducting inspections, and other work. The unions have equal representation with managements in these unique agencies; thus there is a significant chance for organized workers to have an impact on OSH conditions, protective measures, and services. Still, only some 30 percent of workers do belong to unions and unemployment has been high in recent years, so there are many weaknesses in the system. One of the peculiar ones is the limitation of OSH services strictly to preventive efforts—this at the insistence of organized medicine which has sought thereby to keep OSH physicians from invading even the smallest fraction of its patient market. Many, if not most OSH services are offered by GPs and other PHC providers who have had just a two-week introductory training in OSH.

The apparent strength of the workers' movement in the U.K. may be deceiving. The evidence of strength includes the longest tradition of struggle in any of our study countries, some 55 percent of workers currently organized, and a Labour Government in power at various times since WW II, though always as a coalition, never a majority in itself. But unemployment has been exceedingly high—up around 12 percent in recent years—and the very conservative, even repressive Thatcher government has been in power for the last several years. In any case, the U.K. falls toward the weaker end of the spectrum as regards OSH policy and its pursuit, and as regards awareness of, concern for and action about the OSH-PHC connection. Workers simply do not have the say they do in Sweden as regards OSH matters and OSH services are poorly developed and are not made up for by the minimally staffed Employment Medical Advisory Service (EMAS). Not even the NHS (the largest employer in all of Europe) has an OSH service. There seems to be little awareness of OSH among medical educators and no OSH requirements as a general rule (though some schools have some hours) such that PHC providers would become aware of OSH problems and what to do about them. One of the hopeful aspects is the ferment going on around OSH matters in left-oriented activist groups such as the Politics for Health Group and the Work Hazards Groups which is a part of the British Society for Social Responsibility in Science.

The U.S.A. stands clearly at the bottom of our six study countries on this dimension. A history of narrow business unionism (that is, an acceptance of the capitalist system and focus on wages and hours to the exclusion of wider societal concerns and political action)[7] the lack of a labor party, high unemployment (10.8 percent at the high point of the recent recession with some 12 million people ready and wanting to work but unable to find a job), and an atmosphere of union busting with less than 20 percent of the workforce organized, has made it impossible to even keep the gains of OSHA as it first came in. There is not even an OSH services standard in the OSHA law and those (mostly large) plants which do have services generally have poorly trained staffs completely under management's thumb. More than in any other of the study countries, the hope and energy for the future lies outside of official channels in the unions themselves, especially among activist rank and file and in COSH groups (there being forty or so across the country). Finland and the U.K. also have COSH-type groups but the U.S.A. is special in the strength and importance of this element (Sweden did have a network of such groups, but it apparently faded with the success of official organized labor in OSH matters).

There is almost no evidence of concern for the OSH-PHC connection among official circles in U.S.A. Even public health and medical care leaders seem to have ignored the problem. Surveys reveal that about half of U.S.A. medical schools now have an average of four required hours in OSH. Some of the others offer elective OSH courses.

In summary of the policy dimension, I would rank Sweden and the G.D.R. close together at the top, each with a very different system as compared with the other, Finland is next, then some ways down, the F.R.G. and the U.K. close to each other with the U.S.A. at the bottom among these study countries.

SPONSORSHIP AND CONTROL

Both the extent and nature of the sponsorship and control of OSH conditions and services and OSH-PHC relations are to be discussed for each study country. First it should be noted that the laws in all six countries make the managements responsible for protecting workers' health and safety. Beyond this common assignment, there are great variations as to who is in charge of what resources and (the main focus here) the extent and nature of worker control over their own health and safety.

In different ways Sweden and the G.D.R. have strong degrees of worker control of OSH conditions and services. In neither case (nor in any other) have workers or their representatives entered into even a noticeable degree of control over PHC services, personnel or training so as to improve the ability of PHC services to pick up and follow-through on OSH problems.

Considering different levels, Sweden shows the strongest degree of worker control of all study countries at the most important level, that is, on the shop floor. The law of co-determination of production as well as extensive worker education efforts in OSH assure the worker's right to know what hazards may be affecting her or him. Workers have the right to refuse to continue working under hazardous conditions without fear of retaliation. As noted earlier, health and safety reps have a legal right to stop the production process on their word alone if they perceive hazardous conditions. Also, as noted, the union-dominated health and safety committee (and nearly all workshops of any size have one, since some 98 percent of blue-collar workers are organized and probably 80–85 percent of all workers are organized, including 70–75 percent professional and technical workers) has the primary voice in hiring and firing the chief OSH medical and engineering personnel. This committee must also approve the OSH physician's annual report and plan before the NHS will reimburse the employer for one half the costs of the OSH services. At national level too workers have a strong voice through their unions' umbrella organizations (LO and PTK) and the Social Democratic Party which has been in power for over forty-five years. For example, labor representatives are included with at least an equal voice to that of the owners in every official study commission and in the formulation of legislation. It should be specially noted that the unions, particularly LO at the national level, have a significant staff of experts to represent them on such bodies. This even extends to boards overseeing universities and professional schools, including medical education, though organized labor seems not to have raised the issue of OSH-PHC linkages, except in a very recent study and report (see note 6). The special strength of labor's voice is perhaps best symbolized by the fact that the labor and management umbrella organizations first reach agreement on major new programs and directions which only later are confirmed in law. In the U.S.A. by contrast, organized labor often seeks some small advance through lobbying in Congress for new legislation, the provisions of which it could not achieve through collective bargaining alone.

In the G.D.R. the most authentic and ubiquitous control is by politically conscious workers organized through the German Communist Party. This guiding voice touches all things in the society of major social significance, including, very importantly, worker health and safety. The trade unions are of a different character than in our capitalist study countries. They are officially sponsored and encouraged organizations working cooperatively with the management and the state to encourage high productivity as well as protection and benefits and fair treatment for the workers. Thus they are not in an antagonistic or adversarial relationship to the power structure as in capitalist society; instead, they are a part of the power structure. They have the major responsibility, formally speaking, for seeing that the OSH Laws are enforced. This is a watchdog role, for the management of each enterprise (Volkseigener Betrieb—People's Own Undertaking) is legally responsible for protecting the health and safety of workers as in all our study countries. At the shop level there are a number of organizational vehicles involving party representatives, union representatives, and other workers, including a production planning group and the health and safety committee. These have been detailed in Chapter 9 on the G.D.R. While some of the provisions for worker participation and control appear a bit stilted and formalistic, mostly ordained from the top down, it can not be denied that, at least at this formal level, there are an impressive set of structures and procedures giving an opportunity for a major voice to workers and their representatives. The OSH service personnel are neither beholden to management, nor to the unions; they are employees and under the supervision of the NHS or Ministry of Health, though at their local places of work they are responsible to the several worker-management-party bodies mentioned above. In spite of the many opportunities for workers and the unions to exercise a voice in OSH matters, my own experiences in the country led me to see it as a primarily bureaucratically-technocratically-determined system—however, with the major qualification, and it is major, that the overall system presents itself as a workers' state (see also the views of my guide in G.D.R. given in the Appendix to Chapter 9). Finally, as to PHC, there is no evidence of worker control, other than party guidance of the national health service.

Finland does not stand as high on degree of worker sponsorship and control as it did on OSH policy (if it makes sense to think of comparing positions on the different dimensions of the model—something I would not argue for strongly, but use here more as a figurative way of giving a sense of comparative position). And compared to Sweden and the G.D.R. it stands much lower on the dimension of worker sponsorship and control of OSH conditions and services. A history of late industrialization, and, therefore, late development of unions, as well as an inheritance of paternalism and rightist repressions, has meant that labor could really only take hold freely and widely following WW II. In any case, the comparatively high percentage of the workforce which is organized is somewhat misleading as regards the true voice of labor. It is not insignificant; in fact, quite a bit stronger than the U.S.A., but I am unsure of placing Finland above F.R.G. and the U.K. on this dimension. The sense I have is that much in OSH is determined by the renowned National Institute for OSH and by the experts of large firms who are employed by and report to those firms' managements without the significant union voice as in Sweden. Unions themselves lack enough OSH experts of their own. As in other study countries, there is no sign of workers or their representatives having a noteworthy say in the provision of PHC or its relations to OSH or the preparation of PHC personnel. Probably only because worker influence is felt through unions in a higher proportion of shops than in F.R.G. do I place Finland higher on this dimension.

The OSH laws in the F.R.G. do not even mention unions as having a role. They require workers to be included on health and safety committees in every shop of any size, and, in union shops, these will be union representatives, but in a non-union shop they will most likely represent the interests of the boss. The law does place the workplace insurers in a key position with the major responsibility for OSH standards, inspections, and services and the unions have a significant voice on the boards of these agencies. But the OSH services are paid for and under the control of management. And the unions do not have enough expertise of their own. There was no suggestion that the unions had voiced any concern for PHC and its relations to OSH.

Worker sponsorship and control of OSH conditions and services in the U.K. may not be as weak as in U.S.A. but it is surprisingly so, considering what appears to be the comparative strength of the workers' movement. True, the OSH laws are administered through the Ministry of Labor which might be considered a traditionally Labor portfolio, even in a conservative government such as that which has held power for the past several years. But I am not sure much comfort can be taken from this fact. There is no OSH services standard and even where there are OSH services, they are under management's direction. The NHS has not even entered the OSH picture for its own employees to say nothing about other workers. Thus organized labor is particularly suspicious of medical sponsorship and dominance in the OSH field. One strong point is that the labor umbrella organization, the Trades Union Congress (TUC) and several of the larger unions have some excellent OSH experts to help at least keep up with things. There is no evidence of organized labor having voiced its concerns regarding PHC and OSH, though some union OSH experts have begun to gather information and formulate proposals on such matters as OSH as a required part of medical education.

There are several avenues for workers to have a voice in OSH matters in U.S.A., but they are all comparatively weak. At the shop floor level some of the larger unions have achieved significant gains through the contract bargaining process. The UAW-GM contract, for example, provides for a certain number of company-paid, union-selected and trained health and safety reps who are free to walk around and try to identify and remove hazards. But nowhere are unions in control of the hiring and firing of OSH services personnel for the shop as in Sweden, nor can workers leave their jobs in the face of hazards without fear of being fired. In any case, less than 20 percent of workers are organized. This fact relates to the weakness of labor's voice in the legislative process, though the AFL-CIO and some of the larger unions have very good OSH experts who take part in legislative hearings. One of the most important OSH forums in U.S.A. is the courts, where managements' resources and ability to hire lawyers, to say nothing of the management bias of the court system itself, allow the interests of capital to dominate in the testing of new OSHA rulings and other parts of the law. There is no sign of organized labor attempting to influence PHC providers or their educators to have a more effective impact on OSH. The U.S.A. is the weakest of all six study countries as regards worker sponsorship and control of OSH conditions and provisions and relations to PHC.

EDUCATION

Perhaps the most important aspect of this dimension is worker education to allow workers themselves to play a greater role in protecting their own health and safety on the job. The number and complexity of OSH problems and numbers of workplaces make it impossible for any country to hope to control these problems solely through cadres of inspectors and other OSH experts (even if that were an effective approach, which I don't believe it is). Thus a major effort must be made to educate workers about OSH hazards and what to do about them. However, the emphasis here is quite different from the "blame-the-victim" approach taken by industry, particularly in U.S.A.[8] in which attempts are made to focus attention on the, so-called, careless worker. The solution to work-related health and safety problems from this perspective is to convince workers to wear personal protective equipment and "be careful." If not exposed for what it is, such attention takes the heat out of any push to get the owners and management to slow down the dangerous pace of production, undertake sometimes costly engineering controls, or substitute non-hazardous chemicals and materials for hazardous ones. It is this concern which makes the sponsorship and control of the educational process of major importance. Brown and Margo offer an authentic approach[9] which guided us in the New Directions Program in Worker Education for OSH in Connecticut.[10] In this instance (and in about one-half of these OSHA-funded New Directions Programs nationally—others were business or academic programs without significant worker control) organized labor dominated the advisory Board and Executive Committee of the program and adopted policy with regard to overall strategy and emphasis, some specific content, timing, and location of the variety of educational approaches. These included movement from the classroom into the union hall and onto the shop floor where "projects" to correct problems were undertaken by worker students and progress and difficulties reported back to classmates and staff. If suspected individual or work group health problems were uncovered, as happened all too often, the program's ties to the Yale Occupational Health Clinic allowed referral to and follow-up by this independent (non-company) group of OSH clinicians, industrial hygienist, and other prevention-oriented staff. It was an exciting program; "was" because the Reagan regime cut the funding in 1982. In the meantime, fortunately, the Executive Committee contributed importantly to the founding of a COSH in Connecticut (ConnectiCOSH) and worked with an environmental and citizen-action coalition to get legislation passed to fund a Worker Education Division of the Worker Compensation Commission. The former New Directions staff now works for the new Division and fairly generous funding is not dependent upon the political ups and downs of OSHA in Washington D.C.

This brief sketch of one particular experience is given to suggest that this program and other somewhat similar ones across the nation put the U.S.A.

in a higher place on this particular facet of this aspect of this dimension than most other study countries. But this aspect also includes the very important "workers' right to know" what chemicals and other materials they are working with and what the hazards are and what to do about them as well as access to their own company medical records, and reports of any hazard inspections and tests conducted by OSHA (or like agencies in other countries) or other agencies. Unfortunately, the struggle for the right to know has not been won in U.S.A. Some important but small victories have been achieved primarily through state laws, primarily concerning some 400 or more probable human carcinogens. One very hopeful recent development occurred in a New York State court which ruled that a worker's rights were violated when she was fired for refusing to work after her employer failed to provide information on chemicals in her workplace as required by the New York State right to know law (Bureau of National Affairs, "Washington Alert," September 1984). But these laws do not reach workers in every state, do not cover all substances, and are themselves being threatened by a Federal move to adopt a set of regulations emphasizing "trade secrets" which some COSH groups and others have dubbed "right to hide." This situation of inadequate right to know laws has led to educational efforts by unions and COSH groups on how to research chemicals and "crack the company code."[11]

In Sweden, right to know is assured as is the right to refuse hazardous work. Further, although the New Directions efforts in U.S.A. are impressive, they pale in comparison to the forty or more hours of OSH preparation given to some 110,000 union health and safety representatives in Sweden under strong union sponsorship and control. This for a nation of some eight million people. With 230 million population, the U.S.A. would have to have prepared 3,162,500 health and safety reps to an equal level under strong union guidance to be equal to Sweden. Connecticut with three million population has been fairly vigorous and may have some 250 union OSH representatives prepared at the forty hour or more level (equivalent to the "Trainers" of the New Directions Program). If this achievement is generalized to the whole country (probably too generous an estimate) there are fewer than 20,000 worker OSH representatives trained to this level–a deficit of some 3,142,000 as compared with Sweden! But the full sense of difference is only indexed by such arithmetic. Since, the large majority of workplaces of all types are organized in Sweden, the OSH reps have a supportive context within which to function. In U.S.A. with less than 20 percent of the workforce organized (as compared with Sweden's 80 plus percent) only a small proportion of workplaces are organized.

In Finland, there are widespread worker education efforts but they seem to be less under the sponsorship and control of unions than in Sweden and more likely to be offered through company managements or the National OSH Institute. Nor has the right to know been assured by a co-determination law and other measures to the same degree as in Sweden. However, the Socialist Front for Health (a COSH-type group) has reached many union activists and other workers through its conferences and workshops. Thus, on this aspect of the education dimension Finland appears to fall below Sweden but at least on a par with the U.S.A. which I would place above F.R.G. and the U.K.

In F.R.G. there have been extensive worker education efforts organized primarily through the National Institute for Work Protection and Accident Research (BAU) but only a minority of workers (some 30 percent) and workplaces are organized and their voices are not well represented in this education. The picture of the workers' right to know seems as problematic as in U.S.A., perhaps worse, and there are no COSH groups to help generate wide worker awareness and concern.

The situation is somewhat better in the U.K. The larger unions have OSH experts who have organized extensive training programs for several hundred thousand shop stewards and OSH reps who have major responsibilities under the law. This large number has been reached since some 55 percent of workers are organized. Also, there are the COSH-type groups noted earlier. However, the dominance of management perspectives in the current government has not allowed many advances in the workers' right to know.

The G.D.R. is a special case. Extensive efforts have developed a cadre of Red Cross-prepared workers who know what to do in case of an accident or disaster in the shop. These workers have also received some preventive OSH information. Further, every worker must be given instruction by supervisory personnel concerning any hazards on her or his job and the worker must sign off in a personal work hazards diary which is kept on file. While, some of this education has a formalistic, even paternalistic character, and there are obvious large gaps (for example, the Ministry of Health driver who had never heard that asbestos is hazardous) the numbers of workers reached and the existence of a ubiquitous protective worker network are impressive.

In summary of the comparative picture of worker education for OSH, Sweden stands highest; the G.D.R. next, with its own peculiarities; U.S.A. next; then the U.K.; Finland; and finally the F.R.G.

Another aspect of this dimension is the education of service providers. This in turn has several facets—the preparation of OSH experts (of more than one type and at different levels); OSH material in basic education of medical students and other prospective PHC providers; and OSH for practicing PHC providers. It would be too complicated and lengthy a task to attempt here a treatment of every facet of this aspect in full detail. Some details are given in country chapters (7-12). Here I will only present summary assessments with a few illustrations. The focus is more on medical and nursing personnel than engineering, industrial hygiene, ergonomic, social psychologic, and social service personnel because of our special interest in PHC.

As regards specialist medical training for OSH, a trend observable in several of the six study countries is worth special note as it may offer a model for other countries to follow. In G.D.R., Finland, F.R.G., with signs in the same direction in Sweden and the U.K., there are or are developing two levels of OSH training for physicians. The "high" level (or "big" level as it is termed in F.R.G.) is achieved after some three to five years of academic and research work following medical graduation. This is intended to prepare the person for research and teaching and clinical work on OSH problems in academic settings. The "lower" level (or "small" preparation as it is termed in F.R.G.) is intended for OSH physicians based in an in-plant OSH service or an OSH policlinic serving a number of small-to-medium-sized workplaces. This training is usually organized outside of academic centers (though academic people may play a role in the teaching) and may range anywhere from two weeks, spread over half a year, to two or four weeks of concentrated training, to two years of preparation involving about one half class room and one half practical or on-the-job experience under faculty guidance. This strikes me as a generally sensible direction to go—for the U.S.A. which seems less clear in its directions as regards specialist OSH preparation for physicians than the other study countries.

It can be gathered from these observations that the U.S.A. stands at the bottom, comparatively speaking, among study countries as regards preparation of OSH physician specialists. In U.S.A., in Connecticut, the Occupational Medical Association of Connecticut (OMAC) has about 110 members. These are all the OSH physicians for the three million people of this state. Some 80 percent of these are "part-timers," that is, they have contracts to provide OSH services to two or three medium-sized plants. Hardly any of these have any special OSH training. They take time out of their general or family or other practice to do OSH work for a few hours each week. Hardly any of these services are preventive. The remaining 20 percent are "full-timers," that is, they are employed full-time by one or two, usually larger plants. Only some eight or nine of the 110 OMAC members have board certification in Preventive Medicine with a concentration in OSH. Most started out and completed a career in some other branch of medicine and then "retired into" OSH work as a 9-4:30 job with good pay. It is unlikely that other states have a better developed situation. In fact, the overall picture is rather sad. The field itself is not very attractive, since most of necessity become hired hands of the companies and can not exercise independent creative judgment.[12] Further, no OSH services standard has been adopted by OSHA. Thus many many plants, even fairly large ones, are without OSH services of any kind—however, captured and poor they might be in the U.S.A. monopoly capitalist context.

The U.K. is not much better off. There is no OSH services requirement and a professor I spoke with in a renowned school for preventive medicine told me of his part-time OSH work for a battery plant. In the course of the conversation he tried to convince me that some scientific work shows that lead exposure is actually a health-giving, protective factor for workers! He did not succeed. However, specialization at the two levels mentioned above has recently been recognized and accredited training programs are under way.

Although OSH specialization is not especially popular in the G.D.R. (it is also not the lowest specialty in popularity) it is probably more fully developed there—along the two tracks or levels mentioned earlier—than in any other study country. OSH physicians are paid and supervised through the chief OSH physician of each region of the NHS under the Ministry of Health. Thus they are comparatively free of management or union pressures. In any case, there are more well trained in-plant OSH physicians covering a higher proportion of workers than in any other study country. Also, every medical school has its highly trained OSH faculty and the National Institute for OSH in Berlin (ZAM) has a first rate staff.

Perhaps Finland stands next closest to the top among our six study countries on this facet of this aspect of the education dimension. Its newly adopted national law requiring OSH services for every workplace (either in-plant, through municipal OSH policlinics serving several plants, or through private OSH services) has led to at least brief OSH preparation offered by the first-rate staff of the National OSH Institute to hundreds of physicians. Even though many of these may be "part-timers" the general spread of OSH awareness, concern and preparation among physicians is truly a major accomplishment.

The F.R.G. has also adopted a law requiring OSH services. It has led to the workplace insurers and BAU offering two-week preparation to literally thousands of GPs and others. This is not quite the same as the accomplishment in Finland because the "part-time" character of this work and restrictions placed upon it by the entrepreneurial medical care system restrict the efforts largely to "passing by" the plant once a week for pre-employment exams, etc. However, the preparation of some 15,000 GPs with some degree of OSH awareness, if not great OSH competence, may be a major achievement as regards OSH-PHC connections.

Sweden seems to have placed the greatest weight on worker competence and control. Perhaps for this reason its efforts in OSH medical specialist preparation seem to lag behind the G.D.R., Finland, and even the F.R.G. However, various commissions have studied the matter and the National OSH Board (ASV) offers courses and has criteria which must be met before OSH personnel can be considered for open in-plant or policlinic positions. It appears that Sweden too will soon have the high-level university specialists and the workplace OSH services specialists.

Every country depends heavily on specially prepared OSH nurses to complete the OSH services coverage for whatever work settings are covered. This group has not been of central concern in this work, but it should be noted that it is absolutely key to offering adequate OSH services.

OSH preparation of PHC providers at different levels leaves a great deal to be desired in every one of the six study countries. Still, there are important differences.

The G.D.R. stands at the top in comparison with the others because it has a national pattern of at least thirty-six required OSH hours in undergraduate medical education; OSH questions are routinely included on the national qualifying exams for medical practice; and it shows signs of developing programs to reach PHC providers already in practice. None of the other study countries shows this degree of awareness and concern and action as regards the education dimension of the OSH-PHC problem.

Sweden may be next, though far down the line as compared with the G.D.R. This may be said because there have been some deliberate demonstration projects, as in Mora and Skövde, through which GPs have been successfully involved in continuing education as regards OSH problems. The undergraduate OSH medical education picture is no better than in other study countries, but a recent national study of almost all the facets of the need for improved cooperation between the PHC and OSH services (see note 6) tends to put this country ahead of the remaining study countries.

The F.R.G. is probably next in line by the simple fact that it routinely includes OSH questions on the national medical practice qualifying exams (the national board exams as they are called in U. S. A.) though schools vary greatly as to the inclusion of OSH material in their curricula. It appears that almost nothing has been done to reach PHC providers already in practice, though, as noted above, several thousands of GPs have received the "small" (usually two-weeks) preparation in OSH to qualify them for part-time preventive OSH service in one or more medium-sized factories.

The Finnish law requiring OSH services for all workplaces is bringing about at least minimal OSH awareness and competence in increasingly wider medical circles in that country. However, the situation as regards in-service OSH preparation for PHC providers in general seems untouched. And undergraduate medical preparation in OSH varies widely from a generous number of required hours at the medical school in Tampere to almost nothing in some other schools.

The U.K. and U.S.A. stand at the bottom on this aspect and it is hard to distinguish between them. Neither country has an OSH services standard. Little or nothing is done to prepare PHC providers in taking an OSH history or general awareness of problems in the field. And undergraduate medical education varies from no required hours to a few, with some schools offering an elective OSH course. In the U.S.A., about half of the 117 medical schools require some OSH hours (an average of four) some of the others offer elective courses in OSH, and about two-fifths have nothing.

In summary of the comparative picture as regards the medical education facet of the education dimension, the G.D.R. had the best rating, with a sizeable gap before Sweden which is next, then Finland and the F.R.G., with the U.K. and U.S.A. showing the least adequate positions.

ORGANIZATION

This dimension has many facets but for our purposes here two may be high-lighted—the organization of the OSH proctective structure and specialty services and the organization of OSH-PHC linkages.

The G.D.R.'s provision of OSH services through two main forms (in-plant services for large plants, and OSH policlinics serving a number of plants) within a regionalized NHS, under the Ministry of Health, is unique among our six study countries. It has many advantages, especially its independence from management influence. The special OSH policlinics and other units serving high risk working populations such as construction and agricultural workers are another attractive feature of this case. But perhaps the most important and outstanding feature of this system is the 1981 law and its implementation designed to survey, measure and follow-up on the hazards by priority in every workplace of any size in the country. The G.D.R. also offers some examples (though no general pattern) of integrated PHC-OSH service in a few local district policlinics. The G.D.R. has the further advantage (shared only with the U.K. which also has a regionalized NHS) of PHC services targeted toward geographically defined populations. In this way, even in the absence of fully adequate OSH training for PHC providers, the patients' work settings have a greater chance of being taken into account than in fractionated, market-oriented systems as in U.S.A., F.R.G., and the private parts of the Finnish and Swedish systems. (Sweden also has an NHS but the periphery of the system offers four different points of first contact—the local district physician who has a geographically defined area; the hospital, private physicians, and OSH specialty services).

Sweden's pattern of organization of OSH specialty services is particularly good as regards the National OSH Board and the inspection arm (ASV).[13] The organization of in-plant services, especially in large-to-medium-sized plants, is also admirable. The special strength depends as much on the dominant role of worker health and safety representatives and committee members, not simply as individuals, but as representatives of the union collective, as on expert services (OSH physicians, engineers and others). One of the minor problems (also bothersome in the other study countries) is the less than perfect coordination between medical and safety engineering components of the in-plant services. Sweden has not solved the small plant services problem, though some demonstration programs and the recent Wessman Commission report indicate continuing efforts to reach these workers. One of the successful efforts has been a special OSH services for agricultural workers. But Sweden has not passed a law as Finland has to require OSH services for all workers. The organization links to PHC are no better in Sweden than in most other study countries. The country has an NHS but its peripheral input points are fractionated between public district physicians, the hospitals, and private practitioners. Thus there is not the same geographic focus as in the NHS systems of G.D.R. and the U.K. There is a requirement that PHC providers report OSH diseases they see in their practice to the ASV, but the lack of OSH training for PHC providers makes this a weak link. Perhaps the recent very important report (note 6) on the need for greater OSH-PHC cooperation will bring about improvements in this country where workers' concerns are usually acted upon once they are recognized.

Finland's law requiring OSH service coverage for all workers puts it in a unique position among our six study countries. The obvious strength of the National Institute for OSH is also noteworthy. However, the lesser role of workers, as compared with Sweden, does not lead to as high a degree of effective protection for as many workers. There is also the problem of OSH services as well as the general health services being divided between private and public auspices and control. The greater degree of expert determination and paternalistic atmosphere does not encourage as high a rating as for Sweden, but it has a much stronger organization for OSH than in F.R.G., U.K. and U.S.A. As regards links to PHC, the situation is no better than in most of the other study countries.

The U.K.'s provision of OSH protection and services is not particularly impressive. The system falls within the British equivalent of the Ministry of Labor and is organized through the Health and Safety Commission and Executive which oversee the factories inspectorate, the Employment Medical Advisory Service (EMAS) and other branches, including special services for offshore drilling work, mines and quaries, nuclear installations, and other special hazard work. But there is no OSH services requirement for all workplaces and the EMAS can in no way provide adequate coverage for those workplaces without physicians and other personnel. The preparation of OSH specialists to serve in factories is not as well developed as in Finland, the G.D.R., Sweden, or even the F.R.G. Thus where there are services, they may be offered by poorly prepared part-timers as is the case in the U.S.A. Perhaps the one special strength of the U.K. is the geographic focus of the PHC services functioning within the fairly well regionalized NHS. By this means, as noted above for the G.D.R., one can expect PHC providers to become familiar with workplaces and the problems they generate among patients, though this is only one attractive facet which does not make up for lack of OSH training for OHC providers and other measures which would bring about more adequate OSH-PHC links. It is an appalling thing that the NHS has provided no leadership and lacks an OSH service for its own employees.

The F.R.G.'s specialty OSH services and protective structure are characterized by multiple authorities and fractionation. The workplace insurers play the central role in this system (a unique situation among our study countries) but there are quite a number of these and local (Land or state) authorities also play a role in the inspection process. Great efforts have been made to prepare enough GPs and other providers at least minimally in OSH so as to cover more plants with services. This is admirable as far as it goes. But the motivations and mode of functioning of these part-timers have been challenged. In any case, the unique restriction of OSH physicians of all types (full or part-time) strictly to preventive efforts seems unfortunate, for it tends to widen the gap between OSH and PHC. The PHC services themselves have an entrepreneurial cast within a system of National Health Insurance. Thus, there is not generally a geographic focus to the services.

The U.S.A. has neither adequate organization and provision of specialty OSH protection and services, nor geographically-oriented PHC services functioning within a regionalized health service. The details of this assessment are laid out in Chapter 12. Here, perhaps it is enough to observe that OSH and OSH-PHC organization in the U.S.A. is "a many splintered thing," leaving major gaps such as the lack of any OSH services standard in the OSHA law and a general lack of OSH preparation for PHC providers. The current head of NIOSH recently told members of Congress that the reporting of OSH diseases by physicians is seventy years behind the reporting of infectious diseases.[14] The strong points which can be found in U.S.A. are in shops organized by some of the larger unions which have shown special concern for OSH–the Oil, Chemical and Atomic Workers, the United Auto Workers, the United Steel Workers, the United Rubber Workers, and some others. Also the forty or more Councils on Occupational Safety and Health (COSH groups) which are voluntary groups of activist workers, OSH scientists, and others, must be seen as special sources of energy (though not unique–since several other study countries have such groups). Before the conservative Reagan regime took power, the research and training sponsored by NIOSH would have had to be seen as an important strength as would the New Directions Programs sponsored and funded by OSHA. But for the vast number of U.S.A. workers, especially the 81.2 percent who are *not* organized in unions, the OSH situation is bleak.

In summary of the organizational dimension, it can be said that OSH-PHC linkages need great improvement in all six study countries. There are some variations in this important facet, which, taken together with the larger differences in organization of special OSH protective measures and services, lead to an estimate that the G.D.R. shows the best overall organization, Sweden next, then Finland, the F.R.G., the U.K., and finally the U.S.A.

INFORMATION

The attention given above to the workers' right to know and other facets of the education dimension (pp. 423-428) prepare the reader for more of a focus on the systemic aspects of information. Taken as a whole, an adequate OSH information system involves full provision of hazard and hazard-control information to workers, OSH specialists of several kinds, and PHC providers.

The chain should not be broken by the arguments of private capital concerning "trade secrets," inter-organizational jealousies, or other such factors.

The G.D.R. has the most complete and well integrated OSH information system among our six study countries. It is not perfectly complete nor perfectly integrated. For example, the bureaucratic, technocratic center of the system has decided not to alarm workers and citizens generally about hazards such as asbestos, unless a clear plan of action to correct the situation can be announced at the same time as the information is widely shared. This is sad. But, given this kind of major qualification concerning the expert-determination and limitation of the system, there are several strong points to be mentioned. First, the regionalized NHS includes a very complete reporting system for all health problems in relation to clearly defined population denominators; thus providing a very good epidemiologic surveillance system for all diseases. Second, the quite complete coverage of the working population by well-trained OSH services specialists allows for very extensive and systematic follow-up of workers who have been exposed. Third, the 1981 law requiring a survey and follow-up of hazards according to priority for every workplace with more than a few workers will probably yield a more complete OSH hazard information base than for any other country in the world. Fourth, the National OSH Institute (ZAM) carries out an extensive research program. One of the weaknesses (which is even more evident in other study countries) is the inadequate preparation of PHC providers in taking an OSH history and other aspects of the field so as to foster more complete reporting and follow-up of work-related diseases. However, the G.D.R. does have a more adequate OSH preparation for undergraduate medical students than any other study country. Some of the larger plants have computerized systems recording every diagnosis reported back by a physician for every worker who has been off work. These could be used to greater advantage for case and "hot-spot" follow-up, whereas often they are only used for periodic totals, work absence rates, etc.

The OSH information systems of Sweden and Finland appear to me to be very similar to each other and on about the same level. Possibly Sweden has an edge because of the more widespread and authentic involvement of workers themselves in sharing the information—through the law of co-determination of production; the extensive training efforts; and other ways. It has also supported extensive research. However, the OSH-PHC linkages are inadequate as is the preparation of PHC personnel in OSH matters; thus the reporting of OSH diseases and their follow-up (if they have first come to the attention of PHC personnel; which it is supposed they most often do—though this too is an unknown—suggesting another area for improved information) is probably no better than in most other study countries.

The work of Finland's National OSH Institute is so impressive that it deserves special mention. This comment relates to research and publication as well as training of OSH physicians and others.[15] However, the system lacks a significant degree of worker involvement and control and the OSH-PHC connection is no better than in Sweden; in fact, it may be worse as the division between public and private health services is more evident than in Sweden.

The organizational fractionation of the F.R.G. system leads to a lack of shape for the OSH information system. The U.K. is not much better off and perhaps the U.S.A. is in the poorest position, though NIOSH-funded research and extensive OSH epidemiologic studies under a variety of auspices make the special studies aspect of this dimension rather strong for the U.S.A. None of these three countries have a significant degree of worker involvement. The right to know is still being fought for. None of the three has a well worked-out exchange of information between PHC and the OSH inspectorates or OSH services (the very big deficit in reporting of OSH diesases in U.S.A. was cited above–note 14). It is hard to make clear distinctions of accomplishment on this dimension. All three leave a great deal to be desired.

In summary of the information dimension, the G.D.R.'s system could be greatly improved by more authentic worker involvement and less expert determination of the system; otherwise, there is much to learn from it. Sweden's system has the strength of a high degree of worker involvement and clear right of the workers to know what hazards they face and what to do about them, but it shares several of the weaknesses of other systems. Finland's National OSH Institute and its research and educational efforts are impressive. The F.R.G., the U.K., and U.S.A. have such fractionated and partial information systems seriously affected by entrepreneurial interests and "trade secret" ideology that it is difficult to distinguish between them in their inadequacy.

FINANCING

The amount of financing for OSH protection and services and for OSH-PHC linkages generally follows the policy emphasis of a country (see pp. 416-419). Thus it is more the flow and control of funds and the direction of their motivating effect which are in focus here.

In any case, funding figures were not available for most study countries. Only for Finland did I learn that the total allocation to OSH activities approximates some $43-$50 per active worker per year (depending on one's estimate of the exchange rate between U.S.A. dollars and Finnish Marks). My impression is that this is comparatively generous. It may be approached by Sweden and the G.D.R. but it surely surpasses what is spent in the U.S.A., the U.K. and the F.R.G. But, as I said, the amount tends to be a statement duplicative of what has already been observed concerning policy.

The unavailability of information on amounts of funds for OSH does not appear to be a matter of sensitivity of the information, it simply appears that the subject has not achieved recognition as a major "cost center" of concern. Yet the costs of OSH problems to society are large indeed. A now somewhat dated estimate for the U.S.A. put the losses from lost work time, compensation, medical care costs, etc. from occupational injuries and diseases at some $35 billion per year—there is no allowance in this figure for the cost of human anguish.[16] Since most of the problem is at least theoretically preventable, it deserves a great deal more funding than it appears to receive. But this must be funding which motivates to protective action for workers.

The G.D.R.'s inspectorate and other out-of-plant OSH services are state financed through the Ministry of Health. The in-plant OSH personnel are also paid by local levels of the Ministry of Health, while the material resources they need for their work are paid for by the management of whatever firm or undertaking they are assigned to.[17] This tends to be a good arrangement, for it allows OSH personnel to function in an independent but responsible manner. Theoretically it should also allow a primary preventive emphasis, but this sometimes collides with the need for major capital outlays if certain ventilation, heat, noise, or other problems were to be removed and here the management and state may drag their feet at times, leaving the OSH personnel with a secondary preventive emphasis through follow up of individual exposed workers. Funding of PHC services through the NHS does not seem problematic. The lack of adequate links between PHC and OSH seems more attributable to educational and organizational factors.

Sweden's OSH services are paid one half by the NHS and one half by the employer. Theoretically, this is to allow equal emphasis on treatment and prevention. But no one seems too concerned about the exact apportionment of effort in relation to payment source. Probably the key point is that the worker-union-dominated health and safety committee must approve the annual OSH services report and projected plan and budget before the NHS will release its half of the funds. This generally puts the control where it belongs and directs the system toward workers' interests. In this case, as in others, the lack of adequate PHC-OSH ties seems more a matter of education and organization than financing.

Finland has approximately the same financing system as Sweden with a lesser degree of authentic worker control over the OSH specialty services and a greater division between private and public services in the health system generally.

The F.R.G. leaves management in charge of paying for OSH services with little or no worker control. Curiously, and uniquely, the services are restricted to preventive efforts (pre-employment exams, work placement, some screening) at the insistence of community and hospital practitioners who do not want OSH physicians taking in any of the sick slips they get from their patients as proof of services rendered so that they will receive payment. This is a very poor financing arrangement as it tends to drive a wedge between OSH and PHC creating a gulf which would be too wide even without such an arrangement.

The U.K.'s OSH services are paid for by management when they are provided. Workers have no more trust of "company docs" than they do in U.S.A. The one strong point is that the NHS is tax supported; thus, theoretically, good arrangements could be worked out between PHC and OSH services with more of a preventive OSH orientation in the PHC services. This has not happened for educational and organizational reasons.

The U.S.A. has the worst financing system. OSH services, when they are available, a rarity, are company paid and dominated, leading to worker mistrust. The general health services are primarily market oriented and not oriented toward prevention, that is, there are absolutely no rewards to be realized by practitioners for taking the time to inquire into a person's work and follow up, either individually or for the larger similarly exposed work group, if it is a work-related problem.

In summary of the financing dimension, Sweden shows the greatest worker control over the flow of OSH services funds and is strong for this reason. The G.D.R. may be the best in an overall sense because of its NHS sponsorship of both OSH and PHC services. Finland expends an impressive amount on OSH and has a system similar to Sweden's with less worker control and more of a private public split in the health services. The F.R.G.'s system of restricting OSH services strictly to preventive efforts tends to divide OSH and PHC services even more than would otherwise be the case. The U.K.'s OSH services are company paid and dominated, but the tax-supported NHS would theoretically allow preventive OSH efforts on the parts of PHC providers. The U.S.A.'s OSH services are company paid and dominated and there are few redeeming features to the market-oriented (therefore curative oriented) PHC services.

SUMMARY

The following figure (Figure 9) is offered as a brief overview with no pretension as to the certainty of placement of countries on each dimension or aspect of each dimension. I can not even say what measuring units are represented by the numbers 0-10. These are simply used as a device, as is done in questionnaires, to offer a sense of difference and closeness as regards the level of a country's accomplishment. I have not placed any country at 10 or 0 because I don't believe I saw any perfect or absolutely bad situations. There is no defined calculus for combining the several dimensions into the overall rating. This is simply a general estimate of each OSH and OSH-PHC system after the kind of detailed examination presented in each country chapter (7-12).

In general, the working hypothesis that stronger worker movements will be associated with higher accomplishments as regards OSH protection and services is upheld. The G.D.R. and Sweden have the strongest (though different) worker movements and the highest accomplishment. Finland has a quite strong but more recently consolidated workers' movement and more of a paternalistic inheritance than in Sweden; thus it falls at an intermediate position which matches its

	Strength of Workers' Movement	Policy	Worker Sponsorship and Control	Education			Organization		Information	Financing	Overall Rating
				Workers	OSH Providers	PHC Providers	OSH Services	General Health Services			
10											
9	G.D.R. Sweden	G.D.R. Sweden	Sweden G.D.R.	Sweden	Finland G.D.R.	G.D.R.	G.D.R. Sweden	G.D.R. U.K.	Sweden G.D.R.	G.D.R.	G.D.R. Sweden
8		Finland					Finland	Sweden		Sweden	
7	Finland		Finland	G.D.R.	Sweden				Finland	Finland	Finland
6		U.K.	U.K.	U.S.A.	F.R.G.	Sweden	F.R.G.	Finland			
5	U.K.	F.R.G.		U.K.		Finland	U.K.			U.K.	
4	F.R.G.		F.R.G.	Finland	U.K.	F.R.G.		F.R.G.			U.K. F.R.G.
3		U.S.A.		F.R.G.	U.S.A.		U.S.A.		U.K. F.R.G. U.S.A.	U.S.A.	
2	U.S.A.		U.S.A.			U.K. U.S.A.				F.R.G.	U.S.A.
1								U.S.A.			
0											

Figure 9. Study countries on strength of workers' movement and dimensions of the model.

OSH accomplishments. The U.K. would appear to have a slightly stronger worker movement than F.R.G.; yet it is not better than the F.R.G. as regards OSH. It is possible that the declining British Empire, enonomic downturns, and conservative gorvernments have had more of a weakening effect on organized labor and the Labour Party than it at first seemed. In any case, the U.S.A. is in the poorest situation as regards OSH accomplishment and has the narrowest and weakest workers' movement.

In a very general way, the same relationship holds between strength of workers' movement and adequacy of PHC-OSH connections as hypothesized. However, the study countries all leave so much to be desired in this regard that it can only be described as a very weak relationship. A possible interpretation of these results is that workers' organizing and consciousness raising efforts achieve more in relation to conditions directly tied to production because workers and unions have a direct impact on production. The general health services, including PHC lie outside of the productive realm and are more under a traditionally elite work group—physicians. Workers, even though highly organized, as in Sweden, have not succeeded to spread their influence to the PHC services.

NOTES AND REFERENCES

1. All About OSHA, Programs and Policy Series. OSHA Publication No. 2056. Washington, D.C., April 1976 (Revised).
2. Increasing Worker and Community Awareness of Toxic Hazards in the Workplace, *The Nations's Health* (official newspaper of the APHA) pp. 10–12, September 1984.
3. Of course, it depends whose interests are served by this work. Not all such work is hope-giving as far as workers are concerned. Some on the potentials of genetic screening as a measure which might allow capital to avoid the costs of cleaning up, while continuing to expose presumably "immune" workers may be the biggest threat on the horizon. Tony Mazzocchi, A Decade of Genetic Struggle, *Int'l J. Health Services, 14*: 447–451, 1984. Such measures would inevitably lead to new forms of vicious and repressive discrimination (not to mention "race purification" moves) in a capitalist political economy in which misused technology is leading to high levels of chronic unemployment.
4. Molly Joel Coye, Mark Douglas Smith, and Anthony Mazzocchi, Occupational Health and Safety: Two Steps Forward, One Step Back, pp. 79–106 in *Reforming Medicine, Lessons of the Last Quarter Century.* Victor W. Sidel and Ruth Sidel (eds.), New York: Pantheon, 1984. Philip J. Simon, *Reagan in the Workplace: Unraveling the Health and Safety Net.* Washington, D.C.: Center for Study of Responsive Law, 1983.
5. Walter Korpi and Michael Shalev, Strikes, Industrial Relations and Class Conflict in Capitalist Societies, *The British J. of Sociology, 30*:164–187, June 1979.

6. Primärvårdens samverkan med företagshälsovården (Primary health care's cooperation with worker health services) Stockholm: Sjukvårdens och socialvårdens planerings-och rationaliseringsinstitut (Spri) Report No. 143, July, 1983.
7. This history is laid out in Stanley Aronowitz, *False Promises: The Shaping of the American Working Class*. New York: McGraw Hill, 1973.
8. Daniel Berman, *Death on the Job, Occupational Health and Safety Struggles in the United States*. New York: Monthly Review Press, 1978, esp. p. 76ff. A recent variation in this kind of Capitalist ideology is to draw attention to and place greater emphasis on carcinogens and other hazards occurring in nature than those created by industry, for example, aflotoxin found in musty peanuts versus ethylene dibromide. William Havender, Peanut Butter Sandwich Deadlier than Muffins Containing EDB, *The Hartford Courant* (April 4, 1984): B 11.
9. E. R. Brown and G. E. Margo, Health Education: Can Reformers be Reformed? *Int'l J. Health Services, 8*:3–26, January 1978.
10. R. H. Elling, "Worker-Based Education, Organization and Clinical-Epidemiologic Support for Occupational Safety and Health (OSH): The New Directions Program in Connecticut," pp. 171–187 in *Proceedings of the ILO-WHO Conference on Education for Occupational Health* held in Sandefjord Norway, 16–19 August, 1981. Geneva: ILO, 1982.
11. How to Crack the Company's Code: Getting Names of Workplace Chemicals, *UAW Occupational Health and Safety Newsletter, 6*(5):2 November-December 1977; "Researching the Literature for Toxicological Information" (an information packet) New Directions Program, Connecticut, 1982.
12. Susan Mazzocchi, "Training Occupational Physicians, Suppose They Gave A Profession and Nobody Came," pp. 7–10 and 19–24 in *Health PAC Bulletin* (New York, NY, Human Sciences Press) (April, 1977).
13. Navarro details the stronger standards enforcement and other features of the Swedish system as compared with U.S.A. Vicente Navarro, The Determinants of Social Policy. A Case Study: Regulating Health and Safety at the Workplace in Sweden, *Int'l J. of Health Services, 13*:517–560, 1983. He also correctly criticizes Kelman for his inadequate depiction of the differences between U.S.A. and Sweden and lack of understanding of the comparative strength of the workers' movement in Sweden as the major factor generating the more favorable OSH situation for Swedish workers.
14. Bill Keller, "U.S. Health Aide Urges Logging of Job-Injury Data," *The New York Times* (June 21, 1984).
15. Jorma Rantanen, "Occupational health and safety in Finland," *Scandinavian J. of Working Environment and Health, 9*:140–147, 1983.
16. Nicholas A. Ashford, *Crisis in the Workplace: Occupational Disease and Injury, A Report to the Ford Foundation*. Cambridge, MA: MIT Press, 1976.
17. Actually, the labor theory of value tells us that the workers produce the value which pays for these supplies, but I am referring to the management's control of the funds which it supplies to the OSH services.

CHAPTER 14
A Summing Up: Workers' Health and Human Development

As we have seen, workers' health is a complex matter. It is affected by many conditions. Were we to attempt a balance sheet, so to speak, of positive and negative influences on OSH, there would be very long lists on each side. On the negative side, we would have to include many seemingly purely technical matters, such as the tremendous numbers of often untested chemicals already in different workplaces and the new ones entering into production processes each year. Yet, upon closer examination, we would see that the ways in which hazardous materials are brought into and used in the workplace are not purely technical matters. In fact, whether or not workers have the right to know and have been properly informed under acceptable and believable (that is, effective teaching) auspices about the materials they work with, their potential hazards, and how to avoid them–including, perhaps, by an enclosed process, calling for new engineering and capital investment and possibly a slowed production process–are matters of social relations or organization of the production process.

On the positive side too, conditions which might appear to some observers at first glance to be purely technical, for example, a well-trained, worker-sympathetic, but independent (that is, objective) OSH physician and related specialty OSH services, with a proper balance between treatment and prevention, turn out on examination to be determined by policy, sponsorship and control, educational resources, financing, and other organizational matters.

If it is admitted that OSH is not purely a technical matter determined solely by hazard potencies, dose-response relationships, control standards, emission levels, length and concentration of exposures, etc., then a second line of defense, so to speak, of those who wish to somehow keep the problem confined within as narrow, technical boundaries as possible (thereby averting an examination of their own favored interests and privileged positions in relation to the production process) is to focus on psychological factors. That is, even with adequate standards (which these interests usually resist until they are forced upon them by a strong, well informed workers' movement) it has been said historically, and is still said today, that the main problem is "the careless worker."[1] A recent extreme of this ideological position designed to roll back the advances workers have achieved through collective struggle for OSH, asserts that too many safety and health provisions make workers relax and adopt a careless attitude and thereby actually raise accidents and illness above the levels there would be with fewer protective measures.[2] The trouble with this thesis is that

> In order to realistically lay the blame for industrial accidents (or illness) on workers, one has to assume that workers have the ability to act independently to both control work hazards and to react to them in a manner consistent with their own self protection. However, the very nature of most modern workplaces shapes workers into reactors rather than independent actors.[3]

To understand and improve OSH we can not stop either at the technical or personal levels. In fact, the country case studies reported earlier in this volume (Chapters 7-12) and the comparison of them (Chapter 13) suggest that a key force for improvement of OSH is the historical and current strength of the workers' movement. For example, in Sweden where unemployment never reached 3 percent during the most recent recession in the capitalist political-economic world-system, where some 80 percent plus of the workforce is unionized, where the unions follow a broad socialist, social welfare policy, where workplaces are organized on a single-union, closed-shop basis (that is, if a union is voted in, there is just one union and all nonmanagerial personnel belong) where a labor government has been in power for over forty-five years,[4] there are remarkably more protective arrangements for workers than in U.S.A.[5] In U.S.A. only 18.8 percent of the workforce is organized, narrow business unionism prevails, there are often several unions competing in a single workplace, the Taft-Hartley Law forbids closed-shop organizing, union-busting is in full swing, and there has never been a labor party with widespread mass support—to say nothing of a labor-dominated government.[6] As a result, not only are the protective measures for OSH weak and getting weaker,[7] but the results in terms of work accidents compare unfavorably with Sweden and other study countries.[8]

If the strength of the workers' movement has some primacy in determining health and safety on the job, as well as the life chances of working people more broadly and generally,[9] then we must attempt to understand the nexus of factors which contribute to or detract from this force if we are to see a healthier, safer, better life for working people everywhere. I will not pretend to do more than address this matter in this closing chapter. Of necessity, I will do so in a somewhat abstract, disembodied way, citing general concerns and drawing examples from the six study countries with some special attention to U.S.A., but without full examination of the historical context (for this the reader is referred back to country chapters and the more in-depth references concerning the workers' struggle in each, which are given therein). This is a matter of the highest priority for continuing comparative, historical or developmental cross-national study. It will require in-depth country case studies of workers' struggles and the development of the workers' movement. The main focus must be on production and job-related organization. But wider political organization and consciousness raising of the working class and, in dialectic fashion, the responses of capital (including technologies and changing divisions and allocations of labor)[10] and the state, and the replying round of workers' struggles, must be taken fully into account as well. Fortunately, this is not a new idea. Some important work along these lines is already available.[11] However, much of the work is country specific and not cross-national.[12] These are valuable works, but more could be learned by the comparative method—albeit making sure to carry out the description and analysis in terms of the historical dialectic of each country. Also, more could be understood by including a clear examination for each country of the ways in which the course of its workers' movement has been affected by developments in the world system. What I am suggesting is a gigantic task, reaching from 1) the micro level, for example, the ways in which workers' relationships to each other on the job are affected by the cultural hegemony,[13] as it guides self conceptions and images of men and women, or minority ethnic groups; to 2) the intermediate level of union organization, for example, inter-union rivalry or narrowness versus breadth of concern; to 3) the broader level of state action, for example the establishment of open-shop or closed-shop labor organizing laws; to 4) the macro level of the world-system, for example, the transfer or threat of transfer of jobs to peripheral or semi-peripheral countries which can be a way of weakening the solidarity of the unions and workers' movements in core capitalist nations.[14]

Even if our focus were narrowly on OSH, there would be a number of reasons for concerning ourselves with the broader levels of worker influence and control. When capitalist owners prepare to abandon a plant in search of a higher rate of profit elsewhere, in some of these cases, the workers and their union succeed in buying the owners out and take full control of the means of production. While there is a need for further research on OSH conditions in these situations, there is one review of the literature and preliminary report of a study comparing OSH in worker-owned, cooperative plywood factories with conventional capitalist factories producing more or less the same products. The results suggest that productivity rises in the worker-controlled plants, but OSH does not. In summarizing their results and offering tentative conclusions, the authors state:

> . . . preliminary analysis suggests that matched plywood cooperatives may be more productive but also less safe than their conventional counterparts. If this result holds when all the data have been

> collected and analyzed, this will cast serious doubt on the likelihood that the cooperative form of work organization is a realistic alternative for overcoming the contradiction between productivity and safety. Greenberg, in a recent paper examining the wider political implications of workplace democratization, reaches the conclusion that this process is not at all a promising road to take in creating a fully democratic society. In his view, it is a "political dead-end" that fosters "enterprise egoism" and self-exploitation.[15] We cannot consider these issues in this paper (we hope to do so in later work). However, we do agree with him (and others) that the limits of the form may be determined by the external environment in which producer cooperatives operate. Although one can see coop members as worker-owners, the external environment will reinforce a one-sided development of their dual identity. To survive in a capitalist market, they will have to stress the owner's goal of maximizing income and will therefore view themselves as instruments of production, compelled to maximize productivity. True, as collective worker/owners they may outperform firms with antagonistic social relations . . ., but in this instance they have merely found a more effective, and perhaps more satisfying, way to achieve goals that are promulgated by the wider society. However, when there is a conflict between the values and orientations produced by the democratic workplace and those produced by the external environment, the latter may well tend to swamp and submerge the former. If this is indeed a true characterization of what happens, then further research that examines the performance of cooperatives in a more supportive environment is needed before we can conclude that workplace democratization is not a solution to the contradiction between productivity and workers' safety and health.[16]

Braverman enucleates the core of the matter when he distinguishes human labor from other commodities which the capitalist purchases—raw materials, machinery, etc.—in terms of its human qualities and the ability of workers to organize and resist or cooperate in their own exploitation and the ways these alternatives and many variations in between are influenced by the larger context.

> Human labor, . . . because it is informed and directed by an understanding which has been socially and culturally developed, is capable of a vast range of productive activities. The active labor processes which reside in potential in the labor power of humans are so diverse as to type, manner of performance, etc., that for all practical purposes they may be said to be infinite, all the more so as new modes of labor can easily be invented more rapidly than they can be exploited. The capitalist finds in this infinitely maleable character of human labor the essential resource for the expansion of his capital.
>
> It is known that human labor is able to produce more than it consumes, and this capacity for "surplus labor" is sometimes treated as a special and mystical endowment of humanity or of its labor. In reality it is nothing of the sort, but is merely a prolongation of

> working time beyond the point where labor has reproduced itself, or in other words brought into being its own means of subsistence or their equivalent. . . .
>
> The distinctive capacity of human labor power is therefore not its ability to produce a surplus, but rather its intelligent and purposive character, which gives it infinite adaptability and which produces the social and cultural conditions for enlarging its own productivity, so that its surplus product may be continuously enlarged. From the point of view of the capitalist, this many-sided potentiality of humans in society is the basis upon which is built the enlargement of his capital. He therefore takes up every means of increasing the output of the labor power he has purchased when he sets it to work as labor. The means he employs may vary from the enforcement upon the worker of the longest possible working day in the early period of capitalism to the use of the most productive instruments of labor and the greatest intensity of labor, but they are always aimed at realizing from the potential inherent in labor power the greatest useful effect of labor, for it is this that will yield for him the greatest surplus and thus the greatest profit.
>
> But if the capitalist builds upon this distinctive quality and potential of human labor power, it is also this quality, by its very indeterminacy, which places before him his greatest challenge and problem. The coin of labor has its obverse side: in purchasing labor power that can do much, he is at the same time purchasing an undefined quality and quantity. What he buys is infinite in *potential*, but in its *realization* it is limited by the subjective state of the workers, by their previous history, by the general social conditions under which they work as well as the particular conditions of the enterprise, and by the technical setting of their labor. The work actually performed will be affected by these and many other factors, including the organization of the process and the forms of supervision over it, if any.[17]

There is a need for further research on the effects of in-plant worker control both in capitalist and socialist contexts. One of the important reasons to encourage such work has to do with stress-related diseases—anxiety, coronary heart disease, etc. Even though the rush of cooperative production in a capitalist, market-oriented context fails to control accidents, may even increase them, it could be that the sense of control and voluntary engagement in production would cut the stress factor and therefore stress-related disease. But this may be only a minimalist goal and achievement. And the likelihood is that this hypothesis would not be upheld, for the pressures of the competitive capitalist context would probably turn the co-worker/supervisor and the co-worker/manager into bosses with no less stress for all that. In contexts where the workers' movement has achieved a broader social welfare state, even though the fundamentals of production remain capitalist in the sense that the surplus is a privately and not a publicly controlled good (except for the important qualification that there is a progressive, equalizing tax structure), experiments on worker control show significant gains in reduced stress[18] and other work shows that greater worker determination lessens the risk of coronary disease.[19]

Beyond the effects of the wider context on self concepts, relations of production, stress, accidents, and disease, there are the wider control structures and measures directed specifically at OSH. The present study tends to confirm our starting idea that stronger laws, standards, enforcement agencies, more generous support for worker education in OSH, realization of workers' rights to know what hazards they face and to refuse hazardous work without retaliation, and more adequate OSH medical and other services would be associated with stronger workers' movements.[20]

While these and other OSH conditions and measures depend largely on state actions in response to struggles of workers and capitalists', thus the nation state is our primary unit of study, analysis, and action, our concerns as regards OSH extend beyond the nation state to the world-system. If stringent OSH controls are set up in one country, capital may seek to weaken those controls and the workers' movement by threatening to export or actually exporting or setting up hazardous production processes in other countries.[21]

Our concern for the nexus of factors which tend to weaken or strengthen the workers' movement must be similarly broad, ranging from elements of solidarity versus divisiveness between workmates in the shop, to united versus divided political action to influence the nation state, to concerted versus divided working and peasant class consciousness and action on a world scale. As already suggested, our overview here must be illustrative, rather than exhaustive. The point here is to suggest enough of what is involved to lead us toward a concluding statement of action.[22]

Some divisive factors reach across several, even all of the levels and are, therefore, of special importance and concern and require special efforts toward understanding and vigorous, widespread actions to counter them.

Possibly the most ubiquitous, pervasive and debilitating one is sexism—the manifold ways in which the ruling class plays men and women off against each other in order to prevent them from realizing their common humanness and potential for building a socialist world together. On the job, there may be various forms of sexism ranging from gross discrimination which excludes women from higher paying jobs or provides them lower pay for the same job or lower pay for the same amount and quality of work in another job (unequal pay for comparable work) to more subtle things like self concepts in which macho ideas taken over from the cultural hegemony, encouraged by the dominant class, lead men to lift heavier loads, possibly damage their backs thereby, and reinforce the divisions between men and women. At the societal level, as some Feminist-Marxist analysis puts it, patriarchy in the home reinforces capitalist domination at work.[23] Even under socialism, at least in its older versions, this problem was not solved, so heavy was the weight of the sexist cultural hegemony (possibly reaching back to pre-historic times, but reinforced and further exploited under capitalism). In 1904 Mary Garbatt complained that the Socialist man

> . . . does not seem to recognize that the work the housewife does cooking, washing, sewing, mending, nursing and caring for the

> children is in any sense commensurate with the work he does for wages. If he tried it for a week, I am sure he would change his mind. . . . He would come to realize that she is an economic factor in the home as truly as he himself and because of her service entitled to an equal and joint right in the money he earns on the outside, and an equal voice in everything that concerns the home.[24]

Women's work in the home serves to maintain and reproduce the working class—a major service to capital, especially since it is unpaid labor. The continuance of patriarchy in the home, even for women who also work outside the home, means they have a double exposure to occupational hazards—those in the home and those in the factory.[25] Also at the societal or state level, protective legislation, so called, has been used to keep women out of higher paying jobs and thereby continue the divisions between men and women. "Protective legislation helped to eliminate some of the awful conditions imposed on women factory workers (it did not apply to domestics or the growing number of office workers); but it also reinforced the status of women as secondary wage earners whose primary role was in the home."[26] As part of the male- and capitalist-dominated state apparatus, the courts reinforced this division. Thus in 1908 the U.S.A. Supreme Court in *Muller v. Oregon* upheld the law limiting the hours of female factory workers because "women would always be dependent on men for protection."[27] As regards unionization, the unions were dominated by men and generally fell for the capitalist trick of sexism in all our study countries by either keeping women out, or by keeping them out of leadership positions if they were let in. Even in Sweden today, where organized labor generally is very strong, it is weaker than it might be because of this discrimination. I reported in Chapter 7 the example of a factory in Norköpping with several hundred workers, most of them women, with no women among the officers of the union local. A woman I interviewed as she came out of the plant gate said the union "is not very good." On a world scale, some of the most exploitative conditions are being imposed on women workers in underdeveloped countries. In the film, "Controlling Interest"[28] a U.S.A. chief executive officer of a multinational says that women don't mind working sixty hours a week for just $20 assembling his electrical switches in Singapore. "Besides," he says, "they have much greater dexterity than men." One can even feel sorry for this capitalist executive, suffering as he does such mind pollution from his own cultural hegemony.[29]

But it is clear from the history of labor struggles that men and women do not have to be divided against each other. Mother Jones became a bonafide folk hero through her power to inspire and rally both men and women in the coal mining towns where the bosses kept conditions at the sub-human level.[30] In the inspiring "Bread and Roses Strike" in Lawrence Massachusetts in 1912, "The women won the strike," as Big Bill Haywood of the IWW said.[31] The point of course is not really that the women won the strike (though leadership and determination are very important in such situations)[32] the real point is that all the workers won by striking together. A very recent small but significant positive event involved Judy Goldsmith, the President of the National Organization for Women (NOW) and Lane Kirkland, President of the AFL-CIO, appearing in New Haven, Connecticut to offer monetary and organizational as well as moral support to the clerical, and technical workers out on strike against Yale University. Local 34 of the Federation of University Employees is 82 percent women.

> "This strike is not just about better wages," said Kirkland, speaking to a crowd of about 1,000 unionists and strike supporters. "It is about a system that for too long has denied women and black workers equal pay for comparable worth."[33]

It is important to report that the strike at Yale was a success, so the principle of comparable worth will be on the agenda for the future. Racism, as just suggested by Kirkland's statement at Yale, is a similarly ubiquitous, pervasive force of great power in the capitalists' hands. The AFL itself has long been a dupe of the divisive racist ploy.

> In 1905, AFL President Gompers told a union gathering in Minneapolis the "caucasions" were "not going to let their standard of living be destroyed by Negroes, Chinamen, Japs, or any others." In referring to these "others," Gompers meant to include the new immigrants from Eastern and Southern Europe, who came in large numbers to work in a variety of jobs, from the mills of Homestead to the garment shops of New York. The racist ideology of the period extended to include these groups as inferior and not entitled to the full benefits of citizenship—or union membership.[34]

This powerful divisive force takes many forms and is integrally tied up with OSH. If employers can hire lower paid minority group members to do the dirty work and most hazardous work, then they don't have to invest fixed capital to clean up the workplace for all workers. On some occasions veritable slaughters of minority workers have resulted. The blasting, drilling and digging of a hydroelectric tunnel near Gauley Bridge, West Virginia during the Great Depression was one such disaster. The workers were mostly Blacks desperate to have a job. Instead of waiting after blasting for the dust to settle, foremen herded the workers back in the tunnel with pick handles. The mountain through which they were cutting had a high silica content. Raging silicosis took 470 human lives and 1,500 were disabled.[35] Today, when 8½ million are unemployed, the workers over the coke ovens in U.S.A. steel mills are disproportionately black; thus they are more exposed to getting lung cancer than their white brothers and sisters.[36] In other study countries this factor has also had a virulent history. In Nazi Germany, Jews and other minorities were used as scapegoats, persecuted, and "the final solution" of genocide attempted.[37] Today in F.R.G. the foreign "guestworkers" are under pressure to return home to Turkey and elsewhere when unemployment is high and they do the dirtiest, most hazardous, low paying jobs, with the highest accident rates.[38] In the U.K. it is the colored people from the former colonies who are subjected to discrimination and super exploitation in a situation where 3.1 million are unemployed.[39] As Finland was entering the modern capitalist age and for generations thereafter, the bourgeoisie divided the working class by playing the Swede-Finns off against the Finns. Even in Sweden, the most ethnically homogeneous of our study countries, one found discrimination during the most recent worldwide recession in the form of rumblings against the "guestworkers" as in the F.R.G.

There are no cookbook answers to this problem. It is the easiest thing in the world when the capitalist system enters one of its periodic crises and workers' livelihoods and families are threatened through unemployment to turn the blame on one minority group or another saying those "others" want to take your jobs, rather than turning the blame on the capitalist system. We must turn to positive experiences for close study and answers. Again, the "Bread and Roses Strike" of 1912 deserves mention. The textile mill owners and bosses sought to divide the workers by having them clustered by different language and ethnic groups in different work rooms–Poles, Italians, Slovaks, etc.–and attempted to get some groups to scab while others were out on strike. But the workers learned to communicate with each other; realized they were not enemies of each other and that their real enemies were the greedy owners and their bosses and the ruling-class-oriented state apparatus, especially the police; formed representative strike and negotiating committees; and struck together to win the strike.[40] It was nearly a microcosm of the workers of the world who united.

Another periodically widespread, powerful force dividing workers is unemployment and the "reserve army" of the unemployed who stand ready out of desperation to take the jobs of the employed. This condition allows other divisive factors such as racism and sexism to operate with increased force to weaken the workers' movement. It functions between similar workplaces in the same industry within the same town, between similar shops and industries within a nation, and, in the world-system. One of the conditions threatening jobs in core nations is the generally higher unemployment and therefore lower wages and weaker worker organization in semi-peripheral and peripheral countries. In revulsion for the Nazi period, one of our study countries, the G.D.R., has striven hard to combat racism through films, expositions, museums, educational programs etc. But most important, as a measure against many factors which would otherwise tend to weaken the workers' movement, it has adopted the right to a job as a constitutional provision. There is no unemployment! I will come back to this important provision of this socialist-oriented state. This is not to say the situation of working people in the G.D.R. is perfect. I have tried to fill in some of the details to present a fair picture in Chapter 9. It can be easily seen that in this situation a right to work is also a duty to work.

And some of the same pressures for production may operate (in spite of elaborate in-plant mechanisms for workers to voice their concerns–detailed in Chapter 9, as taken primarily from Steele).[41] Such pressures led Braverman to his critical comments on capitalist style production in the soviet union: "Whatever view one takes of Soviet industrialization, one cannot conscientiously interpret its history, even in its earliest and most revolutionary period, as an attempt to organize labor processes in a way fundamentally different from those of capitalism. . . ."[42] This being said, in general, zero level unemployment and even a rather severe labor shortage is a powerful force leading to a very complete set of provisions to protect workers in the G.D.R.

Union organization is one of the keys to the strength of the workers' movement. It could be seen as one of two interwoven components of the organized part of the working class–the other being political party organization and mobilization. At the union level *per se*, the several potentially divisive and weakening factors considered thus far–sexism, racism, the pressures of unemployment–can all severely affect union inclusiveness, solidarity and strength. But there are other factors more entwined with the unions themselves in their intra- and inter-union organizing. There are many topics which could be taken up here. One of the more important is the balkanization of labor in terms of particular crafts and trades versus industrial unions and their more inclusive organizing. Even where there is the potential of division over "raiding" and territorial disputes, strong umbrella organizations from local district on up to national levels can serve to unify organized labor. Historical and contextual factors can be determinative in this as well as other matters affecting union organization and the strength of the workers' movement. In part, Sweden's strong position today relates to its late and more direct entry into industrial capitalism whereby some of the elitism and factionalism stemming from the "labor aristocracy" of the craft unions, as in the U.K., could be avoided in favor of the more open, inclusive, democratic industrial unions.[43] But this and other topics of union organization are more properly treated in specialized works.[44] However, one preeminent concern affecting union strength, as well as other levels of organization, deserves attention here.

If elitism, authoritarianism and autocracy are allowed to reproduce within the union–the very relationships of the wider society to which the worker has been socialized in her or his family, community and school–there will be little chance of unions serving as centers of resistance to help the working class develop a new consciousness and the energy to transform the wider society in a direction which is liberating of human potential. In spite of Michels,[45] this is not a hopeless matter. A comparative approach shows that the, so called, iron law of oligarchy is not so iron. Even John L. Lewis, the archetypical union autocrat, who presided over the United Mine Workers for so many years in U.S.A., starting in 1919, was challenged by a "Save the Union" Committee formed under the leadership of socialist, communist, and syndicalist militants. This move was defeated through the rigging of the election by Lewis' organization.[46] But the fact of this democratizing move and a similar one alive in the Teamsters' Union in U.S.A. today shows that there is a capacity among working men and women to organize for justice and human liberation, sometimes against heavy odds.[47] Cross-national comparisons too suggest that the U.S.A. history of union bossism is not an inevitable thing.

> Commenting in the 1940s on the strength of trade-union autocrats such as Lewis, the English socialist R. H. Tawney explained that special conditions in the United States favored a kind of boss rule which never developed in British trade unions. First, rank-and-filers tended to accept autocratic bosses because they were so dependent upon the union leadership for their jobs. The kind of job control the miners maintained against employers could be used to enhance the power of union officials, who became, in effect, hiring agents. They could function much like a city boss, who kept voters loyal by dispensing jobs. There was a more important source of trade-union autocracy: the violent character of the struggle for control in industries like coal mining. "Trade unionism in the United States had to fight for its existence," Tawney noted. It emerged under far more violent circumstances than British trade unionism. "Systematically persecuted by employers, and half outlawed by the courts, it had recourse, with the consent of its members, to a high degree of centralization of power as a condition of survival. Thus the dictatorship of management was countered by tolerating dictatorial methods in the leaders appointed to mobilize resistance to intolerable conditions." Revolt against the machine occurred periodically, but in a union like the UMW they failed to affect the centralized power of the Lewis organization, which prevailed from the 1920s through the 1960s. Lewis frankly stated throughout his rule that democracy was a luxury the hard-pressed miners could not afford.[48]

This is a complex matter of great import. It can not be left to some kind of automatic, mechanical conception–either inevitable contextual political economic forces (though increasing evidences of crisis in the capitalist world-system are hard to deny)[49] or inherent biological drives (see note 47).

> The adequacy of a theory of socialization that shows the possibility for liberatory self-activity by workers can neither rely on the pressure of eternal events nor on the reduction to biological urges. It must show that inside the structure of social life the activity pattern of the subjects (that is, the workers themselves) can produce values, norms, and ways of interacting that depart from the prescribed patterns of capitalist socialization as well as conform to it. It is the contradiction between the autonomous self-activity of workers in the process of their formation as a class both historically and biographically and the social constraints imposed by the capitalist mode of production (or for that matter, any other system)

> that constitutes the possibility of historical change. If the workers have no elements of their socialization that they can perceive as resistant to the economic role assigned to them by a relatively successful capitalism, neither economic crisis nor biological urges will suffice to change their responses to their life situation.
>
> When Marx asserted the historical role of the working class as the gravedigger of capitalism arising from its unique position in capitalist production, he was only establishing the objective possibility of this process. What is required is to establish its subjective possibility, that is, not the ascribed class consciousness of the working class, but the conditions for its actualization.[50]

To make the future, working people will move within their unions, but also beyond them, to organize politically to affect conditions in the wider society. Workers will move to affect those conditions bearing directly on OSH (for example, the composition and functioning of the state apparatus as it sets and enforces OSH standards) and on PHC (for example, whether there is a national health service, geographically oriented at the periphery and fully integrated with OSH concerns). They will also move on those conditions affecting other aspects of the decency of life–housing, education, work and or creative opportunity for all and a decent income. Perhaps no other factor stands out so clearly from our comparison of six industrialized countries as the breadth of workers' and unions' concerns and their readiness to engage in wider political activity for the achievement of broad socialist goals. U.S.A. "exceptionalism," that is, its narrow "business unionism" contrasts sharply with the picture of working class parties and union organization in the five other study countries. The case of Sweden is particularly notable by contrast.[51] Many of the pitfalls of the U.S.A. experience were at least partially or quite completely avoided in Sweden.[52] Of prime importance was the deliberate focus on the working class and its concerns at large were adopted at the start of union organizing and political formation. Thus, for example, an equity of wage levels was sought avoiding to some extent the divisions between the dominant monopoly capital sector and smaller, competitive capitalist workshops which has left the competitive sector largely unorganized in U.S.A.

Perhaps more important, this class-oriented orientation of the workers' movement in Sweden also supported the building of the Social Democratic Party which held power for over forty-five years, starting in the Great Depression. This led to a progressive tax structure and social welfare, housing, educational, and health provisions, including OSH provisions, second to none. What has occurred in Sweden is a political transformation or "exchange" whereby, rather than being on the defensive, the workers' movement is a central guiding force, perhaps overshadowing, but at least on a par with, the bourgeoisie.[53] Of course there are historical reasons for the U.S.A. experience, but the strategy of unions, and the workers' movement generally, to underplay working *class*-oriented action is terribly significant.

> . . . crucially, given the workers' early enfranchisement in a decentralized polity, their political action was concentrated at the

> community level. Katznelson[54] has argued that under American conditions, ethnic and cultural cleavages have tended to overshadow those of class at this level, thereby further isolating the economic and national-political dimensions of class conflict.
>
> The arguments reviewed above suggest that in the United States a variety of historical forces have shaped a political structure and culture singularly unfavorable to socialist political activity and impermeable to changes which might have facilitated such activity. In our view, however, a factor of even greater significance than those already considered has been the remarkable weakness of the mass (union) base in the industrial arena on which a political strategy of class mobilization and conflict might have rested. . . .
>
> Sweden's longrun experience appears to demonstrate that politics which speak to the broadest and most basic interests of all sellers of labor power—such as employment security—are most effective in achieving a broad base of support.[55] It is evident, therefore, that the discretion of the mobilizers and its impact on their effectiveness are quite considerable.[56]

Any new directions I could suggest would have to include more in-depth contrasting case studies of these matters, particularly union organization and political mobilization which raises working class consciousness and defeats sexism, racism, unemployment, and other divisive and weakening forces.[57] But action should not wait upon studies of the past. Our understandings at this time suggest some broad lines of struggle. These range across the levels of concern already depicted from the in-shop relations of production, to union organizing, to community and wider political action, to linkages between health at work and general health, to working class control of state power, to restructuring of the world-system. In all of these, the raising of working class consciousness and broad unity with supporting forces, wherever they may be found, seems crucial. This may entail a broadened conception of "working class" as Aronowitz suggests in his most recent book to reflect to shifting division of labor and degradation of work, including work performed by secretaries, technical, and service people who may not have thought of themselves as part of the working class.[58] To this broadened consciousness within core capitalist nations, I would add the importance of organized links and consciousness raising to join with the working and peasant classes of peripheral and semi-peripheral nations, for the problems of the workers of the world are shared problems and their opportunities are shared opportunities. Political action should be broadly directed toward assuming control of state power so as to assure working class interests. Most important is the guarantee of a decent income for all who want to work. If workers have to be attracted to work, perhaps it can become the health-giving creative activity it should be.

> . . . The sad, horrible, heart-breaking way the vast majority of my fellow countrymen and women, as well as their counterparts in most of the rest of the world, are obliged to spend their working lives is seared into my consciousness in an excruciating and unforgettable

> way. And when I think of all the talent and energy which daily go into devising ways and means of making their torment worse, all in the name of efficiency and productivity but really for the greater glory of the great god Capital, my wonder at humanity's ability to create such a monstrous system is surpassed only by amazement at its willingness to tolerate the continuance of an arrangement so obviously destructive of the well-being and happiness of human beings. If the same effort, or only half of it, were devoted to making work the joyous and creative activity it can be, what a wonderful world this would be.[59]

Another demand directly related to the concerns of this study should be for comprehensive, regionalized, national health systems which place emphasis on prevention of disease through geographically oriented primary care services fully integrated with the workers' health services. Societies, even so called socialist-societies, which do not apply the highest priority to protecting the health of those who produce the valued products for all to enjoy can not be called a workers' society. Such a demand would have a unifying effect on a broad coalition of workers, the elderly, the poor and minorities.[60]

I am talking of human liberation and development. But this is not something which will be given to the workers of the world, as is pretended when core capitalist institutions seek to tie their exploitative strings ever wider and tighter through "aid."[61] These are human conditions which will be won through organization and political action which speaks "to the broadest and most basic interests of all sellers of labor power."

Walter Rodney pondered these problems before he was brutally murdered in Georgetown Guyana on June 13, 1980 by the very forces from which he sought the liberation of his brothers and sisters. Of course, they didn't kill him. As in the song, "Joe Hill," when the copper bosses shot him dead "he went on to organize." Part of the legacy of Walter Rodney's unsought martyrdom is an understanding of what real development means:

> In all this, a view of what was meant by alternative development to pseudosocialism prevailed. No attempt has yet been made to specify this in full detail by either Rodney or the WPA. And they could not, since in the nature of these things the conception of an alternative society grows out of the process of struggle. But if I may be permitted to sketch such an outline as being constitutive of the thought and practice of Walter Rodney and the WPA, I would argue that as against pseudosocialist development, socialist development, as he spoke of it and as the party of which he was a member speaks of it, means the following things.
>
> First, production to satisfy the basic needs of the masses at large. This implies certain things about the way in which the society is organized. It implies, for example, that in the context of poor societies a systematic attack on poverty as a conscious, deliberate, and planned activity must be engaged upon, rather than sweeping

poverty under the carpet, as is the practice in the Caribbean today. And when you asked him the question "Who are the masses?" the obvious answer would be those "who do not have power in society derived from property, wealth, religion, caste, expertise, or such other sources not widely shared." He recognized that these needs, when we spoke about them, were both personal (food, clothing, housing) and public or collective (health, sanitation, education, water supply, electricity, culture, recreation, etc.) as well as material and non-material. Now the point about this is that the satisfaction of these needs through socialist development had to be generated through the effective right to work. That is, it is not a donation given by the state, but it had to be generated through the effective exercise within the community of the right to work. This meant the right to a job without coercion as to place or type of job, given a person's skills. It also meant that the framework of industrial relations had to permit free collective bargaining and effective representation of workers' interests. The work process, also, had to ensure that effective worker involvement and control was possible, as was protection of health, guarantee of education, and so on, in the context of their work. In other words, what is implied in this concept of alternative development is that work is situated in a self-realization process, so that Rodney came to see work as both an end and a means of development. Indeed, this concept marked a fundamental shift in what was meant by development within the region.

Such development, he also recognized, had to be essentially self-reliant, it had to depend upon internal forces within the society. This is the only sustainable form, not just in a technical sense, but even more important, it is the only form that expresses man's "belief in his capabilities to develop himself." In that sense, therefore, Rodney's history was at the service of humanity, because he saw, even in the context of speaking about work and alternative possibilities of development within the region, that what we are talking about is a reference to a fundamental premise of humanity, that is, the capability of man and his belief in that capability to develop himself. This development he saw as also requiring the democratization of power and effective exercise of fundamental rights such as expression, organization, the abolition of repression and torture, the democratization of decision making, the equitable distribution of income and wealth, and equitable access to the use and management of resources. To him the state control of resources was not enough if it did not lead to equitable access to their utilization and management, which the state that came into existence after a popular social revolution, or as a result of popular consent, would still be subject to these determining conditions in order to make the state socialist.

He also felt it was necessary to see that there was equitable access to information, because without this, effective decisions cannot be made by the working class. The communications that he was making about history, about understanding the development of society, were part and parcel of his fundamental recognition that to be effective within society one has to be informed about

> the nature and processes of the development of the society in which we live.
>
> He was also a globalist and felt responsibility to ensure that in whatever development we spoke about we accepted the global and universal meaning of it and shed all sense of parochialism.
>
> In all of these things it is important to note Rodney's recognition of the interaction of material and other needs and their self-reinforcement; the recognition that these things are to be seen dynamically. People's needs grow and develop as people grow and develop a society. He also understood that there was no single "royal road" towards such a development which can be predetermined at the outset. This road, he came to appreciate, is always socially determined; it did not lie in the heads of any scholars or even the leaders of any political party. Rather, the political party and its leadership were an integral part of an unfolding process that was trying to make the particular road chosen one that was effectively socially determined. In a word, the needs and the methods of determining these within a society were by their very nature culturally specific. Above all, he came to see in all this that the "alienation of man from the process of social reproduction—of the so-called development which is being practiced everywhere," would be brought to an end.[62]

I do not feel I have stretched my own or the reader's concerns beyond the relevant nexus of forces acting upon the health of workers and others by pointing to human development as the goal which will encompass workers' health as well as other desirable human achievements. Health of all persons, workers included, is, after all, as the World Health Organization has stated, "not simply the absence of disease, but is complete physical, mental, and social well being."

NOTES AND REFERENCES

1. Daniel M. Berman, *Death on the Job, Occupational Health and Safety Struggles in the United States*. New York: Monthly Review Press, 1978.
2. K. Vicusi, The Impact of Occupational Safety and Health Regulations, *The Bell J. Of Econ., 10*(1):117–140, 1979; Consolidation Coal Company, Cost Benefit Analysis of Deep Mine Federal Safety Legislation and Enforcement. Pittsburgh, 1980. For a detailed exposure and telling critique of this view, see John Dennis Chasse and David A. LeSourd, Rational Decisions and Occupational Health: A Critical View. *Int'l. J. Health Services 14*:433–445, 1984.
3. R. Sass and G. Crook, Accident Proneness: Science or Non-Science? *Int'l J. Health Services, 11*:175–190, 1981 (quote from p. 189).
4. These static characterizations of some of the indicators of a strong workers' movement in Sweden are given a more historical and dialectic treatment in Chapter 7. For an in-depth study of Swedish labor and its participation

in politics, see Walter Korpi, *The Democratic Class Struggle*. London: Routledge and Kegan Paul, 1983. For a condensed but perceptive comparison of labor struggles and the conditions and responses of capital in Sweden, U.K., U.S.A. and Japan, see Michael Burawoy, Between the Labor Process and the State: The Changing Face of Factory Regimes Under Advanced Capitalism. *American Sociological Review 48*:587–605, October 1983.

5. One observer attempts to minimize the differences in protective OSH provisions as between Sweden and U.S.A.: Steven Kelman, *Regulating America, Regulating Sweden: A comparative Study of Occupational Safety and Health Policy*. Cambridge: MIT Press, 1981. But Navarro details the remarkable differences in standards, for example, a noise standard of 85 dB in Sweden versus 90 dB in U.S.A. and in workers' right to know about and to refuse hazardous work and other provisions (see the long list in Table 9, p. 553 in the following reference) and exposes the inadequate interpretations of Kelman as ideologically oriented toward management interests while ignoring the power of worker consciousness and organization. Vicente Navarro, The Determinants of Social Policy, A Case Study: Regulating Health and Safety at the Workplace in Sweden. *Int'l J. Health Services, 13*: 517–561, 1983.
6. Again, these characterizations of the comparative weaknesses of the worker's movement in U.S.A. are given more detailed treatment in Chapter 12. For more detailed historical analyses, see James R. Green, *The World of the Worker, Labor in Twentieth-Century America*. New York: Hill and Wang, 1980; Stanley Aronowitz, *False Promises, the Shaping of American Working Class Consciousness*. New York: McGraw-Hill, 1973; Philip S. Foner, *Women and the American Labor Movement, From Colonial Times to the Eve of World War I*. New York: The Free Press, 1979. For a condensed historical comparative presentation of labor struggles and the responses of capital and the state in U.S.A., Sweden, U.K. and Japan which also sees a new despotism evolving as the periphery and semi-periphery of the world-system are used to weaken the workers' movements in core capitalist nations see Burawoy, op. cit. (note 4).
7. Molly Joel Coye, Mark Douglas Smith and Anthony Mazzocchi, Occupational Health and Safety, Two Steps Forward, One Step Back, pp. 79–106 in *Reforming Medicine: Lessons of the Last Quarter Century*. Victor Sidel and Ruth Sidel, eds. New York: Pantheon, 1984; Philip J. Simon, *Reagan in the Workplace: Unraveling the Health and Safety Net*. Washington, D.C.: Center for Study of Responsive Law, 1983. Also, Carl Gersuny, *Work Hazards and Industrial Conflict*. Hanover, NH: University Press of New England, 1981; David P. McCaffrey, *OSHA and the Politics of Regulation*. New York: Plenum, 1982; and Ben Bedell, Reagan's OSHA Is Self Destructing, *Guardian* (New York) (January 19, 1983):3.
8. This would probably apply as well to work related illness levels, were adequate comparative data available. For an examination of OSH data and its inadequacies as these are embedded in varying political economies, see Chapter 4. As regards accidents, Witt states that there are half as many job injuries per worker each year in Sweden as there are in U.S.A. M. Witt,

Learning Job-Safety and Health from Europe, *New York Times* (May 7, 1978): A 21. In the U.K. where coal miners are much more widely and solidly organized than they are in U.S.A., in the late 1970s there were an average of fifty-two deaths for a workforce of 246,000 whereas in U.S.A. the coal miner deaths averaged 137 for a smaller workforce of 216,000. P. J. Nyden, Miners, Mr. President, Are Not Slag, *New York Times* 24 (1981): 21.

9. Again using Sweden as the example, there is no question that life chances are better as compared with U.S.A. in the narrow sense of more favorable overall mortality figures. And the picture of social welfare and health care provisions for the population at large is also much more favorable. Richard F. Tomasson, *Sweden: Prototype of Modern Society*. New York: Random House, 1970. Also Richard F. Tomasson, The Mortality of Swedish and U.S. White Males: A Comparison of Experience, 1969–71, *Amer. J. Publ. Health, 66*:968–974, 1976. Also Walter Korpi, *The Working Class in Welfare Capitalism: Work, Unions and Politics in Sweden*. London: Routledge and Kegan Paul, 1978.
10. The shifting division of labor and allocation of labor as between, for example, industrial production and service production and the impacts of such changes on class consciousness and struggle are examined in Harry Braverman, *Labor and Monopoly Capital, The Degradation of Work in the Twentieth Century*. New York: Monthly Review Press, 1974.
11. Burawoy, op. cit. (note 4); Walter Korpi and Michael Shalev, Strikes, Industrial Relations and Class Conflict in Capitalist Societies, *British J. of Sociology, 30*:164–187, June, 1979; Korpi and Shalev, Strikes, Power and Politics in the Western Nations, 1900–1976. *Political Power and Social Theory* I (1979): 299–332; Geoffrey K. Ingham, *Strikes and Industrial Conflict: Britain and Scandinavia*. London: MacMillan, 1974; Ronald Dore, *British Factory–Japanese Factory: The Origins of National Diversity in Industrial Relations*. Berkeley and Los Angeles: University of California Press, 1973; Colin Crouch and Alessandro Pizorno, eds. *The Resurgence of Class Conflict in Western Europe since 1968*. (2 vols.) London: MacMillan, 1978; Harold L. Wilensky, *The Welfare State and Equality*. Berkeley: University of California Press, 1975; Anna DeCormis, Made–and Enslaved–in Japan, *Guardian* (New York) (October 9, 1985): 2.
12. A complete list would be too long even for one country. An interesting piece which attempts to explain U.S.A. "exceptionalism," that is, its narrow business (as opposed to broad, politically-oriented, socialist) unionism with comparative reference to Sweden, the U.K. and other countries is Michael Shalev and Walter Korpi, "Working Class Mobilization and American Exceptionalism," *Economic and Industrial Democracy* (Vol. 1) (1980):31–61. This piece will serve as a key source in our considering the nexus of factors related to the strength of the workers' movement, or, as these authors term it, "working class mobilization." Other key works on the workers' movement in the U.S.A. were cited above (note 6). See also the references for Chapter 12, including Stanley Aronowitz' recent reexamination, *Working Class Hero, A New Strategy for Labor*. New York: The Pilgrim Press, 1983. Korpi's works on the history of workers' struggles in Sweden were cited in

notes 4 and 9. Works by Hobsbawm and others on the U.K. are cited in Chapter 11 and selected key items for other study countries are cited in each country chapter (Finland, Chapter 8; G.D.R., Chapter 9; F.R.G. Chapters 9 and 10).

13. Cultural hegemony was presented as an analytic concept in Chapter 5, drawing on Antonio Gramsci's essay "The Modern Prince" in his *Selections from the Prison Notebooks*. New York: International Publishers, 1971 (original about 1923).
14. Again, a conception of the world-system was offered in Chapter 5 as taken from the works of Wallerstein, Frank, and others. On the worldwide "Taylorization" of labor (that is, breakdown into specific tasks which can be allocated to cheaper non-skilled labor in any "stable" country of the world–often defined by capital as one with a military dictator who represses unions and workers generally) see Susan Jonas and Marlene Dixon, Proletarinization and Class Alliances in the Americas, *Synthesis, 3*:1-13, 1979. Also Immanuel Wallerstein, ed., *The European Workers' Movement* special issue of *Contemporary Marxism* No. 2 (Winter 1980); *The New Nomads; Immigration and Changes in the International Division of Labor* special issue of *Contemporary Marxism* No. 5 (Summer 1982).
15. E. S. Greenberg, Context and Cooperation: Systematic Variation in the Political Effects of Workplace Democracy, *Economic and Industrial Democracy 4*(3):191-223, 1983, (quotes from p. 217).
16. Leon Grunberg, Jerry Everard and Mary O'Toole, Productivity and Safety in Worker Cooperatives and Conventional Firms, *Int'l J. Health Services, 14*:413-432, 1984. (quote from pp. 429-430).
17. Braverman, op. cit. (note 10): 55-57.
18. Bertil Gardell, Scandinavian Research on Stress in Working Life, *Int'l J. Health Services, 12*:31-41, 1982.
19. Robert Karasek, et al. Job Decision Latitude, Job Demands, and Cardiovascular Disease: A Prospective Study of Swedish Men, *Amer. J. of Publ. Health, 71*:694-705, July, 1981.
20. See the summary assessment in Chapter 13. For more specifics on the more adequate provisions for worker health and safety in Sweden and stronger workers' movement as compared with U.S.A. see those country chapters (7 and 12) and the especially detailed and important article by Navarro, op. cit. (note 5).
21. Barry I. Castleman, The Export of Hazardous Factories to Developing Nations, *Int'l. J. Health Services, 9*:569-606, 1979. Leah Levin, Too Young to Work, *World Health* (Jan./Feb., 1984): 24-27; Barbara Ehrenreich and Annette Fuentes, Life on the Global Assembly Line, *Ms* (January, 1981): 53-59 and 71. R. Cantwell, Women Workers of the Asian Rim, *WIN* 15 (March, 1979): 16-19; Ray Elling, The Capitalist World-System and International Health, *Int'l. J. Health Services, 11*:21-51, 1981.
22. I would emphasize again the need for continuing comparative or contrasting cross-national historical case studies of workers' movements and the sets of interwoven factors affecting their strength.
23. Natalie J. Sokoloff, *Between Money and Love: The Dialectic of Women's Home and Market Work*. New York: Praeger, 1980.

24. Mary E. Garbatt, Practical Application in the Home, *Los Angeles Socialist* (March 19, 1904) as quoted in Foner op. cit. (note 6): 273.
25. Wendy Chavkin (ed.), *Double Exposure: Women's Health Hazards On the Job and at Home*. New York: Monthly Review Press, 1983. Reviewed by Kathy Deacon, Women on the Job Put Their Health on the Line, *Guardian* (New York) (October 3, 1984): 19.
26. Green, op. cit. (note 6): 45.
27. Ibid, p. 45.
28. Produced about 1976 by California Newsreel, 630 Natoma Street, San Francisco, CA 94103.
29. He ignores of course that watchmakers have usually been men as have micro surgeons. The matter of "dexterity" is simply an ideological statement encouraging super exploitation. See Linda Y. C. Lim, Women Workers in Multinational Corporations: The Case of the Electronics Industry in Malaysia and Singapore, Ann Arbor: Michigan Occasional Papers in Women's Studies, nd (about 1980); also some of the references in note 21.
30. In 1902 just before she was to be jailed for supposedly violating a court order prohibiting her from addressing striking miners in Clarksburg West Virginia, Mother Jones continued to support the miners by appealing to the Cincinnati Central Labor Council for help. She also took the occasion to promote the cause of working women, saying: "You will never solve the problem until you let in the women. No nation is greater than its women. . . Women are fighters." As quoted in Foner, op. cit. (note 6): 283.
31. This case was recounted briefly in Chapter 12, drawing on the marvelous pictorial history by William Cohn, *Lawrence 1912, The Bread and Roses Strike*. New York: Pilgrim Press, 1980.
32. It was the rank and file women who walked the lines, tried to keep their families fed until the food ran out and children had to be shipped off to the protection (and wider solidarity) of worker families in New York and Boston. They got beat up by the police—even when they were obviously pregnant and some had miscarriages as a result. Among the support which was sent in was the inspiring presence and stirring voice of Elizabeth Gurely Flynn. She had joined the IWW ("The One Big Union,"– the Industrial Workers of the World, otherwise known as the Wobblies) in 1906 as a member of the "mixed" Local No 179 in New York City. "Socialist fervor seems to emanate from her expressive eyes, and even from her red dress. She is a girl (then only fifteen-years old) with a 'mission', with a big 'M'," wrote the *Industrial Union Bulletin* in 1907 concerning a speech she delivered in Duluth. As cited in Foner, op. cit. (note 6): 396.
33. Susan Howard, AFL-CIO Head Backs Strikers in Yale Fight, *The Hartford Courant* (October 12, 1984): C1 and C2.
34. Green, op. cit. (note 6): 46–47.
35. More details of this disaster were given in Chapter 12 as taken from Berman, op. cit. (note 1): 27–28 and from Joseph A. Page and Mary-Win O'Brien, *Bitter Wages*. New York: Grossman Publishers, 1973, esp. pp. 59–63. See also R. Scott, *Muscle and Blood*. New York: E. P. Dutton and Company, 1974, esp. pp. 176–196.

36. Joseph K. Wagoner, Occupational Carcinogenesis: The Two Hundred Years Since Percival Pott, pp. 1–3 in *Occupational Carcinogenesis*, special issue of the *Annals of the New York Academy of Science* 271 (1976).
37. Arthur D. Morse, *While Six Million Died*. New York: Random House, 1968.
38. Hans-Ulrich Deppe, Work Disease, and Occupational Medicine in the Federal Republic of Germany, *Int'l J. Health Services, 11*:191–205, 1981. Associated Press, West Germany Offering Some Foreign Workers Money to Leave Country, *The Hartford Courant* (November 12, 1983).
39. John Downing, *Now You Do Know*. An Independent Report on Racial Oppression in Britain for Submission to a Word Council of Churches Consultation, London, March, 1980.
40. Cohn, op. cit. (note 31).
41. Jonathan Steele, *Inside East Germany; The State That Came in from the Cold*. New York: Urizen Books, 1977.
42. Braverman, op. cit. (note 10): 22. In this connection he cites Marglin who says: "In according first priority to the accumulation of capital, the Soviet Union repeated the history of capitalism, at least as regards the relationship of men and women to their work. . . . The Soviets consciously and deliberately embraced the capitalist mode of production. . . . Now, alas, the Soviets have the 'catch-up-with-and-surpass-the-U.S.A.' tiger by the tail, for it would probably take as much of a revolution to transform work organization in that society as in ours." Stephen A. Marglin, What Do Bosses Do? The Origins and Functions of Hierarchy in Capitalist Production. Cambridge, MA: Harvard University Department of Economics, nd (mimeographed, about 1973).
43. Burawoy, op. cit. (note 4) esp. p. 601.
44. With special reference to union organizing in U.S.A. see the works by Green, Aronowitz and Foner (note 6).
45. Robert Michels, *Political Parties*, translated by E. Paul and C. Paul. Glencoe, Ill.: The Free Press, 1949.
46. Green, op. cit. (note 6): 122.
47. Aronowitz recognizes this too when he contrasts Parson's conservative, static view that people are socialized rather perfectly into set roles and learn to live with the capitalist system (Talcott Parsons, *The Social System*. New York: Free Press, Paperback Edition, 1964) with Marcuse's model of inherent rebellion (Herbert Marcuse, *One-Dimensional Man*. Boston, MA: Beacon Press, 1964). This last is based upon a kind of Freudian concept, "a biologically rooted instinctual structure that provides a material substratum for the liberation of humanity. . . His faith resides in the ultimate inexorability of the liberatory instincts that provide the substratum of rebellion–rebellion that expresses itself as refusal rather than as a genuine alternative." Aronowitz, op. cit. (note 6): 56–57. As we will see, Aronowitz appreciates Marcuse's search for liberating forces but rejects such "biochemical" thinking in favor of something much clearer and yet more complex–the capacity of men and women to organize in pursuit of human justice and potential.
48. Green, op. cit. (note 6): 122.

49. Andre Gunder Frank, *Crisis in the World-Economy*. New York: Holmes and Meier, 1980; Albert Bergesen (ed.), *Crisis in the World-System*. Beverly Hills, CA: Sage, 1983; and E. A. Brett, *International Money and Capitalist Crisis: The Anatomy of Global Disintegration*. Boulder, CO: Westview, 1983.
50. Aronowitz, op. cit. (note 6): 58.
51. Walter Korpi. *The Working Class in Welfare Capitalism: Work, Unions and Politics in Sweden*. London: Rutledge and Kegan Paul, 1978.
52. An attempt was made in Chapter 7 on Sweden to summarize the organizational and political strategies of the unions and their party.
53. Korpi and Shalev, June, 1979, op. cit. (note 12).
54. Ira Katznelson, Considerations on Social Democracy in the United States, *Comparative Politics, 11*:77–99, October, 1978.
55. Here the authors cite Korpi op. cit. (note 51) and Andrew Martin, *The Politics of Economic Policy in the United States: A Tentative View from a Comparative Perspective*. Beverly Hills: Sage Professional Paper in *Comparative Politics* 01-040, 1973.
56. Michael Shalev and Walter Korpi, Working Class Mobilization and American Exceptionalism, *Economic and Industrial Democracy*. Beverly Hills: Sage (Vol. 1):31–61, 1980, quotes from pp. 51–54.
57. The list would be too long to consider again here. See Chapter 4 on the matter of research and ideology. Also in addition to those society-wide conditions related to the cultural hegemony, such as sexism and racism, the history of labor struggles includes violent confrontations with the police and military when state power is under the sway of the bourgeoisie. Laws and the courts and their actions are another set of such forces limiting the power of unions and the workers' movement. As part of the capitalist cultural hegemony, Aronowitz treats technological consumerism and the colonization of leisure, op. cit. (note 6) Chapter 2, pp. 51–134. Navarro has examined the social control functions of the medical care system: Vicente Navarro, *Medicine Under Capitalism*. New York: Prodist, 1976, esp. the last chapter. More directly in the hands of owners and managers are a near multitude of ways to divide and rule, cajole, and influence workers generally. These range all the way from favoritism, to spies in union meetings, to "agents provacateurs" on the picket line so as to bring in the police. One of the more disturbing from the perspective of my own discipline, was Henry Ford's "Department of Sociology" which sent interviewers onto the shop floor, into the workers' neighborhoods and even into their homes to learn how to keep them tranquil! Green op. cit. (note 6): 109–110.
58. Stanely Aronowitz, *Working Class Hero, . . .* op. cit. (note 12).
59. From the Foreword by Paul M. Sweezy to Braverman, op. cit. (note 10): xii–xiii.
60. The Dellums Bill for a National Health Care System has this potential, to choose a specific plan relevant to the U.S.A. Some of the other study countries already have a NHS–G.D.R., Sweden, U.K. However, there is

considerable room for improvement of the linkages between OSH and PHC in all these systems.

61. Andre Gunder Frank, *Dependent Accumulation and Underdevelopment*. New York: Monthly Review Press, 1979; T. Hayter, *Aid As Imperialism*. Harmondsworth, U.K.: Penguin, 1971.
62. Clive Y. Thomas, Walter Rodney and the Caribbean Revolution, pp. 119–132 in *Walter Rodney, Revolutionary and Scholar: A Tribute*. Edward A. Alpers and Pierre-Michel Fontaine (eds.), Los Angeles: University of Los Angeles, Center for Afro-American Studies and African Studies Center, 1982 (quote from pp. 125–127).

Methods Appendix — Origins, Procedures and Limitations

For the past several years I have taught and done research on comparative health systems.[1] For five recent years I have concurrently had the privilege and excitement of working with a first-rate team of people committed to educating workers about hazards at work and ways to organize to prevent damage from such hazards through the New Directions Program, Connecticut.[2] This program has worked mainly because it was worker-based, having strong multi-union sponsorship, as well as having funds from the Occupational Safety and Health Administration (OSHA). Another strong point of the program has been our close connection to the Yale Occupational Health Clinic, a place we could refer workers for assessment and follow-up of suspected occupational disease and, perhaps most important, a place workers felt they could trust because the well trained physicians, industrial hygienist, and other staff had a track record of concern for workers' health and working with organized labor.

During the course of the New Directions Program, I realized how wide the gap is between workers and the general health services. In my own medical school, although I have taught an elective course, "Epidemiology, Carcinogenesis, and Air Pollution at Work," there is not one required hour of OSH for all students. Hardly any of the medical students have themselves worked in factories and very few come from working class families. As Dr. David Wegman, then of the Harvard OSH Program, put it to one of our Training of Trainers classes, "You will have to undertake to educate the doctors in your own communities, since they have not been trained in occupational health and are, for the most part, not motivated to care."

I determined during my sabbatic leave in the Fall of 1981 to bring my interests in comparative health systems and workers' health together by contrasting the appalling disjunctures in the U.S.A. between PHC and OSH with experiences in other countries.

It seemed to me that there would be the most to learn within the time and resources available (essentially 2½ months of actual field work time in study countries) if I could obtain interviews and documents from countries with reputedly well developed specialty OSH services yet reflecting a range of strengths and types of workers' movements as well as variations in health systems. Thus five countries were chosen as main study countries–Sweden, Finland, G.D.R., F.R.G. and U.K. (in addition the U.S.A. is included as a case already known to me).

With the support of WHO-EURO and their sponsorship of me as a Temporary Advisor, it was relatively easy to gain access to officials and official sorts of information in each country. In addition, I made it a point to talk on my own to workers and activist union people in each country, where this was possible. My lack of language capability in Swedish and Finnish limited this kind of data gathering in those countries, but even there I found it possible to make a number of exciting contacts of this sort using English. I have reasonable competence in German and thus was able to pursue this interest to some extent in the two Germanies, and it was easy to pursue this interest in the U.K. I would like to thank WHO-EURO and the many people in each country kind and concerned enough to take the time to be interviewed or provide information in other ways. This expression of thanks to others in no way implies their endorsement of my analyses or views. This research report is entirely my own for which I take full responsibility.

In general, I planned (and more or less succeeded) in each country to gather information from persons representing the following sorts of interests:

- workers and activist unionists
- union officials
- production policy or management persons
- industrial physicians
- ministry of health
- ministry of labor (or labor inspectorate)
- PHC providers
- medical educators

In each country I made it a point to inquire into differences between large, medium and small industries and agriculture. My approach was to introduce myself or ask another person who knew me well to call and introduce me as an American Professor on sabbatic leave serving as a Temporary Advisor to WHO to inquire into occupational health and its relation to the general health system. I can't now recall a single refusal on the part of someone I really wanted to see. There are gaps in the above list of interests in some countries because the best recognized representative had a schedule conflict or I had gone too far in structuring my own time that I could not make the necessary alteration. I interviewed a few people by phone. In general, I spent from ten days to two weeks in each country. In each, I visited places outside the capital. In the interviews themselves I generally took detailed notes. Occasionally, especially in sensitive situations, I waited to do notes by recall as soon after the interview as possible. I found it always important to go over my notes in the evenings or on other proximate occasions to complete unclear words, fill in gaps, etc. Informants were generally told that they would not be quoted by name unless I explicitly sought their permission. For this and other reasons I am not giving a list of all people seen and interviewed in each country as easily identitiable points of view could be associated with the one or two people on each list representing a given interest. This has left me free to report fully the strong points as well as the weak points of each system.

For each study country I had the good fortune of knowing at least one especially well informed key informant with whom I was able to share a draft chapter for that country and receive critiques, comments, suggestions, and added information. Occasionally, feeling a special obligation (but again, implying no endorsement on their part) I have acknowledged especially helpful personal correspondents in country chapters. The commentary on my first draft for the G.D.R. was especially extensive and critical so I have included it as an Appendix to Chapter 9.

While I have accumulated literal mounds of documents, reports, books, articles, notes and correspondence, and have devoted considerable effort to this work, I am but one person with my own biases (very much oriented, as the reader will surely be able to tell, to workers themselves taking greater control of their lives) and my time was limited. Rather than interviewing a representative sample in each interest category listed above, I generally interviewed what might be called a "reputational leadership sample," persons known by other knowledgable people to be among the best informed and leading person in a given interest category.

Thus only one person interviewed certain people in a selection of countries in a rather brief period. *Clearly, this must be considered a preliminary study and report.* If I am so bold as to offer recommendations and draw out general implications, I do so because preliminary as this is, a clear picture seems to emerge; also, no one else has done this kind of thing.[3] If the reader is properly cautioned, as to the tentative and preliminary nature of the whole study, proper use can be made of the recommendations and expressed general implications.

In general, I see a need both to make this work more definitive for present study countries and expand it to other countries, including a variety of underdeveloped countries with varying political, economic, cultural, and health system circumstances. The special need is to understand the dynamics and forces contributing to stronger workers' movements which will help to assure workers' health and human development.

NOTES AND REFERENCES

1. Ray Elling, *Cross National Study of Health Systems, Political Economies and Health Care*. New Brunswick, NJ: Transaction Books, 1980.
2. Ray Elling, "Worker-based education, organization and clinical-epidemiologic support for occupational safety and health (OSH): The New Directions Program in Connecticut." Paper summarized at the ILO-WHO meeting on occupational health education, Sandefjord, Norway, 16–19 August, 1981. Now available in the Conference Proceedings, Geneva: ILO, 1982.
3. The cross-national literature on OSH is either very sparse or non-existent, depending upon what one will accept as a comparative or cross-national study. Ashford included descriptions of some European experience with the main focus on the U.K. and the Robens Report. See Nicholas A. Ashford, *Crisis in the Workplace: Occupational Disease and Injury. A Report to the Ford Foundation*. Cambridge, MA: MIT Press, 1976 (Chapter 11). And there have been a few other descriptive presentations of different systems, usually in collections involving different authors using different unstated frameworks. One valuable collection of this type including descriptions of selected OSH improvement efforts in Sweden, U.K., U.S.A., Australia, Japan, and Jamaica is found as section IX in *Proceedings of the Symposium for Labor Educators on Occupational Health and Safety*. Oct. 15–18, 1979 Madison, Wis.: The School for Workers. I tried to include the literature through 1977 in the second volume of an annotated bibliography: Ray H. Elling, *Cross-National Study of Health Systems: Countries, World Regions and Special Problems*. Detroit: Gale Research, 1980, Sect. 2.14. The study recently conducted of OSH systems in European countries for WHO-EURO by Dr. Sven Forssman should add importantly to our knowledge in this field. The work by Kelman comparing OSH standards in Sweden and U.S.A. and Navarro's telling critique of it are cited in Chapter 7 and elsewhere in this volume. A study by Atkins comparing OSH policy and enforcement in the U.K. and U.S.A. is cited in Chapter 11.

Name Index

Subject Index

List of Abbreviations

AFL – American Federation of Labor (national umbrella union in U.S.A. now joined with CIO)

AFL-CIO – American Federation of Labor and Congress of Industrial Organization (national umbrella union in U.S.A.)

AHF – Air Hygiene Foundation, U.S.A.

AMA – American Medical Association

ASV – Swedish Worker Protection Board

BAU – Bundesanstalt fuer Arbeitsschutz und Unfallforschung (the NIOSH equivalent in F.R.G.)

BNA – Bureau of National Affairs, U.S.A.

BPO – Betriebsparteiorganisationen

CDC – Center for Disease Control, U.S.A.

CDU – Christian Democratic Union, F.R.G.

CNSHS – Cross National Study of Health Systems

COSH – Council on Occupational Safety and Health (voluntary association of activist workers and OSH experts in U.S.A.)

DBCP – A pesticide

DRG system – A diagnostic related group reimbursement system for hospitals in U.S.A.

EMAS – Employment Medical Advisory Service in U.K.

EURO – European Regional Office of the World Health Organization

FDGB – Freier Deutscher Gewerkschaftsbund (Free German Trade Union in G.D.R.)

F.R.G. – Federal Republic of Germany (West Germany)

GDP – Gross Domestic Product

G.D.R. – German Democratic Republic (East Germany)

G.m.b.H. – Gesellschaft mit beschraenkter Haftung (society with limited liability, the F.R.G. equivalent of Ltd. or Inc.)

GNP – Gross National Product

GP – General Practitioner

HSE – Health and Safety Executive, U.K.

IHF – Industrial Health Foundation, U.S.A.
KPD – Kommunistische Partei Deutschlands (Communist Party of Germany)
LO – Landesorganisationen i Sverige (Blue Collar Workers' national umbrella union, Sweden)
NLRB – National Labor Relations Board, U.S.A.
NHI – National Health Insurance
NHS – National Health Service
NIOSH – National Institute for Occupational Safety and Health, U.S.A.
NSC – National Safety Council, U.S.A.
NSV – Swedish National Board of Occupational Safety and Health
OCAW – Oil, Chemical and Atomic Workers, U.S.A.
OSH – Occupational Safety and Health (workers' safety and health)
OSHA – Occupational Safety and Health Administration, U.S.A.
PHC – Primary Health Care
PSRO – Professional Standards Review Organization, U.S.A.
PTK – Privattjanstemannakartellen (White Collar Workers' national umbrella union, Sweden)
SAF – Svenska Arbetsgivareföreningen (Employers Confederation, Sweden)
SAK – Confederation of Unions, Finland
SARB – Swedish COSH-type group
SED – Sozialistische Einheitspartei Deutschlands (Socialist Unity Party of Germany, G.D.R.)
SPD – Sozialdemokratische Partei (Social Democratic Party, F.R.G.)
TLV – Threshold limit value
TUC – Trades Union Congress (National umbrella union in U.K.)
UAW – United Auto Workers, U.S.A.
U.K. – United Kingdom of Great Britain and Northern Ireland
U.S.A. – United States of America
VDT – Video Display Terminal
VEB – Volkseigene Betriebe (People's own undertaking – G.D.R. equivalent of Ltd. or Inc.)
WHO – World Health Organization
WW I – World War I
WW II – World War II
ZAM – Zentral Institut fuer Arbeitsmedizin (Zentral Institute for Work Medicine, G.D.R.)